SUPERCONDUCTORS – PROPERTIES, TECHNOLOGY, AND APPLICATIONS

Edited by **Yury Grigorashvili**

Superconductors – Properties, Technology, and Applications
Edited by Yury Grigorashvili

Published by InTech
Janeza Trdine 9, 51000 Rijeka, Croatia

Publishing Process Manager Sandra Bakic

Technical Editor Teodora Smiljanic

Cover Designer InTech Design Team

Additional hard copies can be obtained from orders@intechopen.com

Superconductors – Properties, Technology, and Applications, Edited by Yury Grigorashvili
p. cm.
ISBN 978-953-51-0545-9

For EU product safety concerns:
IN TECH d.o.o., Prolaz Marije Krucifikse Kozulić 3, 51000 Rijeka, Croatia,
info@intechopen.com or visit our website at intechopen.com.

Meet the editor

Dr. Grigorashvili Yury Evgenievich was born in 1947 in the town of Ussuriysk, in the Far East region of Russia. In 1972, he graduated from the Moscow Institute of Electronic Technology with his Master's degree in Electronic. In 1974, Dr. Grigorashvili received his Ph.D. in Electronic Engineering from the same university. After that, he worked as the lead engineer at the plant, specializing in electronic chip manufacturing. From 1978 to 1988, Dr. Grigorashvili had held a position of senior research associate. Since then, he has taken the role of the director of superconducting microelectronics laboratory at the Moscow National Research University of Electronic Technology (MIET). As a result of completed projects over the years he has published more than 150 papers.

Contents

Preface

Superconductivity is a very interesting phenomenon found in nature. Initially, the superconductivity was characterized by an abrupt drop to zero in the material's electrical resistance. An electric current, induced by the magnetic field in a superconducting loop, can flow indefinitely without any loss. Further research constantly modifies our concepts and understanding of the superconductivity as a more complex phenomenon.

The statement about the zero value of electrical resistance is valid only for direct electric current. In an AC electric field, the superconductor's resistance is different from zero even at the below critical temperatures. This is due to the fact that the sample has not only the superconducting current carriers but also normal current carriers. Normal electrons are the cause of heat allocation in a superconductor when it receives irradiation by high-frequency electromagnetic field. The losses rise with the increase in frequency and become noticeable in the microwave range of frequencies.

The difference between a superconductor and a perfect conductor (resistance of which is also equal to zero) is shown in the Meissner effect. Upon cooling of an ideal conductor in a constant magnetic field to a below critical temperature, the magnetic field remains unchanged in its volume. Under similar conditions the magnetic field found inside the superconductor would be absent after cooling. Even if the external magnetic field remains unchanged, a superconductor has arising electric currents on its surface that compensate for the field present inside the sample. However let's remember that generally the external magnetic field is pushed out only from a partial volume of a superconductor. It depends on the strength of the magnetic field, and the sample's shape and material. In the majority of known superconductors the magnetic field located in limited space, penetrates into the superconductor deeper in the form of a thin thread. The volume of this thread has no superconducting properties. The superconducting material is located outside this zone. Constant electric current finds its way out without any heat loss and the material appears to have zero electrical resistance. Ohmic resistance occurs in sufficiently strong magnetic fields only, when zones with the material in normal state increase and completely cover the path for the movement of the superconducting carriers.

Considerable difference between the superconductor and conductor is found and proven by two Josephson effects. The first effect occurs in a superconductor that is divided by a thin dielectric layer. If the current's value passing through the structure is below critical, the dielectric voltage drop is equal to zero. In the metal-based structures divided by a thin dielectric, the current-voltage characteristics of the tunneling current have no sites with zero voltage found on the dielectric. The second effect occurs when the direct current with the magnitude above critical flows through the superconducting structure. In this case, the voltage drop will occur on the dielectric and the structure will begin to emit electromagnetic waves.

For a long time all discovered superconducting materials had a very low critical temperature. The liquid helium was used to cool the superconducting materials. In 1976, the record for the low-temperature superconductors was achieved by finding the Nb3Ge (Tc = 23,2K) compound. Eminent changes also happened in 1986 when the superconductor – La2-xBaxCuO4. (Tc = 30 K) was synthesized. It took a year for other researchers to prove the existence of the superconductor with Tc > 30 K. In January 1987, the discovery of superconductivity at temperature of 90 K was announced (Y1 Ba2 Cu3 O7-x). One year later, bismuth-based high temperature superconductor was revealed (Tc = 110 K) and in a month, Tl2 Ba2 Ca2 Cu3 O10. (Tc = 125 K) was discovered as well. After the year 2002, there were invented both MgB2 (40 K) superconductors and iron-based superconductors (56 K). Today the record is held by HgBa2Ca2Cu3O10-x (135 K) compound.

Needed progress forward is impossible without thorough understanding of all the features found in superconductivity on both the macro and micro levels. The result of the research taken in this direction may lead to a discovery of a material, which would have high critical temperature and current. The findings of such materials could bring us closer to a breakthrough in the creation of highly efficient electric motors and power transmission lines, levitating transportation, and frictionless bearings. Currently known superconducting materials either have the critical temperature below 77 K, or are very fragile, preventing us from receiving an acceptable economical cost-effectiveness.

Yet, the pace of modern research is extraordinary. Today we already have and use the superconducting SQUID's that have the highest sensitivity to the magnetic field with minimal magnetosensitive zone. We use the Josephson electronic devices in the primary voltage standards. Among the infrared sensors, the superconducting receivers have maximum sensitivity and performance. The wires used in both systems for fusion devices and colliders are also superconductive. It's possible to assume that the quantum computer will be built on the basis of elements of superconducting spintronics. For the high-temperature superconductors the cooling process has already been considerably developed. On today's market we have both the acoustic refrigerators and the refrigerators made based on the thermoelectric effects, which are opening up new horizons for commercial applications of the superconducting devices.

The book gives an overview of major problems encountered in this field of study. A list of countries, in which researchers work, demonstrates global interest to the problem of superconductivity. Thus, this book would be a great reference for further research and development of ideas and applications.

First chapter of the book presents the theoretical results of the research done on superconductivity. The following chapters describe the manufacturing technology of the superconducting materials and structures as well as studied results of their properties. In the end, the book presents the devices currently designed for commercial applications. Besides the superconducting elements, these devices have the cooling systems of various design, electronic control units, information storage and processing units, and electrical elements that run required functions of the device. Technical realization of such devices demands great financial commitment by companies and other investors. Yet, the effect of these investments into the technosphere will pay off in the future, marking new heights in imminent progress and development.

Most of the material presented in this book is the result of authors' own research that has been carried out over a long period of time. A number of chapters thoroughly describe the fundamental electrical and structural properties of the superconductors as well as the methods researching those properties. The sourcebook comprehensively covers the advanced techniques and concepts of superconductivity. It's intended for a wide range of readers.

Dr. Yury Grigorashvili
National Research University, Moscow Institute of Electronic Technology, Moscow, Russia

Role of Antiferromagnetic Fluctuations in High Temperature Superconductivity

Sung-Ho Suck Salk[1,2,3], Jae-Hyeon Eom[1], Seung Joon Shin[1], Jae-Gon Eom[1] and Sung-Sik Lee[4]
[1]*Department of Physics, Pohang University of Science and Technology, Pohang*
[2]*Natural Science Division, The Korean Academy of Science and Technology, Sungnam-si*
[3]*Korea Institute of Advanced Study, Seoul*
[4]*Department of Physics and Astronomy, McMaster University, Hamilton, ON*
[1,2,3]*Korea*
[4]*Canada*

1. Introduction

The two-dimensional layers of cuprate oxides are known to be the systems of strongly repulsive (correlated) electrons as the Mott insulators which have revealed various novel physical properties uniquely different from the conventional low temperature superconductors. They show the antiferromagnetic (AF) infinite-range or long-range order (AFLRO) at and near half-filling. As hole doping increases, the AFLRO diminishes and the short-range (finite-range) AF order takes over with the emergence of d-wave superconductivity. The two-dimensional systems of strongly correlated electrons involved with strong repulsive interactions may favor the spin singlet pairing order (or correlations) of d-wave symmetry over that of s-wave symmetry. Here the spin singlet paring correlations are concerned with the AF spin fluctuations of the shortest possible correlation length among the AF spin fluctuations of all possible correlation lengths which appear in the region of hole doping away from half-filling. In this region of hole doping the cuprate oxides exhibit the novel structure of the high T_C phase diagram characterized by the dome-shaped superconducting transition temperature, T_C and the monotonously decreasing pseudogap temperature, T^*.

Earlier. slave-boson approaches of the t-J Hamiltionian were proposed by researchers in the field[1–3] in an attempt to reproduce the observed high T_C phase diagram. They were successful in reproducing the monotonously decreasing pseudogap temperature T^* in agreement with observation. T^* is shown to be the spin gap temperature at which the spin singlet pairing order or the spin (spinon) pairing correlations begins to open. However, their treatment of single-holon bose condensation has led to a linear increase of the bose condensation temperature T_C rather than the observed dome-shaped T_C[2, 3]. Later we introduced a slave-boson approach which allows the double-holon bose condensation[4] and failed to reproduce the dome-shaped T_C, also yielding the linearly increasing trend of T_C.

Soon after this study we[5] proposed an improved slave-boson theory which fundamentally differs from these approaches in that a term involving coupling between the spin and charge degrees of freedom or simply spin-charge coupling appears in our rigorous slave-boson treatment of the t-J Hamiltonian. The resulting effective mean field Lagrangian reveals coupling between the spin (spinon) paring order, Δ^f and the charge (holon) pairing order, Δ^b. As a consequence the Cooper pairing order is satisfactorily seen to be a composite of these two order parameters, Δ^f and Δ^b to allow for the bose condensation of the Cooper pairs rather than the single-holon bose condensation or the double-holon bose condensation. Accordingly this theory has led to successful reproductions of not only the monotonously decreasing spin gap temperature but also the long-waited dome-shaped structure of the superconducting transition temperature in the phase diagram. Further other important physical observations such as the boomerang behavior of superfluid weight, the peak-dip-hump structure of optical conductivity and both the temperature and doping dependence of spectral functions are reproduced in agreement with observations[6].

For the sake of self-containment we will first review our earlier proposed slave-boson theory[5] of the t-J Hamiltonian which reveals the spin-charge coupling mentioned above. Earlier it was shown by others that inclusion of the t' term in the t-J Hamiltonian leads to satisfactory descriptions of the electronic structure of high T_C cuprates[7–11] and the enhancement of pairing correlation resulting in an increasing trend of T_C in the overdoped region in the phase diagram for the choice of $t'/t < 0$, e.g., $t'/t = -0.3$[12, 13]. It is, thus, of great interest to see how its inclusion affects the entire structure of the phase diagram which includes the pseudogap temperature. At present there has been no study which addresses the role of the diagonal hopping t' on the spin gap temperature, T^*. Such study is needed to find whether there exists any relation between T^* and T_C or the spin gap phase and the superconducting phase. In this regard we would like to draw attention to the fact that the observed phase diagrams of high T_C cuprate samples (e.g., LSCO and BSCCO samples) reveal that higher the T^*, higher the T_C as earlier discussed by Oda et al.[14] This suggests that the two energy or temperature scales, T^* and T_C are no longer independent of each other. Thus one of our main objectives is to study how the pseudogap or spin gap temperature, T^* and the superconducting transition temperature, T_C are correlated and show that such correlation arises owing to the presence of the short-range antiferromagnetic (AF) spin fluctuations of the shortest possible correlation length involved with the spin pairing correlations. For a concerted, self-consistent study, we use a predicted phase diagram to calculate magnetic susceptibility and discuss two important observations made by the inelastic neutron scattering (INS) measurements, namely the temperature dependence of the magnetic resonance peak[17] and the linear scaling behavior between the magnetic peak resonance energy, E_{res} and the superconducting transition temperature[18]. From this study we show that the short-range AF spin fluctuations are directly responsible for the magnetic susceptibility observed by the INS measurements mentioned here.

2. Theory: U(1) slave boson representation of the t-J Hamiltonian

In the present study we limit ourselves to the derivation of the U(1) slave boson representation of the t-J Hamiltonian. We refer details of its derivation to Appendix A. In Appendix B a brief exposure of the SU(2) approach is made in association with the U(1) representation. Here only a rudimentary description is presented by introducing the next-nearest neighbor hopping or diagonal hopping t' term into the $t-J$ Hamiltonian. It is given by,

$$H_{t-t'-J} = -t \sum_{<i,j>} (\tilde{c}^{\dagger}_{i\sigma}\tilde{c}_{j\sigma} + c.c) - t' \sum_{<i,j>'} (\tilde{c}^{\dagger}_{i\sigma}\tilde{c}_{j\sigma} + c.c)$$
$$+J \sum_{<i,j>} (\mathbf{S}_i \cdot \mathbf{S}_j - \frac{1}{4} n_i n_j),. \quad (1)$$

Here $\sum_{<i,j>}$ denotes summation over the nearest neighbor sites i and j, $\sum_{<i,j>'}$, the summation over the next-nearest neighbor (diagonal) sites and $\tilde{c}_{i\sigma}(\tilde{c}^{\dagger}_{i\sigma})$, the electron annihilation(creation) operator with the constraint of no double occupancy at each site i. t is the nearest neighbor hopping integral; t', the next-nearest neighbor hopping integral and J, the Heisenberg coupling constant.

We take the slave-boson representation of electron operator as a composite of spinon (f) and holon (b), that is, $c_{i\sigma} = f_{i\sigma} b^{\dagger}_i$ with the single occupancy constraint at each site i. Following a rigorous slave-boson treatment of $\mathbf{S}_i = \frac{1}{2}\sum_{\alpha\beta} c^{\dagger}_{i\alpha}\sigma_{\alpha\beta}c_{i\beta}$ with $\sigma_{\alpha\beta}$, the Pauli spin matrix in the above equation, the resulting U(1) slave-boson representation of the above $t-t'-J$ Hamiltonian is given by

$$H_{t-t'-J} = -t \sum_{<i,j>,\sigma} (f^{\dagger}_{i\sigma} f_{j\sigma} b^{\dagger}_j b_i + c.c.) - t' \sum_{<i,j>',\sigma} (f^{\dagger}_{i\sigma} f_{j\sigma} b^{\dagger}_j b_i + c.c.)$$
$$-\frac{J}{2} \sum_{<i,j>} b_i b_j b^{\dagger}_j b^{\dagger}_i (f^{\dagger}_{i\downarrow} f^{\dagger}_{j\uparrow} - f^{\dagger}_{i\uparrow} f^{\dagger}_{j\downarrow})(f_{j\uparrow} f_{i\downarrow} - f_{j\downarrow} f_{i\uparrow})$$
$$+i \sum_i \lambda_i (\sum_{\upsilon} f^{\dagger}_{i\sigma} f_{i\sigma} + b^{\dagger}_i b_i - 1). \quad (2)$$

Here λ_i is the Lagrange multiplier field which enforces the single occupancy constraint.

Taking proper Hubbard-Stratonovich transformations and associated algebras by closely following our recently proposed slave-boson theory[5] (see Appendix A for details), one obtains the following effective Lagrangian,

$$\mathcal{L}_{eff} = \mathcal{L}_0 + \mathcal{L}_f + \mathcal{L}_b \quad (3)$$

with

$$\mathcal{L}_0 = \frac{Jx}{2} \sum_{<i,j>} \left\{ |\Delta^f_{ij}|^2 + \frac{1}{2}|\chi_{ij}|^2 + \frac{1}{4} \right\} + \frac{J}{2} \sum_{<i,j>} |\Delta^f_{ij}|^2 (|\Delta^b_{ij}|^2 + x^2), \quad (4)$$

$$\mathcal{L}_f = \sum_{i,\sigma} f^{\dagger}_{i\sigma}(\partial_\tau - \mu^f) f_{i\sigma}$$
$$-\frac{Jx}{4} \sum_{<i,j>,\sigma} \left\{ \chi^*_{ij} f^{\dagger}_{i\sigma} f_{j\sigma} + c.c. \right\}$$
$$-t'x \sum_{<i,j>',\sigma} \left\{ f^{\dagger}_{i\sigma} f_{j\sigma} + c.c. \right\}$$
$$-\frac{Jx}{2} \sum_{<i,j>} \left\{ \Delta^{f*}_{ij} (f_{i\downarrow} f_{j\uparrow} - f_{i\uparrow} f_{j\downarrow}) + c.c. \right\}, \quad (5)$$

for the spin spinon sector and

$$\begin{aligned}\mathcal{L}_b = & \sum_i b_i^\dagger(\partial_\tau - \mu^b)b_i - t\sum_{<i,j>}\left\{\chi_{ij}^* b_i^\dagger b_j + c.c.\right\} \\ & -t'(1-x)\sum_{<i,j>'}\left\{b_i^\dagger b_j + c.c.\right\} \\ & -\frac{J}{2}\sum_{<i,j>}|\Delta_{ij}^f|^2\left\{\Delta_{ij}^{b\,*} b_i b_j + c.c.\right\},\end{aligned} \tag{6}$$

for the charge (holon) sector. Here $\mu^f(\mu^b)$ is the spinon(holon) chemical potential. χ_{ij} is the hopping order parameter and $\Delta_{ij}^f =< (f_{i\uparrow}f_{j\downarrow} - f_{i\downarrow}f_{j\uparrow}) > (\Delta_{ij}^b =< b_i b_j >)$, the spinon (holon) pairing order parameter; x, the hole doping concentration and $J_x = J(1-x)^2$. The last term of Eq. 6 reveals the presence of coupling between the spin (spinon) and charge (holon) degrees of freedom, i.e., simply termed as spin-charge coupling, as seen in the form of the product of the spin (spinon) single pairing order, Δ^f and the charge (holon) pairing order, Δ^b. Thanks to this coupling the Cooper pairing order Δ is, now, properly represented as a composite of these two order parameters, Δ^f and Δ^b. We point out that the holon (charge) sector, Eq. 6 is coupled with the spinon (spin) sector, Eq. 5 owing to the presence of coupling between the spinon paring order Δ^f and the holon pairing order Δ^b as shown in the last term of Eq. 6.

The resulting free energy (see derivation in Appendix A) is given by,

$$\begin{aligned}F &= -\frac{1}{\beta}\ln\mathcal{Z} \\ &= J_x N\left(2|\Delta^f|^2 + \frac{1}{2}|\chi|^2\right) + JN|\Delta^f|^2|\Delta^b|^2 \\ &\quad -(1-x)N\mu_{eff}^f + Nx\mu_{eff}^b + \frac{1}{2}\sum_{\mathbf{k}} E_{\mathbf{k}s}^b \\ &\quad -2Nk_BT\ln 2 - 2k_BT\sum_{\mathbf{k}}\cosh\frac{\beta E_{\mathbf{k}s}^f}{2} \\ &\quad +k_BT\sum_{\mathbf{k}}\ln\left(1 - e^{-\beta E_{\mathbf{k}s}^b}\right),\end{aligned} \tag{7}$$

where $\mathcal{Z} = \int \mathcal{D}f\mathcal{D}b\mathcal{D}\chi\mathcal{D}\Delta^f\mathcal{D}\Delta^b\mathcal{D}\lambda e^{-\int_0^\beta d\tau\mathcal{L}_{eff}}$ is the partition function; $\Delta^f =< (f_{i\uparrow}f_{j\downarrow} - f_{i\downarrow}f_{j\uparrow}) > (\Delta^b =< b_i b_j >)$, the spinon(holon) pairing order parameter; $E_{\mathbf{k}}^f = \sqrt{(\epsilon_{\mathbf{k}}^f - \mu_{eff}^f) + \Delta_{\mathbf{k}}^{f\,2}}$ ($E_{\mathbf{k}}^b = \sqrt{(\epsilon_{\mathbf{k}}^b - \mu_{eff}^b) - \Delta_{\mathbf{k}}^{b\,2}}$), the spinon(holon) quasiparticle energy; x, the hole doping rate;$J_x = J(1-x)^2$ and N, the total number of sites in a square lattice. Here the spinon and holon energies, $\epsilon_{\mathbf{k}}^f$ and $\epsilon_{\mathbf{k}}^b$ are respectively,

$$\begin{aligned}\epsilon_{\mathbf{k}}^f &= -\frac{J_x}{2}\chi(\cos k_x + \cos k_y) - 4t'x\cos k_x\cos k_y \\ \epsilon_{\mathbf{k}}^b &= -2t\chi(\cos k_x + \cos k_y) - 4t'x\cos k_x\cos k_y.\end{aligned} \tag{8}$$

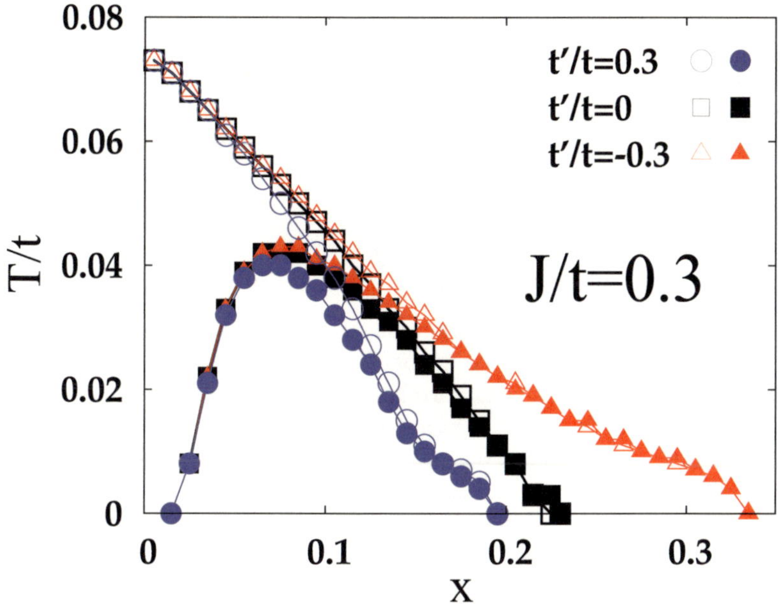

Fig. 1. (color online). Variation of phase diagram with t' for Heisenberg coupling strength, $J = 0.3t$.

The contribution of the next-nearest neighbor hopping or the diagonal hopping is readily understood from the inspection of Eq.8 by noting that the value of $\cos k_x \cos k_y$ is negative at the hot spot $(\pi, 0)$, zero at the cold spot $(\pi/2, \pi/2)$ and positive at $(0, 0)$. From this we see that stabilization (destabilization) of the spinon energy at the hot spot with $t' < 0$ ($t' > 0$) is expected to lead to the enhancement (depression) of AF spin (spinon) pairing correlations or the spin singlet pairing order of d-wave symmetry compared to the case of $t' = 0$), that is, no diagonal hoping. The charge (holon) pairing of s-wave symmetry will be enhanced at the nodal points.

3. Role of next-nearest neighbor hopping on the structure of phase diagram

Here we explore the role of the next-nearest neighbor hopping, i.e., the diagonal hopping on both the pseudogap temperatureT^* and the superconducting transition temperature, T_C and the cause of correlation between these two temperature scales or relatedly the spin gap phase and the d-wave superconducting phase. Earlier the negative value of t' was shown to match well the observed Fermi surface of the hole doped cuprate oxides while its positive value matches that of the electron doped cuprate oxides[7] as mentioned above.

Choosing the two different cases of the diagonal hopping, one for $t' < 0$ (e.g., $t' = -0.3t$) and the other for $t' > 0$ (e.g., $t' = 0.3t$), we examine the dependence of the phase diagram on t' for the hole doped cuprate oxides.

Fig.1 shows the computed results of the phase diagram with the variation of t'/t at a fixed value of the Heisenberg coupling constant, $J = 0.3t$. In the underdoped region both T^* and T_C are predicted to remain largely unchanged despite the considerable change of t'/t

from a positive value (t'/t = 0.3) to a negative ($t'/t = -0.3$) one. On the other hand, in the overdoped region both T* and T_C are seen to simultaneously increase (decrease) for the case of $t'/t < 0$ ($t'/t > 0$) with reference to that of $t'/t = 0$. The predicted superconducting transition temperature at optimal doping concentration did not change appreciably despite the considerable variation of t'/t as shown in the figure. The simultaneous increase (decrease) of T* and T_C with $t'/t = -0.3$ ($t'/t = 0.3$) indicates that the two temperature scales, T* and T_C or the spin gap phase and the superconducting phase are interrelated. To see the cause of such interplay, below we probe the role of the short-range AF spin fluctuations or the spin paring correlations on the determination of the phase diagram.

For the case of $t' < 0$ ($t' > 0$) the spinon energy at the hot spot(π,0) is lowered (raised) with reference that of $t' = 0$, i.e., no diagonal hopping, as can be readily understood from Eq. 8. Thus the spin (spinon) pairing correlation at the hot spot for $t' < 0$ is energetically more stable than the case for $t' > 0$. It is to be recalled that the spin gap temperature is the temperature at which the spin singlet paring order or (correlations) of d-wave symmetry or the spin pairing correlations emerges. The spin paring correlations will be less prone to change in the underdoped region compared to the case of the overdoped region. This is because owing to lower hole concentrations in this region, chances of electron hopping from site to site are reduced and, consequently, the existing short-range AF order is not easily perturbed. Thus the spin paring correlations or the short-range antiferromagnetic spin fluctuations is expected to remain more robust in the underdoped region compared to the case of the overdoped region. Indeed, the predicted T* and T_C is shown to be sensitive to the variation of t', preferentially in the overdoped region. This is displayed in Fig. 1.

It is reminded that the Cooper pairing order can be seen as the composite of the spin (spinon) pairing order Δ^f and the charge (holon) pairing order Δ^b, which results from the presence of the spin-charge coupling shown in the last term of Eq. 6. As a result of the coupling between the two orders, the superconducting phase transition with its onset temperature, T_C may arise owing the short-range AF spin fluctuations involved with the formation of the spin pairing order (correlations) which initiates the onset of the spin gap temperature T*. To put it otherwise, owing to the spin-charge coupling both T* and T_C are simultaneously affected or correlated. Indeed, such simultaneous change is seen to appear by exhibiting the simultaneous increase (decrease) of both T* and T_C with $t'/t < 0$ ($t'/t > 0$) as J increases. Such trend is seen in Fig. 2. Our findings of both the enhancement of the spin pairing correlations and the increasing trend of the superconducting transition temperature for $t' = -0.3t$ which appear in the overdoped region agree well with the variational Monte Carlo, mean-field calculations of Lee and coworkers [13]. However, unlike our present study they did not show a study of the spin gap temperature concerned with the role of the spin paring correlations.

For further verification from a different angle we closely examine the predicted structural dependence of the phase diagram on J in Fig.2. Both T^* and T_C are predicted to simultaneously increase with J as shown in Fig. 2. Needless to say, spin pairing correlations should increase with J. This will, in turn, cause the simultaneous increase of both the spin gap temperature and the superconducting transition temperature with increasing J. Such simultaneous increase with J is predicted as shown in the figure. This clearly demonstrates that the the superconducting transition temperature and the pseudogap temperature or relatedly the spin gap phase and the d-wave superconducting phase are correlated via the spin pairing correlations or the AF spin fluctuations of the shortest possible correlation length.

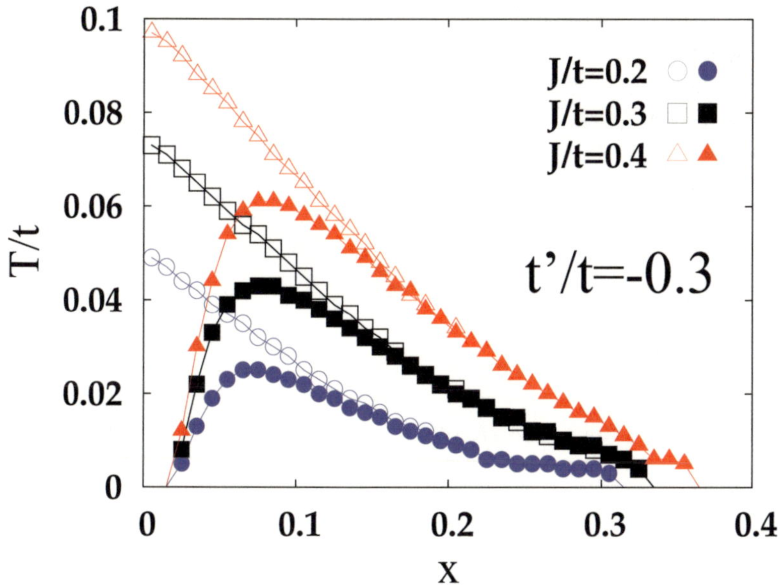

Fig. 2. (color online). Variaton of phase diagram with Heisenberg coupling strength J for $t'/t = -0.3$.

To put it otherwise, the short-range AF spin fluctuations play a key role of causing such inseparable relation between the two temperature scales or relatedly the spin gap phase and the superconducting phase. This finding is consistent with the observed phase diagrams with different cuprate samples which shows higher the T^*, higher the T_C[14] as mentioned earlier. It is then assured that the superconducting phase transition will not arise in the absence of the spin gap phase below T^* which is initiated by the short-range AF spin fluctuations involved with the spin paring correlations

4. Magnetic susceptibility based on the U(1) slave-boson representation

The observed high T_c phase diagrams of cuprate oxides are characterized by the pseudogap or spin gap phase which exists below the monotonously decreasing T^* and the d-wave superconducting phase below the dome shaped T_C[15, 16]. From their inelastic neutron scattering measurements (INS) of the temperature dependence of magnetic resonance peaks for $YB_2Cu_3O_{6+x}$ (YBCO) Dai et al. [17] reported that the magnetic resonance begins to appear at the pseudogap temperature T^* as its onset temperature and continues to exist with an increasing trend of the resonance peak height in the underdoped region as temperature is lowered and that near the optimal doping T^* tends to get closer to T_c. On the other hand, He et al. observed from their INS measurements of the doping dependence of the resonance peak energy, E_{res} for $Bi_2Sr_2CaCu_2O_{8+\delta}$ (BSCCO) that in the underdoped region E_{res} increases with increasing hole concentration x up to optimal doping x_0, and that in the overdoped cuprates it decreases with increasing x, by exhibiting a linear scaling behavior of E_{res} with T_c at all hole concentrations[18]. Most recently Stock et al.[19] observed that spin waves decay above the pseudogap of a heavily underdoped YBCO. Using a time-of-flight neutron spectroscopy for the studies of dynamic spin correlations or spin fluctuations in the overdoped $La_{2-x}Sr_xCuO_4$

(LSCO) sample, Wakimoto et al.[20] showed from their study of the doping dependence of antiferromagnetic spin excitations that the excitations decrease with hole doping above the optimal doping of $La_{2-x}Sr_xCuO_4$ (LSCO) and disappear at $x = 0.3$. Here we discuss the magnetic susceptibility[21, 22] at the wave vector $\vec{Q} = (\pi, \pi)$ in association with our computed phase diagram and focus our attention to the observed linear scaling behavior of magnetic resonance peak energy E_{res} with the superconducting transition temperature T_c. For self-containment we first discuss the U(1) slave-boson representation of irreducible magnetic susceptibility for our calculations of magnetic susceptibility.

Allowing external magnetic field $\vec{h}$, we introduce into the effective Lagrangian $\mathcal{L}_{eff}$ above the Zeeman coupling term,

$$\hat{H}^{ex} = -\sum_i \vec{S}_i \cdot \vec{h}_i, \tag{9}$$

where in the U(1) slave boson representation,

$$\begin{aligned}
\vec{S}_i \cdot \vec{h}_i &= \frac{1}{2}\sum_{k=1}^{3}(c^{\dagger}_{i\alpha}\sigma^k_{\alpha\beta}c_{i\beta})h^k_i \\
&= \frac{1}{2}\sum_{k=1}^{3}(f^{\dagger}_{i\alpha}b_i\sigma^k_{\alpha\beta}b^{\dagger}_i f_{i\beta})h^k_i \\
&= \frac{1}{2}\sum_{k=1}^{3}(1+b^{\dagger}_i b_i)(f^{\dagger}_{i\alpha}\sigma^k_{\alpha\beta}f_{i\beta})h^k_i, \\
&\simeq \frac{1}{2}\sum_{k=1}^{3} < (1+b^{\dagger}_i b_i) > (f^{\dagger}_{i\alpha}\sigma^k_{\alpha\beta}f_{i\beta})h^k_i,
\end{aligned} \tag{10}$$

The associated free energy is formally,

$$F = -\frac{1}{\beta} lnZ, \tag{11}$$

where $\beta = 1/kT$ and the partition function,

$$\mathcal{Z} = \int \mathcal{D}f\mathcal{D}b\mathcal{D}\chi\mathcal{D}\Delta^f\mathcal{D}\Delta^b\mathcal{D}\lambda e^{-\int_0^\beta d\tau \mathcal{L}_{eff}}. \tag{12}$$

Converting the magnetic (spin) susceptibility,

$$\chi_{kk'}(\vec{r}_i - \vec{r}_j, \tau - \tau') = -\beta\hbar \left. \frac{\delta^2 F[\vec{h}]}{\delta h^k_i(\tau)\delta h^{k'}_j(\tau')} \right|_{\vec{h}=0}. \tag{13}$$

into its four momentum space $(\vec{q}, \omega)$expression and allowing isotropic response to the applied magnetic field, the RPA form of magnetic susceptibility is obtained to be,[23]

$$\chi(\vec{q}, w) = \frac{\chi^0(\vec{q}, w)}{1 + J(\vec{q})\chi^0(\vec{q}, w)}, \tag{14}$$

where $J(\vec{q}) = 2J(\cos q_x + \cos q_y)$ and $\chi^0(\vec{q},\omega)$ is the irreducible magnetic susceptibility given by

$$\chi^0(\vec{q},w) = \frac{(1-x)^2}{4N}\sum_k$$
$$\Big[\frac{\left(E^f_{k+q}E^f_k + (\epsilon^f_{k+q}-\mu^f)(\epsilon^f_k-\mu^f)+\Delta^f_{k+q}\Delta^f_k\right)}{2E^f_{k+q}E^f_k}\frac{n^f(E^f_{k+q})-n^f(E^f_k)}{w-(E^f_{k+q}-E^f_k)+i\eta}$$
$$+\frac{\left(E^f_{k+q}E^f_k - (\epsilon^f_{k+q}-\mu^f)(\epsilon^f_k-\mu^f)-\Delta^f_{k+q}\Delta^f_k\right)}{2E^f_{k+q}E^f_k}\frac{n^f(E^f_{k+q})-n^f(-E^f_k)}{w-(E^f_{k+q}-(-E^f_k))+i\eta}$$
$$+\frac{\left(E^f_{k+q}E^f_k - (\epsilon^f_{k+q}-\mu^f)(\epsilon^f_k-\mu^f)-\Delta^f_{k+q}\Delta^f_k\right)}{2E^f_{k+q}E^f_k}\frac{n^f(-E^f_{k+q})-n^f(E^f_k)}{w-(-E^f_{k+q}-E^f_k)+i\eta}$$
$$+\frac{\left(E^f_{k+q}E^f_k + (\epsilon^f_{k+q}-\mu^f)(\epsilon^f_k-\mu^f)+\Delta^f_{k+q}\Delta^f_k\right)}{2E^f_{k+q}E^f_k}\frac{n^f(-E^f_{k+q})-n^f(-E^f_k)}{w-(-E^f_{k+q}-(-E^f_k))+i\eta}\Big], \quad (15)$$

where the quasi-spinon energy is $E^f_k = \sqrt{(\epsilon^f_k-\mu^f)^2+(\Delta^f_k)^2}$ with the effective bare spinon energy, $\epsilon^f_k = -\frac{J_x X}{2}(cosk_x + cosk_y) - 4xt' cosk_x cosk_y$; the spinon chemical potential, μ^f; the spinon gap, $\Delta^f_k = J_p\Delta_f\varphi_k$ with $\varphi_k = cosk_x - cosk_y$ and $n^f(E^f_k) = 1/(e^{\beta E^f_k}+1)$. In the complete expression of the effective Lagrangian Eq. 3, interplay between the two sectors, one for the spinon (spin) sector and the other for holon (charge) sector, namely Eq. 5 and Eq. 6 appears owing to the presence of coupling between the spinon pairing order and holon pairing order as shown in the last term of Eq. 6. Thus it should be noted that the effect of coupling between the two order parameter is embedded in the expression of the above irreducible magnetic susceptibility, Eq. 15, including the effect of the nearest neighbor hopping.

5. Computed results of magnetic susceptibility

Earlier, with the neglect of the next-nearest neighbor (or diagonal) hopping t' term we were able to obtain the generic feature of the dome shaped superconducting transition temperature and the monotonously decreasing pseudogap temperature in the phase diagram[5] in agreement with observaions[15, 16]. Now with the inclusion of the diagonal hopping term, such generic feature is, still, well predicted in the computed result of the phase diagram as shown in Fig. 1 and Fig. 2. As a concerted study we use the predicted phase diagram shown in Fig. 3 to calculate the magnetic spin susceptibility of present interest. As in our earlier study of the magnetic susceptibility[21, 22], we take the negative value[7, 13] of the next-nearest neighbor hopping integral with the choice of $t' = -0.45$ (to conform with the study of Brinckmann and Lee[23]) in the $t - t' - J$ Hamiltonian of interest[21] .

In Fig. 4 we display the variation of magnetic susceptibility at (π, π) with temperature T and transfer energy E at a fixed hole doping, $x = 0.05$. The magnetic resonance peak is shown to

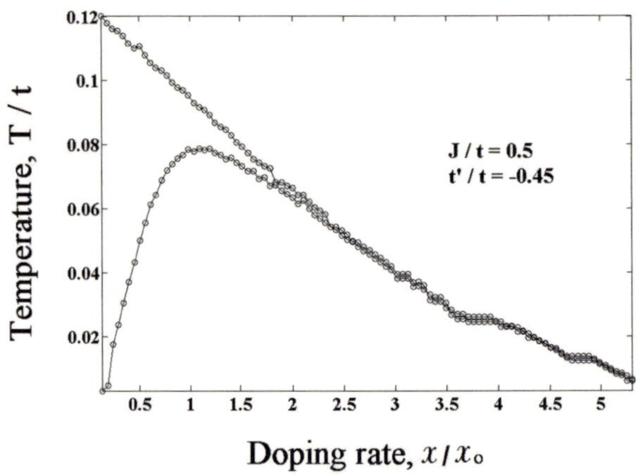

Fig. 3. (color online) Phase diagram with $t'/t = -0.45$, $J/t = 0.5$. Both temperature T and hole concentration x are in reduced units, T/t and x/x_0 respectively.

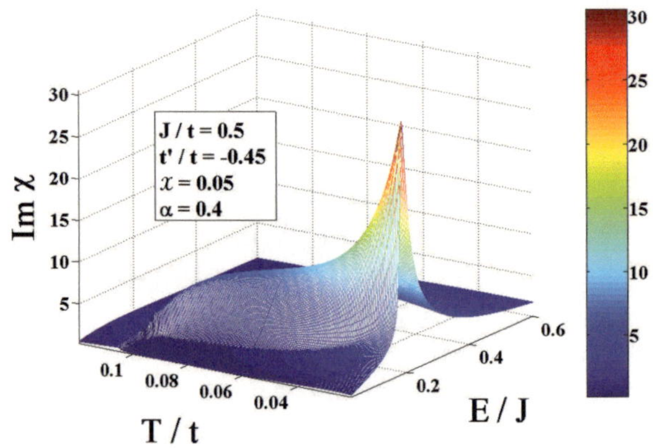

Fig. 4. (color online) Imaginary part of magnetic susceptibility vs. temperature and resonance energy at a fixed hole doping, $x = 0.05$ in the underdoped region

decrease with increasing temperature and disappears at the onset temperature T^*. It shows a steady decrease of the resonance peak peak height with increasing temperature and eventual disappearance at T^* in agreement with observation[17]. This indicates that the short-range AF spin fluctuations involved with the spin pairing correlations or the spin (spinon) singlet pairing order disappears at the onset temperature, T*.

He et al.[18] showed from their INS measurements of $Bi_2Sr_2CaCu_2O_{8+\delta}$ (BSCCO) that in the underdoped cuprates the magnetic (spin) resonance peak energy E_{res} (or ω_{res}) increases with T_c showing a linear scaling behavior between the two energy scales, E_{res} and T_c, i.e., $E_{res}/T_c \simeq const$. In Fig. 5 we show that the predicted E_{res} with $J_{eff} = \alpha J$ (where $\alpha = 0.4$[23]) monotonously increases with increasing T_c, yielding a linear scaling behavior of $E_{res}/T_c \simeq const$. This predicted linear scaling behavior is in agreement with the observations made by He et al.[18]. We note some quantitative differences between the observed value (around 5) and the predicted one (around 3).

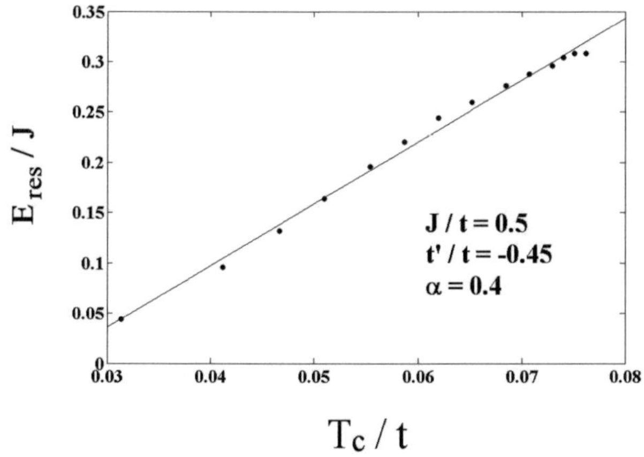

Fig. 5. (color online) Resonance peak energy (E_{res}/J in reduced unit) vs. superconducting transition temperature (T_c/t in reduced unit) with $t'/t = -0.45$.

6. Summary

In this study we applied the recently proposed slave-boson theory[5] in which the spin (spinon) paring order and the charge (holon) pairing order are coupled to result in the generic feature of the dome-shaped superconducting transition temperature and the monotonously decreasing spin gap temperature in the phase diagram. From the present study with the inclusion of the diagonal hopping t' term we also found that such generic feature still holds, as shown in Fig. 1 through Fig. 3. Further we showed that there exists correlation (or interplay) between the two different temperature scales, T^* and T_C, resulting in the increasing T_C with increasing T^*. Relatedly, it can be said that the superconducting phase is correlated with the spin gap phase. We find that such correlation between the two phases is attributed to the short-range AF spin fluctuations involved with spin pairing correlations. The simultaneous increase of the superconducting transition temperature with the spin gap temperature with incrasing J is shown to be consistent with the observed phase diagrams for high T_C cuprate samples (e.g., LSCO and BSCCO samples)[15] which shows that the higher T^* samples always accompany higher T_C. In addition, to achieve a self-consistent, concerted study we used the predicted phase diagram to study the magnetic susceptibility. Specifically, resorting to the computed phase diagram shown in Fig. 3 we found that both the temperature dependence of the magnetic resonace peak and the linear scaling behavior of the magnetic (spin) resonance peak energy E_{res} with the superconducting transition temperature T_c agree with the INS measurements[17, 18]. We showed that this linear scaling behavior is attributed to the short-range AF spin fluctuations. Although not discussed here, such linear relation is found to be invariant with the Heisenberg coupling constant[22], implying high T_C cuprate sample independence. In short, based on the above concerted studies of both the phase diagram and the magnetic susceptibility we find that the short-range (spin dimer) AF spin fluctuations of the shortest possible correlation length involved with the spin pairing correlations are responsible for high T_C superconductivity. We argue that this finding is supported by the reproducibility of both the dome-shaped superconducting transition temperature, T_C in the phase diagram and the linear scaling behavior between E_{res} and T_C, in both of which the T_C

and thus the superconducting phase transition is shown to occur as a result of the short-range AF spin fluctuations in association with the spin-charge coupling.

7. Acknowledgement

One(SHSS) of us greatly acknowledges the Korean Ministry of Education (HakJin Program) and Pohang University of Science and Technology for financial supports at the initial stage of the present work. He thanks the Korean Academy of Science and Technology for continuous encouragements in science research. We are grateful to Professor Ki-Seok Kim at Pohang University of Science and Technology for his earlier contribution.

8. Appendix A: Heisenberg interaction term in the U(1) slave-boson representation

The t-J Hamiltonian of interest is given by,

$$\begin{aligned} H = & -t \sum_{<i,j>} (c^{\dagger}_{i\sigma} c_{j\sigma} + c.c.) + J \sum_{<i,j>} (\mathbf{S}_i \cdot \mathbf{S}_j - \frac{1}{4} n_i n_j) \\ & -\mu \sum_i c^{\dagger}_{i\sigma} c_{i\sigma} \end{aligned} \tag{A1}$$

and the Heisenberg interaction term is rewritten

$$\begin{aligned} H_J &= J \sum_{<i,j>} (\mathbf{S}_i \cdot \mathbf{S}_j - \frac{1}{4} n_i n_j) \\ &= -\frac{J}{2} \sum_{<i,j>} (c^{\dagger}_{i\downarrow} c^{\dagger}_{j\uparrow} - c^{\dagger}_{i\uparrow} c^{\dagger}_{j\downarrow})(c_{j\uparrow} c_{i\downarrow} - c_{j\downarrow} c_{i\uparrow}). \end{aligned} \tag{A2}$$

Here t is the hopping energy and $\mathbf{S}_i$, the electron spin operator at site i, $\mathbf{S}_i = \frac{1}{2} c^{\dagger}_{i\alpha} \sigma_{\alpha\beta} c_{i\beta}$ with $\sigma_{\alpha\beta}$, the Pauli spin matrix element. n_i is the electron number operator at site i, $n_i = c^{\dagger}_{i\sigma} c_{i\sigma}$. μ is the chemical potential.

In the U(1) slave-boson representation[1, 2, 24, 25], with single occupancy constraint at site i the electron annihilation operator $c_{i\sigma}$ is taken as a composite operator of the spinon (neutrally charged fermion) annihilation operator $f_{i\sigma}$ and the holon (positively charged boson) creation operator $b^{\dagger}_i$, and thus, $c_{i\sigma} = f_{i\sigma} b^{\dagger}_i$. Rigorously speaking, it should be noted that the expression $c_{i\sigma} = b^{\dagger}_i f_{i\sigma}$ is not precise since these operators belong to different Hilbert spaces and thus the equality sign here should be taken only as a symbol for mapping. Using $c_{i\sigma} = f_{i\sigma} b^{\dagger}_i$ and introducing the Lagrange multiplier term (the last term in Eq.(A3)) to enforce single occupancy constraint, the t-J Hamiltonian is rewritten,

$$\begin{aligned} H = & -t \sum_{<i,j>} \left((f^{\dagger}_{i\sigma} b_i)(b^{\dagger}_j f_{j\sigma}) + c.c. \right) + H_J \\ & -\mu \sum_i f^{\dagger}_{i\sigma} b_i f_{i\sigma} b^{\dagger}_i \\ & -i \sum_i \lambda_i (b^{\dagger}_i b_i + f^{\dagger}_{i\sigma} f_{i\sigma} - 1) \end{aligned} \tag{A3}$$

with the Heisenberg interaction term,

$$H_J = -\frac{J}{2} \sum_{<i,j>} b_i b_j b_j^\dagger b_i^\dagger (f_{i\downarrow}^\dagger f_{j\uparrow}^\dagger - f_{i\uparrow}^\dagger f_{j\downarrow}^\dagger)(f_{j\uparrow} f_{i\downarrow} - f_{j\downarrow} f_{i\uparrow}). \quad \text{(A4)}$$

The first term represents hopping of a spinon from site j to site i and of a holon (positively charged boson) from site i to site j. In the slave-boson representation a charged fermion (electron or hole) is taken as a composite particle of a 'spinon' and a 'holon'. They can conveniently serve as book-keeping labels to discern physical properties or objects involved with the charge or spin degree of freedom (e.g., spin gap phase, spin singlet pairs, hole pairs, ...). With the single occupancy constraint, electron is allowed to hop from a singly occupied copper site i only to a vacant copper site j. A site of single occupancy in the CuO_2 plane of high T_c cuprates physically represents an electrically neutral site (net charge 0) with an electron of spin 1/2 and the vacant site, a site of positive charge $+e$ with net spin 0. In the slave-boson representation, hopping of an electron (a composite of spinon and holon) from a singly occupied copper site (neutral site) j to an empty site (positively charged site with $+e$) i results in the annihilation of a spinon (a fermion of charge 0 and spin 1/2) and the creation of a positively charged holon (a boson of charge $+e$ and spin 0) at site j while at the copper site i a composite of a spinon (fermion of charge 0 and spin 1/2) and a negatively charged holon is created. It is of note that as a result of electron hopping the newly occupied copper site i in the CuO_2 plane can, also, be labeled as 'spinon' since this is an electrically neutral (charge 0) site with an electron of spin 1/2 and the vacant site j, 'holon' since this is a positively charged site with a single charge $+e$ and the net spin 0 as mentioned above. Thus in practical sense, there is no distinction between the two different cases above. At times, we will call the singly occupied site as 'spinon' and the vacant (empty) site as 'holon' as long as there is no confusion. This is because any site occupied by a spinon is identified as an electrically neutral site occupied by a single electron with spin 1/2 and the site with a positive holon is a positively charged vacant site with spin 0. Thus physical spin-charge separation is not allowed.

The Heisenberg interaction term, Eq.(A4) shows coupling between the charge and spin degrees of freedom. Physics involved with the charge degree of freedom is manifested by the four holon (boson) operator $b_i b_j b_j^\dagger b_i^\dagger$ in the Heisenberg interaction term. Judging from the intersite charge coupling term $\frac{J}{4} n_i n_j$ present in the Heisenberg interaction term $H_J = J \sum_{<i,j>} (\mathbf{S}_i \cdot \mathbf{S}_j - \frac{1}{4} n_i n_j)$, it is obvious that this charge contribution can not be neglected in its slave-boson representation. It is to be noted that the Hubbard Hamiltonian contains repulsive interaction U between charged particles and is mapped into the t-J Hamiltonian H_{t-J} in the large U limit. The Coulomb repulsion, $U n_{i\uparrow} n_{i\downarrow} = \frac{U}{4}(n_{i\uparrow} + n_{i\downarrow})^2 - \frac{U}{4}(n_{i\uparrow} - n_{i\downarrow})^2$ obviously manifests the presence of both the charge (the first term) and spin (the second term) degrees of freedom. Thus, under mapping the charge part of contribution naturally appears in the Heisenberg interaction term.

Let us now take another look at the importance of the charge contribution. In general, uncertainty principle between the number density (amplitude) and the phase of a boson order parameter applies. As an example, arbitrarily large fluctuations of the number density fix the phase, or arbitrarily large phase fluctuations fix the number density of the boson. The conventional BCS superconductors of long coherence length meet the former classification, and thus the phase fluctuations of the Cooper pair order parameter are minimal. For charged bosons, e.g., the Cooper pairs, the number density fluctuations refer to charge density fluctuations. For short coherence length superconductors such as the high T_c cuprate systems of present interest, local charge density fluctuations exist and cause large phase fluctuations. Thus, both the charge and phase fluctuations need to be taken into account to fully exploit the quantum fluctuations .

Let us now consider the importance of the charge and spin fluctuations. In generally, coupling between physical quantities A and B is decomposed into terms involving fluctuations of A, i.e., $(A- < A >)$ and B, i.e., $(B- < B >)$, separately uncorrelated mean field contribution of $< A >$ and $< B >$ and correlation between fluctuations of A and B, that is, $(A- < A >)$ and $(B- < B >)$ respectively; $AB = (A- < A >) < B > +(B- < B >) < A > + < A >< B > +(A- < A >)(B- < B >)$. Setting $A = b_i b_j b_j^\dagger b_i^\dagger$ for charge (holon) contribution and $B = (f_{i\downarrow}^\dagger f_{j\uparrow}^\dagger - f_{i\uparrow}^\dagger f_{j\downarrow}^\dagger)(f_{j\uparrow} f_{i\downarrow} - f_{j\downarrow} f_{i\uparrow})$ for spin (spinon) contribution, the Heisenberg coupling term, Eq(A4) can be decomposed into terms involving coupling between the charge and spin fluctuations separately, the mean field contributions and coupling (correlation) between fluctuations (charge and spin fluctuations). Using such decomposition of the Heisenberg interaction term for Eq.(A3), we write the partition function,

$$Z = \int DbDf_\uparrow Df_\downarrow D\lambda e^{-S[b,f,\lambda]}, \tag{A5}$$

where

$$S[b,f,\lambda] = \int_0^\beta d\tau \Big[\sum_i b_i^\dagger \partial_\tau b_i + \sum_i f_{i\sigma}^\dagger \partial_\tau f_{i\sigma} + H_{t-J}^{U(1)} \Big] \tag{A6}$$

with $\beta = \frac{1}{k_B T}$, the inverse temperature and $H_{t-J}^{U(1)}$, the U(1) symmetry preserved Hamiltonian,

$$\begin{aligned}
H_{t-J}^{U(1)} &= -t \sum_{<i,j>} (f_{i\sigma}^\dagger f_{j\sigma} b_j^\dagger b_i + c.c.) \\
&- \frac{J}{2} \sum_{<i,j>} \Big[\Big\langle (f_{i\downarrow}^\dagger f_{j\uparrow}^\dagger - f_{i\uparrow}^\dagger f_{j\downarrow}^\dagger)(f_{j\uparrow} f_{i\downarrow} - f_{j\downarrow} f_{i\uparrow}) \Big\rangle b_i b_j b_j^\dagger b_i^\dagger \\
&+ \Big\langle b_i b_j b_j^\dagger b_i^\dagger \Big\rangle (f_{i\downarrow}^\dagger f_{j\uparrow}^\dagger - f_{i\uparrow}^\dagger f_{j\downarrow}^\dagger)(f_{j\uparrow} f_{i\downarrow} - f_{j\downarrow} f_{i\uparrow}) \\
&- \Big\langle b_i b_j b_j^\dagger b_i^\dagger \Big\rangle \Big\langle (f_{i\downarrow}^\dagger f_{j\uparrow}^\dagger - f_{i\uparrow}^\dagger f_{j\downarrow}^\dagger)(f_{j\uparrow} f_{i\downarrow} - f_{j\downarrow} f_{i\uparrow}) \Big\rangle \\
&+ \Big(b_i b_j b_j^\dagger b_i^\dagger - \Big\langle b_i b_j b_j^\dagger b_i^\dagger \Big\rangle \Big) \Big((f_{i\downarrow}^\dagger f_{j\uparrow}^\dagger - f_{i\uparrow}^\dagger f_{j\downarrow}^\dagger)(f_{j\uparrow} f_{i\downarrow} - f_{j\downarrow} f_{i\uparrow}) \\
&- \Big\langle (f_{i\downarrow}^\dagger f_{j\uparrow}^\dagger - f_{i\uparrow}^\dagger f_{j\downarrow}^\dagger)(f_{j\uparrow} f_{i\downarrow} - f_{j\downarrow} f_{i\uparrow}) \Big\rangle \Big) \Big] \\
&- \mu \sum_i f_{i\sigma}^\dagger f_{i\sigma} (1 + b_i^\dagger b_i) - i \sum_i \lambda_i (f_{i\sigma}^\dagger f_{i\sigma} + b_i^\dagger b_i - 1).
\end{aligned} \tag{A7}$$

8.1 U(1) mean field Hamiltonian

Noting that $[b_i, b_j^\dagger] = \delta_{ij}$ for boson, the intersite charge (holon) interaction term (the second term) in Eq.(A7) is rewritten,

$$-\frac{J}{2}\Big\langle(f_{i\downarrow}^\dagger f_{j\uparrow}^\dagger - f_{i\uparrow}^\dagger f_{j\downarrow}^\dagger)(f_{j\uparrow}f_{i\downarrow} - f_{j\downarrow}f_{i\uparrow})\Big\rangle b_i b_j b_j^\dagger b_i^\dagger$$
$$= -\frac{J}{2} < |\Delta_{ij}^f| >^2 \left(1 + b_i^\dagger b_i + b_j^\dagger b_j + b_i^\dagger b_j^\dagger b_j b_i\right), \tag{A8}$$

with $\Delta_{ij}^f = f_{j\uparrow}f_{i\downarrow} - f_{j\downarrow}f_{i\uparrow}$, the spinon pairing field. The third term in Eq.(A7) represents the intersite spin (spinon) interaction and is rewritten,

$$-\frac{J}{2}\Big\langle b_i b_j b_j^\dagger b_i^\dagger\Big\rangle(f_{i\downarrow}^\dagger f_{j\uparrow}^\dagger - f_{i\uparrow}^\dagger f_{j\downarrow}^\dagger)(f_{j\uparrow}f_{i\downarrow} - f_{j\downarrow}f_{i\uparrow})$$
$$= -\frac{J_p}{2}(f_{i\downarrow}^\dagger f_{j\uparrow}^\dagger - f_{i\uparrow}^\dagger f_{j\downarrow}^\dagger)(f_{j\uparrow}f_{i\downarrow} - f_{j\downarrow}f_{i\uparrow}), \tag{A9}$$

where $J_p = J(1+ < b_i^\dagger b_i > + < b_j^\dagger b_j > + < b_i^\dagger b_i b_j^\dagger b_j >)$ or $J_p = J(1-x)^2$ with x, the uniform hole doping concentration[27]. The fourth term in Eq.(A7) is written,

$$\frac{J}{2}\Big\langle b_i b_j b_j^\dagger b_i^\dagger\Big\rangle\Big\langle(f_{i\downarrow}^\dagger f_{j\uparrow}^\dagger - f_{i\uparrow}^\dagger f_{j\downarrow}^\dagger)(f_{j\uparrow}f_{i\downarrow} - f_{j\downarrow}f_{i\uparrow})\Big\rangle$$
$$= \frac{J}{2}\left(1+ < b_i^\dagger b_i > + < b_j^\dagger b_j > + < b_i^\dagger b_i b_j^\dagger b_j >\right) < |\Delta_{ij}^f|^2 > . \tag{A10}$$

The intersite spinon interaction term in Eq.(A9) is decomposed into the direct, exchange and pairing channels[25],

$$-\frac{J_p}{2}(f_{i\downarrow}^\dagger f_{j\uparrow}^\dagger - f_{i\uparrow}^\dagger f_{j\downarrow}^\dagger)(f_{j\uparrow}f_{i\downarrow} - f_{j\downarrow}f_{i\uparrow})$$
$$= \frac{J_p}{4}\Big[\sum_{k=1}^{3}(f_{i\alpha}^\dagger\sigma_{\alpha\beta}^k f_{i\beta})(f_{j\gamma}^\dagger\sigma_{\gamma\delta}^k f_{j\delta}) - (f_{i\alpha}^\dagger\sigma_{\alpha\beta}^0 f_{i\beta})(f_{i\gamma}^\dagger\sigma_{\gamma\delta}^0 f_{j\delta})\Big]$$
$$= v_D + v_E + v_P \tag{A11}$$

with $\sigma^0 = I$, the identity matrix and $\sigma^{1,2,3}$, the Pauli spin matrices, where v_D, v_E and v_P are the spinon interaction terms of the direct, exchange and pairing channels respectively ,

$$v_D = -\frac{J_p}{8}\sum_{k=0}^{3}(f_i^\dagger\sigma^k f_i)(f_j^\dagger\sigma^k f_j), \tag{A12}$$
$$v_E = -\frac{J_p}{4}\Big((f_{i\sigma}^\dagger f_{j\sigma})(f_{j\sigma}^\dagger f_{i\sigma}) - n_i\Big), \tag{A13}$$
$$v_P = -\frac{J_p}{2}(f_{i\downarrow}^\dagger f_{j\uparrow}^\dagger - f_{i\uparrow}^\dagger f_{j\downarrow}^\dagger)(f_{j\uparrow}f_{i\downarrow} - f_{j\downarrow}f_{i\uparrow}). \tag{A14}$$

Here σ^0 is the unit matrix and $\sigma^{1,2,3}$, the Pauli spin matrices.

Combining Eq.(A8) and Eq.(A10), we have

$$
\begin{aligned}
&-\frac{J}{2} < |\Delta^f_{ij}|^2 > \left(1 + b_i^\dagger b_i + b_j^\dagger b_j + b_i^\dagger b_j^\dagger b_j b_i\right) \\
&+\frac{J}{2} < |\Delta^f_{ij}|^2 > \left(1 + < b_i^\dagger b_i > + < b_j^\dagger b_j > + < b_i^\dagger b_i b_j^\dagger b_j >\right) \\
&= -\frac{J}{2} < |\Delta^f_{ij}|^2 > b_i^\dagger b_j^\dagger b_j b_i + \frac{J}{2} < |\Delta^f_{ij}|^2 >< b_i^\dagger b_i b_j^\dagger b_j > \\
&-\frac{J}{2} < |\Delta^f_{ij}|^2 > \left[\left(b_i^\dagger b_i - < b_i^\dagger b_i >\right) + \left(b_j^\dagger b_j - < b_j^\dagger b_j >\right)\right]. \qquad \text{(A15)}
\end{aligned}
$$

Collecting the decomposed terms Eq.(A8) through Eq.(A10) in association with Eqs.(A11) through Eq.(A15), we write

$$
\begin{aligned}
H_J &= -\frac{J}{2} \sum_{<i,j>} |\Delta^f_{ij}|^2 b_i^\dagger b_j^\dagger b_j b_i \\
&-J_p \sum_{<i,j>} \left[\frac{1}{2}(f_{i\downarrow}^\dagger f_{j\uparrow}^\dagger - f_{i\uparrow}^\dagger f_{j\downarrow}^\dagger)(f_{j\uparrow} f_{i\downarrow} - f_{j\downarrow} f_{i\uparrow})\right. \\
&+\frac{1}{4}\left((f_{i\sigma}^\dagger f_{j\sigma})(f_{j\sigma}^\dagger f_{i\sigma}) - n_i\right) \\
&\left.+\frac{1}{8}\sum_{k=0}^{3}(f_i^\dagger \sigma^k f_i)(f_j^\dagger \sigma^k f_j)\right] \\
&+\frac{J}{2} \sum_{<i,j>} |\Delta^f_{ij}|^2 < b_i^\dagger b_i >< b_j^\dagger b_j > \\
&-\frac{J}{2} \sum_{<i,j>} |\Delta^f_{ij}|^2 \left[\left(b_i^\dagger b_i - < b_i^\dagger b_i >\right) + \left(b_j^\dagger b_j - < b_j^\dagger b_j >\right)\right], \qquad \text{(A16)}
\end{aligned}
$$

where we considered $< |\Delta^f_{ij}|^2 >= |\Delta^f_{ij}|^2$ and ignored the fifth term in Eq.(A7).

Hubbard Stratonovich transformation for the holon pairing term (the second term of Eq.(A16)) leads to

$$
e^{\sum_{<i,j>} \frac{J}{2}|\Delta^f_{ij}|^2 b_i^\dagger b_j^\dagger b_i b_j} \propto \int \prod_{<i,j>} d\Delta^{b*}_{ij} d\Delta^b_{ij} e^{-\sum_{<i,j>} \frac{J}{2}|\Delta^f_{ij}|^2 \left[|\Delta^b_{ij}|^2 - \Delta^{b*}_{ij}(b_i b_j) - \Delta^b_{ij}(b_j^\dagger b_i^\dagger)\right]}, \qquad \text{(A17)}
$$

and the saddle point approximation yields,

$$
H_P^b = \sum_{<i,j>} \frac{J}{2}|\Delta^f_{ij}|^2\left[|\Delta^{0b}_{ij}|^2 - \Delta^{0b*}_{ij}(b_j b_i) - \Delta^{0b}_{ij}(b_j^\dagger b_i^\dagger)\right], \qquad \text{(A18)}
$$

where $\Delta^{0b}_{ij} =< b_i b_j >$ is the saddle point for the holon pairing order parameter Δ^b_{ij}. Since confusion is not likely to occur, we will use the notation Δ^b_{ij} in place of Δ^{0b}_{ij} for the saddle point. As are shown in Eqs.(A12) through (A14) the spinon interaction term is decomposed into the direct, exchange and pairing channels respectively. Proper Hubbard-Stratonovich transformations corresponding to these channels and saddle point approximation leads to the effective Hamiltonian,

$$
\begin{aligned}
H_{eff} = & \frac{J_p}{4} \sum_{<i,j>} \left[|\chi_{ij}|^2 - \chi_{ij}^* (f_{i\sigma}^\dagger f_{j\sigma} + \frac{4t}{J_p} b_i^\dagger b_j) - c.c. \right] \\
& + \frac{J}{2} \sum_{<i,j>} |\Delta_{ij}^f|^2 \left[|\Delta_{ij}^b|^2 - \Delta_{ij}^{b*} (b_j b_i) - c.c \right] \\
& + \frac{J_p}{2} \sum_{<i,j>} \left[|\Delta_{ij}^f|^2 - \Delta_{ij}^f (f_{i\downarrow}^\dagger f_{j\uparrow}^\dagger - f_{i\uparrow}^\dagger f_{j\downarrow}^\dagger) - c.c. \right] \\
& + \frac{J_p}{2} \sum_{<i,j>} \sum_{l=0}^{3} \left[(\rho_j^l)^2 - \rho_j^l (f_i^\dagger \sigma^l f_i) \right] + \frac{J_p}{2} \sum_i (f_{i\sigma}^\dagger f_{i\sigma}) \\
& + \frac{4t^2}{J_p} \sum_{<i,j>} (b_i^\dagger b_j)(b_j^\dagger b_i) \\
& + \frac{J}{2} \sum_{<i,j>} |\Delta_{ij}^f|^2 < b_i^\dagger b_i >< b_j^\dagger b_j > \\
& - \frac{J}{2} \sum_{<i,j>} |\Delta_{ij}^f|^2 \left[\left(b_i^\dagger b_i - < b_i^\dagger b_i > \right) + \left(b_j^\dagger b_j - < b_j^\dagger b_j > \right) \right] \\
& - \mu \sum_i f_{i\sigma}^\dagger f_{i\sigma} (1 + b_i^\dagger b_i) - i \sum_i \lambda_i (f_{i\sigma}^\dagger f_{i\sigma} + b_i^\dagger b_i - 1),
\end{aligned}
\tag{A19}
$$

where $\Delta_{ij}^b =< b_i b_j >$, $\chi_{ij} =< f_{i\sigma}^\dagger f_{j\sigma} + \frac{4t}{J_p} b_i^\dagger b_j >$, $\Delta_{ij}^f =< f_{j\uparrow} f_{i\downarrow} - f_{j\downarrow} f_{i\uparrow} >$ and $\rho_i^k =< \frac{1}{2} f_i^\dagger \sigma^k f_i >$ are proper saddle points.

We note that $\rho_i^l = \frac{1}{2} < f_i^\dagger \sigma^l f_i >=< S_i^l >= 0$ for $l = 1, 2, 3$, $\rho_i^0 = \frac{1}{2} < f_{i\sigma}^\dagger f_{i\sigma} >= \frac{1}{2}$ for $l = 0$ for the contribution of the direct spinon interaction term (the fourth term). The expression $(b_j^\dagger b_i)(b_i^\dagger b_j)$ in the fifth term of Eq.(A19) represents the exchange interaction channel. The exchange channel will be ignored owing to a large cost in energy, $U \approx \frac{4t^2}{J}$ [25, 26]. The resulting effective Hamiltonian is

$$
H^{MF} = H^{\Delta,\chi} + H^b + H^f, \tag{A20}
$$

where $H^{\Delta,\chi}$ represents the the saddle point energy involved with the spinon pairing order parameter Δ^f, the holon pairing order parameter Δ^b and the hopping order parameter χ,

$$
\begin{aligned}
H^{\Delta,\chi} = & J \sum_{<i,j>} \left[\frac{1}{2} |\Delta_{ij}^f|^2 |\Delta_{ij}^b|^2 + \frac{1}{2} |\Delta_{ij}^f|^2 x^2 \right] \\
& + \frac{J_p}{2} \sum_{<i,j>} \left[|\Delta_{ij}^f|^2 + \frac{1}{2} |\chi_{ij}|^2 + \frac{1}{4} \right],
\end{aligned}
\tag{A21}
$$

H^b is the holon Hamiltonian,

$$H^b = -t \sum_{<i,j>} \left[\chi_{ij}^*(b_i^\dagger b_j) + c.c.\right]$$
$$- \sum_{<i,j>} \frac{J}{2}|\Delta_{ij}^f|^2 \left[\Delta_{ij}^{b*}(b_i b_j) + c.c.\right]$$
$$- \sum_i \mu_i^b (b_i^\dagger b_i - x), \tag{A22}$$

where $\mu_i^b = i\lambda_i + \frac{J}{2}\sum_{j=i\pm\hat{x},i\pm\hat{y}} |\Delta_{ij}^f|^2$ and H^f, the spinon Hamiltonian,

$$H^f = -\frac{J_p}{4} \sum_{<i,j>} \left[\chi_{ij}^*(f_{i\sigma}^\dagger f_{j\sigma}) + c.c.\right]$$
$$- \frac{J_p}{2} \sum_{<i,j>} \left[\Delta_{ij}^{f*}(f_{j\uparrow} f_{i\downarrow} - f_{j\downarrow} f_{i\uparrow}) + c.c.\right]$$
$$- \sum_i \mu_i^f \left(f_{i\sigma}^\dagger f_{i\sigma} - (1-x)\right), \tag{A23}$$

where $\mu_i^f = \mu(1-x) + i\lambda_i$.

As can be seen from Eqs.(A21) through (A23), Eq.(A20) reveals the importance of coupling between the spin and charge degrees of freedom, that is, coupling between the spinon pairing and holon pairing. Thus no spin-charge separation appears in the mean-field Hamiltonian above contrary to other mean field theories[1–3, 24, 26] which pay attention to the single-holon bose condensation. As can be seen from the second term in Eq.(A22) which represents holon pairing contribution it is expected that, owing to the coupling effect, bose condensation (or superconducting phase transition) will occur only in the presence of the non-vanishing spin singlet pairing order, Δ^f owing to the coupling effects mentioned above. Indeed, in high T_c cuprates superconductivity is not observed above the pseudogap (spin gap) temperatures T^* where the spin singlet pairing order disappears.

8.2 U(1) free energy

The diagonalized Hamiltonian for Eq.(A20) above is obtained to be (see Appendix A for detailed derivations),

$$H_{U(1)}^{MF} = NJ\left[\Delta_f^2\Delta_b^2 + \Delta_f^2 x^2\right] + NJ_p\left[\Delta_f^2 + \frac{1}{2}\chi^2 + \frac{1}{4}\right]$$
$$+ \sum_{k,s}{}' E_{ks}^f (\alpha_{ks\uparrow}^\dagger \alpha_{ks\uparrow} - \alpha_{ks\downarrow}^\dagger \alpha_{ks\downarrow}) - Nx\mu^f$$
$$+ \sum_{k,s=\pm 1}{}' E_{ks}^b \beta_{ks}^\dagger \beta_{ks} + \sum_{k,s=\pm 1}{}' \frac{1}{2}(E_{ks}^b + \mu^b) + \mu^b N x, \tag{A24}$$

where E^f_{ks} is the quasispinon excitation energy,

$$E^f_{ks} = \sqrt{(\epsilon^f_{ks} - \mu^f)^2 + (\Delta^f_0)^2} \tag{A25}$$

with the spinon pairing energy (gap), $\Delta^f_0 = J_p \xi_k(\tau^f)\Delta_f$, and E^b_{ks} is the quasiholon excitation energy,

$$E^b_{ks} = \sqrt{(\epsilon^b_{ks} - \mu^b)^2 - (\Delta^b_0)^2}, \tag{A26}$$

where the holon pairing energy, $\Delta^b_0 = J\Delta^2_f \xi_k(\tau^b)\Delta_b$ and with $\phi = \theta, \tau^f$ or τ^b,

$$\xi_k(\phi) = \sqrt{\gamma^2_k \cos^2\phi + \varphi^2_k \sin^2\phi},, \tag{A27}$$

$$\epsilon^f_{ks} = \frac{J_p}{2} s\chi\xi_k(\theta), \tag{A28}$$

$$\epsilon^b_{ks} = 2ts\chi\xi_k(\theta), \tag{A29}$$

with $\gamma_k = (\cos k_x + \cos k_y)$ and $\varphi_k = (\cos k_x - \cos k_y)$. $\sum'$ denotes the summation over momentum k in the half reduced Brillouin zone, and $s = +1$ and -1 represent the upper and lower energy bands of quasiparticles respectively. Here $\alpha_{ks\uparrow}(\alpha^\dagger_{ks\uparrow})$ and $\alpha_{ks\downarrow}(\alpha^\dagger_{ks\downarrow})$ are the annihilation(creation) operators of spinon quasiparticles of spin up and spin down respectively, and $\beta_{ks}(\beta^\dagger_{ks})$, the annihilation(creation) operators of holon quasiparticles of spin 0. ϵ^f_{ks} and ϵ^b_{ks} are the kinetic energies for spinons and holons respectively. The minus sign $(-\Delta^2)$ in the expression of the holon quasiparticle energy $\sqrt{(\epsilon-\mu)^2 - \Delta^2}$ arises as a consequence of the Bose Einstein statistics[28]. From the diagonalized Hamiltonian Eq.(A24), we calculate the total free energy.

Rewriting Eq.(A24) as

$$\begin{aligned} H^{MF}_{U(1)} = & \sum'_{k,s=\pm 1} \left[E^f_{ks}(\alpha^\dagger_{ks\uparrow}\alpha_{ks\uparrow} - \alpha_{ks\downarrow}\alpha^\dagger_{ks\downarrow}) + E^b_{ks}\beta^\dagger_{ks}\beta_{ks} \right] \\ & + H_c \end{aligned} \tag{A30}$$

with

$$\begin{aligned} H_c = & NJ\Delta^2_f\left(\Delta^2_b + x^2\right) + NJ_p\left(\Delta^2_f + \frac{\chi^2}{2} + \frac{1}{4}\right) \\ & + \sum'_{k,s=\pm 1} \frac{E^b_{ks} + \mu^b}{2} - Nx\mu^f + Nx\mu^b, \end{aligned} \tag{A31}$$

the partition function is derived to be,

$$Z = \exp(-\beta H_c) \prod'_{k,s=\pm 1} (2\cosh\frac{\beta E^f_{ks}}{2})^2 (1 - e^{-\beta E^b_{ks}})^{-1}. \tag{A32}$$

Using the above expression, the total free energy is given by

$$F_{U(1)} = NJ\Delta_f^2\left(\Delta_b^2 + x^2\right) + NJ_p\left(\Delta_f^2 + \frac{\chi^2}{2} + \frac{1}{4}\right)$$
$$-2k_BT\sum_{k,s=\pm 1}^{\prime} \ln(\cosh(\beta E_{ks}^f/2)) - Nx\mu^f - 2Nk_BT\ln 2$$
$$+k_BT\sum_{k,s=\pm 1}^{\prime} \ln(1-e^{-\beta E_{ks}^b}) + \sum_{k,s=\pm 1}^{\prime} \frac{E_{ks}^b + \mu^b}{2} + Nx\mu^b. \quad \text{(A33)}$$

The set of uniform phase ($\theta = 0$) for the hopping order parameter, d-wave symmetry ($\tau^f = \pi/2$) for the spinon pairing order parameter and s-wave symmetry ($\tau^b = 0$) for the holon pairing order parameter is found to yield a stable saddle point energy for both the underdoped and overdoped regions. There is another set of order parameters which yield the same energy as the above one; 2π-flux phase ($\theta = \pi/2$) for the hopping order parameter, s-wave symmetry ($\tau^f = 0$) for the spinon pairing order parameter and d-wave symmetry ($\tau^b = \pi/2$) for the holon pairing order parameter. In both cases, the d-wave symmetry of the electron or hole (not holon) pairs occurs as a composite of the d-wave (s-wave) symmetry of spinon pairs and s-wave (d-wave) symmetry of holon pairs. Only at very low doping near half filling, the flux phase[25] becomes more stable. Thus, the phase of the order parameters of present interest are $\theta = 0$, $\tau^f = \pi/2$ and $\tau^b = 0$. Then the d-wave symmetry of the electron or hole (not holon) pairs is a composite of the d-wave symmetry of spinon pairs and s-wave symmetry of holon pairs. Minimizing the free energy with respect to the amplitudes of the order parameters χ, Δ_b and Δ_f, we obtain the self-consistent equations for the order parameters,

$$\frac{\partial F_{U(1)}}{\partial \chi} = NJ_p\chi - \sum_{ks}^{\prime}\left(\tanh\frac{\beta E_{ks}^f}{2}\right)\left(\frac{\partial E_{ks}^f}{\partial \chi}\right)$$
$$+\sum_{ks}^{\prime}\left(\frac{1}{e^{\beta E_{ks}^b}-1} + \frac{1}{2}\right)\left(\frac{\partial E_{ks}^b}{\partial \chi}\right) = 0, \quad \text{(A34)}$$

$$\frac{\partial F_{U(1)}}{\partial \Delta_b} = 2NJ\Delta_f^2\Delta_b$$
$$+\sum_{ks}^{\prime}\left(\frac{1}{e^{\beta E_{ks}^b}-1} + \frac{1}{2}\right)\left(\frac{\partial E_{ks}^b}{\partial \Delta_b}\right) = 0, \quad \text{(A35)}$$

$$\frac{\partial F_{U(1)}}{\partial \Delta_f} = 2NJ_p\Delta_f + 2N\Delta_f(\Delta_b^2 + x^2)$$
$$\sum_{ks}^{\prime}\left(\tanh\frac{\beta E_{ks}^f}{2}\right)\left(\frac{\partial E_{ks}^f}{\partial \Delta_f}\right)$$
$$+\sum_{ks}^{\prime}\left(\frac{1}{e^{\beta E_{ks}^b}-1} + \frac{1}{2}\right)\left(\frac{\partial E_{ks}^b}{\partial \Delta_f}\right) = 0. \quad \text{(A36)}$$

For fixed numbers of spinon and holon at a given hole concentration, we obtain, for the chemical potentials, μ^f and μ^b,

$$\frac{\partial F_{U(1)}}{\partial \mu^f} = \sum_{k,s=\pm 1}' \left(\tanh \frac{\beta E^f_{ks}}{2}\right)\left(\frac{\epsilon^f_{ks} - \mu^f}{E^f_{ks}}\right) - Nx = 0, \quad \text{(A37)}$$

$$\frac{\partial F_{U(1)}}{\partial \mu^b} = -\sum_{k,s=\pm 1}' \Big[\frac{1}{e^{\beta E^b_{ks}} - 1}\frac{\epsilon^b_{ks} - \mu^b}{E^b_{ks}} + \frac{\epsilon^b_{ks} - \mu^b - E^b_{ks}}{2E^b_{ks}}\Big] + Nx = 0. \quad \text{(A38)}$$

Using the five self-consistent equations of Eqs.(A34) through (A38), we determine χ, Δ_b, Δ_f, μ^f and μ^b at each doping and temperature. Both the pseudogap temperature T^* and the superconducting transition (bose condensation) temperature T_c are determined to be the temperatures at which the spin gap Δ^f_0 and the holon pairing energy (gap) Δ^b_0 respectively begin to open.

Appendix B: SU(2) action from the U(1) action

The t-J Hamiltonian is manifestly invariant under the local SU(2) transformation $g_i = e^{i\sigma\cdot\theta}$ for both the spinon and holon spinors with $\begin{pmatrix} f_{i1} \\ f^\dagger_{i2}\end{pmatrix} = g_i \begin{pmatrix} f_{i\uparrow} \\ f^\dagger_{i\downarrow}\end{pmatrix}$, $\begin{pmatrix} f_{i2} \\ -f^\dagger_{i1}\end{pmatrix} = g_i \begin{pmatrix} f_{i\downarrow} \\ -f^\dagger_{i\uparrow}\end{pmatrix}$ and $\begin{pmatrix} b_{i1} \\ b_{i2}\end{pmatrix} = g_i \begin{pmatrix} b_i \\ 0\end{pmatrix}$, satisfying $c_{i\uparrow} = b^\dagger_i f_{i\uparrow} = (b^\dagger_i, 0)\begin{pmatrix} f_{i\uparrow} \\ f^\dagger_{i\downarrow}\end{pmatrix} = (b^\dagger_i, 0) g^\dagger_i g_i \begin{pmatrix} f_{i\uparrow} \\ f^\dagger_{i\downarrow}\end{pmatrix} = (b^\dagger_{i1}, b^\dagger_{i2})\begin{pmatrix} f_{i1} \\ f^\dagger_{i2}\end{pmatrix}$ and $c_{i\downarrow} = b^\dagger_i f_{i\downarrow} = (b^\dagger_i, 0)\begin{pmatrix} f_{i\downarrow} \\ -f^\dagger_{i\uparrow}\end{pmatrix} = (b^\dagger_i, 0) g^\dagger_i g_i \begin{pmatrix} f_{i\downarrow} \\ -f^\dagger_{i\uparrow}\end{pmatrix} = (b^\dagger_{i1}, b^\dagger_{i2})\begin{pmatrix} f_{i2} \\ -f^\dagger_{i1}\end{pmatrix}$ [26]. We introduce additional Lagrange multiplier terms involved with the constraints $f^\dagger_{i\uparrow} f^\dagger_{i\downarrow} = 0$ and $f_{i\downarrow} f_{i\uparrow} = 0$ to write

$$-i\sum_i \lambda_i (f^\dagger_{i\sigma} f_{i\sigma} + b^\dagger_i b_i - 1) - i\sum_i \lambda'_i f^\dagger_{i\uparrow} f^\dagger_{i\downarrow} - i\sum_i \lambda''_i f_{i\downarrow} f_{i\uparrow} \quad \text{(B1)}$$

in order to allow for SU(2) symmetry. Thus writing spinors $\psi^0_{i1} = \begin{pmatrix} f_{i\uparrow} \\ f^\dagger_{i\downarrow}\end{pmatrix}$ and $\psi^0_{i2} = \begin{pmatrix} f_{i\downarrow} \\ -f^\dagger_{i\uparrow}\end{pmatrix}$ for spinon, $h^0_i = \begin{pmatrix} b_i \\ 0\end{pmatrix}$ for holon and the three-component Lagrangian multiplier field $\mathbf{a}^0_i$ with $a^{0(1)}_i = \frac{i\lambda' + i\lambda''}{2}$, $a^{0(2)}_i = \frac{-\lambda' + \lambda''}{2}$, $a^{0(3)}_i = i\lambda_i$, the U(1) action in Eq.(A6) can be rewritten as

$$S_{U(1)}[b,f,a] = \int_0^\beta d\tau \Big[\sum_i h_i^{0\dagger} \partial_\tau h_i^0 + \frac{1}{2} \sum_{i,\alpha} \psi_{i\alpha}^{0\dagger} \partial_\tau \psi_{i\alpha}^0 - t \sum_{<i,j>} \Big((\psi_{i\alpha}^{0\dagger} h_i^0)(h_j^{0\dagger} \psi_{j\alpha}^0) + c.c. \Big)$$

$$-\frac{J}{2} \sum_{<i,j>} h_{i\alpha}^0 h_{j\beta}^0 h_{j\beta}^{0\dagger} h_{i\alpha}^{0\dagger} (f_{i\downarrow}^\dagger f_{j\uparrow}^\dagger - f_{i\uparrow}^\dagger f_{j\downarrow}^\dagger)(f_{j\uparrow} f_{i\downarrow} - f_{j\downarrow} f_{i\uparrow})$$

$$-\mu \sum_i f_{i\sigma}^\dagger f_{i\sigma} (b_i^\dagger b_i + 1) - \sum_i \mathbf{a}_i^0 \cdot \big(\frac{1}{2} \psi_{i\alpha}^{0\dagger} \sigma \psi_{i\alpha}^0 + h_i^{0\dagger} \sigma h_i^0 \big) \Big]. \qquad \text{(B2)}$$

Here the fourth term is the Heisenberg interaction term, $H_J = -\frac{J}{2} \sum_{<i,j>} b_i b_j b_j^\dagger b_i^\dagger (f_{i\downarrow}^\dagger f_{j\uparrow}^\dagger - f_{i\uparrow}^\dagger f_{j\downarrow}^\dagger)(f_{j\uparrow} f_{i\downarrow} - f_{j\downarrow} f_{i\uparrow})$.

We rewrite the spinon part of the Heisenberg interaction,

$$-\frac{J}{2} (f_{i\downarrow}^\dagger f_{j\uparrow}^\dagger - f_{i\uparrow}^\dagger f_{j\downarrow}^\dagger)(f_{j\uparrow} f_{i\downarrow} - f_{j\downarrow} f_{i\uparrow})$$

$$= \frac{J}{4} \Big[\sum_{k=1}^{3} (f_{i\alpha}^\dagger \sigma_{\alpha\beta}^k f_{i\beta})(f_{j\gamma}^\dagger \sigma_{\gamma\delta}^k f_{j\delta}) - (f_{i\alpha}^\dagger f_{i\alpha})(f_{j\beta}^\dagger f_{j\beta}) \Big]$$

$$= \frac{J}{4} \Big[\frac{1}{4} \left(tr \Psi_i^{0\dagger} \Psi_i^0 (\sigma^k)^T \right) \left(tr \Psi_j^{0\dagger} \Psi_j^0 (\sigma^k)^T \right) - (f_{i\alpha}^\dagger f_{i\alpha})(f_{j\beta}^\dagger f_{j\beta}) \Big], \qquad \text{(B3)}$$

where $\Psi_i^0 \equiv \begin{pmatrix} f_{i\uparrow} & f_{i\downarrow} \\ f_{i\downarrow}^\dagger & -f_{i\uparrow}^\dagger \end{pmatrix}$ and $(f_{i\alpha}^\dagger \sigma_{\alpha\beta}^k f_{i\beta}) = \frac{1}{2} tr \left(\Psi_i^{0\dagger} \Psi_i^0 (\sigma^k)^T \right)$[?]. Here $(\sigma^k)^T$ denotes the transpose of the Pauli matrices for $k = 1, 2, 3$.

Realizing $h_i = g_i h_i^0$ and $\Psi_i = \begin{pmatrix} f_{i1} & f_{i2} \\ f_{i2}^\dagger & -f_{i1}^\dagger \end{pmatrix} = g_i \begin{pmatrix} f_{i\uparrow} & f_{i\downarrow} \\ f_{i\downarrow}^\dagger & -f_{i\uparrow}^\dagger \end{pmatrix}$, and using Eq.(B3), the SU(2) symmetric Heisenberg interaction term is given by

$$H_J^{SU(2)} = \frac{J}{4} \sum_{<i,j>} (1 + h_i^\dagger h_i)(1 + h_j^\dagger h_j) \left[\frac{1}{4} \left(tr \Psi_i^\dagger \Psi_i (\sigma^k)^T \right) \left(tr \Psi_j^\dagger \Psi_j (\sigma^k)^T \right) - (f_{i\alpha}^\dagger f_{i\alpha})(f_{j\beta}^\dagger f_{j\beta}) \right]$$

$$= -\frac{J}{2} \sum_{<i,j>} (1 + h_i^\dagger h_i)(1 + h_j^\dagger h_j)(f_{i2}^\dagger f_{j1}^\dagger - f_{i1}^\dagger f_{j2}^\dagger)(f_{j1} f_{i2} - f_{j2} f_{i1}). \qquad \text{(B4)}$$

Taking decomposition of the Heisenberg interaction term above into terms involving charge and spin fluctuations separately, uncorrelated mean field contributions and correlated fluctuations, i.e., correlations between charge and spin fluctuations as in the U(1) case, the SU(2) action is rewritten,

$$S[b_\alpha, f_\alpha, \lambda_i] = \int_0^\beta d\tau \Big[\sum_{i,\alpha=1,2} (b_{i\alpha}^\dagger \partial_\tau b_{i\alpha} + f_{i\alpha}^\dagger \partial_\tau f_{i\alpha}) + H_{t-J}^{SU(2)} \Big], \qquad \text{(B5)}$$

where

$$
\begin{aligned}
H_{t-J}^{SU(2)} = & -\frac{t}{2}\sum_{<i,j>}\Big[(f_{i\alpha}^{\dagger}f_{j\alpha})(b_{j1}^{\dagger}b_{i1}-b_{i2}^{\dagger}b_{j2})+c.c.\\
& +(f_{i2}f_{j1}-f_{i1}f_{j2})(b_{j1}^{\dagger}b_{i2}+b_{i1}^{\dagger}b_{j2})+c.c.\Big]\\
& -\frac{J}{2}\sum_{<i,j>}\Big[\Big\langle(f_{i2}^{\dagger}f_{j1}^{\dagger}-f_{i1}^{\dagger}f_{j2}^{\dagger})(f_{j1}f_{i2}-f_{j2}f_{i1})\Big\rangle(1+h_i^{\dagger}h_i)(1+h_j^{\dagger}h_j)\\
& +\Big\langle(1+h_i^{\dagger}h_i)(1+h_j^{\dagger}h_j)\Big\rangle(f_{i2}^{\dagger}f_{j1}^{\dagger}-f_{i1}^{\dagger}f_{j2}^{\dagger})(f_{j1}f_{i2}-f_{j2}f_{i1})\\
& -\Big\langle(1+h_i^{\dagger}h_i)(1+h_j^{\dagger}h_j)\Big\rangle\Big\langle(f_{i2}^{\dagger}f_{j1}^{\dagger}-f_{i1}^{\dagger}f_{j2}^{\dagger})(f_{j1}f_{i2}-f_{j2}f_{i1})\Big\rangle\\
& +\Big((1+h_i^{\dagger}h_i)(1+h_j^{\dagger}h_j)-\Big\langle(1+h_i^{\dagger}h_i)(1+h_j^{\dagger}h_j)\Big\rangle\Big)\times\\
& \Big((f_{i2}^{\dagger}f_{j1}^{\dagger}-f_{i1}^{\dagger}f_{j2}^{\dagger})(f_{j1}f_{i2}-f_{j2}f_{i1})-\Big\langle(f_{i2}^{\dagger}f_{j1}^{\dagger}-f_{i1}^{\dagger}f_{j2}^{\dagger})(f_{j1}f_{i2}-f_{j2}f_{i1})\Big\rangle\Big)\Big]\\
& -\mu\sum_i(1-h_i^{\dagger}h_i)\\
& -\sum_i\Big(i\lambda_i^{(1)}(f_{i1}^{\dagger}f_{i2}^{\dagger}+b_{i1}^{\dagger}b_{i2})+i\lambda_i^{(2)}(f_{i2}f_{i1}+b_{i2}^{\dagger}b_{i1})\\
& +i\lambda_i^{(3)}(f_{i1}^{\dagger}f_{i1}-f_{i2}f_{i2}^{\dagger}+b_{i1}^{\dagger}b_{i1}-b_{i2}^{\dagger}b_{i2})\Big).
\end{aligned}
\tag{B6}
$$

9. References

[1] G. Kotliar and J. Liu, Phys. Rev. B 38, 5142 (1988); references there-in.
[2] Y. Suzumura, Y. Hasegawa and H. Fukuyama, J. Phys. Soc. Jpn. 57, 2768 (1988).
[3] M.U. Ubbens and P. A. Lee, Phys. Rev. B 49, 6853 (1994); references therein.
[4] T.-H. Gimm, S.-S. Lee, S.-P. Hong and S.-H. Suck Salk, Phys. Rev. B 60, 6324 (1999).
[5] S.-S. Lee and Sung-Ho Suck Salk, Phys. Rev. B 64, 052501 (2001); S.-S. Lee and Sung-Ho Suck Salk, Phys. Rev. B 66, 054427 (2002); S.-S. Lee and Sung-Ho Suck Salk, J. Kor. Phys. Soc. 37, 545 (2000); S.-S. Lee and S.-H. S. Salk, cond-mat/0304293.
[6] S.-S. Lee and S.-H. S. Salk, Phys. Rev. B 71, 134518 (2005); J.-H. Eom and S.-H. S. Salk, Phys. Rev. B 72, 064508 (2005); J.-H. Eom, S.-S. Lee, K.-S. Kim and S.-H. S. Salk, Phys. Rev. B 70, 024522 (2004); S.-S. Lee, J.-H. Eom, K.-S. Kim and S.-H. Suck Salk, Phys. Rev. B 66, 064520 (2002); S.-S, Lee and S.-H. S. Salk, Phys. Rev. B 66, 054427 (2002); S.-S. Lee and S.-H. S. Salk, Phys. Rev. B 64, 052501 (2001); Physica C. 353, 130(2001); J. Kor. Phys. Soc. 37, 545 (2000); Int. J. Mod. Phys. B 13, 3455 (1999); cond-mat/0410229; cond-mat/0212436.
[7] M. Civelli M. Capone, S. S. Kancharla, O. Parcollet, and G. Kotliar, Phys. Rev. Lett. 95, 106402 (2005); references there-in.
[8] K. Tanaka, T. Yoshida, A. Fujimori, D. H. Lu, Z.-X. Shen, X.-J. Shou, H. Eisaki, Z. Hussain, S. Uchida, Y. Aiura. K. Ono, T. Sugaya, T. Mizuno. and I. Terasaki, Phys. Rev. B 70 092503 (2004).
[9] A. Himeda and M. Ogata, Phys. Rev. Lett. 85, 4345 (2000).

[10] F. Ronning, C. Kim, D. L. Feng, D. S. Marshall, A. G. Loeser, L. L. Miller, J. N. Eckstein, I.Bozovic, Z.-X. Shen, Science, 282, 2067 (1998).
[11] E. Gagliano, S. Bacci, and E. Dagotto, Phys. Rev. B 42 6222 (1990).
[12] P. Prelovšek and A. Ramšak, Phys. Rev. B 72 012510 (2005).
[13] C. T. Shih, T. K. Lee, T. Eder, C.-Y. Mou, and Y. C. Chen, Phys. Rev. Lett. 92, 227002 (2004); references there-in.
[14] M. Oda, R. Dipasupil, N. Momono, and M. Ido, J. Phys. Soc. Jpn. 69, 983 (2000); T. Nakano, N. Momono, M. Oda and M. Ido, J. Phys. Soc. Jpn. 67, 2622 (1998).
[15] M. Oda, R. Dipasupil, N. Momono, M. Ido, J. Phys. Soc. Jpn. 69, 983 (2000)
[16] T. Timusk, B. Stattt, Rep. Prog. Phys. 62, 61 (1999).
[17] P. Dai, H. A. Mook, S. M. Hayden, G. Appeli, T. G. Perring, R. D. Hunt and F. Dogan, Science 284, 1344 (1999); references there-in.
[18] H. He, Y. Sidis, P. Bourges, G. D. Gu, A. Ivanov, N. Koshizuka, B. Liang, C. T. Lin, L. P. Regnault, E. Schoenherr and B. Keimer, Phys. Rev. Lett. 86, 1610 (2001); references there-in.
[19] C. Stock, R. A. Cowley, W. J. L. Buyers, R. Coldea, C. Broholm, C. D. Frost, R. J. Birgeneau, R. Liang, D. Bonn and W. N. Hardy, Phys. Rev. B 75, 172510 (2007).
[20] S. Wakimoto, K. Yamada, J. M. Tranquada, C. D. Frost, R. J. Birgeneau and H. Zhang, Phys. Rev. Lett. 98, 247003 (2007).
[21] J.-G. Eom, J.-H. Eom and Sung-Ho Suck Salk, Int. J. Mod. Phys. B, 21 3132 (2007).
[22] Seung-Joon Shin and Sung-Ho Salk, submitted to Phys. Rev. Lett..
[23] J. Brinckmann and P. A. Lee, Phys. Rev. Lett. 82, 2915 (1999); J. Brinckmann and P. A. Lee, Phys. Rev. B 65, 014502 (2001).
[24] P. A. Lee and N. Nagaosa, Phys. Rev. B 46, 5621 (1992).
[25] M. U. Ubbens and P. A. Lee, Phys. Rev. B 46, 8434 (1992).
[26] a) X. G. Wen and P. A. Lee, Phys. Rev. Lett. 76, 503 (1996); b) X. G. Wen and P. A. Lee, Phys. Rev. Lett. 80, 2193 (1998); c) P. A. Lee, N. Nagaosa, T.-K. Ng and X.-G. Wen, Phys. Rev. B 57, 6003 (1998).
[27] Here $b_i b_j b_j^\dagger b_i^\dagger$ represents the occupation number operator of the holon pair of negative charge $-2e$ but not the holon pair of positive charge $+2e$ at intersites i and j. $\left\langle b_i b_j b_j^\dagger b_i^\dagger \right\rangle = 1$ arises where an electron pair occupied at intersites i and j. Otherwise it is zero, that is, when the two intersites or one of the two sites are vacant; $b_i b_i^\dagger$ is equivalent to the electron occupation number operator at site i and from the consideration of uniform electron removal and thus hole doping concentration x, we readily note that $\left\langle b_i b_i^\dagger b_j b_j^\dagger \right\rangle = (1-x)^2$ with $0 \leq \left\langle b_i b_i^\dagger b_j b_j^\dagger \right\rangle \leq 1$.
[28] P. Nozières and D. Saint James, J. Physique, 43, 1133 (1982); references therein.

2

Pairing Symmetry and Multiple Energy Gaps in Multi-Orbital Iron-Pnictide Superconductors

Liang-Jian Zou and Feng Lu
Key Laboratory of Materials Physics, Institute of Solid State Physics, the Chinese Academy of Sciences
People's Republic of China

1. Introduction

Since the new high-Tc superconducting family based on iron pnictides was discovered Kamihara et al. (2008); Chen et al. (2008a;b), and the critical temperature was lifted to 56 K under high pressure Wu et al. (2009) which is considerably larger than the McMillan Limit McMillan (1968), the superconductive pairing mechanism and properties have attracted great interests experimentally and theoretically. These newly-found superconductors also provide a potential application prospect in two respects: on the one hand, simple components and rich resource of the FeAs-based compounds show the most possibility of large-scale applications; on the other hand, extremely large upper critical fields H_{c2}Wen et al. (2008) in FeAs superconductors imply realistic applications in the near future. To find higher T_c FeAs superconductors in further experiments and to improve their critical currrent density, it is essentially important to theoretically understand various normal and superconducting properties of iron pnictides, especially the superconducting pairing symmetry and its microscopic pairing mechanism. Once the pairing symmetry is known, many superconducting properties could be qualitatively understood.

Soon after the finding of the superconductivity in LaFeAsO, the first-principles electronic structure calculations Mazin et al. (2008); Boeri et al. (2008); Cao et al. (2008); Ma et al. (2010) showed that Fe 3*d* orbits contribute the major spectral weight near the Fermi surface: the Fermi surface of LaFeAsO consists of two hole-type circles around the Γ point and two electron-type co-centered ellipses around the *M* point Zhang et al. (2009), indicating that the multi-orbital character is important in the FeAs superconductors. This is in a sharp contrast to the single-orbital character of high-T_C cuprate. In these multi-orbital systems, one of the central problems is the superconducting pairing mechanism and its pairing symmetry. In fact, shortly after the determination of crystal structures of F-doped LaFeAsO, several authors proposed the constraint of the superconducting pairing symmetry of iron pnictides by applying the symmetric operations of the crystal point group on the pairing wavefunctions Mazin et al. (2008). However, such an analysis does not consider the Fermi surface topologies and the pairing potential of the superconductivity, hence can not solely determine the pairing symmetry in superconducting $LaFeAsO_{1-x}F_x$ Lin et al. (2008).

Meanwhile, based on microscopic spin fluctuation interactions in LaFeAsO, some authors proposed various superconducting pairing symmetries range from spin singlet *d*-wave Mu et al. (2008); Millo et al. (2008), *s*-wave Wang et al. (2009a;b); Nomura et al. (2008), the mixture of $S_{x^2y^2}$-wave and $d_{x^2-y^2}$-wave Seo et al. (2008) to spin triplet p-wave Dai et al. (2008); Lee et al. (2008). These suggestions on the pairing symmetry raised critical and hot debates in the literature. A few authors focused on the microscopic origin of the superconducting pairing according to the antiferromagnetic spin fluctuation and the Fermi surface nesting topology through the characteristic wavevector $\mathbf{Q}=(\pi,0)$ of the antiferromagnetic spin fluctuations. They proposed the s_{+-}-wave pairing symmetry Mazin et al. (2008); Kuroki et al. (2008), *i.e.* the phase of the superconducting order parameters of the inner Fermi surface around the Γ point is antiphase to that of the Fermi surface around the M point. The s_{+-}-wave symmetry of the superconducting order parameters seems to receive sufficient support in theory and experiment Mazin et al. (2008), and is consistent with the nesting picture of electron-type and hole-type Fermi surfaces in FeAs-based superconductors. However, the most recently found $K_xFe_{2-y}Se_2$ compounds clearly rule out the presence of a hole-type Fermi surface around the Γ point Xiang et al. (2011), suggesting that an alternative pairing symmetry is possible. Actually, from the researching history of the high-T_c cuperates, we have known that a pairing mechanism based on the Fermi surface nesting is rather delicate, since any finite electron-electron interaction, which usually occurs in high-T_c cuprates and iron pnictides, will destroy the perfect nesting of the Fermi surfaces. These disagreements and debates in the experimental data and theoretical results on the superconducting pairing symmetry of iron pnictides appeal for more efforts to unveil the mysterious nature of the superconducting iron pnictides.

On the other hand, the effect of the electron correlation in iron pnictides should be taken into account, since the bad metallic behavior and the existence of antiferromagnetic spin moments suggest that the iron pnictide is close to a metal insulator transition Haule et al. (2008). In this Chapter, starting with the minimal two-orbital t-t$'$-J-J$'$ model Manousakis et al. (2008); Raghu et al. (2008), we develop a mean-field theory of the multi-orbital superconductors for the weak, intermediate, and strong correlation regimes, respectively. Taking a concrete t-t$'$-J-J$'$ model which has the same topology as the Fermi surface and the band structures of LaFeAsO, we obtain the superconducting phase diagram, the quasiparticle spectra in normal state and superconducting phase, and the ARPES manifestation of the superconducting energy gaps. Our theory is applicable not only for FeAs superconductors, but also for ruthenate and heavy fermion, and other spin-fluctuation mediated multi-orbital superconductors. For realistic iron pnictides, we show that the pairing symmetry $d_{x^2-\eta y^2}+S_{x^2y^2}$-wave is stable in the reasonable parameters region; two superconducting gaps and their weak anisotropy and nodeless qualitatively agree with the observations in ARPES experiments. However, a quantitative comparison between theory and experiment shows a more elaborate theoretical model is necessary.

The rest of this Chapter is arranged as follows: in Sec.II we present the theory and methods for multi-orbital superconductivity; in Sec.III and IV we show the numerical results on the pairing symmetry of multi-orbital iron-pnictide superconductors, and the orbital dependence of superconducting energy gaps; Sec.V is devoted to the comparison between our theory and experimental observations, and finally we make a concluding remarks in Sec.VI.

2. Theory and methods of multi-orbital superconductivity

For the iron-pnictide compounds, the electron-phonon coupling seems to be irrelevant to the superconducting pairing origin Boeri et al. (2008), the antiferromagnetic spin fluctuation is naturally thought to contribute the pairing glue of the superconducting electron pairs, due to the antiferromagnetic ground state in updoped FeAs compounds. Considering the multi-band and electron correlation characters, a minimal model for describing the low-energy physics of the FeAs-based superconductors is the two-orbital t-J model and its extension. Based on the band structures results and theoretical analysis, the twofold-degenerate d_{xz}/d_{yz} orbits are essential for the ironpnictide superconductors. We firstly depict such physical processes with the two-orbital Hubbard model,

$$\begin{aligned}\hat{H} = & \sum_{<ij>\alpha\beta\sigma} t_{ij}^{\alpha\beta} \hat{c}_{i\alpha\sigma}^{\dagger} \hat{c}_{j\beta\sigma} + \sum_{\ll ij \gg \alpha\beta\sigma} t_{ij}^{\prime\alpha\beta} \hat{c}_{i\alpha\sigma}^{\dagger} \hat{c}_{j\beta\sigma} \\ & + U \sum_{i\alpha} \hat{c}_{i\alpha\uparrow}^{\dagger} \hat{c}_{i\alpha\uparrow} \hat{c}_{i\alpha\downarrow}^{\dagger} \hat{c}_{i\alpha\downarrow} + U' \sum_{i\sigma\sigma'} \hat{c}_{i1\sigma}^{\dagger} \hat{c}_{i1\sigma} \hat{c}_{i2\sigma'}^{\dagger} \hat{c}_{i2\sigma'} \\ & - J_H \sum_{i\sigma\sigma'} \hat{c}_{i1\sigma}^{\dagger} \hat{c}_{i1\sigma'} \hat{c}_{i2\sigma'}^{\dagger} \hat{c}_{i2\sigma} \\ & + J_H \sum_{i\alpha\neq\alpha'} \hat{c}_{i\alpha\uparrow}^{\dagger} \hat{c}_{i\alpha'\uparrow} \hat{c}_{i\alpha\downarrow}^{\dagger} \hat{c}_{i\alpha'\downarrow} \end{aligned} \tag{1}$$

where $c_{i\alpha\sigma}^{\dagger}$ creates a d_{xz} (α=1) or d_{yz} (α=2) electron with orbit α and spin σ at site R_i. t and t' denotes the hopping integrals of the nearest-neighbor (NN) and the next-nearest-neighbor (NNN) sites, respectively. U, U' and J_H are the intra-orbital, inter-orbital Coulomb interactions and the Hund's coupling.

In the strongly correlated regime, it is well known that the t-t'-J-J' model can be derived from Eq.(1); meanwhile, even in the weak correlation regime in the atomic limit Manousakis et al. (2008), the two-orbital Hubbard model in Eq.(1) can derive to the t-t'-J-J' model, although not strictly. Thus, we can describe the low-energy processes in iron pnictides with the two-orbital t-t'-J-J' model,

$$H = H_{t-t'} + H_{J-J'}, \tag{2}$$

on a quasi-two-dimensional square lattice. This Hamiltonian consists of the tight-binding kinetic energy $H_{t-t'}$ and the interaction part $H_{J-J'}$. The kinetic energy term reads,

$$H_{t-t'} = \sum_{k\sigma} [(\varepsilon_{kxz} - \mu) c_{k1\sigma}^{\dagger} c_{k1\sigma} + (\varepsilon_{kyz} - \mu) c_{k2\sigma}^{\dagger} c_{k2\sigma} + \varepsilon_{kxy} (c_{k1\sigma}^{\dagger} c_{k2\sigma} + c_{k2\sigma}^{\dagger} c_{k1\sigma})]$$

with the notations

$$\begin{aligned}\varepsilon_{kxz} &= -2(t_1 \cos k_x + t_2 \cos k_y + 2t_3 \cos k_x \cos k_y), \\ \varepsilon_{kyz} &= -2(t_2 \cos k_x + t_1 \cos k_y + 2t_3 \cos k_x \cos k_y), \\ \varepsilon_{kxy} &= -4t_4 \sin k_x \sin k_y,\end{aligned}$$

The intra-orbital components of the nearest-neighbour (NN) hopping integrals $t_{ij}^{\alpha\beta}$ are t_x^{11}=t_1=-1, and t_x^{22}=t_2 = 1.3. The components of next-nearest-neighbour (NNN) hopping integrals, $t_{ij}^{\prime\alpha\beta}$, are $t_3 = t_4 = -0.85$ Raghu et al. (2008). Throughout this paper, all the energies are measured in units of $| t_1 |$. The carrier concentration is equal to 0.18, which is a typical doping concentration in the iron-based superconductors Dubroka et al. (2008).

The interaction term reads,

$$H_{J-J'} = J \sum_{<ij>\alpha\beta} (\vec{S}_{i\alpha} \cdot \vec{S}_{j\beta} - \frac{1}{4} n_{i\alpha} \cdot n_{j\beta}) + J' \sum_{<<ij>>\alpha\beta} (\vec{S}_{i\alpha} \cdot \vec{S}_{j\beta} - \frac{1}{4} n_{i\alpha} \cdot n_{j\beta}) \quad (3)$$

which contains a NN and a NNN antiferromagnetic spin couplings. Here J and J' are the NN and the NNN spin coupling strengths, respectively. $\vec{S}_{i\alpha}$ is the spin operator of the electron in the α-orbit at R_i and $n_{i\alpha}$ is the particle number operator. α, β(=1,2) are orbital index.

To explore the essence of the iron-pnictide superconductors and other multi-orbital superconductors, we discuss the t-t$'$-J-J$'$ model in three different correlation regimes:

2.1 Weak correlation regime

When the kinetic energy of the d_{xz}- and d_{yz}-electrons is much larger than the Coulomb interaction, we adopt the conventional mean-field decoupling approach to study the superconducting pairing symmetry and its orbital dependence. This ansatz is applicable for many FeAs-based and other multi-orbital superconductors with metallic ground states.

Notice that the d_{xz} and d_{yz} orbits are spatially anisotropic, in other words, the intra-orbital hopping integral along the x-direction is not equal to that along the y-direction for each orbital, as one can see from $| t_1 | \neq | t_2 |$. Due to the asymmetry of the different directions in different orbits, the amplitude of the superconducting gap of the local pairing along the x-direction may be not equal to that along the y-direction in each orbit. Thus, the single orbital d-wave or s-wave superconducting order parameter, in which the superconducting energy gap has 4-fold symmetry of rotational invariance in the xy plane, is not suitable for describing the pairing symmetry of the intra-orbital superconducting order parameters in this multi-orbital system. Considering all of the possible kinetic correlations and the superconducting pairings for the NN and NNN sites along different directions, we introduce the following order parameters,

$$\begin{aligned}
P^{\alpha}_{x/y} &= < c^{\dagger}_{i\alpha\sigma} c_{j\alpha\sigma} >, \ (j = i \pm \hat{x}/\hat{y}) \\
P^{1/2}_{xy} &= < c^{\dagger}_{i1\sigma} c_{j2\sigma} >, \ (j = i \pm \hat{x}/\hat{y}) \\
P^{\alpha}_{3} &= < c^{\dagger}_{i\alpha\sigma} c_{j\alpha\sigma} >, \ (j = i \pm (\hat{x} \pm \hat{y})) \\
P^{3/4}_{xy} &= < c^{\dagger}_{i1\sigma} c_{j2\sigma} >, \ (j = i \pm (\hat{x} \pm \hat{y})) \\
\Delta^{1\alpha}_{x/y} &= J < c^{\dagger}_{i\alpha\uparrow} c^{\dagger}_{j\alpha\downarrow} >, \ (j = i \pm \hat{x}/\hat{y}) \\
\Delta^{2\alpha}_{x\pm y} &= J' < c^{\dagger}_{i\alpha\uparrow} c^{\dagger}_{j\alpha\downarrow} >, \ (j = i \pm (\hat{x} \pm \hat{y})).
\end{aligned} \quad (4)$$

Here $P^{\alpha}_{x/y}$ and $P^{1/2}_{xy}$ (P^{α}_{3} and $P^{3/4}_{xy}$) are the kinetic average of the NN (NNN) intra-orbital and inter-orbital hopping integrals. These terms could be decoupled within the framework of the mean-field approximationSeo et al. (2008). $\Delta^{1\alpha}_{x/y}$ ($\Delta^{2\alpha}_{x\pm y}$) is the mean-field amplitude of the local NN (NNN) pairing order parameter in the α-orbit. The inter-orbital pairing parameter $< c^{\dagger}_{i1\uparrow} c^{\dagger}_{j2\downarrow} >$ is very small, hence is neglected Seo et al. (2008).

With these parameters, one can decouple the interaction terms in Eq.(3) within the framework of the self-consistent mean-field approximation, and obtain the mean-field Hamiltonian,

$H_{MF} = \sum_k \Psi(k)^\dagger A(k)\Psi(k) + const$. Here $\Psi(k) = (c_{k,1,\uparrow}, c^\dagger_{-k,1,\downarrow}, c_{k,2,\uparrow}, c^\dagger_{-k,2,\downarrow})$,

$$A(k) = \begin{pmatrix} \epsilon_{k1\uparrow} - \mu & \Delta_1^*(k) & \epsilon_{k12} & 0 \\ \Delta_1(k) & -\epsilon_{k1\downarrow} + \mu & 0 & -\epsilon_{k12} \\ \epsilon_{k12} & 0 & \epsilon_{k2\uparrow} - \mu & \Delta_2^*(k) \\ 0 & -\epsilon_{k12} & \Delta_2(k) & -\epsilon_{k2\downarrow} + \mu \end{pmatrix}, \quad (5)$$

and $const$ is the collection of all the constant energy terms from the mean-field decoupling. The modified intra-orbital and inter-orbital kinetic energy reads,

$$\begin{aligned} \varepsilon_{k1\sigma} &= \varepsilon_{kxz} - 2J(P_x^1 \cos k_x + P_y^1 \cos k_y) - 4J' P_3^1 \cos k_x \cos k_y \\ &\quad -4(J + J')(< n_{1\sigma} > + < n_{2\sigma} >) \\ \varepsilon_{k2\sigma} &= \varepsilon_{kyz} - 2J(P_x^2 \cos k_x + P_y^2 \cos k_y) - 4J' P_3^2 \cos k_x \cos k_y \\ &\quad -4(J + J')(< n_{1\sigma} > + < n_{2\sigma} >) \\ \varepsilon_{k12} &= \varepsilon_{kxy} - 2J(P_{xy}^1 \cos k_x + P_{xy}^2 \cos k_y) \\ &\quad -4J'(P_{xy}^3 \cos(k_x + k_y) + P_{xy}^4 \cos(k_x - k_y)). \end{aligned} \quad (6)$$

The superconducting order parameter $\Delta_\alpha(k)$ of each orbital channel in the momentum space is

$$\begin{aligned} \Delta_\alpha(k) &= -4(\Delta_x^{1\alpha} \cos k_x + \Delta_y^{1\alpha} \cos k_y) \\ &\quad -4(\Delta_{x+y}^{2\alpha} \cos(k_x + k_y) + \Delta_{x-y}^{2\alpha} \cos(k_x - k_y)) \\ &= -4\Delta_x^{1\alpha}[(\cos(k_x) \pm \eta_{1\alpha} \cos(k_y)) \\ &\quad +\xi_\alpha(\cos(k_x + k_y) \pm \eta_{2\alpha} \cos(k_x - k_y))] \end{aligned} \quad (7)$$

Thus the superconducting pairing symmetry of the α-orbit is determined by $(\cos(k_x) \pm \eta_{1\alpha} \cos(k_y)) + \xi_\alpha(\cos(k_x + k_y) \pm \eta_{2\alpha} \cos(k_x - k_y))$. Where $\xi_\alpha = \Delta_{x+y}^{2\alpha}/\Delta_x^{1\alpha}$; the anisotropic factors, $\eta_{1\alpha}=\mid \Delta_y^{1\alpha}/\Delta_x^{1\alpha} \mid$ and $\eta_{2\alpha}=\mid \Delta_{x-y}^{2\alpha}/\Delta_{x+y}^{2\alpha} \mid$, are positive. $\pm$ denotes the relative phase of $\Delta_y^{1\alpha}$ with respect to $\Delta_x^{1\alpha}$ or $\Delta_{x-y}^{2\alpha}$ to $\Delta_{x+y}^{2\alpha}$.

To characterize the complicated superconducting order parameters in different parameter regions, we define the $S_{x^2+\eta y^2}$-wave or $d_{x^2-\eta y^2}$-wave as the pairing symmetry when $\Delta_\alpha \propto cos(k_x) + \eta cos(k_y)$ or $\Delta_\alpha \propto cos(k_x) - \eta cos(k_y)$. It reduces to the conventional $S_{x^2+y^2}$-wave or $d_{x^2-y^2}$-wave symmetry at $\eta = 1$. We also define the $S_{\eta x^2 y^2}$-wave or $d_{\eta xy}$-wave as the superconducting pairing symmetry when $\Delta_\alpha \propto cos(k_x + k_y) + \eta cos(k_x - k_y)$ or $\Delta_\alpha \propto cos(k_x + k_y) - \eta cos(k_x - k_y))$. In this situation, it reduces to the familiar $S_{x^2y^2}$-wave or d_{xy}-wave symmetry at $\eta = 1$.

Diagonalizing the matrix $A(k)$ by an unitary transformation U(k), $U(k)^\dagger A(k)U(k)$, and minimizing the free energy of the system with respect to these parameters in Eq.(5-7), one obtains the self-consistent equations,

$$
\begin{aligned}
n_{(1/2)\uparrow} &= \frac{1}{N}\sum_{k,\alpha} U^*_{(1/3)\alpha}(k)U_{(1/3)\alpha}(k)f(E_\alpha(k)) \\
n_{(1/2)\downarrow} &= \frac{1}{N}\sum_{k,\alpha} U^*_{(2/4)\alpha}(k)U_{(2/4)\alpha}(k)(1-f(E_\alpha(k))) \\
P^{1/2}_{x/y} &= \frac{1}{N}\sum_{k,\alpha} cosk_{x/y}U^*_{(1/3)\alpha}(k)U_{(1/3)\alpha}(k)f(E_\alpha(k)) \\
P^{1/2}_{3} &= \frac{1}{N}\sum_{k,\alpha} cosk_x cosk_y U^*_{(1/3)\alpha}(k)U_{(1/3)\alpha}(k)f(E_\alpha(k)) \\
P^{1/2}_{xy} &= \frac{1}{N}\sum_{k,\alpha} cosk_{x/y}U^*_{1\alpha}(k)U_{3\alpha}(k)f(E_\alpha(k)) \\
P^{3/4}_{xy} &= \frac{1}{N}\sum_{k,\alpha} cos(k_x +/- k_y)U^*_{1\alpha}(k)U_{3\alpha}(k)f(E_\alpha(k)) \\
\Delta^{1(1/2)}_{x/y} &= \frac{J}{N}\sum_{k,\alpha} cosk_{x/y}U^*_{(1/3)\alpha}(k)U_{(2/4)\alpha}(k)f(E_\alpha(k)) \\
\Delta^{2(1/2)}_{x\pm y} &= \frac{J'}{N}\sum_{k,\alpha} cos(k_x \pm k_y)U^*_{(1/3)\alpha}(k)U_{(2/4)\alpha}(k)f(E_\alpha(k))
\end{aligned}
\tag{8}
$$

where $E_\alpha(k)$ is the Bogoliubov quasiparticle eigenvalues obtained from H_{MF}, and $f(E)$ is the Fermi-Dirac distribution function, f(E) = $1/(1+exp(E/k_BT))$. $U_{\alpha\beta}$(k) denotes the (α,β) element of the 4x4 unitary matrix U(k).

With these self-consistent equations, we could obtain not only the groundstate phase diagram, but also the temperature dependence of the Fermi surfaces in normal state and the quasiparticle spectra in the normal and superconducting states. In fact, the intra-orbital hopping integral of the d_{yz} orbit is symmetric with that of the d_{yz} orbit under a coordinate transformation (x,y,z) ← (y,x,z). Due to this symmetry, the superconducting order parameters Δ_2(k) can be obtained from Δ_1(k) under the coordinate transformation. Therefore, we mainly focus on the properties of the superconducting order parameters Δ_1(k) in the first orbit d_{xz}. Nevertheless, the global superconducting pairing order parameter of the two-orbital t-t$'$-J-J$'$ model should be rotationally symmetric in the xy-plane, as we can see from the Hamiltonian Eq.(2).

Within the present scenario, we could obtain not only the groundstate phase diagram, but also the quasiparticle spectra in the normal and the superconducting states. The temperature dependence of the Fermi surface in normal state and that of the spin-lattice relaxation rate in the superconducting state can also be obtained. Among these quantities, the spin-lattice relaxation rate in the NMR experiment is expressed as Matano (2008):

$$
\begin{aligned}
\frac{T_{1N}}{T_{1s}} &= \frac{2}{k_BT}\int\int N_s(E)N_s(E')f(E) \\
&\quad \times[1-f(E')\delta(E-E')dEdE']
\end{aligned}
\tag{9}
$$

Providing $1/T_{1N}$ in the normal state satisfies the Korringa law, the spin lattice relaxation rate $1/T_{1s}$ becomes Xiang (2007): $1/T_{1s} \propto (k_B T) \cdot T_{1N}/T_{1s}$.

2.2 Intermediate correlation regime

When the kinetic energy of the conduction bands becomes small and is comparable with the Coulomb interaction, we need to consider the electronic correlation effect, as one sees in $FeTe_{1-x}Se_x$ and other superconductors. We utilize the Kotliar-Ruckenstein's slave boson approach for some FeAs-based and ruthenate superconductor with intermediate magnetic moments. To reflect the multi-orbital character of iron pnictides, we need to extend the single-orbital Kotliar-Ruckenstein's slave boson approach Kotliar et al. (1988) to the two-orbital Hubbard models for various configurations. In the multiorbital Hubbard model, a few of auxiliary boson field operators representing the possibilities of various electron occupations are introduced, such as e, p, d, b, t, q, which denote the possibilities of none, single, double, triplicate, quaternity occupations. With these auxiliary boson fields, an original fermion operator can be expressed as:

$$
\begin{aligned}
c_{i\alpha\sigma}^{\dagger} = Q_{i\alpha\sigma}^{-\frac{1}{2}} \Bigg(& p_{i\alpha\sigma}^{\dagger} e_i + b_{i\alpha}^{\dagger} p_{i\alpha\sigma} + \sum_{\sigma'} d_{i\sigma_\alpha \sigma'_\beta}^{\dagger} p_{i\beta\sigma'} + t_{i\alpha\sigma}^{\dagger} b_{i\beta} \\
& + \sum_{\sigma'} t_{i\beta\sigma'}^{\dagger} d_{i\bar{\sigma}_\alpha \sigma'_\beta} + q_i^{\dagger} t_{i\alpha\bar{\sigma}} \Bigg) (1 - Q_{i\alpha\sigma})^{-\frac{1}{2}} f_{i\alpha\sigma}^{\dagger} \qquad (10)
\end{aligned}
$$

Here $f_{i\alpha\sigma}^{\dagger}$ is the new slaved fermion operator and $Q_{i\alpha\sigma}$ is an auxiliary particle number operator Kotliar et al. (1988). Projecting the original fermion operators into these boson field and fermion field operators, one could not only obtain an effective Hamiltonian, but also get the groundstate energy in the saddle point approximation with the normalization condition and the fermion number constraints Kotliar et al. (1988). Here we employ a generalized Lagrange multiplier method to enforce these constraint conditions, thus the interorbital hoppings and crystal field splitting can be treated on the same foot. The fermion occupation number is constrained with the penalty function method. To enforce the normalization condition, we have a boundary constrained condition:

$$
1 \geq \sum_{\alpha\sigma} p_{\alpha\sigma}^2 + \sum_{\alpha} b_{\alpha}^2 + \sum_{\alpha\sigma\sigma'} d_{\alpha\sigma\sigma'}^2 + \sum_{\alpha\sigma} t_{\alpha\sigma}^2 + q^2 \qquad (11)
$$

With these projections to the boson states, one can easily obtain an effective t-t$'$-J-J$'$ model subjected to the normalization and fermion number constraints. Following the similar steps above in the weak correlation regime to decouple the spin exchange terms, one could obtain the self-consistent equations similar to Eq.(8) to get various supercoundcting groundstate properties in the intermediate correlated iron pnictides, such as FeSe/FeTe, etc.

2.3 Strong correlation regime

Once the Coulomb interaction is so large that double occupation is excluded on each site, we use the Barnes-Coleman's slave boson approach to discuss the pairing symmetry and orbital-dependent superconducting energy gaps, which is applicable for some FeAs-based superconductors and heavy fermion superconductors with significantly large magnetic moments in the parent phases.

Within the slave-boson representation Barnes (1976); Coleman (1984), Eq.(4) is rewritten in terms of the projected fermion operators $f^{\dagger}_{im\sigma}$ and $f_{im\sigma}$, as well as the slave boson operators at each site, which rule out the double and multiple fermion occupancies. The constrained Hilbert space (S) of each site i is

$$S_i = \{|\, 1,\uparrow>, |\, 1,\downarrow>, |\, 2,\uparrow>, |\, 2,\downarrow>, |\, 0,0>\} \tag{12}$$

including the single-occupied states of the spin-up and spin-down in 1-orbit, and those in 2-orbit, together with the vacancy state, respectively. The present constrained spin-orbital formulation resembles to the 4-fold degenerate state of pseudo-angular momentum j=3/2 proposed by Barnes Barnes (1976) and Coleman Coleman (1984), if we define $|\, f^0>=|\, 0,0>$, $|\, f^1 : 3/2, -3/2> =|\, 1,\uparrow>$, $|\, f^1 : 3/2, -1/2>=|\, 1,\downarrow>$, $|\, f^1 : 3/2, 1/2>=|\, 2,\uparrow>$, and $|\, f^1 : 3/2, 3/2>=|\, 2,\downarrow>$. In this representation, $Pc^{\dagger}_{im\sigma}P = f^{\dagger}_{im\sigma}b_i$. The boson operator $b^{\dagger}_i$ creates an empty occupation state at the ith site, and the fermion operator $f^{\dagger}_{im\sigma}(f_{im\sigma})$ creates (annihilates) a slaved electron at site i with the orbit m and spin $\sigma(=\uparrow,\downarrow)$. After projecting the original fermion representation into the present boson representation, one could obtain an effective t-t$'$-J-J$'$ model subjected to these constraints. Following the similar steps above in Sec.2.1 to decouple the spin exchange terms, one could readily solve the superconducting groundstate properties in strongly correlated $K_xFe_{2-y}Se_2$ compound.

3. Pairing symmetry of multi-orbital iron-pnictide superconductors

We present the pairing symmetry of the two-orbital t-t'-J-J' models for weak-correlated FeAs-based superconductors. The main numerical results in the weak correlation situation are addressed as follows. Also one could easily obtain the numerical results of the intermediate and strong correlation situations.

3.1 Stability of unusual superconducting pairing symmetry

First of all, we determine the stable ground state of the present two-orbital $t - t' - J - J'$ model with the electron filling n=1-δ in a square lattice through comparing the groundstate energies of various pairing-symmetric superconducting states: the isotropic s wave, anisotropic $s_{x^2-y^2}$ wave, $s_{x^2y^2}$ wave, $d_{x^2-y^2}$ wave, and $d_{x^2y^2}$ wave, etc. By minimizing the groundstate energies of various candidates and finding a stablest state, we obtain phase diagrams of the systems for various parameters, such as the hopping integrals t_{ab}, the doping concentration n, the exchange parameters J and J' and so on. Our numerical results show that in the superconducting phase of iron pnictides, the energy of the weakly anisotropic and nodeless $d_{x^2-y^2} + s_{x^2y^2}$-wave-like superconducting state is lower than those of the s-waves and d-wave states for most of the situations we investigated.

3.2 Phase diagram of superconducting pairing symmetry

In this subsection, we first obtain the phase diagrams and mark the pairing symmetry of each stable superconducting phase in the J-J' and t-J planes, and locate the most possible position of the pairing symmetry of iron-pnictide superconductors.

The J$'$-J phase diagram of the t-t$'$-J-J$'$ model at carrier concentration x=0.18 is shown in Fig.1a. Different from Seo *et al.*'s phase diagram Seo et al. (2008), we obtain five stable phases in the present model. The first one is a normal phase in the small J and J$'$ region, denoted

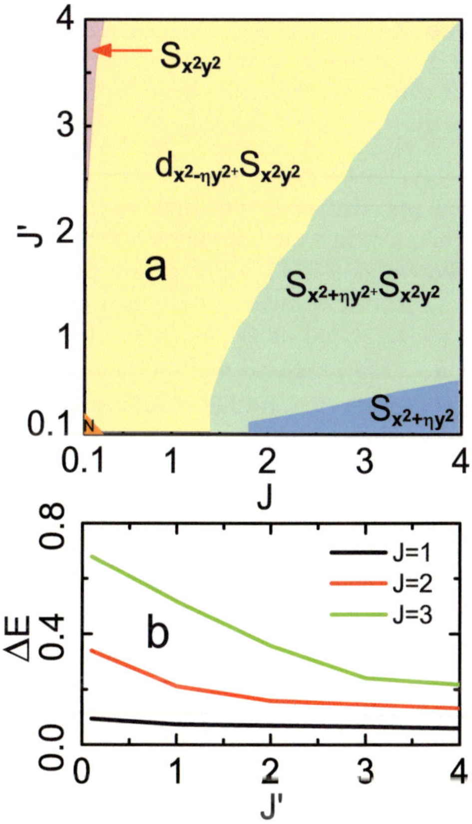

Fig. 1. (a) Superconducting phase diagram of the t-t$'$-J-J$'$ model for the d_{xz}-orbit at the carrier concentration x=0.18. N denotes the normal state, the other four phases are superconductive with different pairing symmetries. (b) The energy difference ΔE between Seo *et al's* Seo et al. (2008) and our ground states vs the NNN spin coupling J$'$ at different J, J=1,2 and 3, respectively.

by N in Fig.1a. Obviously, when the superexchange coupling J and J$'$ are too small to provide the pairing glue, the kinetic energy is dominant, and the electrons stay in the normal state, which is analogous to the single-orbital t-J model Kotliar et al. (1988). Among the four superconducting phases mediated through the spin fluctuations, a large NN spin coupling J and a small NNN spin coupling J$'$, or J>>J$'$, favor the $S_{x^2+\eta y^2}$ (here and below η_1=η) superconducting phase with the gap Δ_1(k) ∝ $cos(k_x) + \eta cos(k_y)$, where the pairing symmetry is the combination of the $S_{x^2+y^2}$-wave and the $d_{x^2-y^2}$-wave components, as seen the pink region in Fig.1a. The $S_{x^2+\eta y^2}$ symmetry arises from the major contribution of the NN spin coupling J term. The NNN spin coupling contributes very little to Δ_1(k) due to J>>J$'$.

On the other hand, small NN spin coupling J and large NNN spin coupling J$'$ favor the $S_{x^2y^2}$ superconducting phase with the symmetry Δ_1(k) ∝ $cos(k_x + k_y) + cos(k_x - k_y)$, as seen in the blue region in Fig.1a, which is mainly attributed to the NNN spin coupling. In this situation, Δ_1(k) is almost isotropic in the xy-plane due to the isotropy of the dominant

NNN hopping integrals in the xy-plane. The superconducting order parameter becomes complicated when J and J$^{'}$ compete with each other. As seen in Fig.1a, the pairing symmetry of the superconducting phase in the green region of Fig.1a is the combination of the $S_{x^2+\eta y^2}$ and the $S_{x^2y^2}$ components, and the symmetry of the superconducting phase in the yellow region of Fig.1a is the combination of the $d_{x^2-\eta y^2}$ and the $S_{x^2y^2}$ components.

It is interesting to ask in which region the realistic parameters of the iron pnictides fall. From the first-principles calculations, Ma *et. al.* suggested that J $\approx$ J$^{'}$ $\approx 0.05\ eV/S^2$ Ma et al. (2010), where S is the spin of each Fe ion. When the hopping integral $|t_1|$ $\approx$0.1 $\sim$0.5 eV, such a set of parameters falls in the yellow region in Fig.1a, implying that the FeAs superconductors should have the $d_{x^2-\eta y^2} + S_{x^2y^2}$ pairing symmetry, and the anisotropic factor η is not equal to 1. Also, some other authors suggested other parameters for the FeAs superconductors, for example, Seo *et al.*Seo et al. (2008) proposed that J=0.25 and J$^{'}$ =0.5; and Si *et al.*Si et al. (2008) thought that J $>$ J$^{'}$/2. This shows that further effort is needed to obtain more accurate interaction parameters in iron pnictides.

We notice that Seo *et al.*'s J-J$^{'}$ parameters also falls in the yellow region in Fig.1a, *i.e.* the pairing symmetry is $d_{x^2-\eta y^2} + S_{x^2y^2}$ type, rather than the $\delta_s\ S_{x^2y^2} \pm \delta_d d_{x^2-y^2}$ type with $\eta = 1$ (here δ_s and δ_d are the weights of the $S_{x^2y^2}$ wave and the $d_{x^2-y^2}$ wave component, respectively). In Fig.1b, we compare the groundstate energy difference between theirs and ours, and find that the groundstate energy in the present superconducting phase, E$_\eta$, is lower than the E$_d$ in Seo *et al.*'s paper Seo et al. (2008). Fig.1b shows the J$^{'}$ dependence of the groundstate energy difference, ΔE=E$_d$-E$_\eta$, between the two superconducting phases at different J, here E$_d$ and E$_\eta$ are the energies of the $\delta_s S_{x^2y^2} \pm \delta_d d_{x^2-y^2}$ symmetric phase and of the symmetric phase in Fig. 1a, respectively. It is clear that in wide J-J$^{'}$ range, the $d_{x^2-\eta y^2} + S_{x^2y^2}$ phase is always more stable than the $\delta_s\ S_{x^2y^2} \pm \delta_d\ d_{x^2-y^2}$ phase. Thus $d_{x^2-\eta y^2} + S_{x^2y^2}$-wave is most likely the superconducting pairing symmetry in iron pnictide superconductors.

3.3 T-dependence of fermi surface and superconducting energy gaps

In the present situation with a weakly broken orbital symmetry, we find that two superconducting energy gaps synchronously approach zero as T is lifted to T$_c$. To concretely discuss the properties of the superconducting state and the normal state, and compare the theory with the ARPES experimental results, in what follows, we focus on two sets of typical superexchange coupling parameters, Case I: J=0.3 and J$'$=0.7, *i.e.*, the NN spin coupling is weaker than the NNN coupling; and Case II: J=0.7 and J$'$=0.3, *i.e.* the NN spin coupling is stronger than the NNN coupling. In both of the situations, the parameters fall in the yellow region in Fig.1a, so the superconducting pairing symmetries are $d_{x^2-\eta y^2} + S_{x^2y^2}$-wave. We present the temperature evolution of the Fermi surfaces in the normal state in Fig.2 for the case I with J=0.3 and J$'$=0.7. The Fermi surface topology for the case II with J=0.7 and J$'$=0.3 is almost identical to Fig.2, hence is not plotted. From Fig.2, one sees that in the large Brillouin zone (BZ) associated with the present t-t$^{'}$-J-J$^{'}$ model with one Fe atom per unit cell, there exist two hole-like Fermi sheets (α_1 and α_2) around the Γ point, and two electron-like Fermi sheets (β_1 and β_2) around the M point. This is in agreement with the ARPES experiment Ding et al. (2008) and consistent with the first-principle electronic structures calculations Mazin et al. (2008); Boeri et al. (2008); Cao et al. (2008); Singh et al. (2008); Lebègue et al. (2007); Ma et al. (2008). Interestingly, the hole-like Fermi sheets expand a little with increasing temperature; in contrast, the electron-like Fermi sheets shrink considerably. This indicates that the

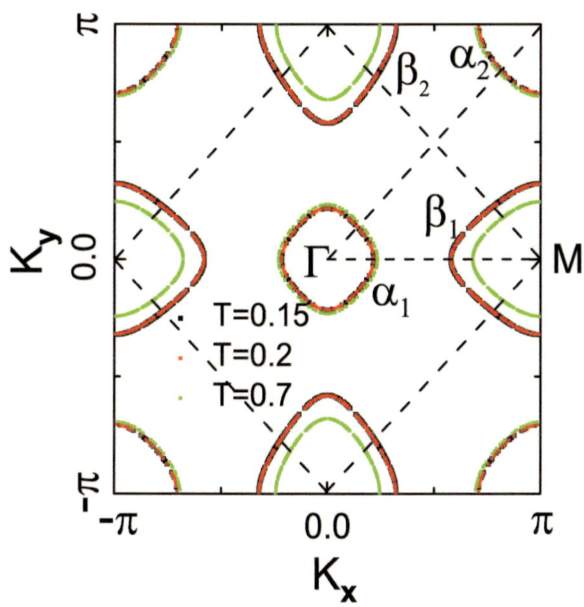

Fig. 2. The Fermi surface topology in the large Brillouin zone at the different temperatures: T=0.15 (black), 0.2 (red) and 0.7 (green). Dashed square outlines the reduced Brillouin Zone. The theoretical parameters: $J = 0.3$, $J' = 0.7$; the other parameters are the same as these in the Fig.1.

electron-like Fermi sheet may play a more important role in the low-energy processes in finite temperatures. This behavior arises from that the electronic thermal excitations increase with the lift of temperature, leading to the chemical potential decreasing with increasing T. Thus the electron-like Fermi surface decreases and the hole-like Fermi surface increases with increasing T.

Fig.3 shows the temperature dependence of the superconducting energy gaps on the two hole-like Fermi sheets along the $\theta = 0^o$ direction in the polar coordinate system for the two sets of parameters. With the increasing temperature, the two energy gaps decrease monotonously and vanish simultaneously, as observed in the ARPES experiments Ding et al. (2008). Obviously, the superconducting-state to normal-state transition is a second-order phase transition. For the case I, the magnitude of the energy gap on the small Fermi srface (α_1) is larger than that on the large Fermi surface (α_2) in the Γ point, in agreement with the ARPES results Ding et al. (2008); Zhao et al. (2008). In contrast, for the case II, the magnitude of the gap on the small fermi surface (α_1) is smaller than that on large Fermi surface (α_2) in the Γ point, which disagrees with the experiment Ding et al. (2008); Zhao et al. (2008). These indicate that the first set of parameters in Case I is more suitable for describing the FeAs superconductors.

From the present theoretical results in Fig.3, we find that in Case I, the ratios of the energy gaps to the transition temperature are $2\Delta_1/k_B T_c$=3.6 for the large gap, and $2\Delta_2/k_B T_c$=2.9 for the small gap, respectively. The ratio of the large energy gap around the small Fermi sheet to the small one around the large Fermi sheet gives rise to Δ_1/Δ_2=1.25. These theoretical

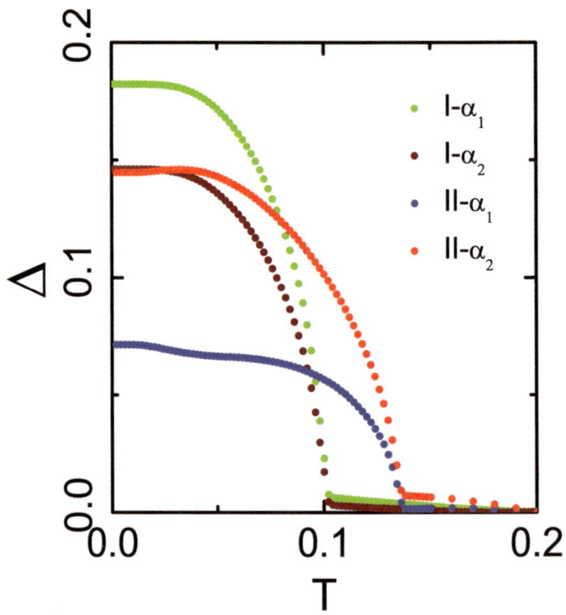

Fig. 3. Temperature dependence of the superconducting energy gaps near the small hole Fermi surface (α_1) and the large hole Fermi surface (α_2) around the Γ point along the $\theta = 0^o$ direction in the polar coordinate system. The doping concentration x=0.18. Theoretical parameters: Case I, $J = 0.3$, $J' = 0.7$ (wine and green circles); and Case II, $J = 0.7$, $J' = 0.3$ (red and blue circles).

results significantly deviate from the ARPES experimental data Ding et al. (2008). In Case II, Δ_1/Δ_2=2, in agreement with Ref.Ding et al. (2008), however, the ratios of these two gaps with respect to T$_c$, $2\Delta_\alpha/k_B T_c$, also strongly disagree with Ref.Ding et al. (2008). These facts demonstrate that there exist some essential shortages in the present t-t$'$-J-J$'$ model or in the self-consistent field method. One also notices that for Case II, the decline of the two superconducting energy gaps with the increasing temperature is not smooth, which comes from the fact that the different local pairing order parameters, $\Delta^{1\alpha}_{x/y}$ and $\Delta^{2\alpha}_{x\pm y}$, interplay with each other, reflecting the anisotropic pairing symmetry in Case II.

3.4 Angle dependence of superconducting energy gaps

In this section we present the dependence of each superconducting energy gap with the d+s-wave pairing on the orientational angle, and show that the anisotropy of the superconducting energy gaps crucially depends on the inter-orbital hopping and the ratio of J$'$/J.

The ARPES experiment provides direct information about the quasiparticle spectra in normal state and the pairing symmetry of the superconducting energy gaps in the superconducting state. Here we present our theoretical results of the angle resolved energy gaps of the two orbits, and compare them with experimental observation. Fig. 4 shows the superconducting energy gap characters of the t-t$'$-J-J$'$ model for the two sets of the parameters in Case I and II. In

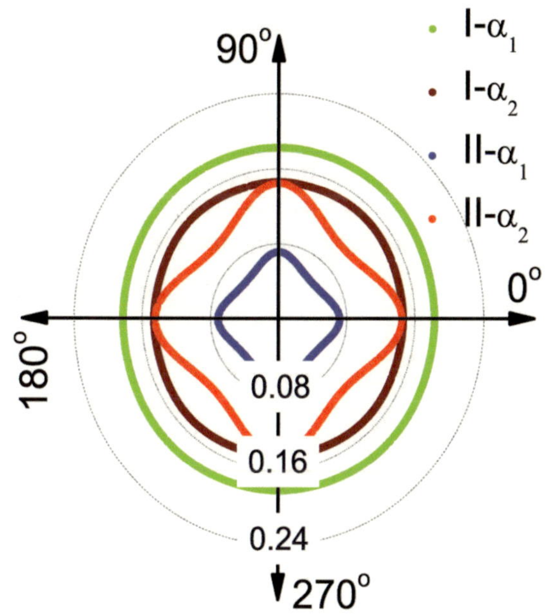

Fig. 4. The angle dependence of the superconducting energy gaps near the small hole Fermi surface (α_1) and the large hole Fermi surface (α_2) around the Γ point in the polar coordinate system. The theoretical parameters are the same as these in Fig. 3, Case I, $J = 0.3$, $J' = 0.7$ (wine and green circles); and Case II, $J = 0.7$, $J' = 0.3$ (red and blue circles).

both cases, two distinct gaps open on the hole-like Fermi sheets, α_1 and α_2, as seen in Fig.2. The presence of two different energy gaps demonstrates the nature of a multi-gap superconductor in the t-t$'$-J-J$'$ model. Our results show that for the case I, the superconducting energy gap structure exhibits a nearly isotropic symmetry with invisible anisotropy, as seen in Fig.4. A large energy gap opens on the small hole Fermi sheet (α_1), and a small energy gap opens on the large hole Fermi sheet (α_2). For the case II, the angular dependence of the energy gaps is visible, exhibiting weak spatial anisotropy. The oscillation amplitude is about 16%. However, the amplitudes of the superconducting energy gaps on the different Fermi surfaces, α_1 and α_2, are contrast to these in Case I, i.e., a small energy gap opens on the small Fermi surface sheet (α_1), and a large energy gap opens on the large hole fermi surface sheet (α_2).

One finds that in Case I, the anisotropy of the superconducting energy gaps is very weak, consistent with Zhao *et al.*'s Zhao et al. (2008) and Ding *et al.*'s Ding et al. (2008) ARPES data. In Case II, the superconducting energy gaps with about 16% anisotropy is in agreement with the ARPES experiment by Kondo *et al.* Kondo et al. (2008). Noticing that in Case II, such spatial anisotropy is still under the resolution of the ARPES experiment, hence does not conflict with Zhao *et al.*'s Zhao et al. (2008) and Ding *et al.*'s Ding et al. (2008) observation. It is the $d_{x^2-\eta y^2}$+$S_{x^2y^2}$-wave pairing symmetry that leads to weakly anisotropy and nodeless superconducting energy gap structures. Although the $d_{x^2-\eta y^2}$-wave pairing has nodes in the line $\cos k_x - \eta \cos k_y = 0$, and the $S_{x^2y^2}$-wave pairing has nodes in the lines $k_x = \pm\pi/2$ or $k_y = \pi/2$. The mixed superconducting pairing symmetry, $d_{x^2-\eta y^2}$+$S_{x^2y^2}$, diminishes the nodes, so the system exhibits weakly anisotropic and nodeless *s*-wave-like energy gaps on the Fermi surface sheets.

3.5 Spin-lattice relaxation rate in NMR

In this subsection one can also obtain the theoretical spin-lattice relaxation data under different temperatures, especially the temperature-dependence of the Knight shift in iron-pnictide superconductors. We attribute the unusual T-dependence of the Knight shift to the multi-gap character.

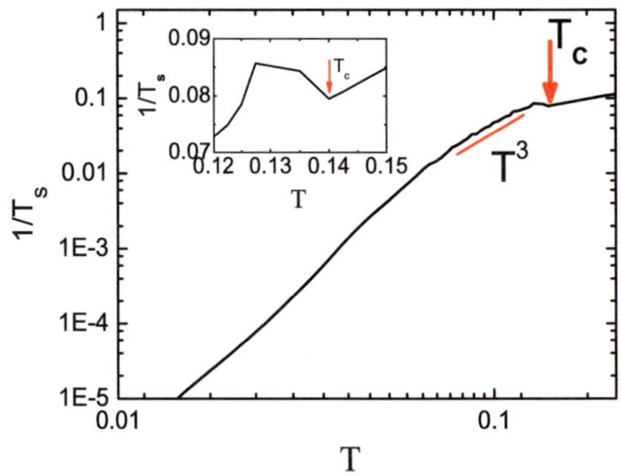

Fig. 5. Temperature dependence of the spin lattice relaxation rate in the t-t$'$-J-J$'$ model. The red arrow indicates the superconducting critical temperature T_c. The red line is the T^3 law for comparison. Inset shows the detail near T_c. The theoretical parameters are the same as case II in Fig.2.

Although many experimental measurements, such as the Andreev reflection Chen et al. (2008), the exponential temperature dependence of the penetration depths and the ARPES Ding et al. (2008); Zhao et al. (2008), observed the nodeless gap function in the superconducting phase of $ReFeAsO_{1-x}F_x$ and $Ba_{1-x}K_xFe_2As_2$ compounds, the line nodes in the superconducting energy gap was also suggested by the NMR experiment Matano (2008). The two characters in the NMR experiment supported the line nodes: lack of the coherence peak and the T^3 behavior in the nuclear spin-lattice relaxation rate, $1/T_1$. Using the gap function obtained in this paper, we calculate the spin lattice relaxation rate $1/T_s$, and the numerical result is shown in Fig. 5. We also plot the T^3 law (the red line) for a comparison. It is found that over a wide temperature range, the spin lattice relaxation rate in the present model can be fitted by the T^3 law, in agreement with the observation of the NMR experiments Matano (2008).

A small coherence peak appears around the critical transition temperature, as clearly seen in the inset of Fig.5. Experimentally, such a small coherence peak may be easily suppressed by the impurity effect or the antiferromagnetic spin fluctuations, similar to the situations in cuprates. This leads to the missing of the Hebel-Slichter coherence peak in the NMR experiment in iron pnictide SC. With decreasing temperature, one finds a drop in the spin lattice relaxation rate, $1/T_{1s}$, consistent with the observation of the NMR experiments Matano

(2008). Such a behavior deviating from the T^3 law may be attributed from the multi-gap character of this system, and such a drop reflects the different energy gaps in different orbits. Surely, more meticulous studies are needed in the near future. We also notice that the extended $s_\pm$ energy gaps found by Parker *etal*. also can give the same NMR relaxation rate in superconducting pnictides Parker et al. (2008). Parish *et al*. Parish et al. (2008) suggested that the deviation of the T^3 law in the spin-lattice relaxation arises from the inter-band contribution.

4. Comparison with other theories and experimental observations

From the preceding discussions, we find that many unusual properties in the normal state and the superconductive phase of newly discovered FeAs compounds could be qualitatively interpreted in the two-orbital t-t$'$-J-J$'$ model, showing that to some extent, this model is a good approximation to describe iron pnictide superconductors. Within this scenario, the mixing pairing symmetry with $d_{x^2-\eta y^2} + S_{x^2y^2}$-wave contributes to the weakly anisotropic and nodeless energy gaps. Such a pairing symmetry assembles the characters of usual *d*-wave and *s*-wave, hence shares the properties of the usual *d*-wave superconductors, like cuprates, and the *s*-wave superconductor, such as MgB_2. Nevertheless, to quantitatively compare the theoretical results with the experimental observation, more subtle band structures of the t-t$'$-J-J$'$ model are expected.

It is of interest that for sufficiently large NNN spin coupling J$'$, $S_{x^2y^2}$ is a dominating pairing state which is the same as the pairing symmetry obtained by Chubukov *et al*Chubukov et al. (2008). It may seem strange that this intermediate coupling theory based upon the proximity to a Mott transition has essentially the same pairing solutions as the Fermi-liquid analysis of Ref.Mazin et al. (2009). But it is not surprising at all because the fermiology and the spin fluctuation wave vector (the structure of magnetic excitations in the reciprocal space) predetermine this symmetry, as is suggested by Mazin et al. (2009). There is, however, an important difference between our results and those of Chubukov *et al* Chubukov et al. (2008). In their case, pairing mechanism is due to the increase in the intra-band pairing hopping term, not necessarily due to spin-fluctuations that is the pairing mechanism in our analysis.

Also, one should keep in mind that a completely quantitative comparison between the theory and experiment is still difficult, since the present two-orbital t-t$'$-J-J$'$ model only describes the topology structure of the Fermi surfaces of the FeAs superconductors, but does not contain all the details of the Fermi surfaces and the band structures in iron pnictide compounds. On the other hand, in the realistic material, the spin couplings, J and J$'$, might be a strong asymmetry Yin et al. (2008), which is not taken into account in the present t-t$'$-J-J$'$ model. Hence, we expect that the more elaborate tight-binding parameters and anisotropic coupling J-J$'$ model will improve the present results in future studies. Also the present constrained mean-field approximation needs to be further improved.

In some FeAs-based superconductors, the weakly anisotropic orbital symmetry makes it very difficult to distinguish which orbitals are involved in the formation of the superconducting state. To further uncover the orbital dependence of the superconducting energy gaps, we study the superconducting properties of a highly anisotropic two-orbital t-J model in the strong correlation regime. We study how the phase diagram evolves with the band asymmetric factor $R = t_{22}/t_{11}$, and the numerical result is shown in Fig. 6 in the strongly correlation regime. Note that we consider only the nearest-neighbor hopping on a square lattice. It is found that at $n = 1.98$, the difference between Δ_1 and Δ_2 increases with R deviating from the unity. The superconducting order parameters exhibit different behaviors: Δ_2 monotonously increases and almost saturates as R $< R_c \approx 0.6$; however, Δ_1 monotonously decreases and vanishes at R_c, indicating the appearance of an orbital dependent superconducting phase, where the superconducting gap in orbit 2 exponentially approaches zero, and the energy gap in orbit remains finite. As the doping concentration increases to $n = 1.95$ in Fig.6b, the two superconducting order parameters behave similarly to Fig.6a. Finally, the TGSC-intermediate superconducting phase transition occurs at $R_c \approx 0.7$. Obviously, with the decrease of R, the bandwidth of 2-orbit considerably shrinks and the pairing coupling of the orbit-2 electrons significantly deviates from that of 1-orbit. Thus, the orbital-dependent intermediate superconducting phase easily occurs when the symmetry of the orbital hopping is broken.

With the increase of the hopping integral asymmetry, the bandwidth of the 2-orbit becomes narrower and narrower, and more and more orbit-2 electrons transfer to orbit-1, hence the amplitude of the superconducting order parameter of 2-orbit gradually decreases to zero. In the same time, the superconducting order parameter of 1-orbit increases. The system enters

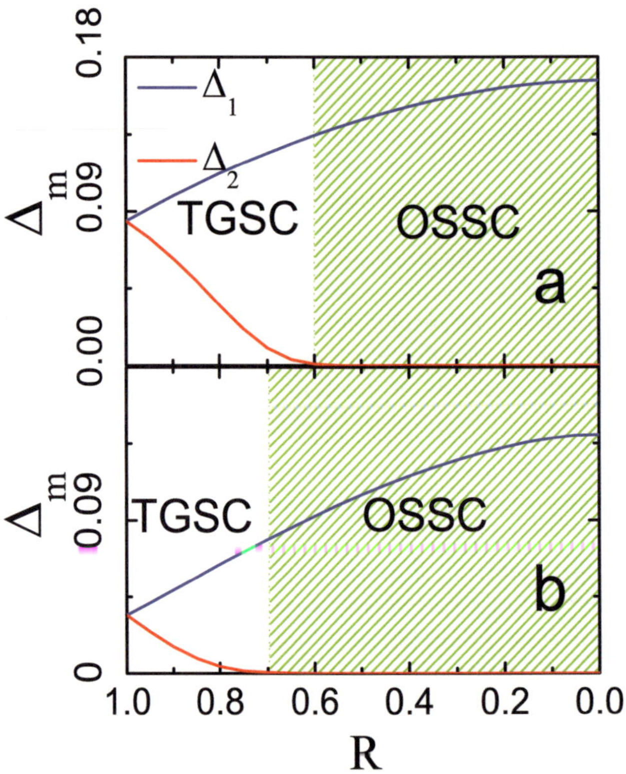

Fig. 6. (Color online) Dependence of the superconducting order parameters Δ_m (m=1,2) on the level splitting E_Δ for (a) $R = 1$ and $\delta = 0.02$, and (b) $R = 0.8$ and $\delta = 0.02$, respectively. Here TGSC and OSSC denote the two-gap and orbital dependent superconducting phases, respectively

the orbital dependent superconducting regime, and the orbital dependent superconducting phase is more robust with the deviation of R from the unity, as we see in Fig.6b. As one expects, when the hopping integral ratio R is larger than the unity, the behavior of Δ_1 is inter-changed with that of Δ_2. The properties in the system with R are analogous to those with $1/R$ in the absence of the crystalline field splitting.

5. Remarks and conclusions

We notice the profound difference of the superconducting properties between cuprates and iron pnictides. Comparing with the copper-based superconductors with a 4-fold rotational symmetry, the inequivalence between the x- and y-direction of the orbit $d_{xz/yz}$ in iron pnictides results in the anisotropic factor, η, and leads to a distinct pairing symmetry. In the present iron pnictide superconductors, the NNN spin coupling contributes an important role to the $S_{x^2y^2}$ pairing symmetry. Further, the multi-orbital character also contributes two weakly anisotropic and nodeless energy gaps, significantly different from the single energy gap in the cuprate superconductors.

Strong next-nearest-neighbour coupling and inter-orbital hopping in iron-pnictide superconductors favor a weak anisotropic and nodeless d+s wave symmetry. From Eq.(3), one could see that the NN interaction J favors the order parameters $\Delta^{1\alpha}_{x/y}$ and the NNN interaction J' favors $\Delta^{2\alpha}_{x\pm y}$. Thus, when the NN interaction J is dominant in the system, the local superconducting order parameters $\Delta^{1\alpha}_{x/y}$ become a dominant term in Eq.(5); and when the NNN interaction J' is considerably larger than J, the local superconducting order parameters $\Delta^{2\alpha}_{x\pm y}$ become dominant in Eq.(5).

In summary, our results have shown that many properties observed in iron-based superconductors could be comprehensively understood in the present model qualitatively. In the reasonable physical parameters region of $LaFeAsO_{1-x}F_x$, the pairing symmetry of the model is nearly isotropic and nodeless $d_{x^2-\eta y^2}$+$S_{x^2y^2}$-wave, mainly originating from the Fermi surface topology and the spin fluctuation in these systems, which is in agreement with the observation of ARPES and the NMR experiments in ironpnictide superconductors.

6. References

Kamihara Y., Watanabe T., Hirano M. and Hosono H. (2008). Iron-based layered superconductor $La[O_{1-x}F_x]FeAs$(x=0.05-0.12) with Tc=26 K, *J. Am. Chem. Soc.*, Vol. 130, 3296-3297, ISSN 0002-7863.

Chen X. H., Wu T., Wu G., Liu R. H., Chen H., and Fang D. F. (2008). Superconductivity at 43 K in $SmFeAsO_{1-x}F_x$. *Nature*, Vol. 453, 761-762, ISSN 0028-0836.

Chen G. F., Li Z., Li G., Zhou J., Wu D., Dong J., Hu W. Z., Zheng P., Chen Z. J., Luo J. L., and Wang N. L. (2008). Superconducting properties of Fe-based layered superconductor $LaO_{0.9}F_{0.1}FeAs$. *Phys. Rev. Lett.*, Vol. 101, 057007, ISSN 0031-9007.

Wu G., Xie Y. L., Chen H., Zhong M., Liu R. H., Shi B. C., Li Q. J., Wang X. F., Wu T., Yan Y. J., Ying J. J., Chen X. H. (2009). Superconductivity at 56 K in samarium-doped SrFeAsF. *J. Phys.: Condens. Matter*, Vol. 21, 142203, ISSN 0953-8984.

McMillan W. L. (1968). Transition Temperature of Strong-Coupled Superconductors. *Phys. Rev.*, Vol. 167, 331, ISSN 0031-899X.

Jaroszynski J., Hunte F., Balicas L., Jo Youn-jung, Raievi I., Gurevich A., Larbalestier D. C., Balakirev F. F., Fang L., Cheng P., Jia Y., and Wen H. H. (2008). Upper critical fields and thermally-activated transport of $NdFeAsO_{0.7}F_{0.3}$ single crystal. *Phys. Rev. B*, Vol. 78, 174523, ISSN 1098-0121.

You W.-L., Gu S.-J., Tian G.-S., Lin H.-Q., (2008). Constrain on possible pairing symmetry in a two-orbital model of FeAs-based superconductors, arXiv:0807.1493.

Mazin I. I., Singh D. J., Johannes M. D. and Du M.H. (2008). Unconventional Superconductivity with a Sign Reversal in the Order Parameter of $LaFeAsO_{1-x}F_x$. *Phys. Rev. Lett.*, Vol. 101, 057003, ISSN 0031-9007.

Boeri L., Dolgov O. V., and Golubov A. A. (2008). Is $LaFeAsO_{1-x}F_x$ an Electron-Phonon Superconductor? *Phys. Rev. Lett.*, Vol. 101, 026403, ISSN 0031-9007.

Cao C., Hirschfeld P.J., and Cheng H. P. (2008). Proximity of antiferromagnetism and superconductivity in $LaFeAsO_{1-x}F_x$: Effective Hamiltonian from ab initio studies. *Phys. Rev. B*, Vol. 77, 220506(R), ISSN 1098-0121.

Ma F. J., Lu Z. Y., and Xiang T. (2010). Electronic structures of ternary iron arsenides AFe_2As_2 (A=Ba, Ca, or Sr). *Front. Phys. China*, Vol. 5, 150 , ISSN 2095-0470.

Zhang H. J., Xu G., Dai X., and Fang Z. (2009). Enhanced Orbital Degeneracy in Momentum Space for LaOFeAs. *Chinese Phys. Lett.*, Vol. 26, 017401, ISSN 0256-307X.

Mu G., Zhu X. Y., Fang L., Shan L., Ren C. and Wen H. H. (2008). Nodal Gap in Fe-Based Layered Superconductor $LaO_{0.9}F_{0.1-\delta}$ FeAs Probed by Specific Heat Measurements. *Chinese Phys. Lett.*, Vol. 25, 2221, ISSN 0256-307X.

Millo O., Asulin I., Yuli O., Felner I., Ren Z. A., Shen X. L., Che G. C., and Zhao Z. X. (2008). Scanning tunneling spectroscopy of $SmFeAsO_{0.85}$: Possible evidence for d-wave order-parameter symmetry. *Phys. Rev. B*, Vol. 78, 092505, ISSN 1098-0121.

Wang Y. L., Shan L., Fang L., Cheng P., Ren C., Wen H. H. (2009). Multiple gaps in $SmFeAsO_{0.9}F_{0.1}$ revealed by point-contact spectroscopy. *Supercond. Sci. Technol.*, Vol. 22, 015018 , ISSN 0953-2048.

Wang F., Zhai H., Ran Y., Vishwanath A., and Lee D. H. (2009). Functional Renormalization-Group Study of the Pairing Symmetry and Pairing Mechanism of the FeAs-Based High-Temperature Superconductor. *Phys. Rev. Lett.*, Vol. 102, 047005, ISSN 0031-9007.

Nomura T., Kim S. W., Kamihara Y., Hirano M., Sushko P. V., Kato K., Takata M., Shluger A. L., and Hosono H. (2008). Crystallographic phase transition and high-Tc superconductivity in LaFeAsO:F. *Supercond. Sci. Technol.*, Vol. 21, 125028, ISSN 0953-2048.

Seo K., Bernevig B. A., and Hu J. P. (2008). Pairing Symmetry in a Two-Orbital Exchange Coupling Model of Oxypnictides. *Phys. Rev. Lett.*, Vol. 101, 206404, ISSN 0031-9007.

Dai X., Fang Z., Zhou Y. and Zhang F. C. (2008). Even Parity, Orbital Singlet, and Spin Triplet Pairing for Superconducting $LaFeAsO_{1-x}F_x$. *Phys. Rev. Lett.*, Vol. 101, 057008, ISSN 0031-9007.

Lee P. A., and Wen X. G. (2008). Spin-triplet p-wave pairing in a three-orbital model for iron pnictide superconductors. *Phys. Rev. B*, Vol. 78, 144517, ISSN 1098-0121.

Kuroki K., Onari S., Arita R., Usui H., Tanaka Y., Kontani H., and Aoki H. (2008). Unconventional Pairing Originating from the Disconnected Fermi Surfaces of Superconducting $LaFeAsO_{1-x}F_x$. *Phys. Rev. Lett.*, Vol. 101, 087004, ISSN 0031-9007.

Yan X.-W., Gao M., LU Z.-Y., Xiang T. (2011). Electronic and magnetic structures of ternary iron selenides AFe_2Se_2 (A=K, Cs, or Tl). *Phys. Rev. B*, Vol. 84, 054502, ISSN 1098-0121.

Haule K., Shim J. H., and Kotliar G. (2008). Correlated Electronic Structure of $LaO_{1-x}F_xFeAs$. *Phys. Rev. Lett.*, Vol. 100, 226402, ISSN 0031-9007.

Manousakis E., Ren J., Meng S, and Kaxiras E. (2008). Effective Hamiltonian for FeAs-based superconductors. *Phys. Rev. B*, Vol. 78, 205112, ISSN 1098-0121.

Raghu S., Qi X. L., Liu C. X., Scalapino D., Zhang S. C. (2008). Minimal two-band model of the superconducting iron oxypnictides. *Phys. Rev. B*, Vol. 77, 220503(R), ISSN 1098-0121.

Dubroka A., Kim K. W., Roessle M., Malik V. K., Liu R. H., Wu G., Chen X. H., Bernhard C. (2008). Superconducting Energy Gap and c-Axis Plasma Frequency of $(Nd,Sm)FeAsO_{0.82}F_{0.18}$ Superconductors from Infrared Ellipsometry. *Phys. Rev. Lett.*, Vol. 101, 097011, ISSN 0031-9007.

Matano K., Ren Z. A., Dong X. L., Sun L. L., Zhao Z. X., and Zheng Q. Q. (2008). Spin-singlet superconductivity with multiple gaps in $PrFeAsO_0.89F_0.11$ *Europys. Lett.*, Vol. 83, 57001, ISSN 0295-5075.

Xiang T., "d-Wave Superconductors", pp226, Science Press Company, (Beijing 2007), ISBN 978-7-03-018920-2.

Kotliar G., and Liu J. L. (1988). Superexchange mechanism and d-wave superconductivity. *Phys. Rev. B*, Vol. 38, 5142, ISSN 1098-0121.

Barnes S. E.. (1976). New method for the Anderson model. *J. Phys. F: Met. Phys.*, Vol. 6, 1375, ISSN 0305-4608.

Coleman P. (1984). New approach to the mixed-valence problem. *Phys. Rev. B*, Vol. 29, 3035-3044, ISSN 1098-0121.

Si Q. M., and Abrahams E. (2008). Strong Correlations and Magnetic Frustration in the High Tc Iron Pnictides. *Phys. Rev. Lett.*, Vol. 101, 076401, ISSN 0031-9007.

Ding H., Richard P., Nakayama K., Sugawara K., Arakane T., Sekiba Y., Takayama A., Souma S., Sato T., Takahashi T., Wang Z., Dai X., Fang Z., Chen G. F., Luo J. L. and Wang N. L. (2008). Observation of Fermi surfaceĺCdependent nodeless superconducting gaps in $Ba_0.6K_0.4Fe_2As_2$. *Europys. Lett.*, Vol. 83, 47001, ISSN 0295-5075.

Singh D., and Du M. H. (2008). Density Functional Study of $LaFeAsO_{1-x}F_x$: A Low Carrier Density Superconductor Near Itinerant Magnetism. *Phys. Rev. Lett.*, Vol. 100, 237003, ISSN 0031-9007.

Lebègue S. (2007). Electronic structure and properties of the Fermi surface of the superconductor LaOFeP. *Phys. Rev. B*, Vol. 75, 035110, ISSN 1098-0121.

Ma F., and Lu Z. Y. (2008). Iron-based layered compound LaFeAsO is an antiferromagnetic semimetal. *Phys. Rev. B*, Vol. 77, 033111, ISSN 1098-0121.

Zhao L., Liu H. Y., Zhang W. T., Meng J. Q., Jia X. W., Liu G. D., Dong X. L., Chen G. F., Luo J. L., Wang N. L., Lu W., Wang G .L., Zhou Y., Zhu Y., Wang X. Y., Xu Z. Y., Chen C. T., and Zhou X. J. (2008). Multiple Nodeless Superconducting Gaps in $(Ba_0.6K_0.4)Fe_2As_2$ Superconductor from Angle-Resolved Photoemission Spectroscopy. *Chinese Phys. Lett.*, Vol. 25, 4402-4405, ISSN 0256-307X.

Kondo T., Santander-Syro A. F., Copie O., Liu C., Tillman M. E., Mun E. D., Schmalian J., Bud'ko S. L., Tanatar M. A., Canfield P. C., Kaminski A. (2008). Momentum Dependence of the Superconducting Gap in $NdFeAsO_{0.9}F_{0.1}$ Single Crystals Measured by Angle Resolved Photoemission Spectroscopy. *Phys. Rev. Lett.*, Vol. 101, 147003, ISSN 0031-9007.

Chen T. Y., Tesanovic Z., Liu R. H., Chen X. H., and Chien C. L. (2008). Superconductivity at 43 K in $SmFeAsO_{1-x}F_x$. *Nature*, Vol. 453, 1224-1227, ISSN 0028-0836.

Parker D., Dolgov O. V., Korshunov M. M., Golubov A. A., and Mazin I. I. (2008). Extended s_{+-} scenario for the nuclear spin-lattice relaxation rate in superconducting pnictides. *Phys. Rev. B*, Vol. 78, 134524, ISSN 1098-0121.

Parish M. M., Hu J. P., and Bernevig B. A. (2008). Experimental consequences of the s-wave cos(kx)cos(ky) superconductivity in the iron pnictides. *Phys. Rev. B*, Vol. 78, 144514, ISSN 1098-0121.

Chubukov A. V., Efremov D. V., and Eremin I. (2008). Magnetism, superconductivity, and pairing symmetry in iron-based superconductors. *Phys. Rev. B*, Vol. 78, 134512, ISSN 1098-0121.

Mazin I. I., and Schmalian J. (2009). Pairing symmetry and pairing state in ferropnictides: Theoretical overview. *Physica* C, Vol. 469, 614, ISSN 0921-4534.

Yin Z. P., Lebĺĺgue S., Han M. J., Neal B. P., Savrasov S. Y., and Pickett W. E. (2008). Electron-Hole Symmetry and Magnetic Coupling in Antiferromagnetic LaFeAsO. *Phys. Rev. Lett.*, Vol. 101, 047001, ISSN 0031-9007.

3

Josephson Current in a Ferromagnetic Semiconductor/Semiconductor/Ferromagnetic Semiconductor Junction with Superconducting Contacts

Y.C. Tao
Department of Physics, Nanjing Normal University, Nanjing
People's Republic of China

1. Introduction

Superconducting spintronics [1–3] is one of the most attractive subjects of spintronics. A variety of spin-injecting tunnelling junctions composed of ferromagnet and superconductor (SC) with different wave pairing symmetries have been explored theoretically and experimentally, where the ferromagnet includes ferromagnetic metals (FM), ferromagnetic semiconductors (FS), and so on.

Recently, the FS arouses great interests due to several important material properties, for example, carrier type (i.e., n or p), carrier densities, and ferromagnetism, can be easily controlled in it. [4] The origin of ferromagnetism in the FS can be explained by applying a picture that uniform itinerant carrier spin mediates a long-range ferromagnetic order between the ions Mn^{+2} with spin. p-type III-V GaMnAs is a prototype FS with a large spin polarization even for small Mn concentrations [4, 5] and recent experiments demonstrate that the Curie temperature of this material can be as high as 150 K, [4, 5] indicating promise for possible technological relevance. [5–8]

On the other hand, current can flow without dissipation through a thin insulating layer or weak link separating two superconductors (SC). A phase difference ϕ between the SCs appears on account of the quantum character of the superflow and the current-phase relationship of the link is periodic function of ϕ. For a tunnel barrier, it is given by $I = I_C \sin(\phi)$ as first found by Josephson [9], where I_C is the critical Josephson current. When the two SCs coupled by a thin ferromagnetic layer [10–29], I_C can change sign. Referring to the Josephson current-phase relationship, the change is corresponding to a π-phase shift of ϕ. And thus Josephson junctions can present a negative coupling, which are usually called π junctions. The transitions from positive to negative coupling (the so-called transition from 0 to π junction) by increasing the FS layer thickness observed by experiments are qualitatively in agreement with those of theoretical calculations. It has been widely accepted that high-critical temperature (T_C) SCs have a $d_{x_a^2-x_b^2}$-wave pairing symmetry, [30–32] in which the energy gap of such a pairing state manifests a sign change at some directions of the Fermi wave vector. This feature is much different from that for a conventional s-wave SC. As a result of this sign change, it was predicted [33] that a sizable areal density of midgap states exists on a {110} surface of a d-wave SC. These midgap states have exerted the influence on the Josephson current of high T_C SC junctions. More recently, the relative Josephson tunnelling junctions composed of FS and SC with s-wave pairing symmetry have being explored and some valuable results have been obtained,[34, 35] however, the works focused on the systems are scarce. Particularly, to our knowledge, we seldom see theoretical and experimental works taking into account the unconventional pairing symmetry, which is relevant to the high T_C superconducting junctions composed of the FS. Therefore, it seems to be timely to review the *Superconducting Spintronics* based on the Josephson junctions consisting of FS and SC with s-wave pairing symmetry and consider the outlook for the future work.

In this chapter, the Josephson current flowing through an FS/semiconductor (SM) /FS trilayer connected to two superconductor (SC) electrodes is studied. We extend a general expression for dc Josephson current to the structure, in which the quasiparticle wave function, satisfying Bogoliubov-de Gennes (BDG) equation, is first obtained and then the temperature Green's function formalism is used. The Josephson current in the junction is find to be strongly dependent on not only the kinds of holes in the FS's and scattering potential strengths at the interfaces but also the relative orientations of the effective exchange field of the two FS's. We also show that in the coherent regime, the oscillatory dependences of the critical Josephson current on the FS and SM layer thicknesses and Josephson current on the macroscopic phase difference for the heavy and light holes in different orientations of the effective exchange field of the two FS's, are much different due to the different mismatches in the effective mass and Fermi velocity between the FS and SC, which is connected with the transition from positive (0) to negative (π) coupling. The effect can be exploited for the design of devices such as superconducting quantum interference devices with improved accuracy and convenience. In Sec. II, we present the theoretical approach. In Sec. III, FS Josephson tunneling junctions are studied. The summary and outlook are given in Sec. IV. Finally, the list of used symbols is presented in Sec. V.

2. Theoretical approach: BDG equation

Applying Bogoliubov-de Gennes (BDG) equation, [36] *Superconducting Spintronics* built on the Josephson junctions composed of GaMnAs FS and s-wave SC are studied, in which the AR ,[34, 35] hole types, mismatches in the effective mass and Fermi velocity of holes between the FS and SC, and strengths of potential scattering at the interface are taken into account. In the next parts of the chapter, the layers are taken to be the $y-z$ plane and to be stacked along the x-direction, and the FS/SC interface, for example at $x=a$, is described by a δ-type potential $V(x)=U\delta(x-a)$ with U dependent on the product of the barrier height and width.

As in FMs, a spin-polarized subband model with the spins parallel and antiparallel to the local magnetization is extended to the heavy and light hole bands. The hole Hamiltonian in the FS is simply given by

$$H_{FS}(\boldsymbol{r}) = H_0(\boldsymbol{r}) - \boldsymbol{h}(\boldsymbol{r}) \cdot \boldsymbol{\sigma}. \quad (1)$$

Here $H_0(\boldsymbol{r}) = -\hbar^2\nabla_{\boldsymbol{r}}/2m_p + V(\boldsymbol{r})$ is the kinetic energy with m_p the mass of a hole plus the usual static potential, which is also the hole Hamiltonian of the SM GaAs, $\boldsymbol{h}(\boldsymbol{r})$ is the effective exchange field with the magnitude equal to $\Gamma/2$, and $\boldsymbol{\sigma}$ the conventional Pauli spin operator.

At finite temperatures, the spin splitting energy Γ different for different kind of holes, is given by $\Gamma = J_{pd} < S^z >$, where J_{pd} stands for the $p-d$ exchange coupling strength between itinerant holes and Mn^{2+} ion impurity spins connected with the densities of Mn impurity and heavy or light holes, and $< S^z >$ implies the thermal average of the impurity Mn ion spins. To model the subband spin splitting and the relative spin-polarized degree P, we define the parameter $P \equiv \Gamma/E_F$. The SC is assumed s-wave pairing and described by a BCS-like Hamiltonian, in which the excitation energy is $\zeta_k = \sqrt{\epsilon_k^2 + \Delta^2}$ with $\epsilon_k = \hbar^2 k^2/2m_e - E_F$ the one-electron energy relative to E_F and Δ the energy gap. The energy gap Δ is determined by the self-consistent equation [34, 35]

$$\ln\left(\frac{\Delta_0}{\Delta}\right) = \int_0^{\hbar\omega_D} \frac{d\epsilon_k}{\zeta_k} \frac{2}{1+e^{\beta\zeta_k}}, \tag{2}$$

where ω_D refers to Debye frequency and Δ_0 denotes energy gap at zero temperature.

The quasiparticle wave functions, in the Blonder-Tinkham-Klapwijk model, BDG equation [36]

$$\begin{bmatrix} H_0^{SC}(\boldsymbol{r}) - \eta_\sigma h(\boldsymbol{r}) & \Delta(x) \\ \Delta^*(x) & -H_0^{SC*}(\boldsymbol{r}) + \eta_{\bar{\sigma}} h(\boldsymbol{r}) \end{bmatrix} \begin{bmatrix} u_\sigma \\ v_{\bar{\sigma}} \end{bmatrix} = E \begin{bmatrix} u_\sigma \\ v_{\bar{\sigma}} \end{bmatrix}, \tag{3}$$

where u_σ and $v_{\bar{\sigma}}$ are the two-component wavefunction, $H_0^{SC}(\boldsymbol{r})$ has the same expression as $H_0^{FS}(\boldsymbol{r})$ except for the effective mass replaced by the mass of electron m_e, $\Delta(x)$ is pair potential, E is the quasiparticle energy relative to E_F, $\bar{\sigma}$ is the spin opposite to σ with $\uparrow$ and $\downarrow$, $\eta_\sigma = 1$ for $\sigma = \uparrow$, and $\eta_\sigma = -1$ for $\sigma = \downarrow$. In Eq. (3), as a result of that the spin-flip process has been neglected in the SC near interfaces in Refs. 34 and 35, the spin-dependent (four-component) BDG equation may be decoupled into two sets of two-component equations: $u_\sigma, v_{\bar{\sigma}}$ describing the spin-σ electronlike quasiparticle (ELQ) and spin-$\bar{\sigma}$ holelike quasiparticle (HLQ), respectively.

3. FS Josephson tunnelling junctions

3.1 Model and theory

A SC/FS/SM/FS/SC Josephson tunnel junction as in Fig. 1 is considered [34, 35]. The left and right electrodes are made up of the same SCs: they are separated from a FS/SM/FS trilayer with the same FS GaMnAs layers of thickness b and SM GaAs of thickness a. Either the left and right FS/SC interfaces or SM/FS interfaces, can be approximately described by four δ-type barrier potentials $V(x) = U\delta[z \pm (b + a/2)]$ and $V'(x) = U'\delta(z \pm a/2)$, respectively and assumed to be symmetrical due to the same SCs and the same FS's, where U and U' depend on the product of the barrier height and width. For the ferromagnetic (F) alignment of the magnetizations of the two bilayers, $\boldsymbol{h}(-b - a/2 \le z \le -a/2) = \boldsymbol{h}(a/2 \le z \le b + a/2)$, while for the antiferromagnetic (A) alignment, $\boldsymbol{h}(-b - a/2 \le z \le -a/2) = -\boldsymbol{h}(a/2 \le z \le b + a/2)$. For the injection of an ELQ from the left SC with Andreev reflection as a HLQ (a) and normal reflection (b), the wave function are given by [34, 35]

$$\psi_L^{SC}(x) = [e^{ik^+x} + be^{-ik^+x}] \begin{pmatrix} ue^{i\phi_L/2} \\ ve^{-i\phi_L/2} \end{pmatrix} + ae^{ik^-x} \begin{pmatrix} ve^{i\phi_L/2} \\ ue^{-i\phi_L/2} \end{pmatrix}, \tag{4}$$

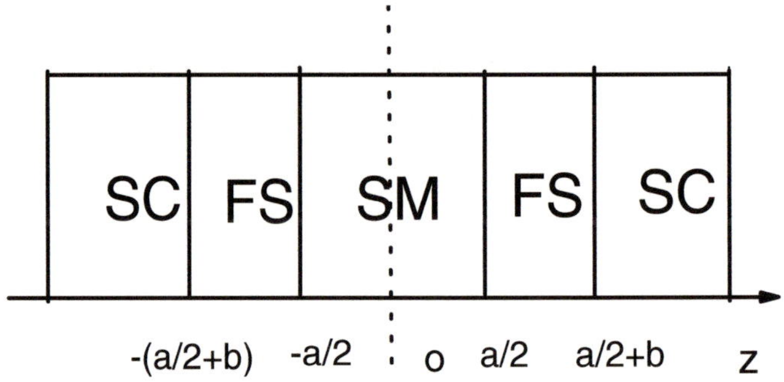

Fig. 1. The SC/FS/SM/FS/SC system (from Ref. 34).

for $x \leq -b - a/2$,

$$\psi_1(x) = [c_{11}e^{iq_{\sigma}^{L+}x} + c_{12}e^{-iq_{\sigma}^{L+}x}]\begin{pmatrix}1\\0\end{pmatrix} + [c_{13}e^{iq_{\bar{\sigma}}^{L-}x} + c_{14}e^{-iq_{\bar{\sigma}}^{L-}x}]\begin{pmatrix}0\\1\end{pmatrix} \tag{5}$$

for $-b - a/2 \leq x \leq -a/2$,

$$\psi_2(z) = [c_{21}e^{iq'^{+}x} + c_{22}e^{-iq'^{+}x}]\begin{pmatrix}1\\0\end{pmatrix} + [c_{23}e^{iq'^{-}x} + c_{24}e^{-iq'^{-}x}]\begin{pmatrix}0\\1\end{pmatrix} \tag{6}$$

for $-a/2 \leq x \leq a/2$,

$$\psi_3(x) = [c_{31}e^{iq_{\sigma}^{R+}x} + c_{32}e^{-iq_{\sigma}^{R+}x}]\begin{pmatrix}1\\0\end{pmatrix} + [c_{33}e^{iq_{\bar{\sigma}}^{R-}x} + c_{34}e^{-iq^{R-}{}_{\bar{\sigma}}x}]\begin{pmatrix}0\\1\end{pmatrix} \tag{7}$$

for $a/2 \leq x \leq a/2 + b$,

$$\psi_R^{SC}(x) = ce^{ik^{+}x}\begin{pmatrix}ue^{i\phi_R/2}\\ve^{-i\phi_R/2}\end{pmatrix} + de^{-ik^{-}x}\begin{pmatrix}ve^{i\phi_R/2}\\ue^{-i\phi_R/2}\end{pmatrix}, \tag{8}$$

for $x \geq a/2 + b$, where $q_{\sigma}^{L(R)\pm} = \sqrt{(2m_p/\hbar^2)(E_F \pm E + \eta_\sigma h_0) - k_{\parallel}^2}$, $q'^{\pm} = \sqrt{(2m_p/\hbar^2)(E_F \pm E) - k_{\parallel}^2}$, $k^{\pm} = \sqrt{(2m_e/\hbar^2)(E_F \pm \Omega) - k_{\parallel}^2}$, $u = \sqrt{(1+\Omega/E)/2}$, and $v = \sqrt{(1-\Omega/E)/2}$ with $\Omega = \sqrt{E^2 - \Delta^2}$ and the parallel component of the wave vector $k_{\parallel}$ which is conserved. In Eqs. (4)-(8), all coefficients will be determined by the usual matching conditions of the wave functions.

From Eqs. (4)-(8), the transmission and reflection coefficients can be obtained. [34, 35] In the derivation, two dimensionless parameters $Z = 2m_eU/(\hbar^2 k_F^{SC})$ and $Z' = 2m_eU'/(\hbar^2 k_F^{SC})$, and χ are respectively introduced to describe the barrier strengths at the interfaces and ratio of the masses m_p and m_e, the phase difference between the SCs $\phi = \phi_L - \phi_R$, $|k_{\parallel}| = k_F^{SC}\sin\theta$, where the corresponding Fermi vector in the SC is $k_F^{SC} = \sqrt{2m_eE_F}/\hbar$. Analogously, one can easily obtain the Andreev reflection (a') as an ELQ for the injection of a HLQ into the right FS from the left SC.

After having obtained the coefficients a and a', we can express the dc Josephson current in terms of the Andreev reflections amplitudes by using the temperature Green's function formalism [37]

$$I = \frac{e\Delta}{2\hbar} \sum_{\sigma,k_{\parallel}} k_B T \sum_{\omega_n} \frac{1}{2\Omega_n} (k_n^+ + k_n^-) \left(\frac{a_n}{k_n^+} - \frac{a_n'}{k_n^-} \right), \tag{9}$$

where k_n^+, k_n^-, a_n, and a_n' are obtained from k^+, k^-, a, and a' by the analytic continuation $E \to i\omega_n$, the Matsubara frequencies are $\omega_n = \pi k_B T(2n+1)$ with $n = 0, \pm 1, \pm 2, \ldots$, and $\Omega_n = \sqrt{\omega_n^2 + \Delta^2}$. As for the same E, σ, and θ in the approximation with $\Omega/E_F \ll 1$ and $E/E_F \ll 1$, the amplitudes of two Andreev reflections $a(\phi)$ and $a'(\phi)$ are simply connected by $a'(\phi) = a(-\phi)$. Performing integration over $k_{\parallel}$, we can deduce [34, 35]

$$I = \frac{2\pi k_B T \Delta}{eRk_F^{SC}} \int_0^{\pi/2} d\theta \sin\theta \cos\theta \sum_{\omega_n,\sigma} [a_n(\phi) - a_n(-\phi)], \tag{10}$$

where $R = 2\pi^2\hbar/Se^2(k_F^{SC})^2$ with S the area of junction. In accordance with Eq. (10), we can calculate the Josephson currents in the F and A alignments.

3.2 Calculations and results

In Fig. 2 are illustrated the numerical results for I_C as a function of the FS thickness b with $a = 40/k_F^{SC}$ at different scattering potential strengths Z and Z' for the heavy and light holes in the F and A alignments. [34] The damped oscillations of the critical Josephson currents I_C for the heavy and light holes are all found to be as a function of b, which is also the characteristic feature of the ballistic SC/FM/SC junctions and always related to the crossovers between 0 and π states [24]. Microscopically, Andreev reflections play an important role in the transfer of Cooper pairs through the FS. An electronlike quasiparticle in FS with energy lower than the superconducting energy gap Δ, cannot enter into the SC, but is reflected at the FS/SC interface as a hole and reflected back as an electron at the opposite SC/FS interface. The constructive interference of the electronlike and holelike excitations gives rise to Andreev bound states in FS, which carry the supercurrent. However, in the FS, the spin splitting of spin band exerts an influence on the Andreev reflection when the spin of the quasiparticles is reversed by the Andreev reflection, which means that the exchange energy is gained or lost by a quasiparticle Andreev-reflected at the FS/SC interface. It follows that the superconducting pair amplitude induced in FS by the proximity to the SC is spatially modulated and the phase increases linearly with the distance from the SC/FS interface. Since the Josephson critical current I_C is proportional to the pair amplitude in the FS, I_C follows the sign of the order parameter in the FS. Whether it is either positive or negative is determined by the phase accumulating in the FS. Furthermore, it is found that as shown in Fig. 2, at $Z = 0$ and $Z' = 0$, for the heavy holes, the general tendency of variations of the oscillation amplitude with increasing b is to first decrease, then increase, and last gradually reduce to zero in the F and A alignments, while for the light holes, gradually decrease to zero all the time. We also find that for either the heavy or light holes, the behaviors in the F and A alignments are very similar in spite of the values of Z and Z'. However, for Z and Z' unequal to zero, it is noticeable that the peak values for heavy holes all have bigger difference between the F and A alignments, whereas for the light holes, no much bigger difference, which is the same as that for $Z = 0$ and $Z' = 0$. Besides, with the increase of the barrier strengths Z and Z', the peak values all decrease for the two kinds of holes in the F and A alignments, however, the number of periodicity is decreased for

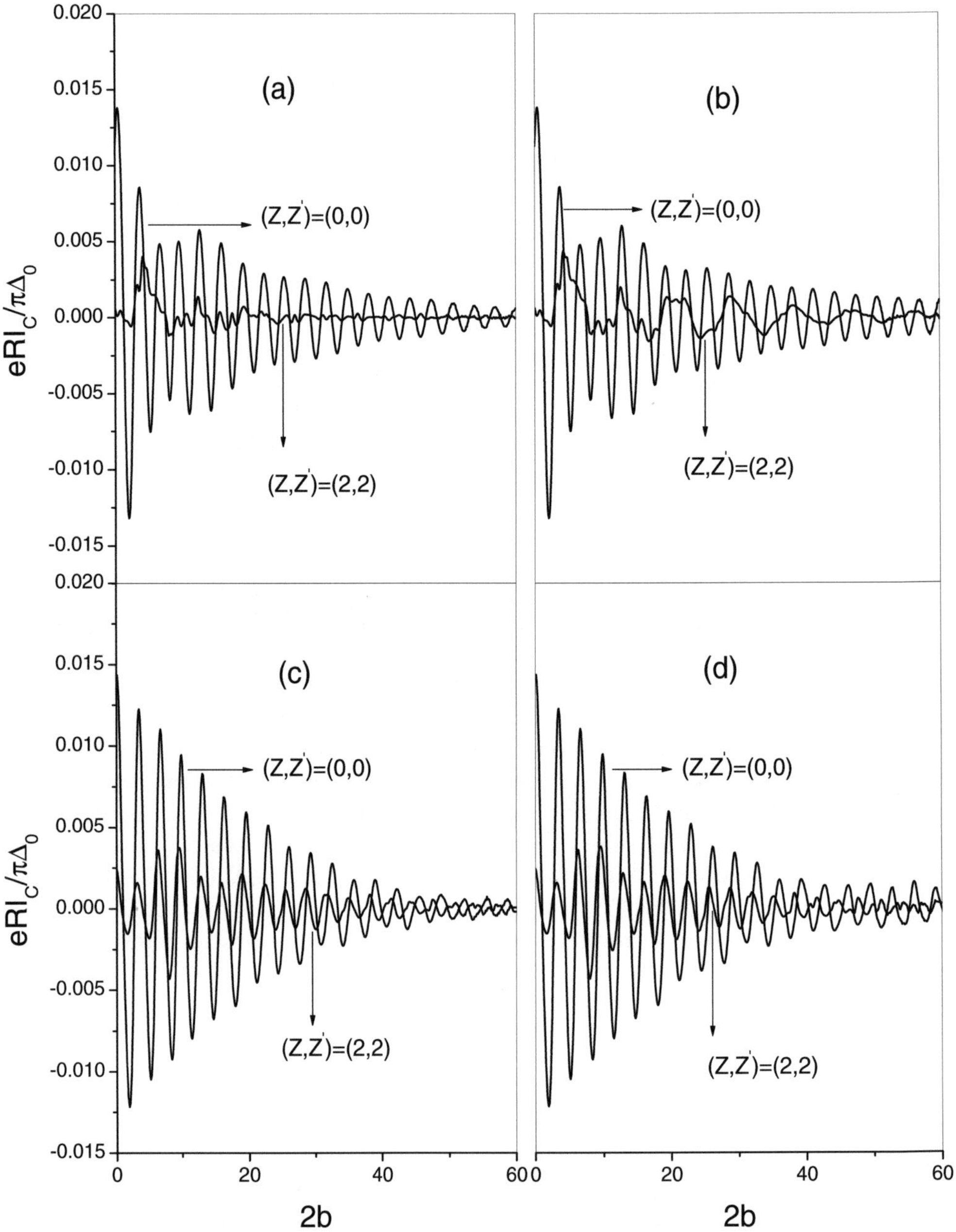

Fig. 2. Dependences of I_C on the FS thickness b at different interface scattering potential strengths (Z, Z') with $(0, 0)$ (solid line) and $(2, 2)$ (dotted line). Here, $E_F = 100.0$meV, $\Delta_0 = 1.4$meV, $k_B T = 0.05\Delta_0$, $m_e = 1.0$, $m_p = 0.45 m_e$ for the heavy holes in the F (a) and A (b) alignments, and $m_p = 0.08$ for the light holes in the F (c) and A (d) (from Ref. 34).

the heavy holes while that for the light is increased. The occurrence of periodic behavior can be explained by the two terms $exp^{-i[(k^{\pm} \pm q'^{\pm})a/2 + (k^{\pm} \pm q_{\sigma}^{L\pm})b - \phi]}$ in Eqs. (A4)-(A13) of Appendix

A, [34] including the product of b and wave vector $q^{L(R)\pm}$ and varying periodically. However, the Fermi vector q_F for the heavy holes is much bigger than that for the light because of effective mass bigger than that for the light, as a result, the number of the periodicity for the heavy is also slightly bigger than that for the light as shown in Fig. 2 at $Z = 0$ and $Z' = 0$. More importantly, when b is smaller, $exp^{-i[(k^{\pm}\pm q'^{\pm})a/2}$ have the bigger effect on the peak value for the heavy holes than that for the light holes, so the behavior of the periodicity for the two holes are different at the smaller b. In fact, the effect of the coefficients of the above two terms corresponds to a physical phase for each equation, therefore, when Z and Z' are unequal to *zero* and vary, the phase shift appears and the peaks all shift with the b. The phase shifts for heavy and light holes are different on account of different mismatches in the Fermi velocity between the FS and SC, resultantly, the behaviors for the variations of the number of the periodicity are different with Z and Z'. For the effect of Z and Z' on the critical Josephson current, physically, for the heavy or light holes, the exchange energy is not easily or lost by a quasiparticle Andreev-reflected at the the FS/SC interface with bigger z, therefore, Josephson current can not easily pass the junction and thus decreases with increasing the Z and Z'.

Fig. 3 shows the SM thickness a dependence of Josephson critical current I_C for the heavy and light holes in the F and A alignments at different Z and Z'. [34] We find that there are same features as in Fig. 2 except that the behaviors and values of the critical Josephson currents I_C are slightly different, which can be similarly easily interpreted as in Fig. 2.

In Figs. 4 (5) are illustrated the phases of the Josephson current $I(\phi)$ of the heavy (light) holes for different Z and Z' in the F and A alignments. [34] It is shown that, for the dependence of the Josephson currents on the phase, if the junction is the 0 or π state or transition from the 0 to π state is determined by not only the magnetic alignment but also kinds of the carriers in the FS. No matter what P, Z and Z' are, for the heavy hole, the junctions in the F and A alignments are only the 0 state, otherwise these are π state, while for the light holes, the situations are thoroughly different from those for the heavy. In the F alignment, regardless of P, the minimum values of Z and Z' at the same time when there is a transition of the junction changing from the 0 to π state are respectively 1.5 and 1.7 or so although the transition is not too obvious for the bigger P. If one of Z and Z' is increased, the other is at least the above value and P is only decreased for the existence of the transition. In the A alignment, ii spite of P, the minimum values of Z and Z' at the same time when there exists a transition of the junction changing from the 0 to π state are basically the same as those in the F. However, in the A alignment, with the increase of Z , Z' can be decreased in small magnitude and P is only enhanced for the existence of the transition, yet both of Z and Z' can be increased in the meantime and can be bigger for the bigger P. At the half metallic limit, they are respectively approximate to 2.2 and 2.1. It can be concluded from these results that the different mismatches in the effective mass and Fermi velocity between the FS and SC and the relative orientations of the effective exchange field of the two FS's play an important role in transition of the junction changing from the 0 to π state. The interpretation can be given as follows. In Eqs. (A4)-(A7) of Appendix A, [34] there are two terms, including $exp^{-i[(K^{\pm}-q_{\sigma}^{L\pm})b-\phi]}$ and $exp^{-i[(K^{\pm}+q_{\sigma}^{L\pm})b-\phi]}$. If the junction is the 0 or π state or transition from the 0 to π state is mainly dependent on the contribution of the two terms. For the heavy holes, due to the effective mass bigger than that in the SC, Fermi vectors in the F and A alignments are not much smaller compared with those in the SC, therefore the sum of the two terms can not easily change the sign with ϕ increased, therefore, there only 0 or π state.

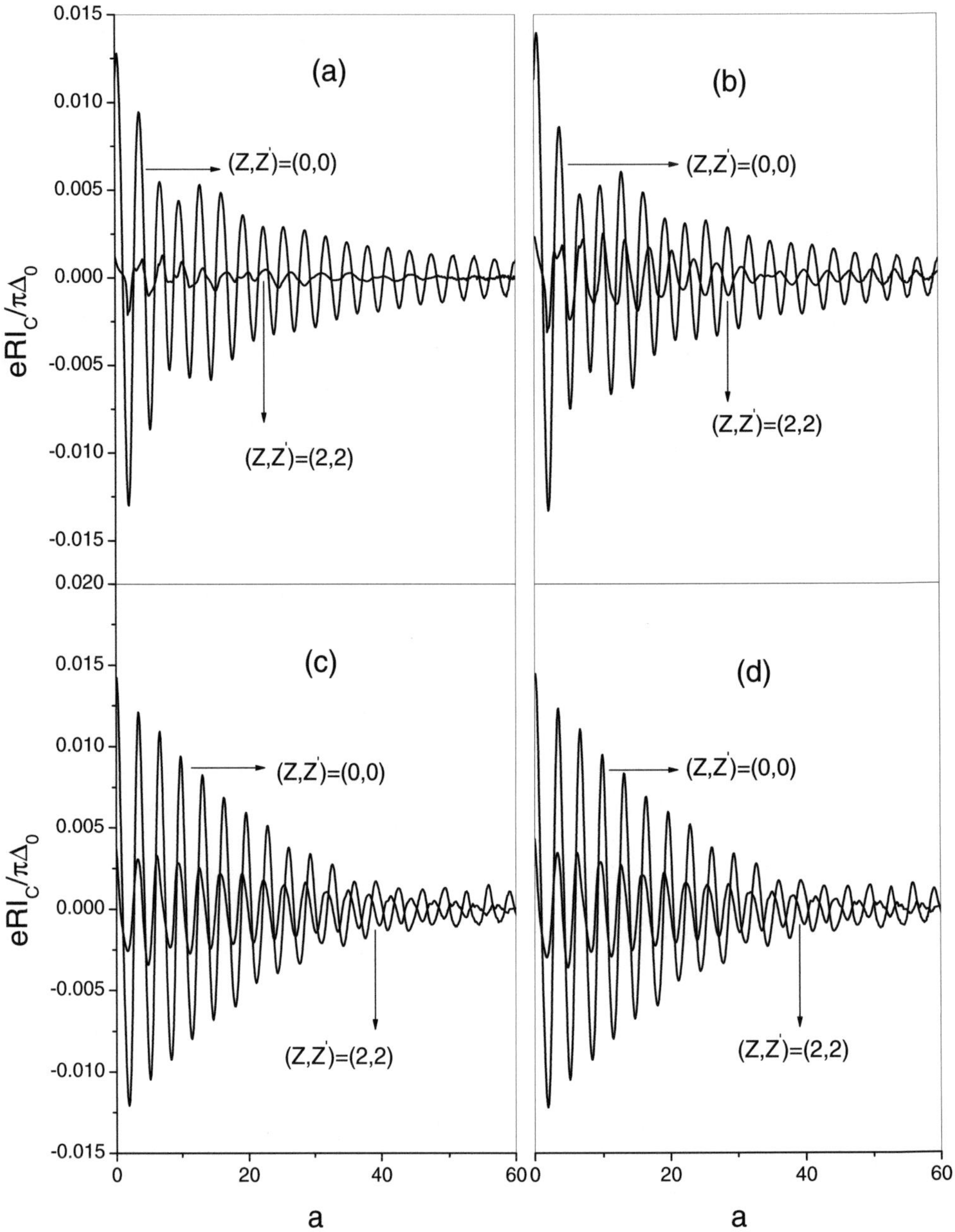

Fig. 3. Dependences of I_C on the SM thickness a at $b = 20/k_F^{SC}$ with the same situations as in Fig. 2 (from Ref. 34).

However, as concerned about the light holes, for the F alignment in which the spin of the majority (minority) holes is still that of the majority (minority) holes in the right FS, since the effective mass is much smaller than that for the heavy hole, the Fermi vector is much smaller than those for the heavy and in the SC, and thus $exp^{-i[(K^{\pm}-q_{\sigma}^{L\pm})b-\phi]}$ and $exp^{-i[(K^{\pm}+q_{\sigma}^{L\pm})b-\phi]}$ can easily change the sign, particularly for the minority holes. It follows that the general tendency is that smaller P more easily gives rise to the sign variation of the sum of the two terms. In the A alignment, a majority (minority) spin in the right FS will be regard as a minority (majority) spin in the right FS, therefore the increase of P is favorable for the sign change of the sum of the two terms. Furthermore, the barrier strengths in coefficients of the two terms in Eqs. (A4)-(A7) also exert an important influence on the sign of the value. When the two barrier strengths Z and Z' or one of them is much bigger or smaller, one of the two coefficients is zero, as result, the increase of ϕ can not cause the change of sign of the sum.

In Fig. 6, the normalized critical currents I_C at different scattering potential strengths Z and Z' in the F alignment for the heavy and light holes are shown as a function of P' [35], where P' is define as $P' \equiv \Gamma/\Delta_0 = PE_F/\Delta_0$ [28]. It is found that the critical Josephson current I_C for the heavy holes, at Z = 0 and $Z' = 0$, increases with P', while for the light, decreases. For Z or Z' unequal to zero, the situations are very different. As $Z' = 0$, for the heavy hole, I_C always increases with P' at the small Z and decreases at the big Z, however, as Z = 0, at some Z', I_C increases with P', and at other Z', decreases. For the light hole, the critical Josephson currents I_C are found, at $Z' = 0$, to decrease all the time with P' regardless of Z, while at Z = 0, there are the same features of I_C with variation of P' as those for heavy holes.

In Fig. 7, are illustrated the P' dependences of Josephson critical current I_C for the heavy and light holes in the A alignment at different Z and Z'. [35] It is shown that there exist some features similar to those in Fig. 6, for example, as $Z(Z') = 0$, for the heavy hole, I_C increases at some Z' with P' and decreases at other Z'. Simultaneously, it is also found that some behaviors and values of the critical Josephson currents I_C with P' increased are remarkably different from those in Fig. 6, which are obviously due to the existence of spin splitting energy in the FS's and resultant different mismatches in the Fermi vectors between the FS and SC for the F and A alignments. The critical Josephson current in the A alignment I_C for the heavy holes, at Z = 0 and $Z' = 0$, decreases with P', while for the light, basically has no change. In addition, I_C in the A alignment, at $Z(Z') = 0$, always increases with P' for the light holes regardless of $Z'(Z)$.

Fig. 8 shows the Z (Z') dependence of Josephson critical current I_C for different Z' (Z) in the F alignment for the heavy and light holes. [35] The critical Josephson currents I_C for the heavy holes, is found to exhibit oscillations with $Z(Z')$ enhanced, finally reduce to zero and there are two peaks, however, at $Z' = 0$, firstly increase with the increase of Z, whereas at Z = 0, firstly decreases and then increases with increasing Z'. For the light hole, the critical Josephson currents I_C is also found to display oscillations with $Z(Z')$ increased. In the meantime, we find that as $Z' = 0$, I_C firstly decreases with the enhancement of Z', while as Z = 0, I_C firstly increases and then decreases with the increase of Z'. At Z and Z' unequal to zero, the varied behaviors of I_C with Z' and Z are also very similar, and I_C basically decreases, then increases, and lastly decreases to zero with increasing Z' or Z, which indicates that there is only a peak.

In Fig. 9, the variations of Josephson critical current I_C for the heavy and light holes with Z (Z') at different Z' (Z) in the A alignment are plotted. [35]It is found that although there are the same features as those in Fig. 8, the values of the critical Josephson currents I_C with increasing Z or Z' are a little different, this can be also ascribed to the same reason as in the explanation of the same and different properties between Figs. 6 and 7. Here, the differences

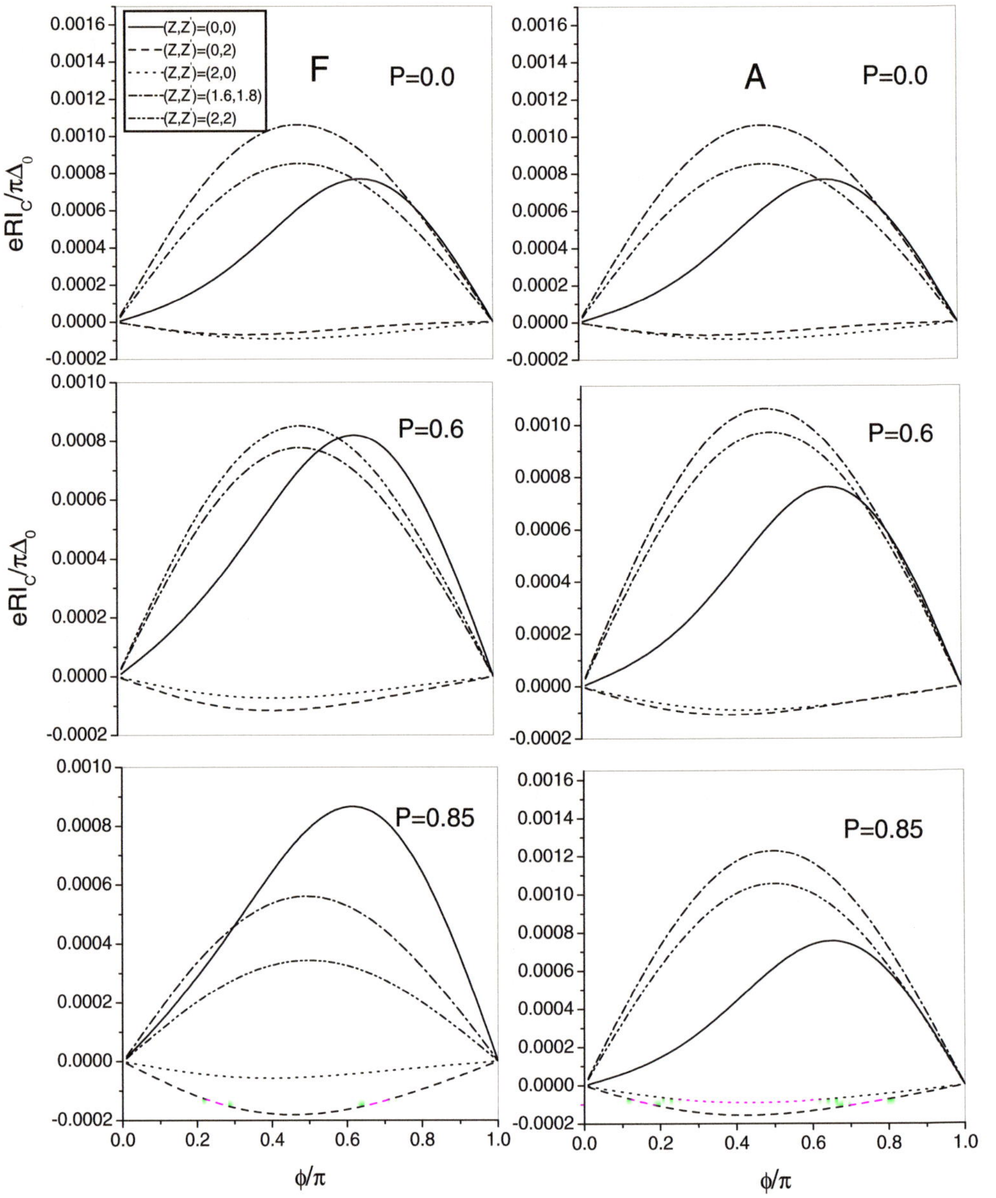

Fig. 4. The phases of the Josephson current $I(\phi)$ for the heavy holes at different (Z, Z') in the F and A alignments. Here, $a = 40/k_F^{SC}$, $b = 20/k_F^{SC}$, Z and Z' are given in the inset of Fig. 4, and the other parameters are the same as in Fig. 2 (from Ref. 34).

of the results between the heavy and light holes in Fig. 9, together with Figs. 6-8, can be all owing to the fact that there are the different mismatches in the effective mass and Fermi velocity between the FS and SC for the two kinds of holes.

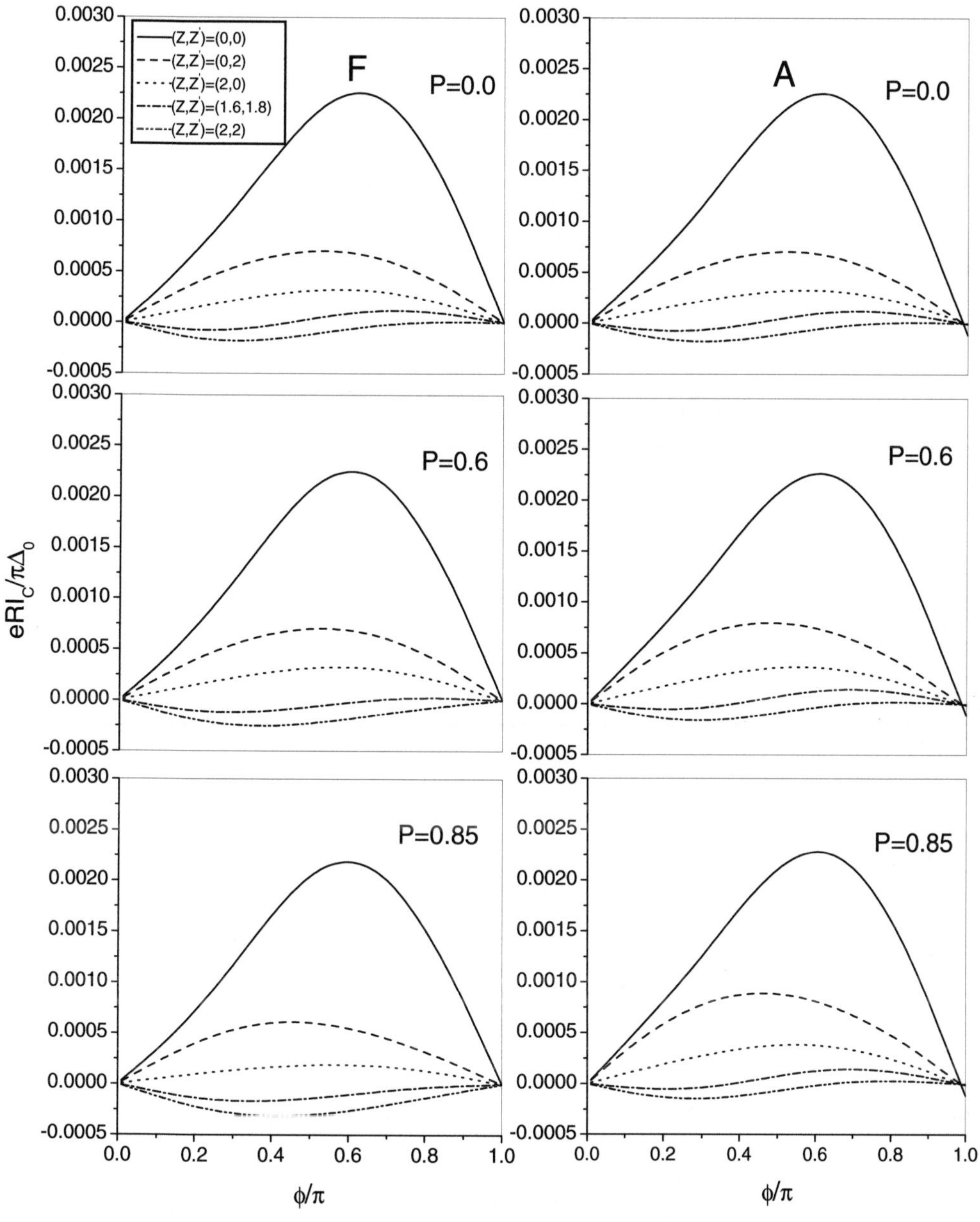

Fig. 5. The same dependence for the light holes as in Fig. 4 (from Ref. 34).

Last, I want to briefly mention that if at finite voltage, there also can exist the dc current which exhibits the so-called subharmonic structure [38, 39], i.e., current singularities at voltages $V = 2\Delta/en$ with $n =,1\ 2,...,$ where Δ the energy gap in the SC. Therefore, in such situation, the decoherence due to the subgap currents would have to be taken into account as pointed by Ref. [39].

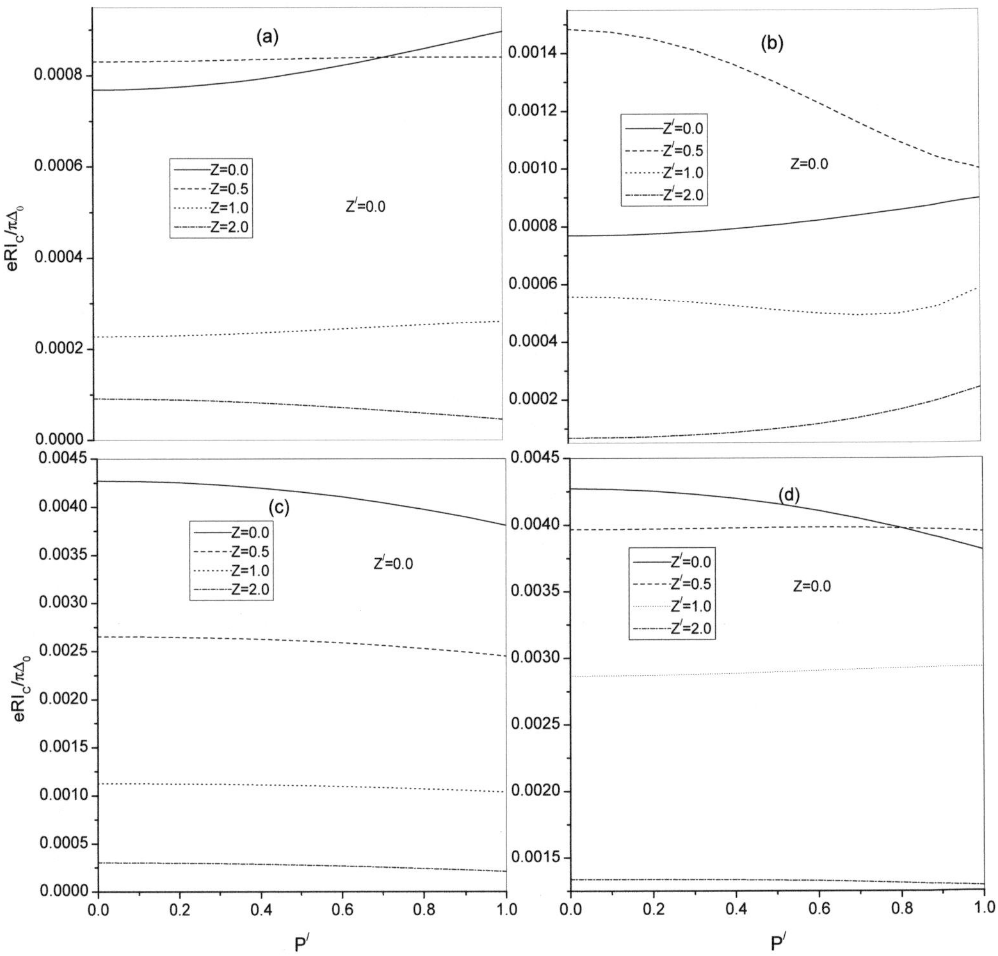

Fig. 6. Dependence of the normalized critical current I_C on P in the F alignment at different scattering potential strengths: Z (a) and Z' (b) for the heavy holes; Z (c) and Z' (d) for for the light. Here. the parameters are the same as in Fig. 4 (from Ref. 35).

4. Summary and outlook

In this chapter, we study the Josephson current flowing through an FS/semiconductor (SM) /FS trilayer connected to two superconductor (SC) electrodes. It is shown that the Josephson current in the junction is strongly dependent on not only the kinds of holes in the FS's and scattering potential strengths at the interfaces but also the relative orientations of the effective exchange field of the two FS's. We also find that in the coherent regime, the oscillatory dependences of the critical Josephson current on the FS and SM layer thicknesses and Josephson current on the macroscopic phase difference for the heavy and light holes in different orientations of the effective exchange field of the two FS's, are much different due to the different mismatches in the effective mass and Fermi velocity between the FS and SC,

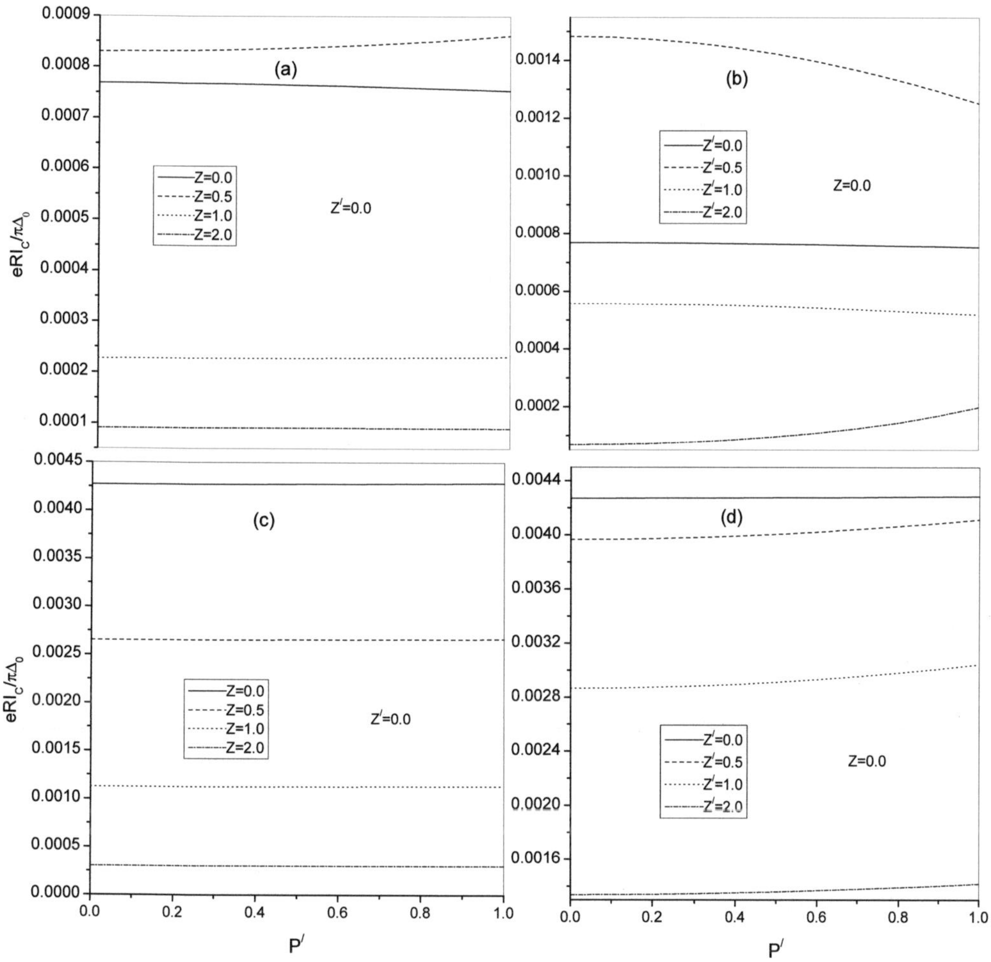

Fig. 7. The same dependence as in Fig. 6 in the case of the A alignment (from Ref. 35).

which is connected with the transition from positive (0) to negative (π) coupling. The effect can be exploited for the design of devices such as superconducting quantum interference devices with improved accuracy and convenience.

Up now, we hardly find theoretical and experimental works taking into account the unconventional pairing symmetry, which is relevant to the high T_C superconducting junctions comprising the FS. Due to the interplay between the various pairing symmetry in the unconventional SC and hole types in the FS, there will exhibit substantial physical content, which will have not only great intrinsic scientific importance but also potential device applications in the future. We will focus our attention on the *superconducting Spintronics* based on the junctions consisting of FS and SC with unconventional pairing symmetry in the future work.

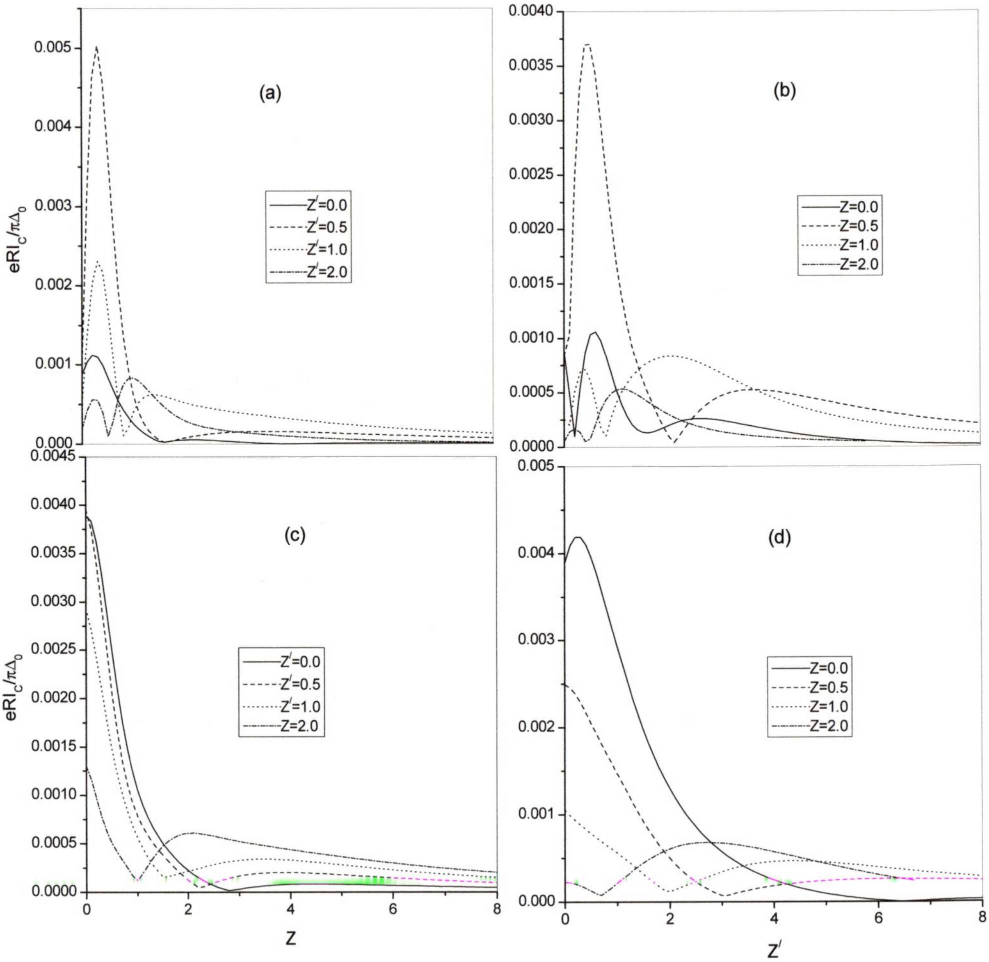

Fig. 8. Dependence of I_C on Z and Z' at $P = 0.9$ with the same situations as in Fig. 6 (from Ref. 35).

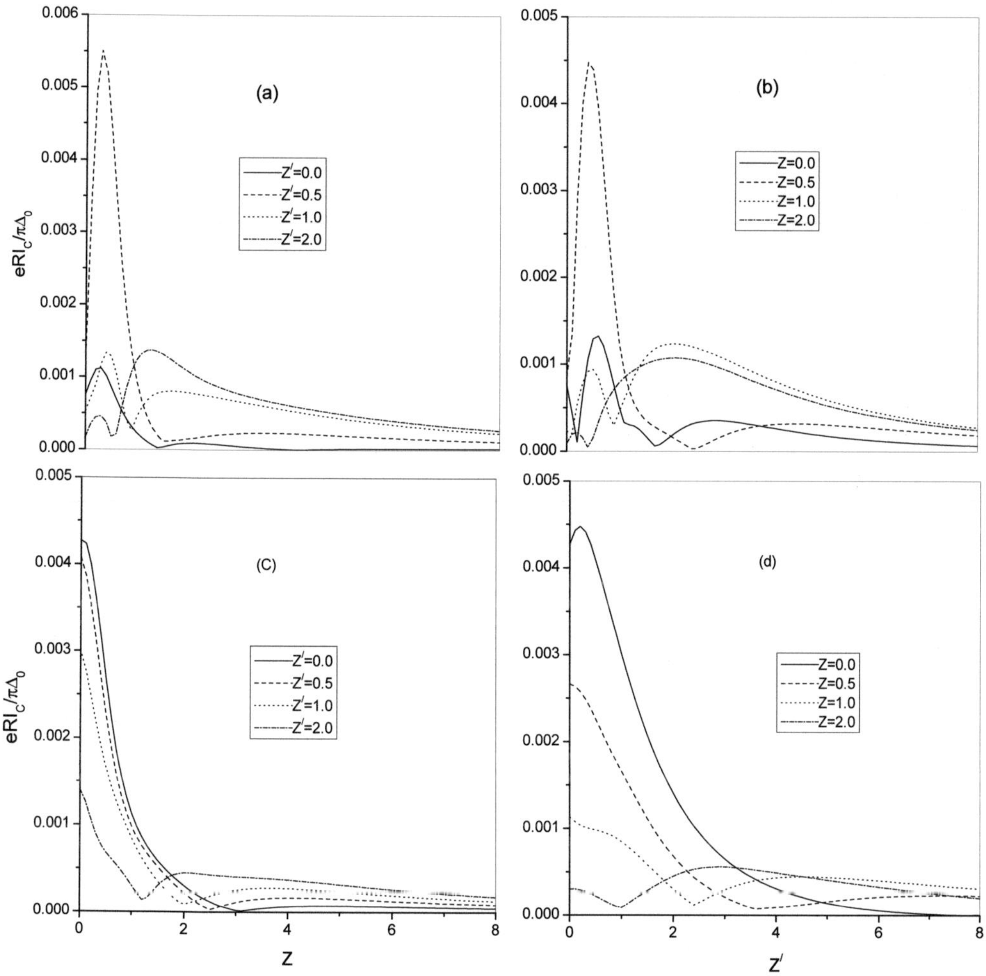

Fig. 9. Dependence of I_C on Z and Z' at $P = 0.9$ with the same situations as in Fig. 7 (from Ref. 35).

5. Symbol list

SC ··· superconductor

FS ··· ferromagnetic semiconductor

SM ··· semiconductor

BDG ··· Bogoliubov-de Gennes (BDG) equation

ELQ ··· electronlike quasiparticle

HLQ ··· holelike quasiparticle

n ··· n-type semiconductor

p ··· p-type semiconductor

ϕ ··· a phase difference between the SCs

I ··· the dc Josephson current

I_C ··· the critical Josephson current

$V(x)/V(\boldsymbol{r})$ ··· the usual static potential

U ··· the barrier height

$\delta(x)$ ··· δ-type function

$H_{FS}(\boldsymbol{r})$ ··· the hole Hamiltonian in the FS

$H_0(\boldsymbol{r})$ ··· the hole Hamiltonian of the SM GaAs

$\hbar$ ··· Planck constant

m_p/m_e ··· the mass of a hole/electron

$\boldsymbol{h}(\boldsymbol{r})$ ··· the effective exchange field

$\boldsymbol{\sigma}$ ··· the conventional Pauli spin operator

Γ ··· the spin splitting energy

J_{pd} ··· the p-d exchange coupling strength

$< S^z >$ ··· the thermal average of the impurity Mn ion spins

p ··· the relative spin-polarized degree

E_F ··· the fermi-energy

ζ_k ··· the excitation energy

ϵ_k ··· the one-electron energy

Δ_0 ··· energy gap at zero temperature

Δ ··· energy gap at non-zero temperature

$\Delta(x)$ ··· the pair potential

$u_\sigma, v_{\bar{\sigma}} \cdots$ **the two-component wavefunction**

$\psi(x) \cdots$ **the wave function**

$a, b, c, d \cdots$ **transmission and reflection coefficients**

$q_\sigma^{L(R)\pm} \cdots$ **the wave vectors in the left (right) FS**

$q'^{\pm} \cdots$ **the wave vectors in the SM**

$k^{\pm} \cdots$ **the wave vectors in the SCs**

$k_{\|} \cdots$ **the parallel component of the wave vector**

$k_F \cdots$ **the Fermi wave vector**

$Z/Z' \cdots$ **two dimensionless parameters**

$k_B \cdots$ **boltzmann factor**

$T \cdots$ **tempreture**

$\omega \cdots$ **the Matsubara frequencies**

6. References

[1] B. M. Andersen. Y. S. Barash, S. Graser, and P.J. Hirschfeld, Phys. Rev. B 77, 054501 (2008).
[2] I. Zŭtic, J. Fabian, and S. Das Sarma, Rev. Mod. Phys. 76, 323 (2004).
[3] A.I. Buzdin, Rev. Mod. Phys. 77, 935 (2005)
[4] T. Jungwirth, J. Sinova, J. Mašek, J. Kučera, and A.H. MacDonald, Rev. Mod. Phys. 78, 809 (2006); H. Saito, S. Yuasa, and K. Ando, Phys. Rev. Lett 95, 086604 (2005).
[5] H. Ohno, A. Shen. F. Matsukura, A. Oiwa, A. Endo, S. Katsumoto, and Y. Iye, Appl. Phys. Lett 69, 363 (1996); H. Ohno, D. Chiba, F. Matsukura, T. Omiya, E. Abe, T. Dietal, Y. Ohno, and K. Ohtani, Nature (London) 408, 994 (2000); K.C. Ku, S.J. Potashnik, R.F. Wang, S.H. Chun, P. Schiffer, N. Samarth, M.J. Seong, A. Mascarenhas, E. Johnston-Halperin, R.C. Myers, A.C. Gossard, and D.D. Awschalom, Appl. Phys. Lett. 82, 2302 (2003).
[6] H. Ohno, Science 281, (1991); J. Magn. Magn. Matter. 200, 110 (1999); T. Dietl, H. Ohno, F. Matsukura, J. Cibert, and D. Ferrand, Science 287, 019 (2000); J. Konig, H.-H. Lin, and A.H. MacDonald, Phys. Rev. Lett. 84, 5682 (2000); J. Konig, T. Jungwirth, and A.H. MacDonald, Phys. Rev. B 64, 184423 (2001).
[7] Y.C. Tao, J.G. Hu, and H. Liu, J. Appl. Phys. 96, 498 (2004)
[8] Y.C. Tao and J.G. Hu, Int. J. Mod. Phys B 16, 2247 (2004).
[9] B.D. Josephson, Phys. Lett. 1, 251 (1962).
[10] V.V. Ryazanov, V.A. Oboznov, A. Yu. Rusanov, A.V. Veretennikov, A.A. Golubov, and J. Aarts, Phys. Rev. Lett. 86, 2427 (2001).
[11] V.V. Ryazanov, V.A. Oboznov, A.V. Veretennikov, A. Yu. Rusanov Phys. Rev. B. 65, 020501 (2001).
[12] T. Kontos, M. Aprili, J. Lesueur, and X. Grison, Phys. Rev. Lett 86, 304 (2001).
[13] J.M.E. Geers, M.B.S. Hesselberth, J. Aarts, and A.A.Golubov, Phys. Rev. B 64, 094506 (2001).

[14] O. Bourgeois, P. Gandit, J. Lesueur, A. Sulpice, X. Grison, and J. Chaussy, Eur, Phys. J. B 21, 75 (2001).
[15] T. Kontos, M. Aprili, J. Lesueur, F. Genetêt, B. Stephanidis, and R. Boursier, Phys. Rev. Lett. 89, 137007 (2002).
[16] M. Fogelström, Phys. Rev. B 62, 11812 (2000).
[17] Yu. S. Barash and I. V. Bobkova, Phys. Rev. B 65, 144502 (2002).
[18] F.S. Bergeret, A.F. Volkov, and K.B. Efetov, Phys. Rev. B 64, 134506 (2001).
[19] V.N. Krivoruchko and E. A. Koshina, Phys. Rev. B 64, 172511 (2001).
[20] N.M. Chtchelkatchev, W. Belzig, Yu. V. Nazarov, and C. Bruder, Pis'ma Zh. Éksp. Teor. Fiz. 74, 357 (2001) [JEPT Lett. 74, 323 (2001)].
[21] A.A. Golubov, M. Yu. Kupriyanov, and Ya. V. Fominov, Pis'ma Zh. Éksp. Teor. Fiz. 75, 709 (2002) [JEPT Lett. 75, 588 (2002)].
[22] Yu. S. Barash, I. V. Bobkova, and T. Kopp, Phys. Rev. B 66, 140503 (2003).
[23] A. Yu. Zyuzin, B. Spivak, and M. Hruška, Europhys. Lett. 62, 97 (2003).
[24] Z. Radović, L. Dobrosavljević-Grujić, and B. Vujičić, Phys. Rev. B 63, 214512 (2001); M. Božović and Z. Radović, Phys. Rev. B 66, 134524 (2002); Z. Radović, N. Lazarides, and N. Flytzanis, Phys. Rev. B 68, 014501 (2001).
[25] K. Halterman and O.T. Valls, Phys. Rev. B 65, 014509 (2002); M. Zareyan, W. Belzig, and Yu. V. Nazarov, Phys. Rev. Lett. 86, 308 (2001).
[26] A. Brinkman and A.A. Golubov, Phys. Rev. B 61, 11297 (2000).
[27] I. Baladie and A. Buzdin, Phys. Rev. B 64, 224514 (2001).
[28] F.S. Bergeret, A.F. Volkov, and K.B. Efetov, Phys. Rev. B 65, 134505 (2002).
[29] J.S. Jang, D. Davidović, D.H. Reih, and C.L. Chien, Phys. Rev. Lett. 74, 314 (1995); Y. Obi, M. Ikebe, T. Kubo, and H. Fujimori, Physica C 317-318, 149 (1999).
[30] D.J. van Harlingen, Rev. Mod. Phys. 67, 515 (1995).
[31] J.X. Zhu, B. Friedman and C.S. Ting, Phys. Rev. B 59, 9558 (1999).
[32] J.X. Zhu and C.S. Ting, and A.V. Balatsky Phys. Rev. B 66,
[33] C.R. Hu, Phys. Rev. Lett. 72, 1526 (1994).
[34] Y.C. Tao, J. Appl. Phys. 100, 063901 (2006)
[35] Y.C. Tao, Eur. Phys. J. B. 56, 83 (2007).
[36] P.G. de Gennes, *Superconductivity of Metal and Alloys* (Benjamin, New York, 1996).
[37] A. Furusaki and M. Tsukada, Solid State Commun. 78, 299 (1991).
[38] W.M. van Huffelen, T.M. Klapwijk, D.R. Heslinga, M.J. de Boer, and N. van der Post, Phys. Rev. B 47, 5170-5189 (1993).
[39] M.J. Storcz and F.K. Wilhelm, Appl. Phys. Lett. 83, 2589 (2003).

4

Investigating Superconductivity with Electron Paramagnetic Resonance (EPR) Spectrometer

D. Shaltiel et al*
The Hebrew University of Jerusalem
Israel

1. Introduction

Shortly after the discovery of high T_c superconductors there was an influx of publications where EPR technique was used to investigate their properties (Srinivasu et al., 2001). As no substantial information was obtained it was soon almost abandoned. The fact that EPR technique provided a large number of experimental data in the superconducting regime, without significantly contributing to the deeper understanding of high T_c superconductors was puzzling. Meanwhile, our continuous effort to further investigate superconductivity using EPR technique yielded important experimental and theoretical results related to properties of high T_c superconductors. Furthermore, some of these results are unique as they were not obtained with conventional experimental techniques.

The work shows that an electron paramagnetic resonance spectrometer is a powerful tool to investigate properties of superconductors. As superconducting systems are not conventional paramagnetic substances, the EPR signals do not originate from the usual resonant absorption, namely magnetic dipolar transitions between spin levels. Thus, it is essential in analyzing the experimental results, to explore the mechanism that induces the EPR signals. Indeed, by detailed experimental and theoretical studies of the EPR spectra of different high T_c superconductors, it has been possible to apprehend the mechanism that induces the observed microwave absorption. Based on these results, new unknown properties of superconductors were obtained. As the EPR spectrometer is the central experimental tool in the present work, we start by describing the EPR spectrometer in details, to facilitate the reader understanding its various functions. This step is necessary, as it turned out that among experts in superconductivity many are not familiar with the elements of the EPR spectrometer, how and what it measures. We will then continue by presenting experimental

*H.A. Krug von Nidda[2], B.Ya. Shapiro[3], A. Loidl[2], T. Tamegai[4], T. Kurz[2], B. Bogoslavsky[1], B. Rosenstein[5,6] and I. Shapiro[3]
[1]*The Hebrew University of Jerusalem, Israel*
[2]*Experimental Physics V, Center for Electronic Correlations and Magnetism, University of Augsburg, Augsburg, Germany*
[3]*Bar-Ilan University, Ramat-Gan, Israel*
[4]*The University of Tokyo, Hongo, Japan*
[5]*National Chiao Tung University, Department of Electrophysics, Hsinchu, Taiwan, R.O.C.*
[6]*Ariel University Center, Israel*

results obtained in high T_c superconductors subjected to various factors such as DC and AC magnetic fields, temperature, sample orientation and to different sample treatments such as zero-field-cooling or field-cooling. This overview will allow a systematic classification of the measurements and will provide the theoretical background (Shaltiel et al., 2008) necessary to analyze these experiments and to understand the inherent physics. The presentation is based on publications by the authors over a period from just after the discovery of high T_c superconductors to date.

2. Experimental

An EPR spectrometer, Fig. 1, consists of a microwave source that feeds a cavity where a sample is installed. The sample is exposed to three magnetic fields; a DC field, H_{DC}, a collinear low frequency AC field, h_{AC}, and a transverse microwave field, H_{rf}. The microwave and the AC frequencies used in the present investigation were 9.3 GHz and 100 kHz, respectively. At zero DC field the coupling of the waveguide to the cavity is adjusted critically in such a way that the reflection from the cavity is zero. Regarding a usual EPR experiment on a paramagnetic sample, the variation of the DC field results in a detuning from the critical adjustment on passing the resonance condition for magnetic dipolar transitions between the Zeeman levels ($\nu=\gamma H_{DC}$, with microwave frequency ν and gyromagnetic ratio γ) and in turn yields microwave reflection from the cavity. The reflected microwave power is rectified by a detector, i.e. a microwave diode, yielding a signal proportional to the absolute value of the reflected power P at H_{DC} and modulated by the AC frequency (note that $h_{AC} \ll H_{DC}$). This signal is further fed into a lock-in-detector **(LID)** which samples the signal by convolution with a periodic signal of the same AC frequency over a large number of AC periods, thus yielding amplitude and phase while suppressing statistical noise. Provided that the width of the resonance absorption dependent on H_{DC} is larger than the amplitude of the AC modulation field h_{AC}, the LID output submits a signal proportional to the field derivative of the paramagnetic resonance signal dP/dH.

An accepted assumption in analyzing experimental results obtained by an EPR spectrometer is that the role of the AC field is to introduce variations in the magnitude of the DC magnetic field; the AC field is usually referred to as the modulating magnetic field. However, as will be shown below, the AC field does not just result in a modulation of the microwave signal dependent of the modulated static field like in a usual EPR experiment, but in superconductors it may also induce an additional mechanism of microwave power dissipation.

Therefore, one should be perceptive to the different dissipation process, when analyzing the EPR results in the superconducting phase. Here it is important to note that a complication in the measuring process has been observed. It is related to variations in the signal phase that may develop through the measuring process. The reference AC phase of the LID is adjusted for usual EPR measurements to yield the optimum paramagnetic resonance signal. However in case of the microwave dissipation observed in anisotropic superconductors this adjustment turns out to be not appropriate. Variations in the signal phase were observed when varying the magnitude of the variables involved such as DC magnetic field, temperature, sample orientation and others. This effect necessitates a procedure where the lock-in detector phase has to be adjusted, throughout the measuring process, to be in-phase with the AC phase. To overcome this problem, conducted during the measuring process, the measurements were performed in steps of the variables involved, and at each step the lock-in detector phase had to be adjusted accordingly. This procedure results in an extremely cumbersome measurement, where the possibilities offered by the modern EPR Bruker spectrometer ELEXSYS E500 are extremely useful. A feature of this spectrometer is its possibility to measure the signal intensity in steps of two variables. It allows obtaining the measured signal intensity at each step, where the lock-in phase is varied from zero to 180^0. Repeating the measurement, by changing, in steps, the magnitude of the variable involved (such as magnetic field, temperature, angle of the oriented crystal) gives the measured signal intensity as function of the variable at the full 180^0 range of the lock-in phase. Then the measurements were analyzed using a proper computer program. It allowed obtaining two curves, the measured signal amplitude, and the corresponding signal phase, as a function of the variable involved such as: DC magnetic field, AC magnetic field, sample orientation angle and temperature. Thus, the actual "calculated" signal was derived from the measured signal amplitude and its corresponding signal phase (see below).

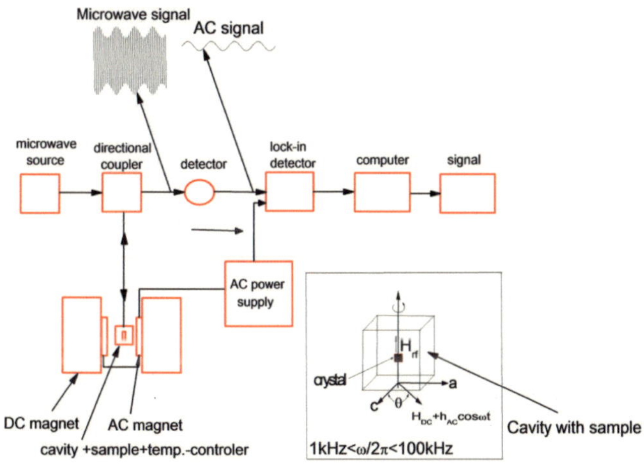

Fig. 1. Block diagram of the EPR spectrometer.

Fig. 2 presents such a result, for a $Bi_2Sr_2CaCu_2O_{8+\delta}$ (Bi2212) single crystal, Zero Field Cooled **(ZFC)** to 4K with the DC magnetic field applied parallel to the a-b plane. The upper curve shows the measured signal amplitude, and the lower curve shows its corresponding signal phase (with respect to the lock-in AC phase) as function of DC magnetic field. The measured signal amplitude exhibits two maxima followed by an exponential decay towards zero at high fields. The signal phase is close to zero at low fields and drops steeply towards -180^0 at high fields. It indicates that the signal is in phase and out of phase with the AC field at low and high fields, respectively. The details of the phase shift, i.e. deviations (±20°) from 0° at low fields and from 180° at high fields have not been investigated so far and need further attention. As the allover phase shift from low to high fields is evidently less than 180° it cannot be explained just by a constant shift of the calibration.

Fig. 3 presents EPR results similar to those shown in Fig. 2, for temperatures, 80K, 30K and 4K. It indicates that the measured signal amplitude and signal phase have similar functional behavior as in Fig. 2. It implies that the EPR responses as function of field are similar in the temperature range from T_c down to 4K.

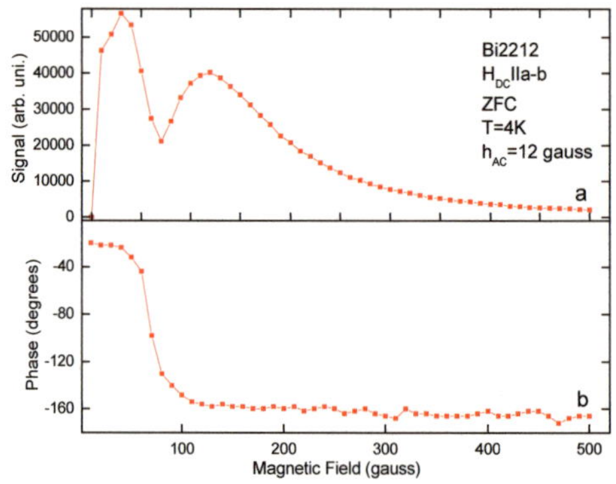

Fig. 2. AC induced microwave dissipation signal amplitude (a) with the corresponding phase value (b) of a Bi2212 single crystal, ZFC to 4K, as function of DC magnetic field aligned parallel to the a-b plane.

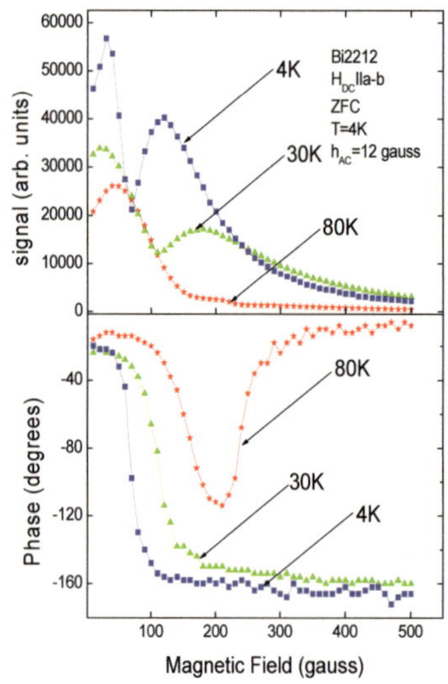

Fig. 3. AC induced microwave dissipation signal amplitudes with their corresponding phase values, of the same Bi2212 single crystal shown in Fig. 2, ZFC to 80K, 30K, and 4K, as function of DC magnetic field aligned parallel to the a-b plane.

Understanding the mechanism that induces the signal, the origin of the spectra, and its corresponding signal-phase as function of field, is an important key in studying superconducting properties with the EPR spectrometer.

Fig. 4 shows the actual calculated signal as function of field in Bi2212, derived by multiplying the measured signal amplitude of Fig. 2 with the cosine of its phase. It shows a sharp increase in the intensity at very low magnetic field followed by a steep decrease with change of sign to negative values and a final increase that asymptotically tends to zero. The theoretical considerations presented in (Shaltiel et al., 2008) showed that this corresponds to the field derivative of the dissipated microwave power due to shaking of JV away from the zero crossing, but is subject to strong nonlinearities close to the zero crossing.

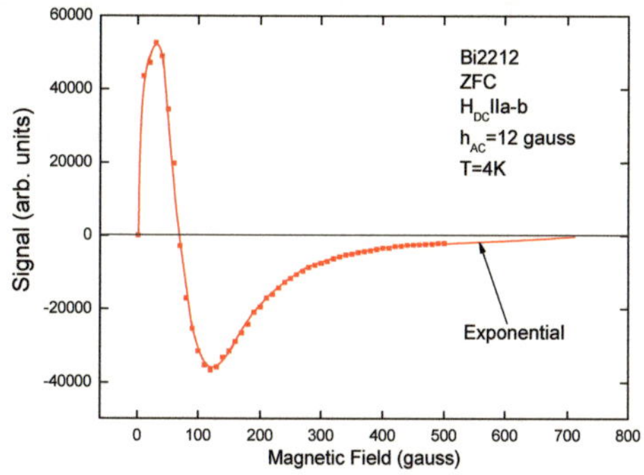

Fig. 4. Actual calculated signal in the Bi2212 crystal as function of field parallel to the conduction plane at 4K, derived by multiplying the measured signal amplitude with the cosine of its corresponding phase using the values shown in Fig. 2.

Fig. 5 presents the measured signal amplitude and the corresponding signal phase as function of magnetic field applied parallel to the a-b plane for a $Bi_2Sr_2Ca_2Cu_3O_{10+x}$ (Bi2223) single crystal ZFC to 4K. It shows a response similar to the observed signal in Fig. 2 for Bi2212. As both Bi2212 and Bi2223 are high-anisotropy superconductors, we deduce that a high anisotropy is a necessary condition to observe similar results in the high-T_c superconductors. This conjecture is also confirmed by our measurements in the high-T_c, low-anisotropy YBCO ($YBa_2Cu_3O_{7-\delta}$) superconductor: no significant EPR signal was observed in optimally dopes YBCO, when the DC magnetic field is applied either parallel or perpendicular to the a-b plane, while sizable signals appeared for overdoped, i.e. more anisotropic YBCO with the magnetic field applied within the a-b plane. These recent results are intended to be published elsewhere. Fig. 6 presents the actual calculated signal as a function of field in Bi2223 derived from the measured signal amplitude and from its corresponding signal phase of Fig. 5. It shows a similar DC-field dependence like that observed in Fig. 4 for Bi2212. It confirms again the earlier conclusion that high-anisotropy is a necessary condition for obtaining a similar behavior in different superconductors.

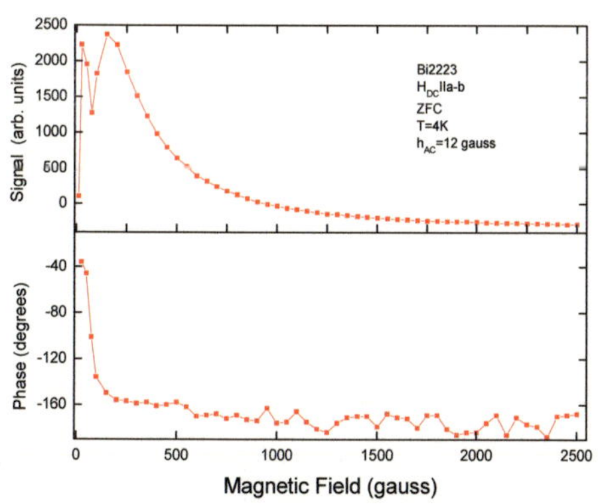

Fig. 5. AC induced microwave dissipation signal amplitude with the corresponding phase value of a Bi2223 single crystal, ZFC to 4K, as function of magnetic field aligned parallel to the a-b plane. The temperature behavior of the signals in Bi2223 is similar to that observed in Bi2212.

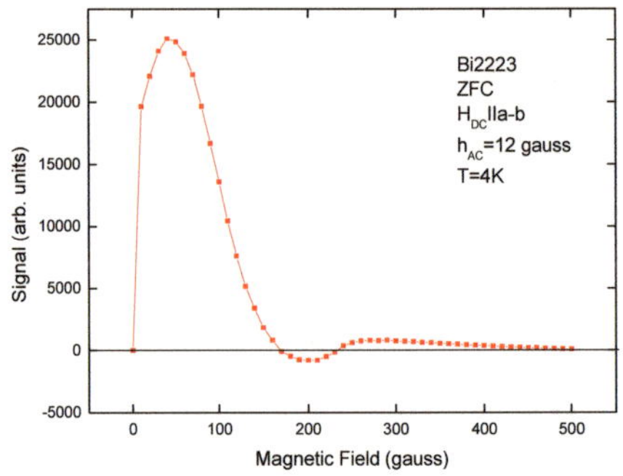

Fig. 6. Actual calculated signal in the Bi2223 crystal as function of field parallel to the conduction plane at 4K, derived by multiplying the measured signal amplitude with the cosine of its corresponding phase using the values shown in Fig. 3.

The experimental results presented in Figs. 2-6 show the EPR signal as a function of DC magnetic field using a constant AC field. An additional picture is obtained by interchanging the roles of the variables, namely the DC field is kept constant and the EPR signal is measured as function of AC field. Such a result is illustrated in Fig. 7; it shows the signal amplitude and signal phase as function of AC magnetic field for a Bi2212 crystal. The measurements were performed by zero-field cooling the crystal down to 30K, followed by applying a DC field of 20 gauss or 150 gauss, prior to sweeping the AC field. The figure shows striking differences in the signal amplitude and in the signal phase as function of AC field when either of the two DC fields is applied. The 20 gauss signal amplitude increases strongly and linearly with increasing AC field. In contrast, the 150 gauss signal amplitude - after a slight increase - remains low and constant. At low AC field the two signals are out of phase with respect to the AC phase. At high AC field the signals are in phase and out of phase respectively when either the low or the high DC field are applied.

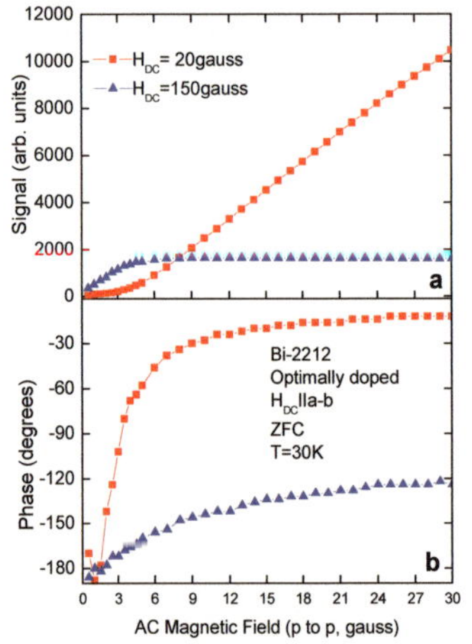

Fig. 7. Signal amplitude (a) and signal phase (b) in Bi2212 vs. AC magnetic field, obtained by zero-field cooling the crystal to 30K, followed by applying a DC field of 20 gauss or 150 gauss prior to sweeping the AC field.

Similar results are observed for a Bi2223 superconductor as shown in **Fig. 8**. They were obtained by zero-field cooling the crystal down to 30K, followed by application of a DC field of 30 gauss, 500 gauss or 5000 gauss prior to sweeping the AC field. Here again, as with the Bi2212 crystal, the Bi2223 signal amplitude, obtained when the low magnetic field of 30 gauss was applied, increases strongly and linearly with increasing AC field, whilst the signal amplitude observed, when 500 gauss was applied, is low and remains constant with increasing AC field. The signal obtained, when the much stronger 5000 gauss cooling field was applied, shows negligibly small signal amplitude for all AC field values. It confirms the previous conclusion regarding the similarity of EPR response in different high anisotropy superconductors.

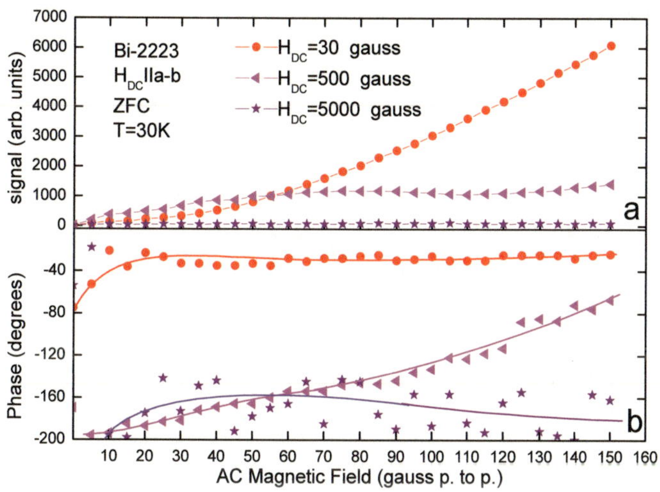

Fig. 8. Signal amplitude (a) and signal phase (b) in Bi2223 vs. AC magnetic field, obtained by zero-field cooling the crystal to 30K, followed by applying a DC field of 30 gauss or 500 gauss or 5000 gauss prior to sweeping the AC field.

3. Understanding the physics of the AC-field-induced signals

Fig. 9 presents an extremely important result vital to understand the physics, and to comprehend the mechanism of the AC field induced microwave dissipation results in the highly anisotropic superconductors. The figure shows an unexpected effect where the signal intensity as function of temperature in the high anisotropy Bi2212 crystal for the DC magnetic field aligned parallel to the c-axis is zero from T_c to 4K, whilst at DC magnetic field parallel to the a-b plane, the signal intensity increases sharply already just below T_c and continues to increase sharply towards lower temperatures. The small signal observed close to T_c in Fig. 9 for the DC magnetic field applied parallel to the c-axis results from the thermally activated flux flow resistivity (FFR) (Shaltiel et al., 2001). (See Fig. 6 of Shaltiel et al., 2001)

A reasonable conclusion to the extreme difference in the superconducting response under the parallel and the perpendicular DC magnetic field, is derived from Fig. 10. It shows that, when a magnetic field is parallel to the a-b plane, only Josephson vortices (JV) are formed and, while applying a magnetic field parallel to the c-axis, only Abricosov vortices, i.e.

pancake vortices (PV), are formed. This indicates that the AC magnetic field induces microwave dissipation, when interacting with JV, but none when interacting with PV.

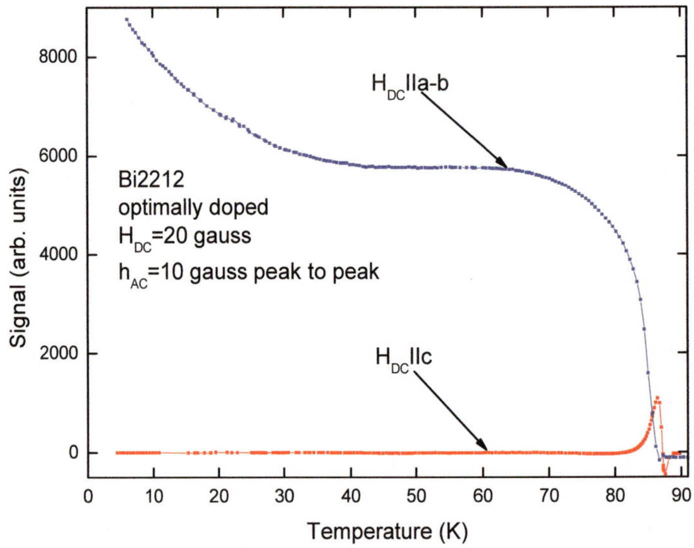

Fig. 9. Signal amplitude as function of temperature in the high anisotropy Bi2212 crystal for DC magnetic field aligned parallel to the a-b plane or to the c-axis. Note the large different response for the two orientations. The small signal observed close to Tc for magnetic field parallel to the c-axis results from thermally activated flux flow resistivity.

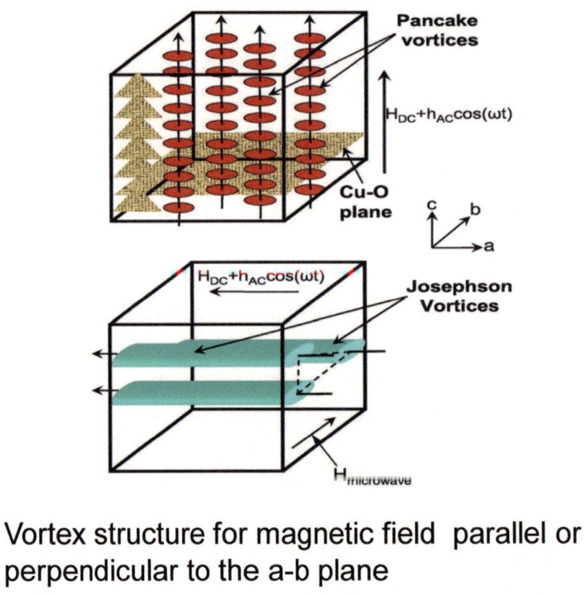

Fig. 10. The figure shows that applying a magnetic field parallel to the a-b plane forms only JV, whilst, applying a magnetic field parallel to the c-axis forms only PV.

This conclusion is corroborated by the fact that microwave dissipation results similar to those shown in Fig. 7, were observed also in Bi2223 which like in Bi2212 forms JV and PV for the magnetic field applied parallel and perpendicular to the a-b plane, respectively. Moreover, we recall that, unlike in Bi2212 or in Bi2223, no significant microwave dissipation was observed in optimally doped $YBa_2Cu_3O_{7-\delta}$ for any direction of the external DC magnetic field besides the signal at T_c from the thermally activated flux flow resistivity. As only PV are formed in optimally doped $YBa_2Cu_3O_{7-\delta}$ for all magnetic field orientations, this again confirms the conclusion that no microwave dissipation is induced by the AC magnetic field, when interacting with PV.

A mechanism that explains the induced microwave dissipation by the AC field, when interacting with JV, is derived from soliton propagation in Josephson transition lines observed in itinerant Josephson junction (IJJ) mesas (Hechtfischer et al., 1997). Such a mesa is given in Fig. 11[1] : A DC bias current applied parallel to the a-b plane induces a Lorentz force on the JV **(fluxons)** that reside between the conduction layers of the superconductor. Extremely high JV velocities, that may reach the speed of light, are obtained when the DC current is applied (Ustinov, 1998). The JV motion induces electromagnetic oscillations and, when interacting with the microwave field, microwave dissipation may result.

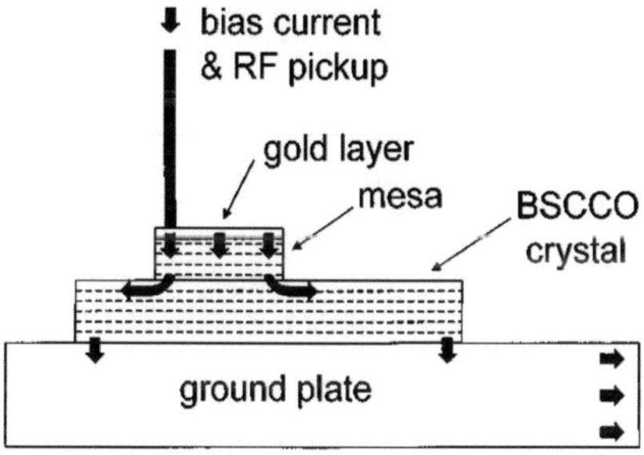

Fig. 11. Mesa structure of a BSCCO crystal. The bias current applies a Lorentz force on the JV that are present between the conduction planes.

Fig. 12 shows that the AC magnetic field applied parallel to the a-b plane induces oscillating Eddy currents in the conduction plane in a Bi2212 compound. As with the current in IJJ mesas, the Eddy currents give rise to a Lorentz force on the JV whose oscillating intensity induces oscillating interaction with the microwave field that generates the EPR microwave losses. This explains why EPR signals are observed at H||a-b plane but none at H||c-axis, as JV are formed only when the magnetic field is parallel to the conduction plane and are not formed when the field is parallel to the c axis. A straight forward conclusion in this analysis is that EPR dissipation signals in high anisotropy superconductors are observed only due to the motion of

[1] Permission to publish this figure (taken from Hechtfischer et al., 1997) was granted by APS, that keeps its copyrights).

JV. Therefore JV pinning plays an essential role on the signal shape. As JV pinning depends on the variables such as temperature and sample orientation, the analysis of the signal as a function of the variables involved enables the microscopic study of JV pinning.

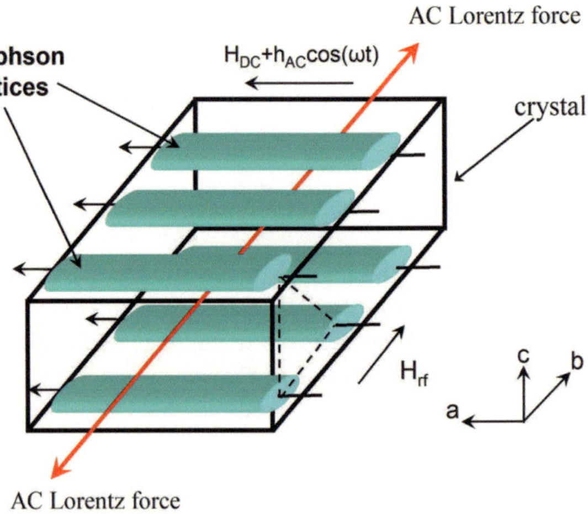

Fig. 12. Schematic picture that interprets the formation of AC-induced oscillating Eddy currents. The currents apply oscillating Lorentz forces on the Josephson vortices.

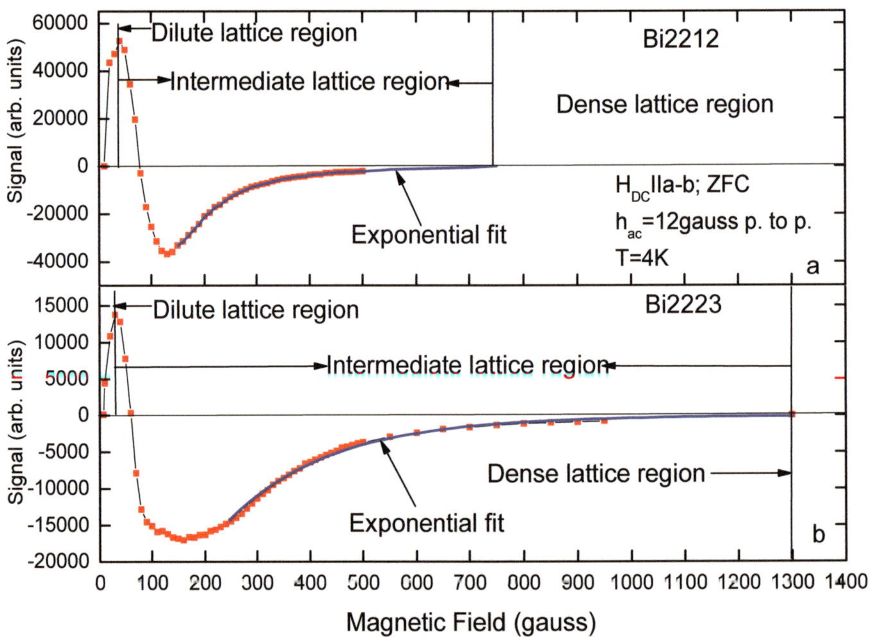

Fig. 13. Calculated signal in Bi2212 (a) and in Bi2223 (b) as a function of magnetic field parallel to the a-b plane at T=4K. It is obtained by multiplying the signal amplitude, with its corresponding phase. It shows formation of the dilute and dense region at low and high magnetic field respectively and an in-between intermediate region, in both compounds.

Fig. 13 presents the calculated signal of Bi2212 and Bi2223 as a function of magnetic field applied parallel to the a-b plane at T=4K. It is obtained by multiplying the measured signal amplitudes shown in Figs. 2 & 3, with their corresponding phase. It indicates the presence of three regions: A low field region, from zero to around 30 gauss, where the signal intensity increases almost linearly with field up to a maximum value; a high field region, where the signal intensity is zero; and an intermediate transition region. At low JV densities, the JV form the unpinned "dilute lattice region" due to negligible interaction between the JV. As the JV density increases linearly with the magnetic field, the Lorentz force on the JV induces a signal in the low-field region proportional to the DC field, as observed. At higher fields the inter JV interaction is strong and the JV form the "dense lattice region" where the JV are strongly pinned. The intensity of the AC field is not strong enough to induce motion of the JV in this region; therefore the signal intensity is zero. The inbetween intermediate region is characterized by a sharp decrease in the signal intensity and by an exponential decrease towards zero with increasing field. The field widths of the intermediate regions observed in Bi2212 and Bi2223 are different. This indicates a different distribution of the JV within the two compounds. The evolution of the JV system in the intermediate region has not been reported elsewhere.

4. Study of superconducting properties with EPR spectrometer

The theory which discusses the EPR experimental results was published earlier (Shaltiel et al., 2008). It shows that pinning of Josephson vortices effects strongly on their behavior. This section presents EPR results obtained in high anisotropic superconductors, which document the strength of this method in the investigation of the properties of Josephson vortices. It will also discuss the possibility to study small and nano-scale superconducting systems with EPR.

4.1 Interaction between Josephson vortices and Abricosov vortices

We have shown that the absence or presence of JV pinning is vital when investigating superconducting properties with the EPR technique. For example, the interaction of pancake vortices (PV), also known as Abrikosov vortices (AV), with JV induces pinning. Pinning of JV can be studied by varying the density of the Josephson vortices and Abricosov vortices in a controlled way and measuring the effect on the EPR signal. This can be achieved by tilting the orientation of a Bi2212 crystal, exposed to a given magnetic field H_{DC}, from magnetic field parallel to the a-b plane, where only JV are formed, towards the c-axis where only AV are formed. The density of the JV and PV as a function of the tilting angle, θ, of the magnetic field (θ being zero at magnetic field parallel to the a-b plane) varies as cosθ and sinθ, respectively. Thus, the signal amplitude being proportional to the JV density should vary as cosθ, provided the JV remain unpinned. Deviation from cosθ dependence would result from pinning of JV induced via the JV-AV interaction. This is demonstrated in **Fig. 14** for H_{DC}=20 gauss.

The figure shows that the signal amplitude decreases close to cos θ at θ values, 0^0 to 40^0. But at higher angles the signal amplitude decreases faster than cos θ. The increase in deviation indicates an increase in pinning above 40^0 due to the increase of the JV-AV interaction. As the pinning strength increases with angle, at angles below 40^0 the JV-AV interaction is not strong enough to induce an observable effect. The intensity of the interaction depends on H_i.

Increasing H_{DC} increases both JV and AV densities for all tilting angles. Therefore, the deviation from cos θ at higher H_{DC} should start at a lower tilting angle. This is observed in Fig. 15 for H_{DC} strengths of 20 gauss, 50 gauss, and 90 gauss where the deviation angle starts at 40^0, 20^0, and 5^0, respectively, indicating that the higher H_{DC} the lower the angle where the deviation starts. The systematic analysis of the signal amplitude as a function of the tilting angle for different H_{DC} values can be used to study the JV-AV interaction.

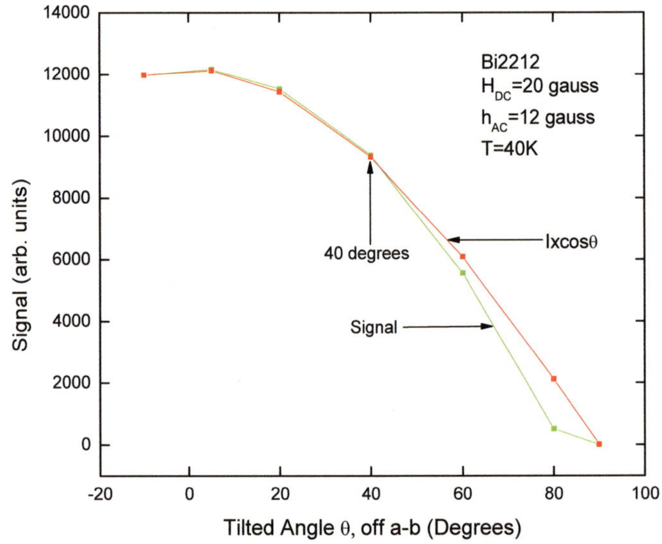

Fig. 14. Bi2212 signal amplitude as function of tilted angle θ in an applied field of 20 gauss.

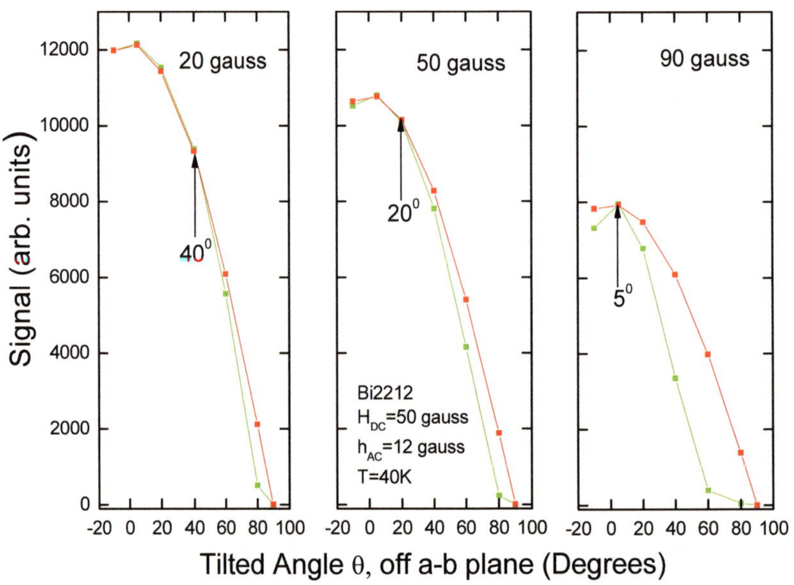

Fig. 15. Bi2212 signal ampltude as function of tilted angle θ in applied fields of 20 gauss, 50 gauss and 90 gauss.

4.2 Comparison of EPR signals, derived from either the field cooling or from the zero field cooling protocol using identical measurements

The maximum signal amplitude as function of temperature, derived from measurements of signal amplitude vs. magnetic field applied parallel to the a-b conduction planes in Bi2212, for zero field cooled (ZFC) protocol and for 500 gauss field cooled (FC) protocol is shown in Fig. 16. The maximum signal amplitudes were obtained from the corresponding maxima in the amplitude values similar to that shown in Fig 2. The figure shows that in both procedures, the maximum signal amplitudes have similar values from Tc down to 40K. Below 40K the maximum signal amplitude strongly increases for the ZFC measurement and sharply decreases for the FC measurement. The sharp decrease of the signal below 40K indicates that field cooling introduces pinning of JV. The increase of the ZFC signal indicates the absence, or only a very small pinning effect, of the JV below 40K.

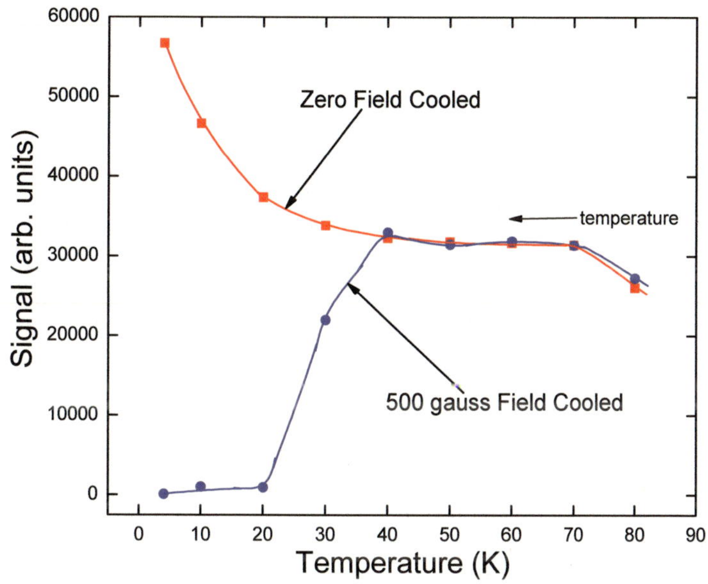

Fig. 16. Maximum signal amplitude as function of temperature in Bi2212 obtained when the sample is field cooled at either zero magnetic field or at 500 gauss magnetic field.

4.3 Memory effect

A memory signal is observed in the high anisotropy superconductors, Bi2212 and Bi2223. It is obtained by field cooling the superconductors in a magnetic field H_m, applied parallel to the conduction planes down to 4K, the lowest available temperature of the spectrometer. After setting the DC field to zero and then measuring the microwave dissipation on increasing DC field, a signal is observed slightly above H_m, indicating a **memory effect**. Memory signals are shown in Fig. 17 for Bi2212; with H_m values that range from 1500 gauss to 17000 gauss. (17000 gauss was the maximum available field with the operating spectrometer).

A detailed theoretical analysis of the memory effect was presented in an earlier publication (Shaltiel et al., 2010). It indicates that by adiabatic field cooling from T_c to a low temperature, the JV explore the available phase space that nucleates into the deepest valley of the landslide potential and are confined inside deep pinning landslide-potential minima in a quasi equilibrium glassy state. As the field is subsequently increased the fluxons are immobile and prevent penetration of new JV sites and cannot absorb microwave energy up to H_m. No signal was observed in this field region. When the magnetic field is increased above H_m new unpinned vortices start to penetrate.

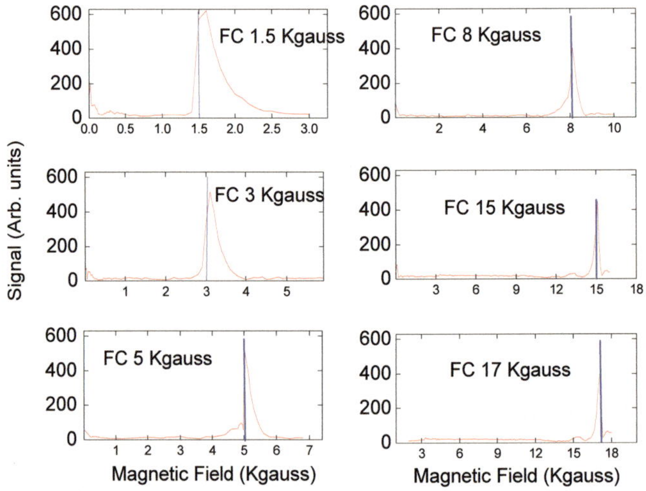

Fig. 17. Memory signal observed at magnetic fields, for H_m values 1.5 kgauss to 17 kgauss. Note that the memory signals appear at field values slightly above H_m.

Further analysis of the Bi2212 memory results presented in Fig. 17 demonstrates additional interesting properties. They indicate that the memory signals hold the following features, common to all H_m values: (a) the signals exhibit the same shape and retain similar intensities, and similar widths for all H_m values. (b) The field position H_{fp} of each memory signal occurs at a field slightly above H_m with similar H_{fp}-H_m value. (c) A narrow single-peak, small-intensity signal is observed at H=0 for all H_m values that decays strongly towards zero. d) In the field region from above H=0 to H=H_m, only a very weak signal is observed indicating the presence of a very small amount of unpinned Josephson vortices along this field region. Analyzing the memory features described in (a) to (d), implies the following additional properties: The signal shape of the memory signals described in (a) indicates similar features as those obtained in the signal shape of Fig. 2 presented earlier. The signal in Fig. 2 was obtained by zero field cooling the crystal (with the field oriented parallel to the a-b plane) followed by increasing the intensity of the magnetic field. In this process the AC field interacts with the JV that penetrate the crystal from H=0 and above, inducing the observed signal. The similarity between the signal shapes of Fig. 2 with the memory signals indicates that the same mechanism induces the memory signals in both cases. Thus as in Fig. 2, unpinned JV start to penetrate the crystal at magnetic fields above H_m, generating the memory effect. This conclusion is in agreement with the theoretical analysis given in reference (Shaltiel et al., 2010).

The analysis presented above in the field region where the memory signal is formed implies unexpected and intriguing consequences. It indicates that JV are present in the magnetic field region from zero to H_m. It predicts the formation of JV remanent magnetization in this field region. To check this conjecture, the following procedure to measure the magnetization in a Bi2212 crystal was carried out. The magnetization was measured in a commercial SQUID (superconducting quantum interference device) magnetometer MPMS5 (Quantum Design) working in a temperature range from $1.8 < T < 400K$ and in magnetic fields up to 50 Kgauss. The Bi2212 single crystal was aligned within the a-b plane. To obtain the field dependence of the ZFC or FC magnetization at 4K, the sample was cooled down to 4K from above T_c at the field indicated at the corresponding hysteresis loop, of 0 Kgauss, 5 Kgauss and 10 Kgauss. Then the field was set to zero and the magnetization measured with increasing field up to 50 Kgauss, then in decreasing field down to -50 Kgauss and again up to the highest field to close the hysteresis loop. The magnetization results are shown in **Figs. 18** a and b. Indeed, as expected, no magnetization is observed at zero magnetic field when a zero magnetic field is applied. In contrast, a certain magnetization is observed at zero magnetic field when the sample is field cooled at 5 Kgauss. An even larger magnetization occurs at a cooling field of 10 Kgauss. A magnetization ratio of about 2 is obtained from Figs. 18 at H=0 in these field cooling processes of 5 Kgauss and 10 Kgauss, as expected.

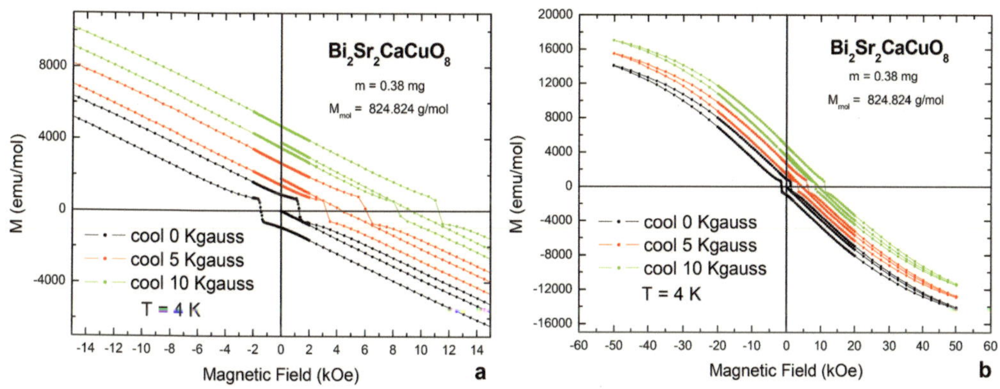

Figs. 18a and 18b. Magnetization as function of magnetic field in a Bi2212 crystal, field cooled at DC fields of 0, 5 and 10 Kgauss. Note that non zero magnetization is observed at zero magnetic fields when the compound is field cooled at 5 Kgauss or 10 Kgauss. It indicates the formation of JV remnant magnetization.

The memory results presented so far, did not involve the effect of temperature or time on the microwave-dissipaton signals. A limited experimental data related to time effects was achived as follows. After obtaining a H_m memory signal, the magnetic field was decreased to zero and kept at this field for 2 hours. Then the field was increased to a value above H_m to disclose the memory signal. The observed signal intensity and signal shape were similar to those observed earlier. It indicates that, if a time effect is present, it should be much longer than two hours.

The dependence of the maximum signal amplitude of a memory signal on the temperature was obtained as follows: A Bi2212 crystal aligned parallel to the a-b plane is cooled in a H_m field to 4K. After decreasing the field to zero, it was increased to a field that yields the maximum memory signal. Keeping the field constant the signal amplitude is measured by increasing temperature from 4K to 30K. The result is shown in Fig. 19. The figure illustrates that the amplitude of the memory signal remains constant in the temperature range 4K to 10K and decreases at higher temperature. This indicates that the JV remain frozen up to 10K and loose the equilibrium state with increasing temperature.

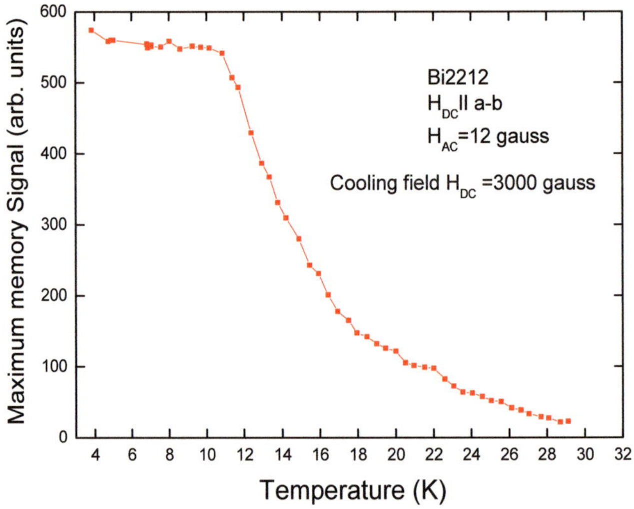

Fig. 19. Dependence of the maximum memory signal ampitude on temperature, derived from the memory signal after field cooling the sample to 4K and measuring the signal amplitude at the DC field of maxium memory signal on increasing the temperature continuously to higher values.

4.4 JV phase diagram

The magnetic field of the JV phase diagram, as function of temperature, derived from the results of the memory effect, is shown in Fig. 20. It indicates the presence of a glass state at the lowest temperature followed by a cross over state, an Anderson glass state, and a liquid state at higher temperatures.

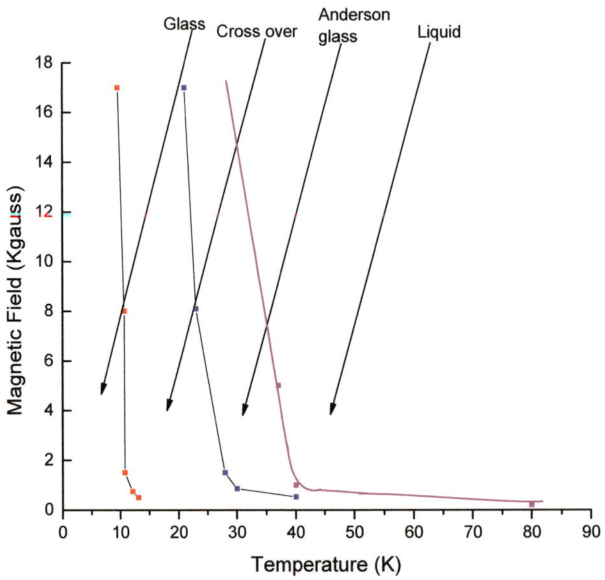

Fig. 20. Magnetic field-temperature phase diagram of JV in Bi2212.

4.5 Evaluating the possibility to study small or nano-scale superconducting systems with EPR spectrometers

The microwave-dissipation signal of a high T_c Bi2212 superconductor, shown in Fig. 21, is obtained from Fig. 2. The sample size was 10x10x0.1 mm^3. Its maximum intensity is of the order of 10,000 arbitrary units. The remarkably high intensity signal promises the possibility to investigate nano-scale superconductors and sizable small samples obtained by reducing the sample thickness to a few superconducting layers.

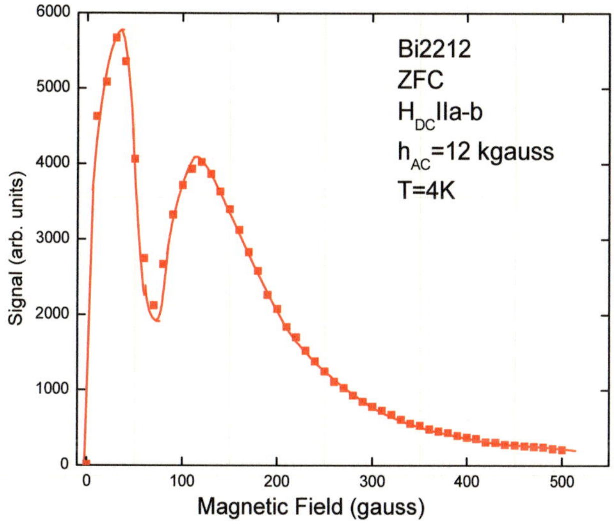

Fig. 21. Signal amplitude as function of magnetic field in a 10x10x0.1mm^3 Bi2212 crystal. The remarkable high intensity signal promises investigation of EPR properties in nano crystals.

Extremely thin superconductors with a limited number of layers are required when studying sample size effects on their properties. EPR technique can be considered in investigating their properties, by comparing results obtained in small layered nano crystals of similar structure with those obtained in crystals with a larger layer number. In samples with a small number of conducting layers the following EPR measurements are anticipated: a) Differences in the interaction between JV and AV vortices due to the small number of neighboring conducting layers. b) Results presented so far have shown that the observed properties depend on sample treatments, such as field cooling or zero field cooling procedure. Therefore the effect of cooling on thin samples may show different response that would depend on the sample thickness. c) The magnetic field memory effect, being by itself a very interesting effect, is explained using the hierarchical model of glassy state, commonly used in spin glass theory (Shaltiel et al., 2010). As the glassy state may depend on the thickness of the sample, it would be interesting to investigate the memory effect in small layer samples. Nano superconducting samples needed to investigate their properties are readily obtained using cleavage methods. In contrast the preparation method of thin superconductors for investigating IJJ properties, when regular methods are applied, is cumbersome, the reason being that conducting elements should be installed on the mesas' surface. Here we quote such a sample preparation in mesas that has been used to study teraherz-wave emission from Intrinsic Josephson Junctions in high T_c superconductors (Kadowaki et al., 2008): "The experiments were performed with the conventional methods commonly used in performing similar measurements and include soldering of conducting wires on the sample's surface. Preparation of the mesas needs skilful techniques as can be deduced from the following description. "The samples were prepared from a piece of single crystal grown by the traveling floating zone method. A cleaved thin single crystal with a thickness of several µm was glued on a sapphire substrate by U-vanish (polyamide resin), and a silver thin layer and a gold film are sputtered on the cleaved sample surface. Then, the sample mesa and two contact pads with a desired size were patterned by the photolithography and Argon ion milling techniques. After removing unnecessary resist, CaF_2 is evaporated on the part of mesa for the electrical isolation purpose, and Au is finally deposited for the contact pad. This Au is again patterned by photolithography and Argon milling techniques again." In contrast to this difficult and cumbersome sample-preparation method in preparing thin mesas, preparation of samples to be used in investigating nano samples with the EPR technique can be described as follows: "A cleaved thin single crystal of conventional size should be glued on a sapphire with proper dimensions to fit the sample holder of the spectrometer". The contrast between the extremely complicated preparation method, where the so called "conventional measurement method" is used, and the simplicity in sample preparation in performing EPR nano measurements is striking. The EPR method should therefore be preferred in studying nano superconductors, when similar properties are to be investigated. Microwave emission from Intrinsic Josephson Junctions in high T_c superconductors can readily be investigated using a device where an AC modulating technique is applied, and whose features are similar to those of the EPR spectrometer. With such a device the microwave output will be modulated at AC frequency.

4.6 Searching for superconducting compounds, with EPR spectrometers

Fig. 22, derived from Fig. 1 of Shaltiel et al., 1991, plots, in schematic representation, the resistivity R(T,H) in cooling a superconductor from above to below T_c at different DC magnetic fields. The figure shows that an ESR signal is obtained when sweeping the temperature from above to below T_c. The signal is induced by the AC field. It results from variations of the sample resistivity during the cooling process that affects the quality factor Q of the cavity of the spectrometer. Thus, EPR spectrometers can be used to search for superconducting materials, to determine their T_c, to study resistivity in superconductors, and to investigate the effects of various paramenets on the sample resistivity, without applying wiring contacts.

4.7 Determination of the anisotropy in superconductors

A description of a procedure that measures the anisotropy in superconductors using an EPR spectrometer was reported earlier (Shaltiel et al., 1992). The anisotropy was derived by analyzing EPR results of the signal-peak intensity, obtained when varying the temperature across T_c, as function of the orientation-angle θ of the magnetic field with respect to the crystal axis (Shaltiel et al., 1992). The detailed description of the measuring procedure including analysis of the results is presented in the above article. The results derived from these measurements, agree with a theory proposed by Tinkham, where the superconducting anisotropy constant ε is involved. Values of $\varepsilon = 5$ and $\varepsilon = 40\text{-}60$ were obtained for YBCO and Bi2212, respectively (Shaltiel et al., 1992). The results described in Shaltiel et al., 1992, which show the dependence of the signal intensity as function of the orientation angle, are a guide that can be used to orient a superconducting crystal with respect to an externally applied magnetic field. The orientation of the superconducting crystal through all the measurements presented in this work were performed by this procedure.

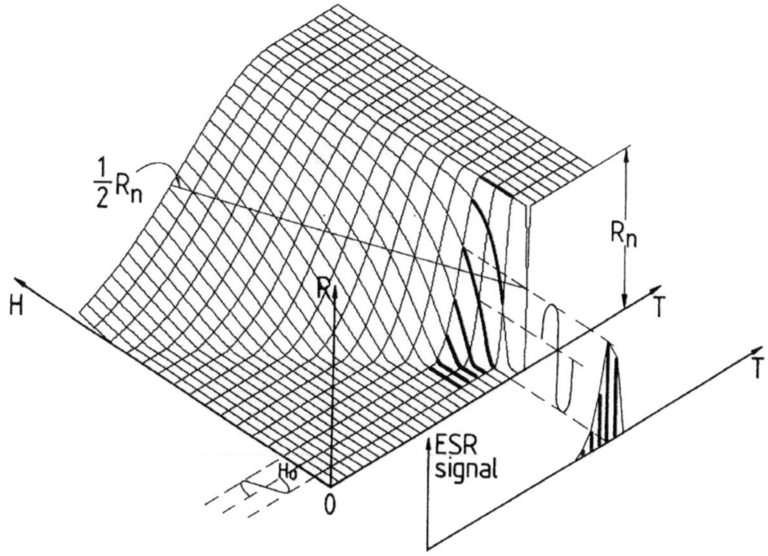

Fig. 22. Schematic, three dimensional presentation of the resistivity as function of magnetic field and temperature in a superconductor. Application of an AC field induces variation in the resistivity. The figure shows that an EPR signal is observed at the superconducting transition temperature.

5. Conclusion

The title of the present article: "Investigating the superconductivity with EPR (Electron Paramagnetic Resonance) spectrometer", also known as ESR (Electron Spin Resonance) spectrometer, is to a certain sense misleading but was intentionally chosen to put attention to an unconventional possibility to use the EPR technique. It gives the impression that the spectrometer investigates properties related to paramagnetic states and transitions between spin states in superconductors. As spin states and paramagnetic properties have not been an important subject in investigating properties in superconductors, various physicists ignore publications related to EPR technique. However, studying superconductors with EPR spectrometers generates a rich spectrum of experimental results of different nature. Hence, questions appeared regarding the mechanism that gives rise to the EPR signals and what do the results indicate. The present work provides answers to both questions. Investigating mesas shows that applying current in Josephson transmission lines (JTLs) induces motion of the fluxons that reside between the conduction planes and that the fluxons interact with the environment (Ustinov, 1998). Fig. 12 shows that in the case of EPR technique the AC modulation field induces Eddy currents that apply a Lorentz force that induces the motion of fluxons in the high anisotropy superconductors Bi2212 and Bi2223. Their motion interacts with the microwaves and induces the observed signal. This is the mechanism that provides the experimental results. The detailed theoretical derivation in the framework of Josephson phase electrodynamics (Shaltiel et al., 2008) enables to model the observed dependence of the microwave absorption on the DC magnetic field and the strong increase of the signal with increasing anisotropy. The analysis of the results yields information related to the properties of Josephson vortices in high anisotropy superconductors, which are otherwise not or only difficult to access. Here we recall two prominent achievements: (a) by means of field cooling and zero-field cooling experiments a memory effect was discovered and a corresponding phase diagram of the Josephson-vortex system was derived. (b) From angular dependent measurements pinning of the Josephson vortices by Abricosov vortices was established. Hence, the EPR spectrometer is the experimental device that measures these properties. We note that it was originally designed to investigate different properties than those probed presently. Experimentally the spectrometer investigates properties of superconductors, where Josephson transmission lines are present. If such an instrument did not exist, it would have to be designed to enable the investigation of such properties.

6. References

Hechtfischer, G.; Kleiner, R.; Schlenga, K.; Walkenhorst, W. & Müller, P. (1997). Collective Motion of Josephson Vortices in Intrinsic Josephson Junctions in $Bi_2Sr_2CaCu_2O_{8+y}$. *Physical Review B*, Vol.55, No.21, (June 1997), pp. 14638-14644.

Kadowaki, K.; Yamaguchi, H.; Kawamata, K.; Yamamoto, T.; Minami, H.; Kakeya, I.; Welp, U.; Ozyuzer, L.; Koshelev, A.; Kurter, C.; Gray, K.E. & Kwok, W.-K. (2008). Direct Observation of Tetrahertz Electromagnetic Waves Emitted from Intrinsic Josephson Junctions in Single Crystalline $Bi_2Sr_2CaCu_2O_{8+\delta}$. *Physica C*, Vol.468, (April 2008), pp. 634-639.

Shaltiel, D.; Bill, H.; Grayevsky, A.; Junod, A.; Lovy, D.; Sadowski, W. & Walker, E. (1991). Microwave Absorption across T_c: Determination of the Angular Dependence $H_{c2}(\theta)/H_{c2}$. *Physical Review B*, Vol.43, No.16, (June 1991), pp. 13594-13597.

Shaltiel, D.; Ginodman, V.; Golosovsky, M.; Katz, U.; Boasson, H.; Gerhouser, W. & Fischer, P. (1992). Investigation of the Magnetic Field-induced Microwave Absorption across T_c in Single Crystals of $YBa_2Cu_3O_{7-\delta}$ and BiSrCaCuO Anisotropy and Magnetic Field Dependence. *Physica C*, Vol.202, (November 1992), pp. 303-320.

Shaltiel, D.; Bezalel, M.; Revaz, B.; Walker, E.; Tamegai, T. & Ooi, S. (2001). Dynamic Response of Microwave Absorption Triggered by Magnetic Modulation in High Anisotropy $Bi_2Sr_2CaCu_2O_x$ Crystals. *Physica C*, Vol.349, (January 2001), pp. 139-149.

Shaltiel, D.; Krug von Nidda, H.-A.; Loidl, A.; Rosenstein, B.; Shapiro, B.Ya.; Shapiro, I.; Tamegai, T. & Bogoslavsky, B. (2008). Sharp Increase of Microwave Absorption Due to Shaking of Josephson Vortices in BiSCCO Superconductor. *Physical Review B*, Vol.77, (January 2008), pp. 014508-1-014508-7.

Shaltiel, D.; Krug von Nidda, H.-A.; Shapiro, B. Ya.; Rosenstein, B.; Loidl, A.; Bogoslavsky, B.; Shapiro, I. & Tamegai, T. (2008a) Interaction between Josephson and pancake vortices investigated by the induced microwave dissipation by ac magnetic field technique, *Phys. Rev. B* 77, (June 2008), pp. 214522-1-214522-5.

Shaltiel, D.; Krug von Nidda, H.-A.; Rosenstein, B.; Shapiro, B.Ya.; Golosovsky, M.; Shapiro, I.; Loidl, A.; Bogoslavsky, B.; Fujii, T.; Watanabe, T. & Tamegai, T. (2010). Field Cooling Memory Effect in Bi2212 and Bi2223 Single Crystals. *Superconductor Science and Technology*, Vol.23, No.7, (July 2010), pp. 75001-75006.

Srinivasu, V.V.; Ken-ichi, I.; Hashizume, A.; Sreedevi, V.; Kohmoto, H.; Endo, T.; da Silva, R.R.; Kopelevich, Y.; Moehlecke, S.; Masui, T. & Hayashi, K. (2001). Nonresonant Microwave Absorption in Bi2212 Single Crystal: Second Peak and Microwave Power Dependence. *Journal of Superconductivity: Incorporating Novel Magnetism*, Vol.14, No.1,(February 2001), pp. 41-46.

Ustinov, A.V. (1998). Solitons in Josephson Junctions. *Physica D*, Vol.123, (November 1998), pp. 315-329.

5

Non Resonant Microwave Absorption (NRMA) Anomalies in High Temperature Superconductors (HTS) Relevance of Electromagnetic Interactions (EMI) and Energy Stabilized Josephson (ESJ) Fluxons

G.K. Padam

Materials Physics and Engineering Division, National Physical Laboratory, New Delhi, India

1. Introduction

High temperature superconductors (HTS) due to their higher operating temperatures, 2-dimensional (2D) layered structure and very short coherence length (ξ) are known to show some interesting properties especially magnetic field dependent microwave (MW) absorption effects at very low magnetic fields of the order of zero gauss; Kadam et al.1987; Blazy et al.1987, 1988, 1989;) while carrying out energy /absorption studies at MW frequencies. One striking example of this is the appearance of very strong non-resonant microwave absorption (NRMA) signal centered around zero magnetic field in the superconducting state at near below T_c in cuprate based HTS :YBCO, reported by Bhat et al.(1987) using conventional electron spin resonance (ESR) spectrometer. This was followed by reports on other cuprate based HTS (Portis et al.,1987, Dulcic et al.1987,1988,1989,1990; Pakulis & G.V.Chandershekar,1989; Sugawara et al., 2000;Hashmizume et al.2001; Padam et al.,1999) and also on a different variety of superconductors like: borocarbides (Kadam et al.,1994), oxyflorides (Kadam et al.,1995) alkali fullerides (Roberts,1998-II), granular conventional low-T_c typeI [Pb, Al, Sn, In etc.]/type II [Nb, Nb_3Sn, Nb_3Ge/$PbMoS_6O_8$ etc.] (Kheifetsand et al., 1990; Blazey et al.1988,1989; Suss et al,1989;Kataev et al.,1991;Baranowsky et al.1991;Moorjani et al.1990; Bele et al.1994; Blazey et al 1987;Veinger et al. 1995),MgB_2(S.Sarangi & S.V.Bhat, Joshi et al.,2002) including newly discovered iron pniticide (Panarina et al. 2010; Pascher et al.2010 etc.).Before coming to the actual problem, what follows is a brief introduction of the NRMA studies done before the advent of HTS.

It is to note that the NRMA response using ESR technique was reported for the first time in 1970s by Indovina et al. in the case of granular thin films of conventional type I superconductor: Pb. Similar results were then reported in other type I superconductors: Sn (Kim et al., 1972) and Al (Muller et al., 1980) at magnetic fields varying from few hundred gauss to many thousand gauss. These reports strongly pointed out that the NRMA signal appears only if Josephson junctions (JJs) are present and their thickness (t) is ≥ ξ and it does not appear in JJs free superconductors and/or superconductors having very thin JJs. These JJs are the non-superconducting regions separating (i) microscopic metal superconducting grains and/or (ii) superconducting regions within the grain. The main mechanism of MW power absorption was assigned to: resistance of JJs (Kim et al.,1972; Muller et al.1980) and/or viscous motion of fluxons penetrated at JJs (Kim et al., 1972; Gittleman & B.Rosenblum,1966) It is to be pointed out that in both these models, JJs of intergranular type (here after called as JJ_{inter} are considered.

Despite these reports on the observation of NRMA revealing an important and direct information related with manifestation of JJs_{inter} in granular conventional superconductors which cannot be rendered by resistivity and specific heat measurements, not much literature on NRMA studies was available at that time. It was until the advent of HTS which are intrinsically granular in nature and showed a very strong NRMA signal at nearly zero magnetic field, NRMA became a valuable, informative and indispensable tool for examining all kinds of granular superconductors existing till today from conventional (both type I and type II) to recently discovered iron pniticide etc. particularly for the detection, variety and behavior of JJs and to know the mechanism of MW absorption.

Further, to explain the origin of NRMA signal in HTS (Portis et al., 1988) and Dulcic et al.1889) extended the previous models for conventional granular superconductors in which JJ_{inter} play the dominant role (Indovina et al.,1970;Kim et al.1972; Muller et al.1980;Gittleman & B.Rosenblum, 1966).This was done by including the contribution from intragranular JJs (JJ_{intra}) also. As is known that unlike in the granular conventional superconductors where, the effect of JJ_{intra} will be suppressed due to large ξ, in HTS due to very short ξ, the effect of JJ_{intra} can be easily observed. Although these models considered contribution from both the JJs (JJ_{inter} and JJ_{intra}), however, they associated to different mechanism of absorption. That is, Portis et al. associated NRMA to damped fluxon (intergranular fluxons at JJ_{inter} / intragranular fluxons at JJs_{intra}) motion, whereas Dulcic et al. associated to JJ_{inter} and JJ_{intra} resistance to the induced boundary currents (Josephson currents).Further, it was pointed out by Portis et al. (1988) "although we associate the microwave absorption with damped fluxon motion, it may be equivalent to associate the absorption with Josephson currents as others have done".

The continuous growth in this field of NRMA studies, along with the extreme sensitivity and expedition of the NRMA technique apart from its ability to detect the phase transition to the superconducting state, it has opened a new window (i) to unambiguously discover/detect new superconducting phases present even in quantity as low as of the order of sub micrograms, (ii) to distinguish multiple superconducting phases in a single matrix, (iii) to tell something about the quality of the superconductor, (iv) to study a new type of device for example: a single chip comprising of a combination of insulating, semiconducting, and superconducting layers, (v) to study the phase diagram of vortex matter through temperature dependent NRMA signal amplitude and to (vi) reveal important information on the anomalous nature of magnetic hysteresis etc. This provides new insights to both the fundamental and technological aspects of HTS.

There are two very important observed features which need further elucidation as their origins are still not very clear though there is a consensus on the origin of NRMA – a manifestation of JJs. Firstly, how could the amplitude of the NRMA signal which is a manifestation of the JJ_{inter} / JJ_{intra} with markedly different temperature dependences originate in such JJs? Secondly, how could the anomalous nature of NRMA hysteresis in contrast to the normal nature of DC magnetic hysteresis under similar conditions of temperature and magnetic field is generally observed? A critical analysis of the existing understanding to explain these observed features reveals that the full explanation of these anomalies are still far from clear understanding and thus needs modifications.

Recently (Padam et al. 2006, 2010) these two aspects have been taken seriously and analyzed in a different perspective, i.e. (i) by including the role of electromagnetic interactions (EMI) for explaining the reason for undulatory temperature dependence of the NRMA signal amplitude, and (ii) by giving a special reference to energy stabilized Josephson fluxons (ESJ) for explaining the anomalous magnetic hysteresis as seen by NRMA.

The main objective of this article is to further elucidate these two problems and the importance of NRMA technique in understanding their origin in a broader and deeper perspective with inclusion of studies made on recently discovered iron pniticides. The first problem has been analyzed in more generalized way by including some more data (i.e. temperature dependence under magnetic field) so that the confusion related with the different explanations for the origin of the temperature dependence of the NRMA signal can be minimized. Similarly, the second problem based on the existing status of NRMA hysteresis and their comparison with DC magnetization hysteresis.

The samples of BSCCO based HTS used in the present study are $(Bi, Pb)_2Sr_2Ca_2Cu_3O_{10+x}$ which were prepared from spray dried route in addition to the earlier studied $Bi_2Sr_2Ca_1Cu_2O_{8+x}$ samples with some more data and were characterized. The analysis of the results leads us to emphasize the role of EMI to explain a temperature-dependent JJs structure and to assert the role of ESJ fluxons to explain the anomalous nature of the hysteresis.

With this objective, in the section 2 of this article, the experimental details of the sample preparation and the characterization are given. In section 3, first results of work already done related with different temperature dependences of NRMA signal amplitude and the explanations given so far (3.1) are presented, then report results of present NRMA studies on sintered pellets of BSCCO samples (3.2) followed by discussion asserting the role of EMI (3.3). Similarly, section 4, will be first devoted to a brief summary of the results reported earlier on the nature of magnetic hysteresis in granular superconductors while carrying out DC magnetization and NRMA measurements (4.1) and the explanations for its origin, then report the present results (4.2) followed by discussion and emphasizing the role of energy stabilized Josephson fluxons (ESJ) in subsection 4.3. Finally, in section 5, we will summarize and give some conclusions.

Based on this exhaustive study, it is shown in this article that the inclusion of EMI and ESJ can further confirm their roles in the observed different temperature dependent NRMA signal amplitude and magnetic anomaly respectively.

In this article, the additional NRMA data showing the anomalous nature of hysteresis in some other variety of superconductors including the recently discovered iron superconductors which further confirms/supports the earlier reports makes the scenario more interesting and intriguing.

2. Experimental details

The bulk pellets (diameter=12mm, thickness= 2mm) of $Bi_2Sr_2CaCu_2O_{8+x}$: Ag (5wt %) [Bi-2212] were synthesized by the matrix reaction method and the details are described earlier (Padam et al., 1999)

Whereas, the bulk pellet samples of $Bi_{1.84}Pb_{0.4}Sr_2Ca_{2.2}Cu_3O_{10+y}$: Ag (0 to 10wt %) [(Bi, Pb)-2223] were synthesized by using spray drying method. For this, high purity powders of Bi $(NO_3)_3.5H_2O$, Pb $(NO_3)_2$, Sr $(NO_3)_2$, Ca $(NO_3)_2.4H_2O$ and Cu $(NO_3)_2.3H_2O$ were used. These powders were taken in a nominal composition of 1.84:0.4:2:2.2:3 and dissolved separately in double distilled water. Each of the individual solutions was continuously stirred with mild heating to clear all solutions and a few drops concentrated nitric acid were added, whenever needed, to fasten the process. Then all the solutions of individual powders were mixed in a beaker and labeled 'pure solution'. Different silver nitrate solutions were also made in double distilled water to obtain a silver content that corresponds to the composition of (Bi,Pb)-2223 by weight percentages of 0,1,5 and 10. They were mixed, one by one, with the same amount of 'pure solution'. These final solutions were then kept separately in a beaker and spray dried separately to prepare the desired powders (0-10wt% Ag) of the above mentioned compositions. The powders so obtained were calcined separately first at 800^0C and then at 820^0C for one hour with one intermediate grinding. Then pellets having 12mm diameter and 1mm thickness, were made of each 0-10wt% Ag doped (Bi,Pb)-2223 calcined powders and sintered at 835^0C in air for 100 hours with one intermediate grinding.

Since the properties of the non- resonant microwave absorption depend to some extent on the phase purity and microstructure of the sample etc, which may vary from sample to sample even made under identical conditions, therefore, to avoid any ambiguity all the characterization studies: NRMA, X-ray diffraction (XRD), scanning electron microscope (SEM), EDAX, electrical and ac susceptibility for (T_c) were done on a single sample.

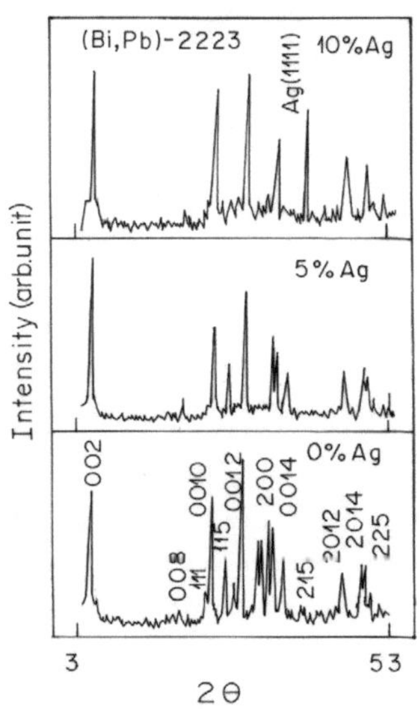

Fig. 1a. XRD patterns of (0-10wt%Ag) added (Bi, Pb)-2223 samples showing improvement in grain alignment on increasing Ag content.

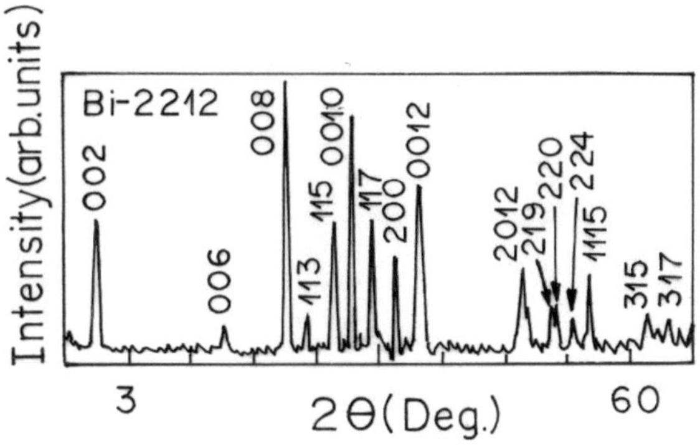

Fig. 1b. XRD pattern of powder of Bi-2212 (5wt%Ag) sintered pellet.

Room temperature X-ray powder diffraction (Fig.1a) of the material 0, 5 and 10 wt% Ag doped $Bi_{1.84}Pb_{0.4}Sr_2Ca_{2.2}Cu_3O_{10+y}$ samples, showed all these products consisted of (Bi, Pb)-2223 phase. However, on addition of Ag, the samples was found to improve in terms of crystallinity as revealed from the sharper peaks and directionally oriented as revealed from the increased 00ℓ reflections. This indicated these contents of Ag addition improved the quality of the samples. Similarly, room temperature powder X-ray diffraction (Fig.1b) for the $Bi_2Sr_2CaCu_2O_{8+x}$ (5wt% Ag) samples established that the products consisted of (Bi-2212). Furthermore, the absence of lines associated with secondary phases like Ca_2PbO_4, CaO, CuO etc. in the Bi-2212 and in the (Bi, Pb)-2223 samples indicated that the materials used in the present experiments were single phase.

A scanning electron microscopic studies of three (Bi, Pb)-2223 (0wt%, 5wt% and 10wt% Ag) samples revealed that in pure samples, there is a broad distribution of grain size and in Ag doped samples, there is a uniform grain size distribution with better oriented well connected plate like grains (not shown). Similarly, $Bi_2Sr_2CaCu_2O_{8+x}$ (5wt% Ag) revealed a reasonably good microstructure with well aligned and well connected grains (not shown). Superconducting transition temperature (T_c) was measured by both the electrical resistivity and ac susceptibility technique for these samples and the results are shown in (Fig.2). For the (Bi, Pb)-2223 samples T_c increases from 105 to 117K on increasing Ag content from 0 to 10wt%. The normal state DC resistance is lowest for the 5wt% doped sample and is highest for the 10wt% sample. In 5wt% Ag doped sample, it appears that most of the silver goes to the grain boundaries and hence is available for conduction. Therefore, minimum normal state resistance is observed in this sample. However, for the 10wt% Ag doped sample, most of the silver ions go into the lattice; therefore less silver is available for normal electrical conduction, hence responsible for highest normal state resistance

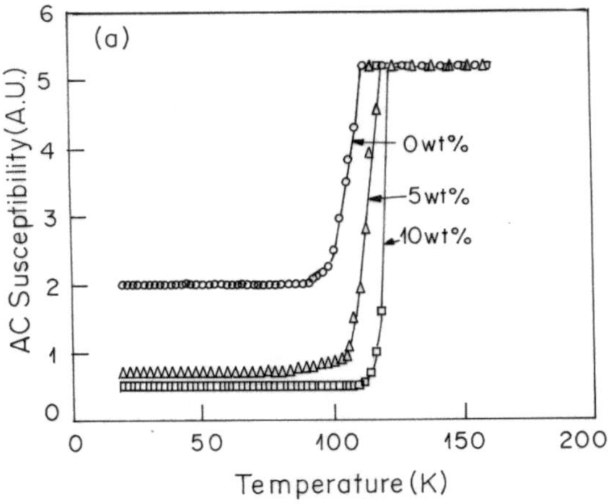

Fig. 2a. AC susceptibility dependence on temperature for silver added (0-10wt%) samples.

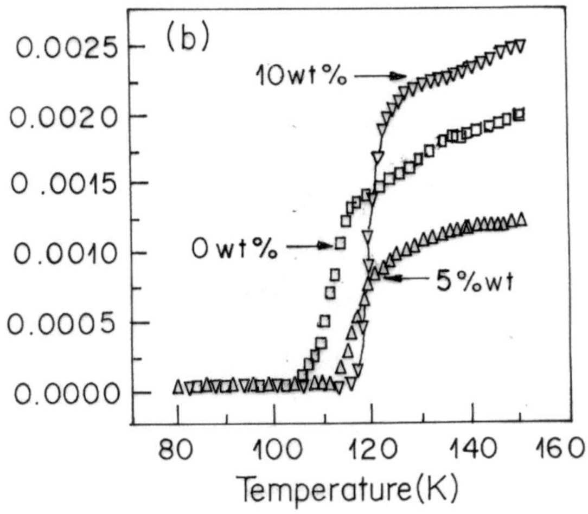

Fig. 2b. Temperature dependence of DC resistance of silver added (0-10wt%) samples.

For the Bi-2212 sample, T_c of 96K was obtained (Fig .3).

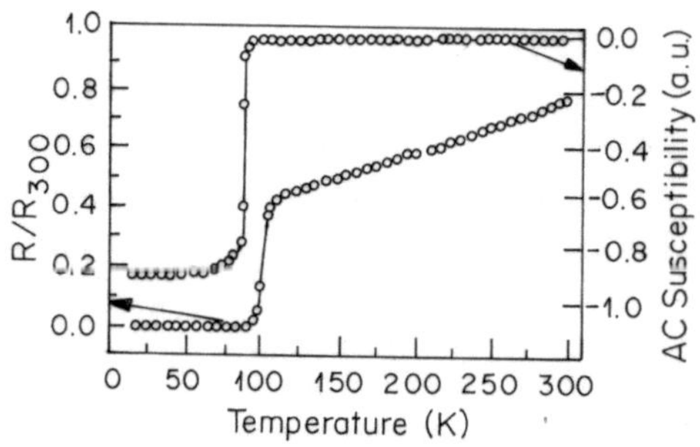

Fig. 3. Resistance and Susceptibility vs. Temperature characteristics of the Bi-2212(5wt%Ag) sample.

The quality test was also confirmed by measuring ESR of the samples at room temperature and observed no ESR signal due to most expected Cu^{2+} ions in all the samples was observed (Fig.4). For comparison, ESR spectra of (Bi, Pb)-2223 under sintered sample (containing binary phases) which is showing Cu^{2+} is also depicted.

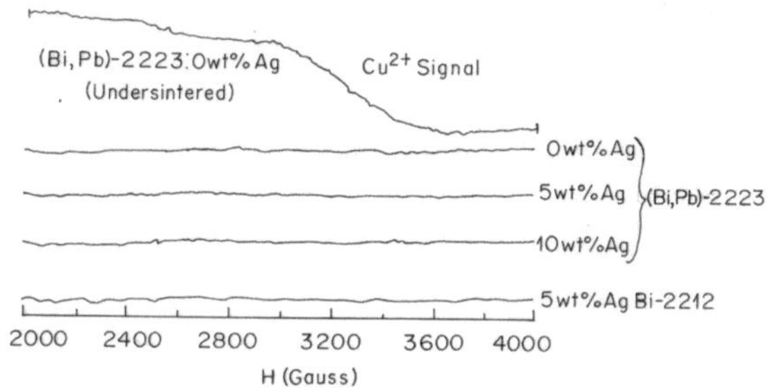

Fig. 4. EPR of (Bi, PB)-2223 and Bi-2212 samples showing absence of any signal to Cu^{2+} ions (100 mW, x1.25x10^3)

All these results showing nearly single phase nature of these samples with good microstructures and sharp superconducting transitions suggested that both the samples are of a reasonably good quality. For NRMA studies, an electron paramagnetic resonance (EPR) spectrometer Bruker ER 200D X-band (9.47GHz) equipped with an Oxford ESR900 continuous helium gas flow temperature variation accessory was used. Details are given in reference [7]. The measurements were done at modulation frequency at 100 KHz for different values of temperature (5-300K) by cooling the samples in zero field and then sweeping the static magnetic field. To avoid any possibility of experimental conditions (modulation amplitude, MW power, magnetic field scan range etc.) dependent shifting, shape, phase reversal and hysteresis or other artifacts of the signal, all the measurements were done under identical conditions for each study. That is, moderate modulation amplitude of typically 4G and the microwave power of 20mW were kept the same. Similarly, the magnetic field scan ranges were also kept the same for both the forward and reverse scans: the field was varied between the field was varied between -50 (+ 50) to +50 (-50) G. Some of the measurements were also done for high field scans from -50 to + 1050G. It was checked and confirmed that the signal shapes are not distorted.

To see the actual variation of the NRMA signal amplitude as a function of temperature, the NRMA signal amplitude is recorded by the temperature scanning method. The NRMA signal amplitude was monitored by warming the samples from 5K to 300K. For the Bi-2212 sample, these measurements were also done at four different fixed magnetic field values of 0, 50 and 100G for each scan in order to see the effect of magnetic field on the temperature dependent NRMA signal amplitude behavior.

Signals were recorded in the derivative form [i.e. field derivative (H) of microwave power absorbed (P): dP/dH] resulting from magnetic field modulation of typically 4G peak-to-peak at 100 KHz and phase-sensitive detection technique when H_{mw} (magnetic field component of microwaves) is perpendicularly to DC magnetic field H_{DC}. This provides information on extremely small temperature and magnetic field dependent microwave absorption (P) changes.

3. Results and explanations on temperature dependent behavior of NRMA signal amplitude

3.1 Temperature dependent behavior of NRMA signal amplitude in different superconductors reported by other researchers

Fig. 5 shows temperature dependence of NRMA signal amplitude reported in for different superconductors: cuprates (Blazey et al., 1987; Moorjani et al.1987; Glarum et al.1988) organic superconductors (Haddon et al., 1991) conventional superconductors[14,17] (Suss et al., 1989; Moorjani et al., 1987)etc. It is evident that markedly different variations in the NRMA signal amplitude in all these superconductors are seen on increasing temperature i.e. going from superconducting state to normal state. These different behavior mostly observed are: peak, undulatory and monotonous fall:

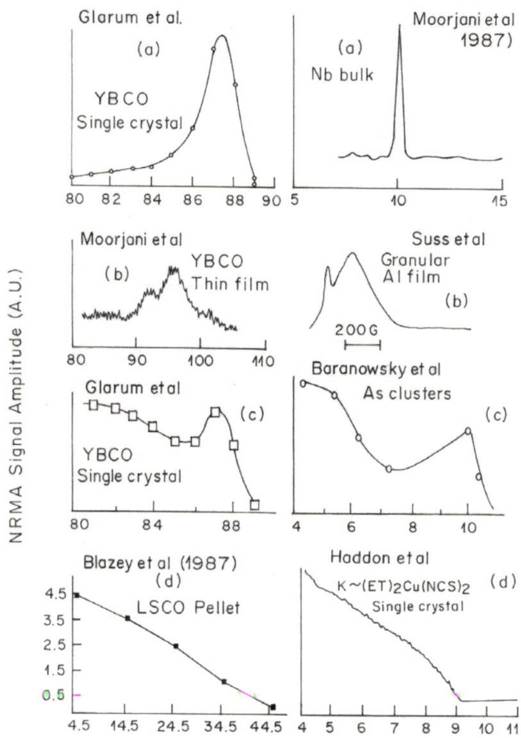

Fig. 5. Temperature dependence of NRMA signal amplitude showing different behavior irrespective of class of granular superconductor (data from literature).

i. Observation of peak/peaks in temperature vs. NRMA signal amplitude i.e. on increasing temperature, an initial exponential rise in the amplitude reaching a maximum with a subsequent exponential fall near below T_c (Fig.5a), or there can be more than one peak as shown in Fig.5b.
ii. Undulatory i.e. an initial fall then a rise reaching maxima followed by a final fall (Fig.5c) on increasing temperature.
iii. A monotonous fall (Fig.5d) on increasing temperature.

It is interesting to note that these different behavior are independent of the sample form (i.e. whether polycrystalline, thin film or single crystal) and types of superconductors. However, to realize, the full understanding of these different behavior it is necessary to clarify a considerable amount of confusion in the literature regarding the different origins of the temperature dependence of the NRMA signal amplitude by different researchers given so far which probably is due to some missing parameter. In the following, a brief account of various approaches and explanations has been given.

3.1.1 Peak type temperature dependence

Single peak type temperature dependence in HTS which is a characteristic feature of a sample having single superconducting phase has been well studied and its origin generally has been attributed to the different coupling strength of different JJs. So, throughout the temperature range in the superconducting state, some junctions decoupled, flluxons (JFs) penetrated and form a flux line lattice (FLL) which is controlled by Josephson interactions (JI) and dissipate power. That is, on increasing temperature, coupling strength of JJs and strength of (JI) among JFs decreases this result into an increase in number density and mobility of the fluxons penetrated within JJs. Consequently, an initial rise in the NRMA signal amplitude reaching maxima in the superconducting region at a few degrees below T_c, until most of the junctions are broken and showed weak superconductivity, hence an exponential fall in the superconducting-to-normal transition region. Above this temperature, all the JJs are broken and the material becomes normal, where the signal disappears. Details can be seen in earlier references (Portis et al., 1988, Dulcic et al., 1989)

Further, width and position of the peak have been related to a different variety of JJs. This different variety of JJs are: (i) JJ_{inter} which dominates in sintered pellets and/or dense powders can be due to misorientation of anisotropic grains, oxygen depleted grain boundaries or secondary phases at the grain boundaries etc. and are generally thick, and (ii) JJ_{intra} which dominates in isolated grains can be due to twin boundaries, stacking faults or point defects etc. and are generally thin. Narrow peaks near below T_c have been assigned to a sample having a small variety of JJs which are generally thin type (i.e.JJ_{intra}). On the other hand, broader peaks at temperatures lower than T_c have been cited as samples having a larger variety of JJs and are more of thick type like JJ_{inter}.

It is to be noted that in contradiction to the above explanation given by majority of the researchers regarding the origin of peak type behavior assigned to temperature dependent coupling strength of different kinds of JJs, however, there are some researchers for example Moorjani et al.(1990) who still have a different opinion. That is, according to them: (i) the peak response is due to the intrinsic superconducting transition, (ii) if any NRMA signal (i.e. rising base line) appears below this peak then that is due to the presence of different JJ_{inter}. Drawback of this report is that it considers only the role of JJs_{inter} and neglects that of JJ_{intra} even in HTS which are known to have inevitable presence of JJ_{intra}. Considering the role of JJ_{intra} e.g. due to the formation of a thin oxide layer on their Nb samples as reported by other researchers, can be one of explanations for the peak behavior in their Nb samples. Another possible reason of formation of superconducting droplets near T_c giving peak response can not be ruled out.

Like single peak which is a characteristic feature of a single superconducting phase, multiple peak behavior is a signature of multiple superconducting phases with different T_c (Kataev et al., 1991; Moorjani et al., 1987) in the same sample. Like the origin of single peak behavior assigned to temperature dependent coupling strength of different JJs and strength of JI in a single superconducting phase, the origin of multiple peaks is attributed to temperature dependent coupling strength of different JJs present in each superconducting phase.

3.1.2 Undulatory and monotonous type temperature dependence

With regard to the origin of undulatory and monotonous fall type behaviors, following different explanations have been given.

a. For the undulatory type behavior, different sources (Dulcic et al., 1990; Moorjani et al., 1990; Bohandy et al. 1988) have been reported to responsible. For example: (i) impurities, (ii) high density of defects and dominance of a variety of thick JJs etc.,(iii) to a mixture of phases with a distribution of superconducting transition temperatures and lower modulation frequency/amplitude.
b. Similarly, for a monotonous fall has been attributed to (i) impurities(Glarum et al., 1988) (ii) a deteriorated T_c superconducting phase (Sastry et al.,1987), (iii) presence of imperfections such as stacking faults (Bohandy et al.,1991)and (iv) a decrease in the viscosity of the fluxons on increasing temperature (Haddon et al., 1991).

It is clear that as compared to the origin of peak type response, origins of the undulatory and monotonous type behaviors are uncertain. That is, whether it is due to some impurities, imperfections, multiple superconducting / non-superconducting phases or variations in flux dynamics etc. still not clear. Further, all these reports consider the role of JI interactions only but none of them give reason for the initial fall in the signal amplitude on increasing temperature.

All these observations irrespective of the class of superconductors make this problem intriguing and motivate to investigate this discrepant data more carefully.

Recently in our earlier work (Padam et al., 2006) an attempt has been made to analyze this problem by considering the effect of electromagnetic interactions (EMI) in addition to JI and found that involving EMI can satisfactorily explain particularly the undulatory behavior.

Before going into further details, lets us see the temperature dependence of the NRMA signal amplitude in the present samples.

3.2 NRMA signal amplitude vs. temperature behaviors in present BSCCO samples – Relevance of EMI

3.2.1 (BI, Pb) samples

Fig.6 depicts the peak behavior for Ag (0, 5 and10 wt %) added $(Bi, Pb)_2Sr_2Ca_2Cu_3O_{10+x}$ [here after called as (Bi, Pb)-2223] bulk samples the temperature dependence of the signal amplitude for of constant magnetic field of 0 G by varying the temperature (45-160K). In these experiments equal amounts of each sample and identical spectrometer operating conditions were used to facilitate comparison of the resulting spectra. It is clear that all the samples showing a peak behavior indicating that all the samples have a different variety of JJs. Since these results are similar to as reported by others[1-7,10,15,] therefore, its origin can also be explained on the bases of temperature dependent coupling strength of JJs / strength of JI.

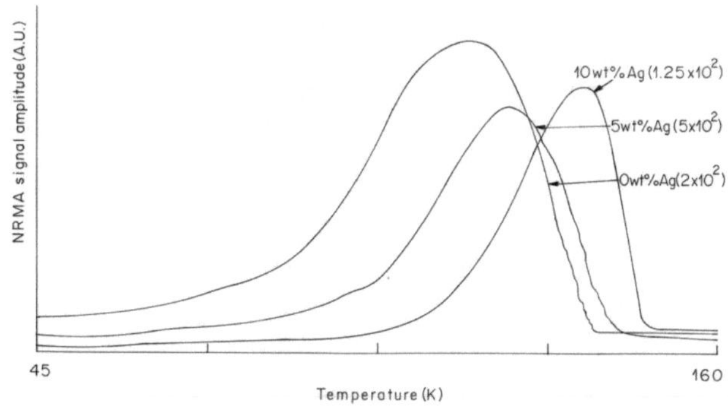

Fig. 6. Temperature of NRMA signal amplitude of Ag added (Bi, Pb)-2223 sample showing narrowing and shifting of the peaks towards higher temperature on increasing Ag content.

However, width and position of the peak was found to depend upon the amount of Ag. That is, narrowest peak at the highest temperature of 135K for the 10wt% Ag doped sample, whereas the broadest peak at the lowest temperature of 118K for the pure sample was observed. This narrowing and shifting towards higher temperature of the peak on increasing Ag content may be due to the improvement in the quality of the sample in terms of decreasing number probably of JJ_{inter}. These results are supported by the sharper XRD peaks with enhanced (ooℓ) reflections (Fig.2) and sharper superconducting transition and increased T_c (Fig.1) on increasing Ag content. Similar results related with observation of narrow peaks at high temperatures for good quality samples and broader peaks at lower temperatures for poor quality samples have been reported by others (Bhat et al.,1987; Portis et al.,1988; Dulcic et al.,1989; Pakulis & G.V.Chandershekar, 1989; Sugawara et al.,2000; Kataev et al., 1991,Padam et al.,1999).

3.2.2 Bi-2212 samples

Fig.7a depicts an undualtory temperature dependence of the NRMA signal amplitude for the Bi-2212 (5wt %) bulk samples kept at constant magnetic field of 0G and varying the temperature from 15K to 105K rather than the peak behavior as shown by all the three (Bi,Pb)-2223 samples. This Bi-2212 sample showed showing three distinct regions, i.e. an initial decreasing curve ABC (15-71.8), narrow flat CD (71.8-76.5K) and then a rise DE (76.5-82.5K) with maxima at around 82.5K followed by subsequent exponential fall EF.

This undulatory behavior may be discussed first in the frame work of the explanation used for peak behavior. That is, on increasing temperature, strength of JI decreases (due to increasing decoupling of JJs) which results into an increase in number density and mobility of JF. This leads to an initial rise in the NRMA signal amplitude reaching a maxima at a few degrees below T_c, until most of the JJs are broken and showed weak superconductivity, hence an exponential fall in the superconducting - to - normal region. Above this temperature, all the JJs are broken and the material becomes normal, where the NRMA signal disappears.

From this it is very clear that, the undulatory behavior of an initial fall in the present samples can not be explained by the temperature dependent strength of JI alone as it will lead to an initial rise with increasing temperature.

Other explanations which relates with impurities and to a mixture of phases with a distribution of superconducting transition temperatures are unlikely as the Bi-2212 samples are single phase as revealed from the observations of: (i) sharp superconducting transition from resistivity/AC susceptibility data, (ii) no reflections, corresponding to the Bi-2201 structure within the accuracy of 3% from even the powder XRD measurements and (iii) no impurity signal from EPR data (Fig.4).

Similarly another explanation which relates to the presence of high density of defects and dominance of a variety of thick JJs etc. considering only the JI is also ruled out for the obvious reason that JI alone can only explain the initial rise but not the initial fall as in the present case. This indicated that there is a need to look for some other alternative explanation and/or to include the role of some other interaction in addition to JI.

Recently the present authors (Padam et al.,2006)have given an alternative explanation for the origin of undulatory behavior by including in addition to JI, the role of electromagnetic interactions (EMI) which are not only significant but are also dominant over most of the low magnetic field and low temperature regimes, where generally the NRMA measurements are carried out. The combined effect of JI and EMI controls the FLL behavior and play an important role in absorption of microwaves. Based upon the inclusion of EMI in addition of JI, the undulatory behavior in the present Bi-2212 (5wt %) sample has been explained as follows:

Appearance of NRMA signal at 15K, suggested the presence of thicker JJs (most likely JJ_{inter}) which allow fluxons to penetrate at such low temperatures. Since density of fluxons is low at 15K, therefore there is a large separation among them. This leads to a weak attractive EMI. However, at this temperature, JI among fluxons is still finite. Thus combined effect of finite JI but feeble attractive EMI results into a loose packing of fluxons to form, a quasi-3D dilute liquid like FLL (i.e. disordered phase) in which the fluxons are mobile. An increase in temperature causes breaking of more number of JJ_{inter} and allows more number of fluxons to penetrate, thus reduces the inter-fluxon distance. This in turn decreases strength of JI but increases that of the EM attractions among the fluxons. Thus are able to suppress their even increasing thermally activated fluxon motion as a result of raising temperature. This trend continues up to 71.8K and results in the decrease of MWA, hence, a decrease in the signal amplitude.

On further increasing temperature, due to continuation of the above trend, probably, the EM attraction becomes which minimize the thermally activated fluxon motion and results in weak temperature dependence of the amplitude in the narrow temperature range (71.8-76.5K). In this range, a phase transition from quasi-3D dilute liquid like FLL to a quasi-3D solid like FLL (ordered phase) probably also occur.

However, above 76.5K fluxon density increases to an extent that the inter-fluxon EM repulsion overpowers inter-fluxon EM attraction and the system now becomes more sensitive to increasing thermal fluctuations. Due to dominance of both the EM repulsion and increasing thermal fluctuations, fluxon motion is more prevented and the FLL starts melting and transits into a quasi-2D dense liquid. This leads to an increase in MWA, hence a rise in the amplitude with a peak at around 82.5K, where it melts and leads to a fall in MWA in 82.5-105K.

From the above one can say that with the inclusion of EMI the undulatory particularly the initial fall portion has been explained which otherwise was unexplainable with taking JI alone. Based upon this explanation of inclusion of EMI in addition to JI, the undulatory temperature dependence of NRMA signal amplitude as reported by others in superconducting sample having thick JJ_{inter} can also be explained. To confirm, the measurements were done by keeping the sample at a constant magnetic field as is known that the magnetic field suppresses JJ_{inter} response. The results obtained for two fields of 50G and 100G are shown in Figs.7b, 7c. Interestingly, for both the fields, the undulatory behavior which was obtained at 0G disappeared and instead a peak response obtained and the peak gets narrowed on increasing magnetic field. This clearly suggested that the initial fall is related with the presence of thicker JJ_{inter} which play a dominating role at low temperatures.

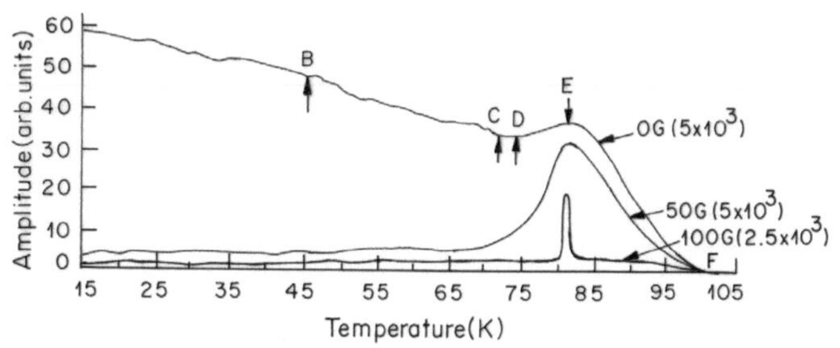

Fig. 7. NRMA signal amplitude vs. temperature curve for Bi-2212 (5wt%Ag) sample in OG, 10G and 50G.

Based upon this explanation, all the results reported in literature showing undulatory can also be explained on the basis of presence of thick JJs which dominate at low temperature and lead to rising strength of EMI initially, hence an initial fall in the NRMA signal amplitude on increasing temperature.

Similarly, the monotonous fall in the NRMA signal amplitude might also be due to the dominance of thicker JJ_{inter} (which can be due to impurities, impurity phases, defects etc.) which allows fluxons to enter at low temperatures and increase the probability of EMI among fluxons, thereby decreasing the mobility of the fluxons hence the NRMA signal amplitude with increasing temperature.

From the above one can say that if a superconductor has more of multitude of thick JJ_{inter}, then the strength EMI overpowers that of JI and generally show undulatory or monotonous response on increasing temperature.

In the next section the second equally important results related with anomalous nature of NRMA hysteresis are given.

4. Results and explanations on anomalous nature of magnetic hysteresis behavior of NRMA

It would be appropriate to give a brief account of earlier reports related particularly with the nature of NRMA hysteresis in comparison to that of the DC magnetization method and its universality in different granular superconducting systems including the present results before any elaborative discussion related to its origin.

4.1 A comparative data on the nature of DC and NRMA magnetic hysteresis reported by other researchers

Fig.8 shows a comparison of set of data available in literature in single crystals of YBCO (Dulcic et al. 1990; Pakulis & G.V.Chandershekar, 1989) and Bi-2212 (Pradhan et al., 1995) of these two measurements under almost similar experimental conditions of temperature, magnetic field scan and orientation of the magnetic field with respect to crystal axis. It is worth noting that for both the YBCO and Bi-2212 systems, the decreasing H curve lying above the increasing H curve observed in DC magnetization, the decreasing H curve lies below the increasing H curve nature of hysteresis shown by NRMA method. This indicated that in contrast to the normal nature of hysteresis observed by DC method, the nature of hysteresis shown by NRMA is anomalous under similar experimental conditions of temperature, magnetic field.

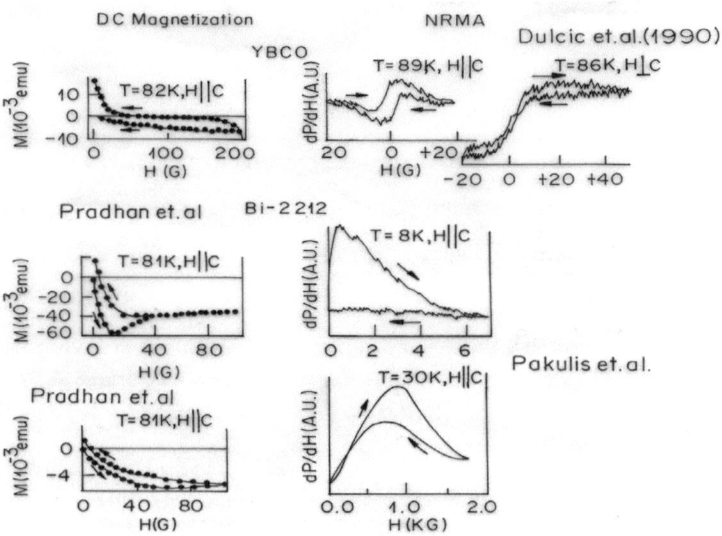

Fig. 8. Comparison of NRMA and DC magnetic hysteresis in YBCO/Bi-2212 single crystals showing the anomalous hysteresis as seen in NRMA in contrast to normal in the DC magnetization (data from literature).

4.2 Magnetic hysteresis of NRMA in different superconductors from unconventional to conventional observed by other researchers

Fig.9 exhibits the results of NRMA hysteresis reported earlier in different superconducting systems like: unconventional intrinsically granular cuprates (Blazey et al.,1988),borocarbides (Kadam et al.,1994) fullerene (Zakhidov et al.1991) MgB_2(Joshi et al.,2002) Iron pniticides (Panarina et al. 2010)and granular conventional type I/type II superconductors (Baranowsky et al., 1991; Mahel & S.Benacka,1992)) It further confirms that irrespective of the system of the superconductor, in all the NRMA curves, the decreasing H lies below the increasing H curve at temperatures much below T_c and in low magnetic field regions. This clearly indicated anomalous hysteresis nature at LT/LH in different superconducting systems from unconventional to conventional while carrying out NRMA measurements.

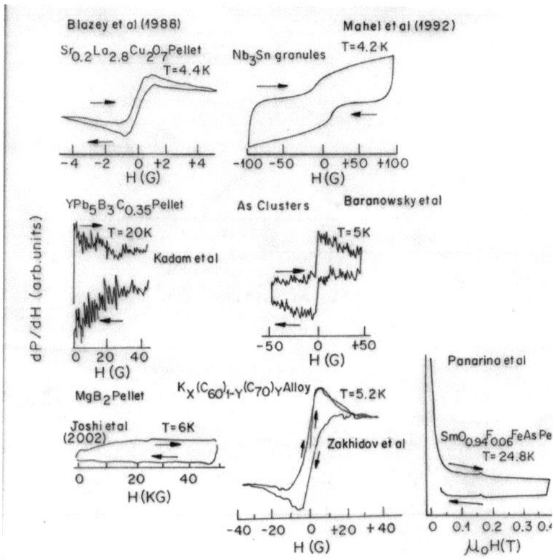

Fig. 9. NRMA hysteresis in different superconductors (data from literature) showing anomalous hysteresis irrespective of class of granular superconductor.

4.3 NRMA hysteresis in the present BSCCO samples

dP/dH vs. H curves of NRMA signals recorded for the present Bi-2223 samples in the high field scans for both forward (-50 to +1025G) directions at temperatures in the range of 5.5 to 135K were taken and the results for 90K are shown in Fig.10a. Two signals: one the usual near zero field signal and another a hump like feature (encircled) which varies in amplitude and width with increasing temperature along with the former signal have been observed. Similar observations of zero field signal along with this kind of feature (encircled) have been reported earlier by some researchers including present authors (Padam et al. 2010) and are also shown in Fig. 10b for Bi-2212 single crystal (rf) (Baginskii et al. 1994, rf absorption studies) and in Fig.10c for Tl-2212 single crystal (NRMA studies) (Joshi et al. 2000, MW absorption studies)The former well known zero field signal is due to viscous motion of intergranular (IGBJ)/intragranular (IG) fluxons. Whereas the latter hump like feature is due to fluxons generated due to tilt motion of flux lattice (Brandt, 1991; Kleiner et al. 1992) It is these fluxons which have been considered to be responsible for the general anomalous magnetic hysteresis and are named as ESJ in our earlier publication (Padam et al. 2010)

Similar results were obtained in Bi-2212 samples. Fig.11 depicts low field forward (-50 to +50G) and reverse field (+50 to -50G) magnetic sweeps of the NRMA signals in the both the Bi-2223 and Bi-2212 samples showing hysteresis at LT (5.5K, 20K) and at HT (110K, 80K). It can be seen that the decreasing field curve lies below the increasing field curve for both the samples not only at HT but also at LT. This indicated that the nature of NRMA magnetic hysteresis is anomalous in the present samples even at low temperatures as low 5.5K.

Thus all the above data reported earlier and in the present work confirms the anomalous nature of hysteresis in different granular superconductors including recently discovered iron pnictides under LT/LH conditions as seen by the NRMA method.

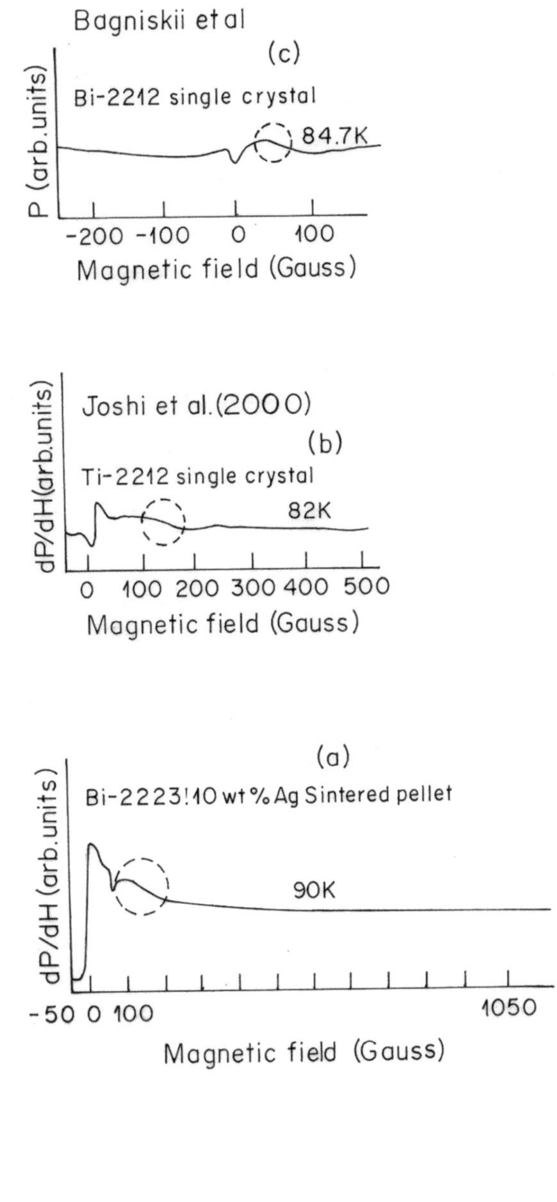

Fig. 10. dP/dH as a function of magnetic field showing signal (encircled) due to ESJ fluxon

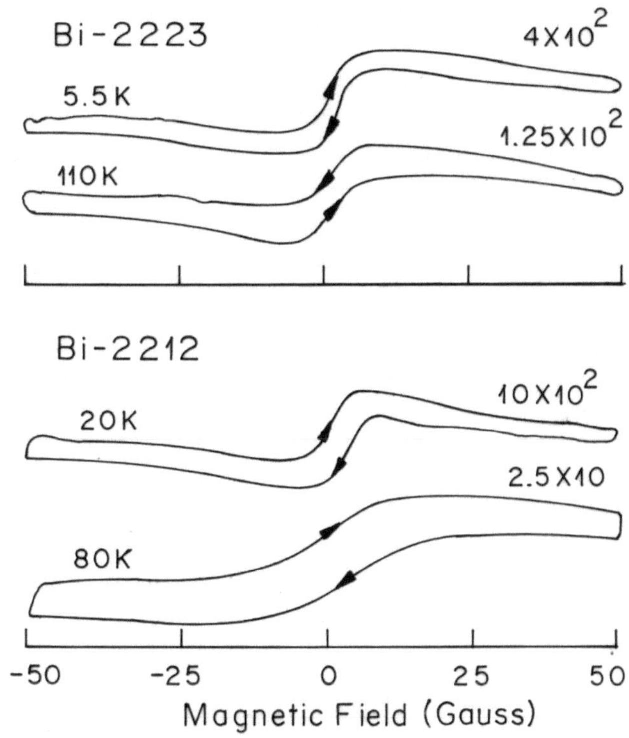

Fig. 11. NRMA hysteresis in present BSCCO sintered pellets of LT and at HT.

4.4 Origin of NRMA anomalous magnetic hysteresis and the relevance of ESJ fluxons

Before elaborating more on the origin of anomalous hysteresis, let us give a brief account of the development of the understanding related with the observations of different nature of hysteresis depending upon the (i) class of superconductors (i.e. conventional or unconventional), (ii) experimental conditions of temperature (T) and magnetic field (H) and (iii) the sensitivity of the method (DC magnetization methods: DC magnetic field (H_{DC}) variation of (i) flux density (B), (ii) microwave surface resistance (Rs) and (iii) transport critical current density [J_c] (Ji et al. 1993)and of modulated magnetic field variation of non-resonant microwave power absorption (P) NRMA technique(Czyzak & J.Stankowski, 1991).

That is, (i) normal nature of hysteresis in conventional superconductors at all temperatures and/or magnetic field by any measurement method, (ii) normal hysteresis at low temperatures (LT) and/or low magnetic field (LH) and anomalous hysteresis at high magnetic field (HH) and/or high temperatures (HT) by in granular conventional / unconventional superconductors by DC magnetization and microwave methods and (iii) anomalous hysteresis not only at HH /HT but also at LH/LT in granular conventional /unconventional superconductors while carrying out NRMA measurements.

The normal nature of hysteresis under any conditions conventional superconductors has been very well explained by the Beans model (Beans, 1962) which considers the role of intragranular Abrikisov/Pancake fluxons (IG). The anomalous nature of hysteresis at HT/HH in granular conventional/ unconventional superconductors has been well explained by Ji et al model by modifying Beans model on considering the role of intergranular Josephson (IGBJ) fluxons in addition to IG.

It is surprising that despite the voluminous literature on the NRMA hysteresis; only a few groups (Sugawara et al. 2000; Padam et al. 1991; Czyzak & J.Stankowski, 1991; Mahel et al., 1992; Ramachandran et al. 1994) have mentioned about its anomalous nature and related it to Ji et al model and its variant by Mahel et al. and Ramachandran et al. but they have their own drawbacks as will be clear below.

The present/ others results on the presence of anomalous magnetic hysteresis can be discussed in the frame work of the most successful model by Ji et al.which explain HT/HH anomalous hysteresis in granular superconductors. According to this model, there are two kinds of fluxons: intergranular (IGBJ) and intragranular (IJ) in granular superconductors. Further, at HT/HH, due to the higher number density and higher mobility of the IGBJ fluxons to those of the IJ fluxons, the magnetization during increasing field curve is lower than that during decreasing field curve and gives anomalous hysteresis. Although this model successfully explains the anomalous hysteresis at HT/HH, however the observed anomalous hysteresis at LT/LH in the present samples and reported by others can not be explained.

Further, these results also cannot be explained even by using variant of Ji et al. model by Mahel et al. and Ramachandran et al. particularly used for explaining NRMA anomalous hysteresis. This is because, the observation of anomalous magnetic hysteresis in the present samples as well as reported by others is not limited by the conditions of imposed by these models. For example, the former model includes the effect of modulation amplitude threshold field on the distribution of both the intergranular Josephson (IGBJ) and intragranular (IJ) fluxons). However, this model is applicable only for a limited modulation amplitude range. Similarly, the latter model which includes the effect of modulation amplitude threshold field on the distribution of flux suffers from its stringent requirement of threshold field value.

In our recent work (Padam et al., 2010) an attempt has been made to solve this problem by using another variant of Ji et al. model by including the role of Energy Stabilized Josephson (ESJ) fluxons in addition to IGB/IG fluxons and highlighting the features of the extremely sensitive NRMA method which makes it unique in creating and detecting a situation for the appearance of anomalous hysteresis. It was found that by combining the role of ESJ fluxons with the uniqueness of the NRMA technique, it is possible to explain the observation of LT/LH anomalous hysteresis too.

It is known that apart from IGBJ fluxons and IJ fluxons, there exists another kind of Josephson fluxons which are generated as a result of thermal/Lorentz forces induced tilt motion of flux lattice (Brandt, 1991; Kleiner et al. 1992) It is these fluxons are ESJ fluxons which not only increase the total number of more mobile fluxons at LT and/ LH, etc but also are detected due to the extra sensitivity of the NRMA method (Khachaturyan et al., 1987), thereby revealing the anomalous hysteresis.

Before proceeding further, it is essential to essential to highlight the experimental features of NRMA method, which makes it unique. For example, its ability (i) in creating a situation for enhanced ESJ fluxons,(ii) in detecting even minute changes in the fluxon density and (iii) in distinguishing between different types of fluxons,where all other DC magnetization techniques unable to do so.

In this context it is known that NRMA technique uses ESR spectrometer, where DC magnetic field (H_{DC}) is modulated with a modulation field (H_m) superimposed parallel on H_{DC}. Due to this H_m of magnetically modulated hysteresis are observed unlike direct field hysteresis measurements. This modulation field in addition to H_{DC} (used in B vs.H and Jc vs. H methods) and H_{DC}+H_{mw} (both used in Rs vs. H method) gives a few advantages. For example, it provides extra energy (due to a combined effect of three fields: H_{DC}+H_{mw}+H_m) required to overcome the strength of the Josephson junctions; as a result, IGBJ/IGJ fluxons enter at much lower temperatures/low magnetic fields. Another is a stronger Lorentz force, particularly due to H_m induced oscillatory currents, which tilt the IGBJ/IGJ fluxons, and probably becomes a volume source of generation of ESJ fluxons. Further, a combination of the modulation field in conjunction with phase sensitive makes ESR an extremely sensitive and powerful tool to detect any feeble change and to distinguish between different types of fluxons (Portis et al.1988; Dulcic et al.1897, 1988, 1989). Furthermore, the sensitivity of the NRMA method, which uses ESR, is even five orders of magnitude greater than the limit of the ESR spectrometer (Khachaturyan et al. 1987).

All these parameters make ESR technique an extra sensitive and unique in comparison of other techniques. Therefore, any change however minute and at times seemingly intractable, related with density/distribution/motion and different types of fluxons depending upon temperature/magnetic field/sample form etc. can be easily detected in NRMA. Furthermore, NRMA is the only method (Portis et al.1988; Dulcic et al.1897, 1988, 1989) which is able to distinguish between above mentioned different types of fluxons, i.e.IGBJ, IGJ, IG and ESJ etc.

Using this explanation, the observed anomalous nature of hysteresis in the present samples (which also shows the signal due to ESJ fluxons) and reported in a variety of superconductors can be explained.

5. Summary and conclusions

The present NRMA studies made on Bi-2223 / Bi-2212 samples and similar studies by other researchers in a variety of superconductors from conventional to unconventional have been reported. Main focus has been given on two most important aspects. First is the different temperature dependences of the NRMA signal amplitude observed by us and by others. The second one is on the observation of anomalous magnetic hysteresis in the present samples and in reports of others. These studies showing different types of behaviors: peak type undulatory and monotonous temperature dependent signal amplitude behavior and the anomalous magnetic hysteresis at all temperatures below T_c have been analyzed and discussed by taking into consideration of parameters/interactions not envisaged before.

From the above studies and discussion, it comes out that the undulatory type/ monotonous temperature dependent behavior of NRMA signal amplitude in a variety of granular superconductors can be very well explained if the role of EMI is included with JI.

In addition, the presence of additional fluxons (ESJ) in the present Bi-2223 / Bi-2212 samples and in other granular superconductors at all temperatures which probably formed due to a combined effect of additional fields like oscillatory MW field and modulation field are responsible for the anomalous hysteresis while carrying out NRMA measurements.

6. Acknowledgements

The author is grateful to Prof.R.C.Budani, Director, and National Physical Laboratory for his constant encouragement during the course of this work. The author is highly indebted to Prof.S.V.Bhat (IISC, Banglore) and Prof.M.R.Tripathy (Dept.Of Electrical and Communications Engng. ASET, Noida) for NRMA experiments. The author is highly grateful to Dr.S.N.Ekbote for his guidance and valuable suggestions.

7. References

Baginskii, K.V., Berezin, V.A., Govorkov, S.A. and Tulin, V.A. (1994).Sov.Phys.JETP, 160, 60.

Baranowsky, J.M.,Lilientel, W.Z., Yan W.F. and Weber, E.R. (1991).Phys. Rev. Lett. 66, 3079.

Beans, C.P. (1962) Phys.Rev.Lett, 8, 253.

Bele, P., Brunner, H., Schweitzer, D. and Keller, H.J. (1994) Solid State Communications 92,189.

Bhat, S.V., Ganguly, P and Rao, C.N.R. (1987) Pramana-J. Physics 28, 2425.

Blazey, K.W., Muller, K.A., Bednorz, J.G., Berlinger, W., Amoretti, G., Buluggiu, E., Vera A.and Matacotta, F.C. (1987) Phys. Rev. B36, 7241.

Blazey, K.W., Portis, A.M.and Bednorz, J.G. (1988) Solid State Commn.56, 1153.

Blazey, K.W., Portis, A.M.and Holtzberg, F.H. (1989) Physica C157, 16.

Bohandy,J.,Kim, B.F., Adrianand F.J., Moorjani, K. (1989)Phys.Rev.B 39,2733.

Bohandy, J., Kim, B.F., Adrianand F.J., Moorjani, K., Arcangelis S.D. and Cowan, D.O., (1991) Phys.Rev.B43, 3724.

Brandt, E.H. (1991) Phys.Rev.Lett.67, 2219.

Czyzak B. and Stankowski, J. (1991) Phys. Stat. Sol. (b) 166, 219.

Dulcic, A., Leontic, B., Peric, M.and Rakvin, B. (1989) Europhys.Lett.,10,593 Dulcic,A.,Crepeau,R.H.,Freed,J.H.,Schneemeyean,L.H.and Waszezak,J.V.(1990) Phys.RevB.42,2155.

Gittleman J.I.and Rosenblum, B. (1966) Phys. Rev.Lett. 16, 734.

Glarum, S.H., Marshall, J.H.and Schneemeyer, L.F. (1988) Phys.Rev.B 37, 7491.

Haddon, R.C., Glarum, S.H., Chichester, S.V., Ramirez, A.P. and Zimmerman, N.M. (1991) Phys.Rev.B. 43, 2642.

Hashizume, A., Yamada, J., Kohmoto, H., Yamada, Y. Endo T. and Shahabuddin, M. (2001) Physica C 357-360,481.

Indovina, P.L., Onoti, P. and Tabet, E. Solid State Commun. (1970) 8, 1721.

Ji, L., Rzchowski, M.S., Anand, N. and Tinkham, M. (1993) Phys.Rev.B47,470.

Joshi, J.P., Anand, A., Bhat, S.V., Chowdhary P.and Bhatia, S.N. (2000) Physica C 341-348, 1229.

Joshi, J.P., Sarangi, S., Sood, A.K., Pal, D.and Bhat, S.V. (2002) Parman, J.Phys. 58,361.

Kadam, R.M., Sastry, M.D., Hassain, Z., Mazumdar, C., Nagarajan, R., Gupta, L.C.,Godart,C.andVijayaraghvan,R.(1994)PhysicaC,232,359.

Kadam, R.M., Wani, B.N., Sastry M.D.and Rao, URK, (1995) Physica C, 246,262.

Kataev, V., Knauf, N., Buchner B.and Wohlleben, D. (1991).Physica C 184, 165.

Khachaturyan, K., Weber, E.R., Tejedor, P., Stacy Aand A.M.Portis, (1987) Phys.Rev. 36, 8309

Kheifets, A.S. and Veinger, A.I. (1990) Physica C 165, 491.

Kim, Y.W., deGraaf, A.M., Chen, J.T., Friedmanand E.J., Kim, S.H. (1972).Phys, .Rev.B 6,887

Kleiner, R., Steinmeyer, F., Kunkel, G. and Muller, P.Phys.Rev.Lett. (1992) 68, 2394.

Mahel, M.and Benacka, S. (1992a) Solid State Commun. 83,615.

Mahel, M., Adam, R., Darula. M., Chromik S. and Benacka, S. (1992b) Physica C, 202,243.

Moorjani, K., Bohandy, J., Adrian, F.J., Kim, B.F., Shull, R.D., Chiang, C.K., Swartzendruber L.J.and Bennet, L.H. (1987) Phys. Rev.36, 4036.

Moorjani, K., Bohandy, J., Kim B.F.and Adrian, F.J. (1990). Solid State Commun. 74,497.

Muller, K.A., Pomerantz ,M.and Knoedler, C.M. (1980) Phys.Rev.Lett.45,832.

Padam, G.K., Ekbote, S.N., Tripathi, R.B., Sharma, M., Suri, D.K., Rao, S.U.M., Rao, Nagpal, K.C.and Das, B.K. (1991).Solid State Commun,40,271.

Padam, G.K., Ekbote, S.N., Tripathi, M.R., Srivastava, G.P. and Das, B.K. (1999) Physica C 315, 45.

Padam,G.K., Ekbote, S.N., Sharma, M., Tripathi, M.R., Srivastava, G.P.and Das, B.K. (2006) Jap.J.Appl.Phys., 45,No.1A,84.

Padam, G.K., Arora N.K.and Ekbote, S.N. (2010) Materials Chem.Phys.123, 752.

Pakulis E.J.and Chandershekar G.V., (1989) Phys. Rev.B39, 808.

Panarina,N.Y.,Talanov, Y.I., Shaposhnikova, T.S., Beysengulov, N.R., Vavilona,G.,Behr, E., Kondrat, A., Hess, C., Leps, N., Wurmehl, S., Klinger, R., Kataev, V.and Büchner, B. (2010) Phys. Rev. B 81, 224509.

Pascher, N., Deisenhofer, J., Krug von Nidda, H.-A., Hemmida, M. , Jeevan, H., Gegenwart, S. P. and Loid, A. (2010) Phy. Rev.B 82, 054525.

Peric, M.,Rakvin,B. Prester, M., Brnicevic, N.,Dulcic, A(1988) Phys.Rev. B37, 522.

Portis, A.M., Blazey, K.W., Muller K.A.and J.G.Bednorz (1988)Europhys. Lett. 5, 467.

Pradhan, A.K., Roy, S.B., Chaddah, P., Chen, C. and Wanklyn, B.M. Phys.Rev.B (1995) 52, 6215.

Ramachandran, J.S., Huang, M.X., Bhagat, S.M.(1994) Physica C,234,173.

Roberts, J., Petit, P., Yildirim, T and Fischer, J.E. (1998-II) Phys. Rev B 157, 1226.

Saharangi , S.and Bhat, S.V.http://arxiv.org/fH/cond-Mat/papers/0511701.pdf

Sastry,M.D., I.Dalvi, A.G., Babu, Y., Kadam, R.M., Yakhmi, J.V. and Iyer, R.M. (1987) Nature, 330, 49.

Sugawara, K., Arai, N. Kouzuki, A., Ichimura, S., Naoi, H., Hotta K.and H.Hirose, (2000) Inter.J.Modern PhysicsB14, No.16, 1633.

Suss, J.T., Berlinger, W., Portis, A.M., Muller, K.A., Jeanneret B.and Martinoli, P. (1989) Solid State communications, 71,929.

Veinger, A.I., Zabrodskii A.G. and Tisnek, T.V. (1995).Supercond.Sci.Technol. 8,368.
Zakhidov, A.A., Imaeda, K., Ugawa, A., Yakushi, K., Inokuchi, H., Iqbal, Z., Baughman, R.H., Ramakrishana B.L.and Achiba, Y. (1991) Physica C 185,411.

6

Superconductivity in Nanoscale Systems

Meenakshi Singh[2], Yi Sun[1] and Jian Wang[1,2*]
[1]*International Center for Quantum Materials & State Key Laboratory for Mesoscopic Physics, School of Physics, Peking University, Beijing,*
[2]*The Center for Nanoscale Science & Department of Physics, The Pennsylvania State University, University Park,*
[1]*China*
[2]*USA*

1. Introduction

According to the BCS theory, charge transport in superconductors is carried out by paired electrons named 'Cooper pairs'. The distance between a pair of electrons is called the superconducting coherence length (ξ). If superconductivity is made possible by electrons 'holding hands' to form Cooper pairs, what happens to the superconductivity when one or more dimensions of the sample are smaller than the size of the cooper pair (ξ)? The answer to this question is of both fundamental interest and practical importance. The fundamental interest stems from the fact that in these reduced dimensionality systems, fluctuations play a very important role, even at low temperatures. In addition, because of the large surface to volume ratio, surface effects become important. Therefore, many novel phenomena not seen in bulk manifest in reduced dimensionality systems. From a practical standpoint, reduced dimensionality systems are becoming important because of rapid miniaturization of electronic circuits. With electronic devices getting smaller by the day engineers have to design smaller transistors. For comparison, the first chip had about 2,200 transistors on it whereas today, hundreds of millions of transistors can fit on a single chip. There are already computer chips that have nanometer sized transistors and future transistors will have to be even smaller. This makes understanding electronics in general and superconductivity in particular in reduced dimensionality samples very relevant.

Let us begin by introducing the specific scales involved and systems of interest. A superconductor with one, two or three dimensions smaller than ξ is in the quasi-two-dimensional (2D), quasi-one-dimensional (1D) or quasi zero dimensional (0D) regime respectively. For most conventional superconductors, ξ is of the order of microns. Therefore, systems falling in the 2D, 1D or 0D category are nanoscale systems. According to Hohenberg-Mermin-Wagner theorem (Hohenberg, 1967; Mermin & Wagner 1966), in these reduced dimensionality systems, fluctuations should destroy superconducting order even at low temperature. In 2D samples, the Berezinski-Kosterlitz-Thousless transition occurs, enabling superconducting order to exist at low temperature. However, the existence, limits

*Corresponding author

and nature of superconductivity in 1D remain a subject of interest. Fortunately, the advent of advanced synthesis techniques like electron beam lithography has enabled scientists to probe the nature of superconductivity in quasi-1D samples. In these systems, superconductivity exhibits many novel phenomena that shed light on the nature of superconductivity. In addition, the aforementioned miniaturization of electronic devices is opening up new horizons for applications of 1D nanowires, and need for understanding superconductivity at this scale. These quasi-1D systems therefore form the basis of the studies described in this chapter.

In this chapter, we hope to provide a comprehensive picture of superconductivity in nanoscale systems starting from synthesis techniques currently in use to examples of the new physics these systems are helping in uncovering. The systems that we focus on are quasi-1D nanowires and nanobelts. In the synthesis section, we will give a brief overview of the common techniques currently in use for making nanoscale samples, especially nanowires. The synthesis and manipulation of these samples can be extremely challenging. Many of the more advanced techniques are continuously evolving and have merited chapters and books on their own. We will include references to some of these more detailed descriptions of the techniques. In the section 3, we will discuss some unexpected discoveries in quasi-1D nanowires and nanobelts. Some of the novel phenomena discussed include macroscopic quantum phenomena, extremely long range proximity effects, the counter-intuitive anti-proximity effect and superconductivity in new materials amongst others. We hope to provide a glimpse of the measurements performed as well as some physical insight into the mechanism behind these phenomena. Finally and most importantly, we will discuss the potential of these new discoveries to be used in technology. The potential for application has not been the least of the reasons why superconductivity continues to capture the imagination of scientists and engineers alike even 100 years after its discovery.

2. Synthesis

The first controlled fabrication of nanowires started in the late 1980s. Early experiments used a top-down approach where a large piece of the material was taken and whittled down using an electron beam to the required size (Knoedler, 1990). Most approaches to nanowire synthesis used today are however bottom-up: where the nanowire is assembled from smaller particles.

2.1 Electrochemical synthesis

Electrochemical synthesis is an inexpensive and versatile method that yields high quality nanowires in a large number. Its limitation is that it is only useful for materials that can be electrochemically deposited from solution (like metals) and its application to semiconductor and non-elemental nanowires is possible but limited. It is a bottom-up approach in which wires are assembled atom by atom from an electrolytic solution with or without an applied potential. Electrochemical synthesis can further roughly be categorized into restrictive template based electrodeposition and active template assisted electrodeposition (Bera, 2004).

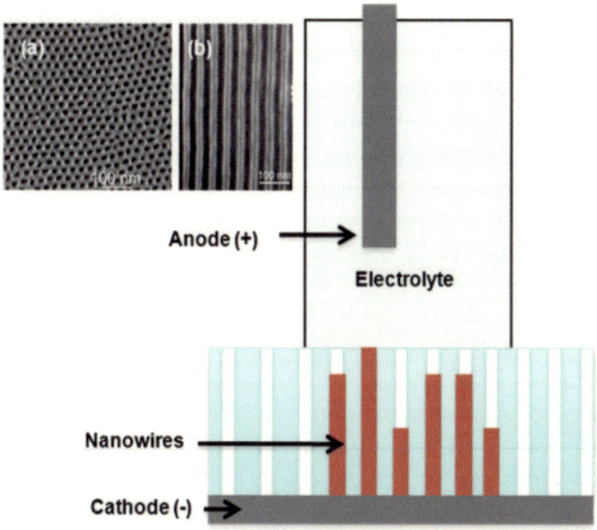

Fig. 1. A schematic of the setup used to electrodeposit nanowires into the pores of a tracke-etched polycarbonate or anodized oxide (AAO) membrane. The inset shows scanning electron micrographs of (a) a cross section of an AAO membrane and a (b) longitudnal section of an AAO membrane. The regularly arranged pores meant for deposition of nanowires can be seen.

For restrictive template based electrodeposition, commercially available track etched polycarbonate or anodic aluminum oxide (AAO) filtration membranes are used as templates. The pores in these membranes are created by irradiating the membrane with energetic heavy ions and anodizing in acid respectively (Lee, 2006). The resultant pores are parallel to each other and perpendicular to a cross section of the membrane (insets of figure 1). The length and diameter of the nanowire can be controlled by changing the thickness and the pore size of the membrane. The typical pore density varies from 10^8 to 10^{11} pores/cm^2 which even with low filling fractions yields a large number of nanowires. A schematic of the electrochemical cell typically used is shown in figure 1. For use in electrodeposition, silver or gold is thermally evaporated on one surface of the membrane closing off the pores on that side. This evaporated layer acts as the cathode for electrodeposition. The membrane is then immersed in an electrolyte containing cations of the material(s) to be electrodeposited. A counter electrode and sometimes a reference electrode are put in the circuit and a potential is applied. Since the membranes are insulating, the metal evaporated on one surface acts as the cathode through which the current flows. As the open side of the template is exposed to the electrolyte, positively charged ions move towards the metal at the bottom of the pore. When they reach the metal, they accept electrons and get reduced to their metallic form. When the pores are filled with material, the deposition is halted. The membrane can be dissolved away and the nanowires are released for further use. Variations of this process are used to deposit multilayered nanowires and nanotubes. Some metals like gold can be deposited without an applied potential in an electroless process (Menon & Martin, 1995). In another variation, the pores of the membrane are wet by a precursor followed by thermal decomposition leading to formation of either nanowires or nanotubes in the pore (Nielsch et al., 2005).

The formation of nanostructures in the active template-based synthesis results from preferential electrodeposition of atoms or molecules at hole and defect sites on the electrode surface. Once the initial layer of atoms is crystallized on the defect site, subsequent atoms also preferentially deposit at the site leading to formation of different nanostructures. The preparation of the electrode surface plays a major role in determining the morphology and structure of the electrodeposited nanostructure. Other nanosynthesis techniques like photolithography have been used in conjunction with this technique. Photolithography is used to create nanostructures on the electrode surface into which the required material is then electrodeposited (Yang et al., 2008).

2.2 Lithography

Lithography translates to 'stone writing' and originally referred to the art of writing text on stone. Modern lithographic techniques find a wide variety of uses in the printing process. Our foci of interest for this section are, however, micro-lithographic processes like photo-lithography and electron beam lithography (EBL) with which most top down nanofabrication processes start. Photo-lithography is the more popular of the two processes and is used for manufacturing of semiconductor microchips. EBL, although more expensive, offers significantly improved resolution, sometimes going down to a few nanometers. Figure 2 shows a schematic of the lithography process.

The substrate (most often doped silicon coated with an insulating layer of silicon oxide or silicon nitride) is spin coated with a photo resist or e-beam resist. The choice of resist depends on the resolution required, spacing of features and thickness of metal to be evaporated post lithography. The resist is baked for hardening. In case of photo-lithography a hard mask is used with the required pattern made on it. In case of EBL, a 'soft' mask drawn using computer software like L-Edit is used. The areas within the pattern are exposed to either UV light for photolithography or an electron beam for EBL. The exposure time and beam-current used depend on the resolution required and the feature size. The exposure of the resist causes it to undergo a chemical reaction. The substrate is then soaked in a developer. In case of positive resist (as depicted in figure 2), the parts of the resist exposed to the beam and chemically altered are dissolved by the developer. For negative resist, the parts not exposed to the beam are dissolved away. The substrate is now left with resist everywhere except for places where the pattern should be. The required material is now evaporated on the substrate. This material fills the blanks left by the dissolved resist and also covers up the remaining resist. The substrate is then taken for the 'lift-off' process where it is dipped in a solvent that dissolves away the remaining resist. The material covering the resist is stripped away with the resist, and only the material in the pattern remains on the substrate. Lithography has the advantage of being an extremely flexible technique that can be used to create a large variety of shapes and sizes of nanostructures. It is also a technique most likely to become scalable to uses in chip design and manufacture in the future. The disadvantage is that lithography, especially EBL tends to be very expensive and the resulting nanostructures are most often granular (this depends on the evaporation conditions). A more in-depth description of lithography can be found in (Madou, 2002).

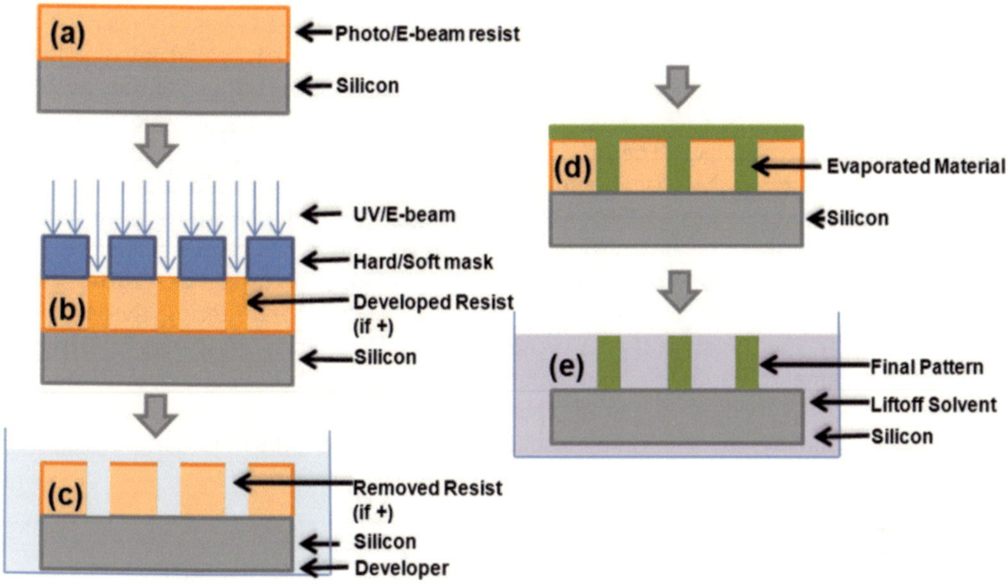

Fig. 2. A schematic of the lithography process for making nanostructures (a) a layer of resist is spin-coated on the substrate, (b) the resist is exposed to either UV radiation in case of photolithography or an electron beam in case of electron-beam lithography through a mask which has the required pattern cut out from it, (c) the substrate with the resist is soaked in a developer which removes the parts of the resist that were exposed to radiation earlier (in case of pisitive resist) or parts that were not exposed to radiation earlier (in case of negative resist), (d) material is evaporated onto the substrate with the resist acting as a mask, (e) the substrate is soaked in a solvent that dissolves the remaining resist and also the layer of evaporated material on the resist. This 'lift-off' process only leaves behind the material evaporated directly on the substrate in step (d).

2.3 Chemical vapor deposition

The technique of Chemical Vapor Deposition (CVD) is mostly used for the synthesis of thin films. The process is very versatile and allows for growth of amorphous, epitaxial, monocrystalline and polycrystalline films. The CVD equipment consists of a tube-like, chamber in which the substrate for synthesis is placed, a mechanism to heat the substrate, a gas handling system which is used to introduce vapors into the chamber and an exhaust from which the waste vapors escape.

For the synthesis, a precursor dissolved in a solvent or in the form of an emulsion is vaporized and introduced into the chamber where some of the precursor mixture deposits on the substrate. The substrate is then heated causing the precursor mixture to decompose and the solvent to evaporate away. A layer of the required precursor is thus left on the substrate. If a metal organic compound is used as the precursor, the method is referred to as Metal-Organic Chemical Vapor Deposition (MOCVD). A slight variation of this technique is also used for nanowire and nanotube growth under the name of Vapour Liquid Solid method (VLS). This method was developed in 1964 for the growth of Silicon whiskers (Wagner & Ellis, 1964). In the first step, a layer of a solid impurity such as gold is thermally evaporated on the substrate. The substrate is then heated so that liquid droplets of an alloy of the impurity and the substrate form on the surface of the substrate. In the next step, like the CVD method, a vaporized solution of the desired material is introduced into the chamber. The liquid droplets formed on the substrate in the earlier step act as catalysts in the reaction to deposit atoms of the required material on the substrate. By a continuation of this process, a nanowire on the substrate is formed. A schematic of this process is shown in figure 3. This technique has recently been used for successful synthesis of topological insulator nanoribbons (Peng et al., 2010). Using lithography in conjunction with this technique also allows for ordering the nanowire growth. Recently a variation of the CVD method has been used to grow very large aspect ratio (length:diameter) nanowires in microstructured optical fibers (Sazio et al., 2006). The advantage of this method is its adaptability to a wide range of materials. The disadvantages mostly involve safety and contamination as poisonous and pyrophoric (ignite in contact with air) precursors are used.

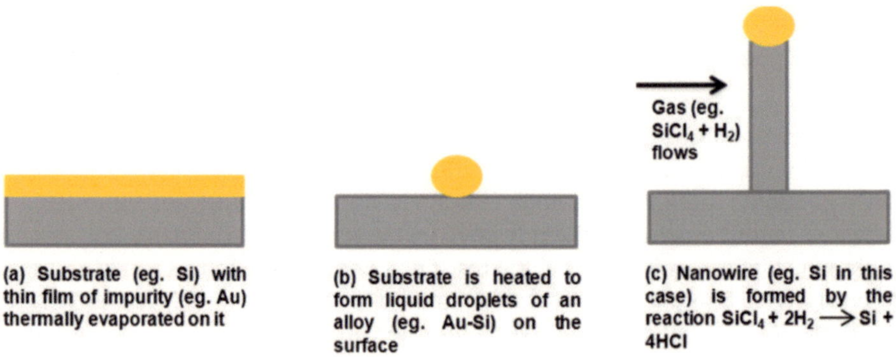

Fig. 3. A schematic of the chemical vapor deposition process modified to the vapor-liquid-solid method for nanowire growth.

2.4 Molecular beam epitaxy

Molecular Beam Epitaxy (MBE) is a refined form of thermal evaporation. In MBE, directed neutral beams of atoms or molecules of a substance being evaporated get deposited on a substrate. The major difference from thermal evaporation is that the deposition is extremely slow, allowing for atomic layer by atomic layer epitaxial growth. The samples grown by MBE are high quality single crystals. The very slow growth rate requires the evaporation to be done in ultra-high vacuum to keep the contamination levels in check. The apparatus to fulfill these requirements is considerably complicated (schematics of a typical MBE chamber are shown in figure 4). In addition, the substrates for sample growth also have to be of very high quality. This technique generates extremely smooth and thin samples with a high degree of compositional control and crystal perfection. The payoff is its high cost. The method is largely used in the synthesis of thin films but, can also have applications to nanowire growth (section 3.6). Further details of this method of sample synthesis can be found in (Arthur, 2002).

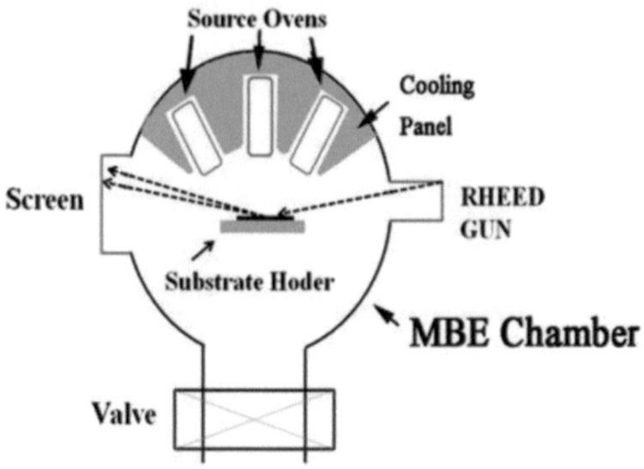

Fig. 4. A molecular beam epitaxy chamber.

In addition to these commonly used techniques, several other ingenious methods have been used for nanostructure synthesis. These include DNA (Stoltenberg & Woolley, 2004) and carbon nanotube template nanowire synthesis (Bezryadin et al., 2000) and microbial nanowires (Reguera et al., 2005) amongst others.

3. Novel phenomena – What they imply and their potential applications

When a material transits from a normal to a superconducting state, electrons with opposite spins pair up. As mentioned in the introduction, this pair is called a 'Cooper pair'. Once part of a Cooper pair, the two electrons can be described by a single quantum mechanical wave function $\psi = |\psi|e^{i\varphi}$. The spatial extent of this wavefunction, or the 'size' of a cooper pair, is known as the coherence length (ξ) and φ is the phase of the wavefunction. If the superconducting state is made possible by electrons 'holding hands' to form Cooper pairs, what happens to the superconductivity when the diameter of the nanowire the Cooper pairs are supposed to go through is smaller than ξ? Do the Cooper pairs break up and is superconductivity destroyed in these quasi-1D systems? This dilemma lies at the root of why superconductivity in quasi-1D systems is different. It is also responsible for the plethora of novel phenomena seen on studying superconductivity in quasi-1D wires as described below.

3.1 Phase slips – Thermally activated and quantum

In the superconducting state, the phase φ is spatially coherent. This means that if the phase at any one point is known, the phase at any other point can be predicted. Fluctuations (thermal and quantum) lead to loss of phase coherence in superconducting samples from time to time. The region where the coherence is lost becomes temporarily normal (φ becomes ill defined). The spatial extent of such a region is given by ξ. For 3D or 2D samples, these normal regions without phase coherence do not affect the transport measurement as the charge carriers can bypass them. In a quasi-1D sample however, since the diameter (d) < ξ, the normal region encompasses the entire cross section of the wire and cannot be bypassed. Therefore, phase slips result in a loss of superconductivity in quasi-1D systems. Mathematically, this can be understood using Josephson relation:

$$V = \frac{d(\Delta\varphi)}{dt} \tag{1}$$

Here, V is the voltage across the system, t is the time, φ is the phase, and Δφ is the phase difference across the system. In a phase slip event, the phase at some point in the sample becomes ill defined. When it recovers, the value of Δφ is changed by 2π causing a finite voltage V to appear across the system and hence, a loss in superconductivity. So, in a phase slip event, the system essentially goes from a superconducting state with some value of Δφ to a superconducting state with a value of Δφ different from the previous value by 2π. To go from one state to another, the system needs to cross a potential barrier (figure 6). The energy to cross the barrier can be provided by thermal energy (thermally activated phase slips) or by quantum fluctuations (quantum phase slips). Extensive experiments attempting to see both kinds of phase slips have been done.

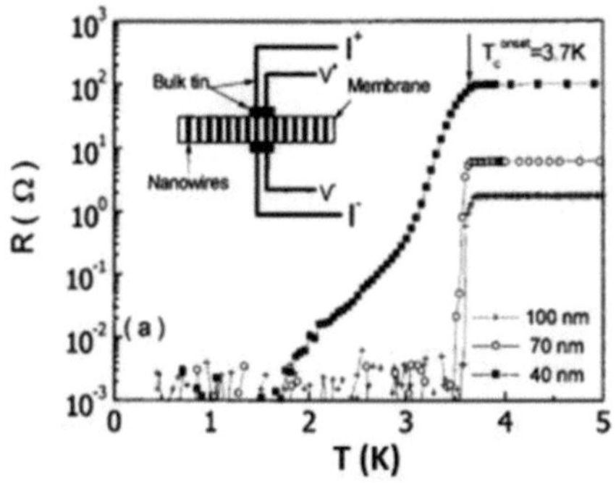

Fig. 5. (Tian et al., 2003) Resistance vs. Temperature measurements on tin nanowire arrays embedded in track-etched polycarbonate membranes and squeezed between bulk tin electrodes. The schematics of the measurements are shown in the inset. The measurements shown here are performed on three different arrays which the wires are 40 nm, 70 nm and 100 nm in diameter. For the 40 nm diameter tin nanowires rounding of the superconducting transitions and a long resistance tail can be seen.

Transport measurements of resistance (R) as a function of temperature (T) of quasi-1D samples differ from bulk samples in two ways. The first is a rounding of the superconducting transition and the second is the presence of a superconducting tail in the quasi-1D samples. These two features can be seen to appear in figure 5. The figure shows R vs. T measurements performed on single crystalline tin nanowire (SnNW) arrays embedded in polycarbonate membranes contacted with bulk tin electrodes (Tian et al., 2003). A schematic of the measurement is shown in the inset of figure 5. Measurements are performed on 3 different sets of nanowires of diameters 100 nm, 70 nm and 40 nm. The transition for the 100 nm and 70 nm diameter nanowires is sharp and the resistance falls from the normal state value to zero in a temperature interval of 90 mK. The transition for the 40 nm nanowire on the other hand, is rounded and also has a long resistive tail stretching from 1.8 K to 3.7 K, a range of 1.9 K. These two features are attributed to thermally activated (TAPS) and quantum phase slips (QPS) respectively.

In TAPS, the system crosses the energy barrier between the superconducting states whose values of Δφ differ by 2π with the help of thermodynamic fluctuations (figure 6). The resistance due to TAPS below the critical temperature of the superconducting transition (T_c) can be written as (Langer & Ambegaokar, 1967; McCumber & Halperin, 1970):

$$R_{TAPS} \propto e^{-\Delta F/k_B T} \tag{2}$$

where ΔF is the free energy barrier between the two states and the proportionality constant is related to the attempt frequency and the temperature. For low excitation currents $\Delta F = \sqrt{2}\sigma H_c^2 \xi / 3\pi$ where σ is the cross sectional area, H_c is the critical field and ξ is the coherence length. As the temperature decreases, the size of the free energy barrier relative to $k_B T$ increases, leading to a rapid reduction in the thermal-activation rate. At very low temperature this rate becomes negligible. TAPS is therefore, only important close to T_c, and is associated with the rounding of the superconducting transition seen in the SnNW. However, it cannot explain the resistive tail that persists to low temperatures. To explain the resistive tail, some other process by which the phase can relax has to be taken into account. QPS, or the tunneling through the energy barrier (figure 6) is the dominant process at low temperatures (Giordano, 1988, 1991; Giordano & Schuler 1990). A heuristic argument (Giordano, 1994) suggests that the resistance given by the QPS will follow the same form as R_{TAPS} except for the energy scale being given by $\hbar\omega = \hbar / \tau_{GL}$ instead of $k_B T$. Here, τ_{GL} is the Ginzburg Landau relaxation time given by $\tau_{GL} = \pi\hbar / 8k_B(T_c - T)$. Therefore,

$$R_{QPS} \propto e^{-a\Delta F \tau_{GL}/\hbar} \tag{3}$$

where α is possible numerical factor. QPS has fascinated experimentalists and theorists alike for almost 2 decades now because of its tremendous implications (Altomare et al., 2006; Lau et al., 2001). The observed resistance tail is an example of macroscopic quantum tunneling because a macroscopic, measurable quantity (resistance) is changing in response to a quantum tunneling event. It has been suggested that QPS in nanowires can be used to realize a superconducting quantum bit that can have much lower 1/f noise than existing quantum bits using Josephson junctions (Mooij & Harmans, 2005). These quantum bits not using Josephson junctions can form a new basis for superconducting quantum computing.

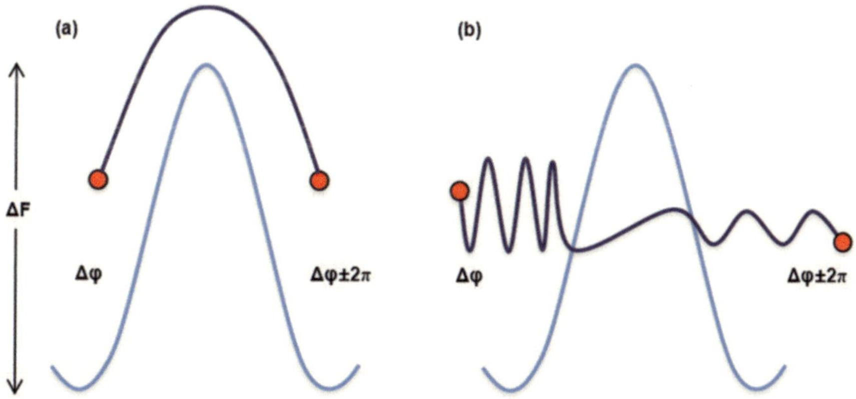

Fig. 6. A representation of the (a) the thermally activated phase slip wherer thermal energy allows the system to go over the energy barrier separating two superconducting states, (b) quantum phase slips that allow the system to tunnel from one superconducting state to the other even at low temperatures. The two superconducting states in question have the phase difference between the two ends of the superconductor (Δφ) different by 2π.

3.2 Long range proximity effect in ferromagnetic nanowires

In a singlet (s-wave) superconductor, the spins of the electrons making a Cooper pair are antiparallel (the spin part of the wavefunction is given by $|\uparrow\downarrow\rangle - |\downarrow\uparrow\rangle$). The spins of electrons in a ferromagnet, on the other hand, are aligned parallel to each other. This incompatible spin order between singlet superconductors and ferromagnets makes for new and interesting physics when the two are placed in contact with each other. When a superconductor is placed in contact with a normal metal, the Cooper pairs in the superconductor tunnel into the normal metal, making it partially superconducting. This is known as the superconducting proximity effect. The distance to which Cooper pairs survive in the non-superconducting material is the range of the proximity effect. This range, for a bulk normal metal can be ~ 1μm. For a bulk ferromagnet however, the incompatible spin order ensures that the singlet Cooper pair cannot survive beyond a few nanometers. This expectation has been confirmed in macroscopic (Fe, Ni)–In junctions (Chiang et al., 2007) and submicrometre Ni–Al structures (Aumentado & Chandrasekhar, 2001) where the spatial range of the proximity effect is limited to ~ 1 nm. In mesosopic ferromagnet-superconductor hybrid structures, a surprisingly long range proximity effect has been found (Bergeret et al., 2005; Giroud et al., 1998; Keizer et al., 2006; Pena, V. et al., 2004; Petrashov et al., 1999; Sosnin et al., 2006). In particular, a recent experiment detected super-current in half-metallic ferromagnet CrO_2 thin film sandwiched between two singlet superconducting electrodes separated by 1 μm (Keizer et al., 2006). To account for the long range proximity in these systems, the superconductivity was attributed to the spin triplet (spin wavefunction given by one of $|\uparrow\uparrow\rangle$, $|\downarrow\downarrow\rangle$ *or* $|\uparrow\downarrow\rangle + |\downarrow\uparrow\rangle$ which do not preclude parallel spins) rather than the spin singlet state. The formation of the spin triplet state requires the presence of a region of non-collinear magnetization. This region, in the case of an experiment on a Co thin film was provided by a thin weakly superconducting CuNi or PdNi layer between the superconducting Nb and the ferromagnetic Co film (Khaire et al., 2010).

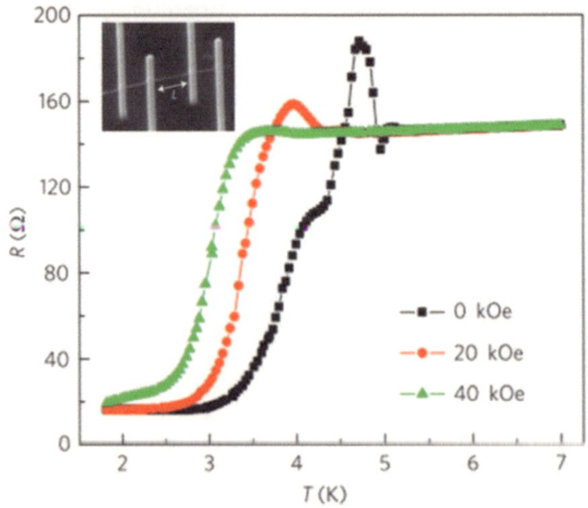

Fig. 7. (Wang et al., 2010a) Resistance vs. Temperature measurements for a 40 nm diameter, 1.5 μm long cobalt nanowire at different applied fields. The applied fields are perpendicular to the wire. The inset shows an scanning electron micrograph of the cobalt wire contacted with 4 focused ion beam deposited superconducting tungsten electrodes.

The occurrence of this long-range proximity effect was also reported in ordinary, hard ferromagnetic crystalline nanowires in standard 4-probe measurements (Wang et al., 2010a). The Co and Ni nanowires (CoNW and NiNW) were synthesized using template based electrodeposition into the pores of a track-etched polycarbonate membrane. The CoNW were single crystalline while the NiNW were polycrystalline. The nanowires were released from the membrane and dispersed on a Si/SiN substrate and contacted by a focused ion beam (FIB) system with superconducting tungsten (W) electrodes. The R vs. T measurements at different applied fields (H) on a 40 nm diameter, 1.5 μm long CoNW are shown in figure 7 (the inset shows the measurement geometry). A large resistance drop (to 11% of the normal state value) at the T_c of the W electrode suggests a very long range (~ 500 nm) proximity effect in the CoNW. The range of the proximity effect is smaller (~ 375 nm) in thicker (80 nm) diameter nanowires. The expected singlet coherence length in Co nanowires is given by $\xi_F = \sqrt{\hbar D / k_B T_{Curie}}$ (Keizer et al., 2006), where T_{Curie} is the Curie temperature and $D = (1/3)v_F\, l$ is the electron diffusion constant (here, v_F is the Co Fermi velocity ~10^8 cm s^{-1} and l is the mean freepath). The experimentally measured resistivities yield a singlet coherence length of 3.3 nm for the sample shown in figure 7, which is much smaller than the observed coherence length of 500 nm. The superconductivity in this system also may be triplet in nature. It has been suggested that the region between the W electrodes and the CoNW is damaged by the FIB process and may thus be largely inhomogeneous and act as a seeding ground for triplet pairing. Another explanation may be that the surface of the CoNW and NiNW is not ferromagnetic and the superconducting proximity effect is only on the non-ferromagnetic surface of the nanowires. In (Konschelle et al., 2010) it is suggested that a phase sensitive measurement, experimentally testing the Josephson current-phase relationship can be used to discriminate between the two scenarios.

Other than the extremely long-range proximity effect, another striking feature visible in figure 7 is the very large resistance peak (25% of the normal-state resistance) found between 5 K and 4.5 K, just above the temperature of the superconducting resistance drop. This 'critical peak' is preceded by a small dip on the high-temperature side of the peak. Measurements in both warming up and cooling down scans show an absence of hysteretic behavior. Both the peak and the superconducting drop are stable on cycling the sample to room temperature. This peak does not conform to charge imbalance peaks typically seen in superconducting mesoscopic systems since a very large magnetic field is required to destroy it. Charge imbalance peaks on the other hand are destroyed by very small applied fields (Santhanam et al., 1991). Another mechanism used to explain resistance peaks in Fe-In junctions is spin accumulation (Chiang et al., 2007). In that case, however, the absolute peak value (ΔR) was many orders of magnitude smaller ($10^{-8}\Omega$), as was the relative change from the normal-state resistance ($\Delta R/R$ ~ 0.05%). Furthermore, spin accumulation would assume that the induced superconductivity is singlet in nature, which seems inconsistent with the long range proximity effect. A satisfactory explanation of this peak is yet to be found.

The long range proximity effect in ferromagnets offers the possibility of combination of the zero-resistance supercurrents of superconductors with the spin alignment of ferromagnets. This has tremendous implications for spintronics (Eschrig, 2011). Superconducting spintronic devices offer elements of non-locality (if the superconducting coherence length is larger than the nanoscale device), coherence (electrons in a cooper pair are coherently coupled even if they go in separate electrodes) and entanglement (stemming from coherence), all of which are essential for quantum computing. The availability of triplet supercurrents may also revolutionize the field of spintronics with new spin valves, pumps, switches, transistors and filters.

3.3 Anti-proximity effect

When a superconductor is placed in contact with a normal metal, signs of superconductivity appear in the normal metal. This 'proximity effect' is a much documented and well-studied phenomenon. However, a number of recent experiments have reported an unexpected 'anti-proximity effect' (APE) in zinc nanowires (ZnNW) contacted with bulk superconducting electrodes (Chen et al., 2009, 2011; Singh et al., 2011; Tian et al., 2005, 2006). In the original experiment, (Tian et al., 2005) ZnNW 6 μm in length embedded in track-etched polycarbonate membranes were squeezed between superconducting bulk electrodes for two-electrode transport measurements (the schematics of the measurement are similar to those shown in the inset of figure 5). In this configuration, the resistance of the long nanowires dominates the overall series resistance, and the contact resistance is minimized as in a 4-electrode measurement. In 70 nm diameter nanowires contacted with bulk tin (Sn) electrodes, the superconducting transitions of Sn at 3.7 K and of the ZnNW at 1 K were both seen, consistent with expectation. However in 40 nm diameter nanowires, the transition for Sn was seen but the superconducting transition for Zn was absent. When the Sn electrodes were driven normal by a magnetic field of 300 Oe, the superconducting transition of Zn reappeared. The APE in the 40 nm Zn wires was replicated with indium (In) and lead (Pb) electrodes (Tian et al., 2006a). The experiment with In electrodes showed that the APE in the 40 nm Zn wires was switched off precisely when the magnetic field was increased above the critical field of In (Tian et al., 2005). With Pb electrodes, the strength of APE was weaker and there was no obvious suppression of superconductivity in the ZnNWs in resistance measurements as a function of temperature or magnetic field. The APE however, could be detected in the critical current (I_c) of the Zn wires. Specifically the I_c of a Pb/Zn/Pb sample showed a dramatic increase when the magnetic field was increased towards the critical field (H_c^{Pb}) of Pb. Only at fields higher than H_c^{Pb} did the I_c of the ZnNWs show the 'normal' behavior, namely a decrease with increasing field (figure 8). Similar I_c behavior vs. magnetic field was also confirmed in aluminum nanowire (AlNW) arrays embedded in anodic aluminum oxide membrane contacted with In electrodes (Singh et al., 2011) and in four electrode measurements on e-beam assisted evaporated granular ZnNWs (Chen et al., 2009, 2011). The nanowires and the electrodes for this experiment were evaporated in a single step using an e-beam lithographically fabricated mask. The authors attributed the increase in I_c to the creation of quasi-particles as the bulk electrodes are driven toward the normal state by the external magnetic field. In all these experiments, a magnetic field was used to see the APE. It is therefore natural to question whether the APE is caused by the magnetic field. To answer this question an experiment was performed where two single crystalline AlNW grown using electrodeposition (Singh et al., 2009) of the same diameter (70 nm) and length (2.5 μm) were contacted with superconducting W and normal platinum (Pt) electrodes using the FIB process. The I_c of the wire contacted with the superconducting electrode was found to be significantly smaller than the I_c of the wire contacted with the normal electrode (figure 9) indicating suppression of superconductivity of the AlNW because of the bulk superconducting electrode. As there were no applied magnetic fields in this experiment, this served to definitively decouple the APE seen in the 70 nm wire from negative magnetoresistance caused by a weak applied magnetic field seen in other experiments (Herzog et al., 1998; Rogachev et al., 2006; Xiong et al., 1997). The phenomenon also shows a strong dependence on the diameter and length of the sample and the nature of the measuring electrodes. It is not seen in 200 nm AlNW and is very weak in 50 μm long AlNW. It is also weaker in ZnNW with Pb electrodes compared to ZnNW with In electrodes.

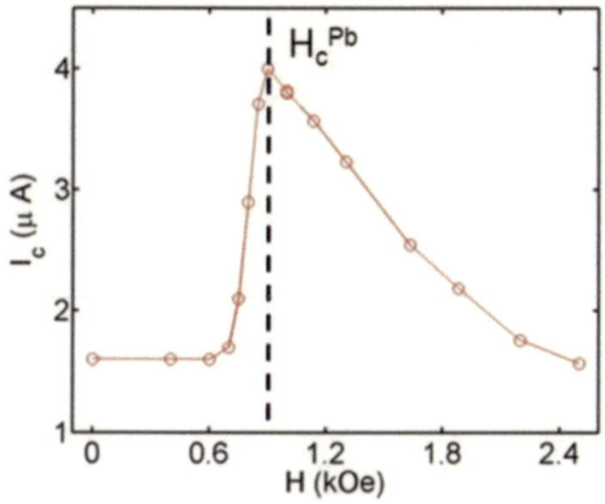

Fig. 8. (Tian et al., 2005) An ehancement in the critical current (I_c) of a zinc nanowire array squeezed between superconducting lead electrodes is seen when the magnetic field is increased toward the critical field of the bulk superconducting lead electrodes contacting the nanowires. The critical current peaks at the critical field H_c^{Pb} of the electrodes.

A theoretical model applicable to this system was given by Fu et al. (Fu et al., 2006) generalizing the resistively shunted Josephson junction to include superconducting nanowires. In this model the nanowire array connected to the two bulk superconducting electrodes is modeled as a nanowire in series with two resistors and in parallel with a capacitor. When the electrodes are normal, the resistors in series provide a medium for dissipation of quantum phase slips, thus stabilizing the superconductivity. When the electrodes are superconducting, this dissipation path disappears and the wire becomes normal. A similar explanation was proposed to understand the results of the experiments on granular Zn nanowires. Although the model agrees with the results qualitatively, it does not explain how the strength of the effect depends on the contacting electrode in AlNWs or ZnNWs. In addition to the material of the electrodes, other important parameters in these experiments are the 'characteristic' diameter and the length of the nanowires that define the presence or absence of APE. It is difficult to obtain quantitative predictions for these quantities from this model. A more quantitative mechanism using the time independent Ginzburg-Landau equations was proposed by Vodolazov (Vodolazov, 2007). The qualitative behavior of the system is reproduced at temperatures close to T_c and the authors expect traces of this behavior to exist at low temperatures also. The mechanism of Vodolazov uses the fact that the diameters of the wires are smaller than the superconducting coherence length to model them as a 1D system. The coherence length for the AlNWs in these experiments can be estimated using the value of $\xi_0.l$ (where l is the mean free path) as 4×10^{-16} Ω m^2, the dirty limit coherence length $\xi(0) = 0.855(\xi_0 l)^{0.5}$ and 1600 nm as the value for ξ_0. $\xi(0)$ is estimated to be ~ 50 nm. In comparison, APE was seen 70 and 80 nm diameter AlNW but, not in 200 nm diameter nanowires. In ZnNWs, the $\xi(0)$ was estimated to be ~ 150 nm and the APE was seen in 40 nm diameter nanowires but not in 70 nm diameter nanowires (Tian et al., 2005). The model also predicts a weakening or absence of the APE in nanowires with length L > Λ_Q (the charge imbalance length). The charge imbalance length of the Zn NWs was calculated to be 22 µm (Vodolazov, 2007) and that of AlNW to be 19 µm (Singh et al., 2011), and this was used to explain the relatively weak APE in long ~ 35 µm Zn and ~ 50 µm AlNWs. Lastly, this mechanism uses an external magnetic field to model the critical current enhancement seen in the APE and the APE has now been seen in the absence of an external magnetic field.

The APE is a counterintuitive phenomenon that challenges our understanding of superconductivity. The phenomenon is qualitatively consistent with the Vodolazov model, but the presence of the phenomenon without an applied field shows that further theoretical work is needed to gain a more complete and quantitative understanding of the APE. A complete picture of the mechanism behind this behavior will offer great insight into superconductivity in nanoscale systems and perhaps open doors to new applications.

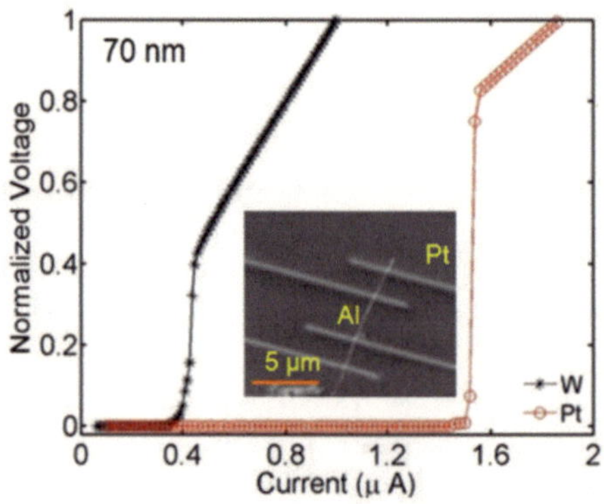

Fig. 9. (Singh et al. 2011) Normalizes voltage vs. applied current for two 70 nm diameter, 2.5 µm long single aluminum nanowires at a temperature of 0.1 K. One of the nanowires is measured using normal focused ion beam deposited normal Pt electrodes and the other is measured using superconducting W electrodes. The inset shows a scanning electron micrograph of the aluminum nanowire contacted with the normal platinum electrodes. The nanowire contacted tungsten also looks the same under a scanning electron microscope.

3.4 Mini-gap state

In superconducting nanowires contacted with superconducting electrodes, the counterintuitive APE is seen. What happens when a normal nanowire is contacted with superconducting electrodes? In this situation, in addition to the expected and well-studied phenomenon of the proximity effect, some surprising new phenomena are uncovered. In a recent experiment (Wang et al., 2009a), crystalline gold nanowires (AuNW) 70 nm in diameter were contacted with superconducting W electrodes using FIB. As expected, the superconducting proximity effect caused the short wires (1 µm and 1.2 µm long) to go completely superconducting whereas the longer nanowires (1.9 µm) retained some residual resistivity to the lowest measured temperatures. This gives the range of the proximity effect to be ~ 600 nm. A surprising feature however, was that the superconducting transition of the 1.2 μm long wire occured in two steps at two distinct T_cs (figure 10). In the R vs. H measurements two distinct H_cs were also seen. The two-step drop in resistance is suggestive of two distinct transitions. The higher T_c and H_c are almost same as the corresponding values for the W electrodes. The results are interpreted with proximity induced superconductivity in the nanowire with a small superconducting "mini-gap" δ near the centre of the wire which is different from the gap elsewhere in the wire. The gap near the electrodes in the wire corresponds to the bulk W gap (Δ). This "mini-gap" state is a novel superconducting state not seen in bulk samples.

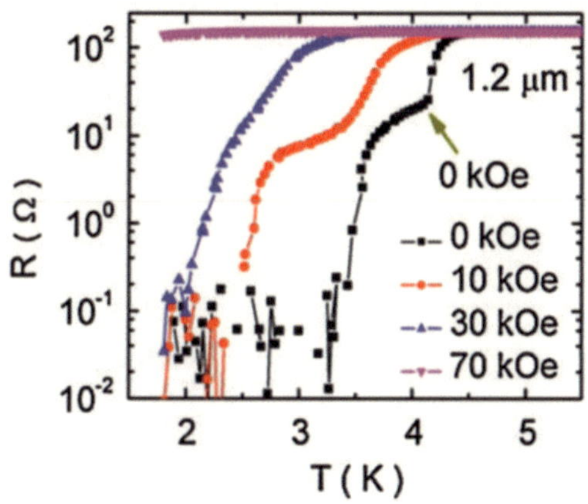

Fig. 10. (Wang et al., 2009a) Resistance vs temperature plots for a 70 nm diameter, 1.2 μm long single crystaline gold nanowire at different magnetic fields applied perpendicular to wire axis. The gold nanowire is contacted with superconducting tungsten electrodes using a focused ion beam system and the measurement shown has been done in the standard four-probe configuration. The superconducting transition can be seen to occur in two steps at two distinct 'critical' temperatures.

The R vs. H measurement of a 1.0 μm long nanowire and 1.2 μm nanowire are shown in figure 11 (a) and 11(b) respectively. A small resistance valley is more apparent in the 1.2 μm long nanowire. This resistance valley is associated with the mini-gap state. The details of the temperature evolution of the mini-gap in the R vs. H measurements can be seen in the magnified graph in figure 11 (c). The H_c of the mini-gap is ~ 3000 Oe in comparison to the larger transition H_c corresponding to the W electrode H_c of ~ 80000 Oe (figure 11(b)) . The curves have been offset for clarity. At 2.4 K, there is no sign of the mini-gap. At 2.5 K, the resistance suddenly jumps and then quickly drops back to zero at 2.5 kOe. At 2.6 K, this magneticfield-symmetric resistance fluctuation becomes more clearly developed and the baseline of the fluctuating resistance at fields above the mini-gap region is seen to be smoothly increasing with field. With increasing temperature the mini-gap valley continues to shrink in width, disappearing at T = 4 K. Evidence for the mini-gap in proximity structures was also revealed in recent scanning tunnelling microscopy studies (Gupta et al., 2004; Le Sueur et al., 2008; Vinet et al., 2001) and another transport measurement (Lucignano et al., 2010). In addition to the mini-gap state, resistance plateaus can be seen in the R vs. H measurements in the transition region (figure 11(a) and 11(b)).

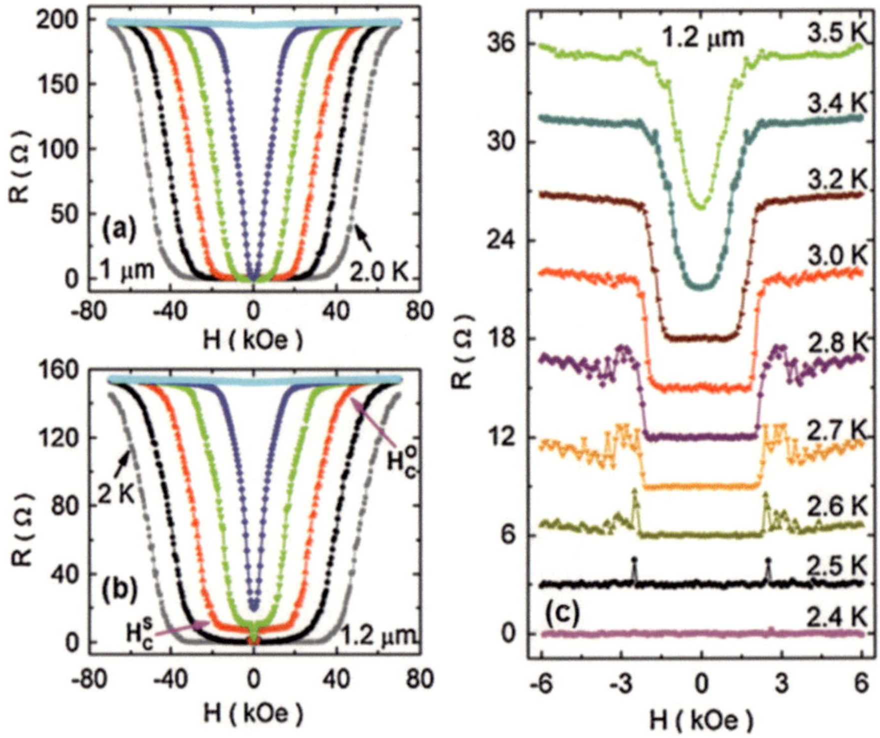

Fig. 11. (Wang et al., 2009a) (a) Magnetoresistance of the 1 μm gold nanowire, from bottom to top, at 2.0 K (gray), 2.5 K (black), 3.0 (red), 3.5 K (green), 4.0 K (blue), and 5.5 K (cyan). (c) Close-up view of (b) near zero magnetic field at different temperatures for the 1.2 μm long nanowire. The curves are offset for clarity; except for 3.5K, the resistance in all plots is zero at zero field. The mini-gap state seen as the small resistance valley is apparent in the 1.2 μm long nanowire but, not in the 1 μm long nanowire.

In figure 12, plots of the numerical derivative dR/dH of the 1 μm and 1.2 μm nanowires are shown as a function of H at 3K (He & Wang, 2011). The applied field is perpendicular to the wire. This differential magnetoresistance shows uniform oscillations with increasing field with a period of $\phi_0/(2\pi r^2)$ (where, $\phi_0 = h/2e$ is the superconducting flux quantum, r = 35 nm is the radius of the nanowire). The period of the differential resistance plateaus can be explained by theorizing that each step in resistance is caused by a phase slip. The phase slips are created by vortices moving across the wire that are generated by the perpendicular magnetic field. The extraordinary feature in this experiment is that the nature of the gap in the proximity induced superconductivity in the AuNW allows for generation of individual vortices one at a time along the nanowire. The gap in the middle of the wire is the smallest and is the first site for single vortex creation. As the field is increased the vortex moves to higher gap regions and a new vortex is created in the center of the wire. As a result, the resistance of the Au wire increases with increasing field step by step so that the generated vortices move continuously across the wire. This is very different from the well-known behavior of superconductors where when the field is increased beyond a critical value, vortices proliferate all at once and the superconductivity is destroyed. In figure 12, the numbers assigned to each peak are theorized to be associated with the number of vortices in the nanowire. An interesting fact is that if the spatial extent of each vortex is taken to be limited by the diameter of the nanowire, the total number of vortices that will fit in the 1 μm nanowire is 1um/70nm = 15 which in fact corresponds to the number of peaks seen before the wire becomes normal.

The implications of the range of new phenomena discovered in proximity induced superconducting nanowires are profound. First of all, the fact that AuNW up to 1.2 μm long act as a superconductor provides us a large number of 'new' superconductors that can be used in nanoscale circuitry. The 'mini-gap' state is a new superconducting state that opens new avenues for exploration of the physics of nanoscale superconductors. The ability to introduce and manoeuvre vortices one by one in the proximity induced superconducting system is a first. If scalable synthesis for such samples can be developed, the manoeuvrable vortices can be utilized as memory units in quantum computing.

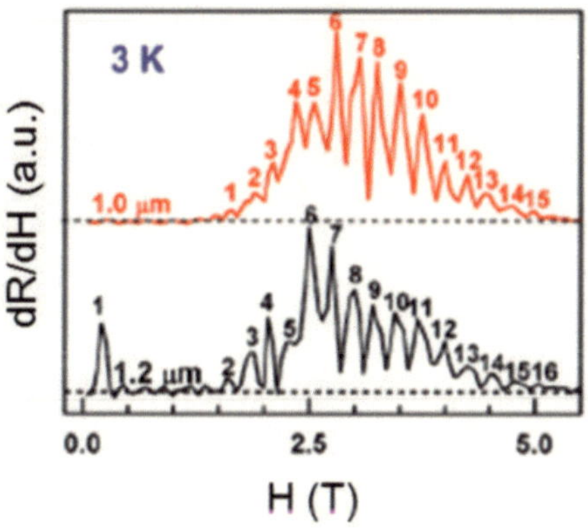

Fig. 12. (He & Wang, 2011) The differential magnetoresistance as a function of magnetic field of the 1.0 and 1.2 μm long gold nanowires measured at T = 3K. The dashed lines represent dR/d|H| = 0 for the two wires. There are small oscillations of the value of dR/d|H| around zero H, possibly due to the limited resolution in the R and H readings in low magnetic field. The clearly resolved peak at about 1.65 T was picked as the first peak of the 1.0 μm wire. The peak at the differential magnetoresistance shows uniform oscillations with B of 0.25 T. The number on the peak was theorized to match with the number of vortices in the nanowire.

3.5 New superconducting materials – Bismuth

Bi has an extremely small Fermi surface (FS) (just 10^{-5} of the Brillouin zone), low carrier density (~ 3×10^{17} /cm^3 at 2 K), small effective carrier mass ($m_e^* < 0.003$ free electron mass for electrons along the trigonal direction) (Liu & Allen, 1995) and very long electron mean free path (exceeding 2 μm at room temperature) (Hartman, 1969). These properties distinguish Bi from other metals and make it particularly suitable for studying quantum phenomena. A most striking example of superconductivity behaving differently in nanoscale systems was observed in granular bismuth nanowires (BiNW) electrodeposited in track-etched polycarbonate membranes (Tian et al., 2006b). Bulk rhombohedral Bi is a semimetal with no superconducting transition observed down to 50 mK under ambient pressure. In granular BiNW however, 70 nm and 100 nm in diameter, superconductivity was found under ambient pressure in 18 out of 38 samples. Careful transmission electron microscopy (TEM) studies revealed that the samples that exhibit superconductivity had a uniform granular morphology with all the rhombohedral Bi grains (~10 nm) aligned with a preferred orientation along the entire length of the wire. For the non-superconducting granular Bi samples, the rhombohedral Bi grains showed random orientation and both wire diameter and the grain size also showed considerable variations along the length of the wire. The superconductivity in the BiNW was attributed to small regions of high pressure Bi phase formed at the boundaries of partially aligned grains. These high pressure regions with the strained lattice probably cannot be detected by X-ray diffractometry because the strain regions are only a few atomic layers thick and do not contribute much to the signal from the sample.

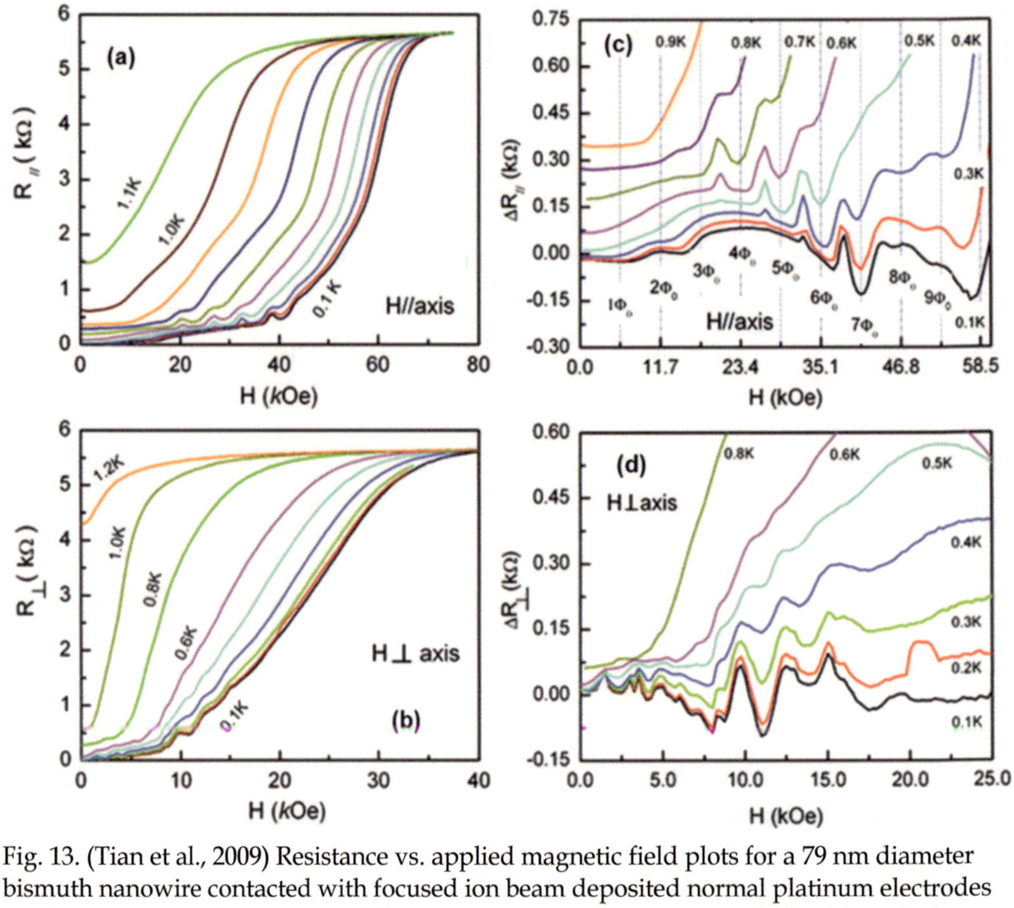

Fig. 13. (Tian et al., 2009) Resistance vs. applied magnetic field plots for a 79 nm diameter bismuth nanowire contacted with focused ion beam deposited normal platinum electrodes in a four-probe configuration at different temperatures for magnetic fields: (a) applied parallel to the wire axis; (b) applied perpendicular to the wire axis. (c) $\Delta R_{||}$-H and (d) $\Delta R_{\perp}$-H plots, obtained by substracting smooth background from the data in Figures 13(a) and (b) respectively. Periodic oscillations of R were clearly seen in parallel H, and the dashed lines in panel (c) indicate the positions of fluxoid quantization as predicted by $H_{(\pi d^2/4)} = n\Phi_0$ with d = diameter of the nanowire = 67 nm. In pependicular H, the R oscillations are more complex and their periodicity is with 1/H.

Bi under high pressure is known to turn superconducting due to the formation of high-pressure metallic polymorphs, namely, monoclinic Bi-II at 2.55 GPa (Brugger et al., 1967; Degtyareva et al., 2004; Nédellec et al., 1974), a complex tetragonal Bi-III at 2.7 GPa (Degtyareva et al., 2004; Haussermann et al., 2002), and a body centered cubic (bcc) Bi-V at 7.7 GPa. (Chen et al., 1969; Degtyareva et al., 2004). The transition temperatures of these polymorphic phases are respectively 3.9, 7.2, and 8.3 K (Brandt & Ginzburg, 1969). T_cs of 8.3 K and 7.2 K are also seen in the superconducting transition of the granular BiNW further strengthening the idea that small regions of high pressure Bi phase exist in the nanowire. As discussed in the next paragraph, crystalline BiNW also remarkably displayed superconductivity with a T_c of 1.3 K.

In a recent experiment (Tian et al., 2009), quantum oscillations of the resistance as a function of applied field (H) were found in crystalline BiNW in a standard four-probe measurement with FIB deposited normal Pt electrodes. The crystalline BiNW were superconducting with a transition temperature of 1.3 K. In parallel field, below the critical field ($H_c \sim 7T$), a long resistance tail was present (figure 13(a)). On subtracting a uniform background from the resistance, and plotting the remainder $\Delta R_{\parallel}$, the oscillations are clearly seen (figure 13 (c)). The oscillations are periodic with H and the period corresponds to the superconducting quantum flux ($\varphi_0 = {}^{h}/_{2e} = 2 \times 10^{-7} Gcm^2$). These oscillations are reminiscent of the Little-Parks (LP) oscillations caused by fluxoid quantization in hollow superconducting cylinders when the thickness of the shell is comparable to ξ (Little & Parks, 1962; Parks & Little, 1964). The standard LP oscillations in a small hollow cylinder are characterized by the integer multiples of the superconducting flux quantum, Φ_0, that is, $H(\pi d^2/4)) = n\Phi_0$, where d is the diameter of the hollow cylinder and n is an integer. Using this model, the period of the oscillations seen in BiNW, ΔH = 5.85 kOe results in d = 67.0 nm. The diameter of the nanowire determined by TEM is 79 nm with a ~ 3.7 ± 0.5 nm oxidation layer on the surface leaving the nanowire with an effective diameter of ~ 72 nm. This is close to the diameter calculated using LP oscillations. The LP-like oscillations unambiguously suggest that the observed superconductivity in BiNW originates from the cylindrical shell between the inner core of the Bi nanowire and its surface oxide.

When the field is aligned perpendicular to the wire axis, the residual resistance below 0.8 K still shows oscillations with varying H (figure 13(b)). On subtracting the residual resistance and plotting the remainder $\Delta R_{\perp}$, (figure 13(d)), it is seen that these oscillations are periodic with $1/H$ rather than with H. The $1/H$ periodicity of the residual resistance oscillations is suggestive of the Shubnikov-de Haas (SdH) effect due to the Landau quantization of the conduction electrons in a metal. The SdH behavior in the superconducting state is not expected; therefore it is tempting to attribute it to the normal core. This explanation however, is not correct, because in the presence of a superconducting shell, the current will bypass the normal core and the normal core will not contribute to the resistance. In the paper (Tian et al., 2009) it is speculated that the SdH oscillations also originate from the superconducting shell. The finite residual resistance is a result of the phase slips in the shell. The normal region created in the shell because of these phase slips, behaves like a normal metal and shows SdH oscillations in the presence of a strong field. Both LP-like oscillations in parallel H and SdH-like oscillations in perpendicular H being observed in the same system suggest a completely unexpected and novel coexistence of superconducting and metallic states in the surface shell of the Bi nanowire below T_c. The presence of superconductivity in ambient pressure BiNW is in itself a surprising result that brings to mind possibilities of other materials known to be non-superconducting in bulk showing superconductivity in the nanoscale. In addition, the fact that the superconductivity seems to be present only on the surface while the core continues to be normal, indicates that this superconductivity is an interface related phenomenon. The presence of SdH oscillations indicates metallic states on the surface. Metallic surface states are a prerequisite for the formation of a topological insulator (Fu et al., 2007). These new findings suggest that this old material may have potential application in developing next-generation quantum computing devices.

3.6 Pb nanobelts

Superconducting films on semiconductor substrates have attracted much attention since the derived superconductor-based electronics have been shown to be promising for future data processing and storage technologies. Current semiconductor technology is mainly based on silicon wafers. The superconductivity and applications of low dimensional Pb nanostructural superconductors have been extensively studied in recent decades (Wang et al., 2007, 2008a, 2008b, 2009b, 2009c, 2010b) because of the relatively higher transition temperature (T_C ~ 7.2 K) of bulk Pb compared to other conventional superconductors. The structure formed by two dimensional (2D) superconducting Pb films on Si substrate provides an excellent research system and has a wide application field. Qi-Kun Xue et al. have carried out a series of systematic experiments in high quality 2D single crystalline Pb film systems by utilizing scanning tunneling microscope (STM) and have achieved many outstanding results (Bao et al., 2005; Guo et al., 2004; T. Zhang et al., 2010; Y. F. Zhang et al., 2005). The Pb films are synthesized by using an MBE chamber. Based on previous work (Jalochowski et al., 1995; Upton et al., 2004), Q. K. Xue's group made great achievements in fabricating crystalline Pb films by MBE. They have made ultrathin Pb films from one atomic layer to tens of layers on Si (111) substrates. The Si substrate was firstly heated at 400°C for hours to get rid of the molecular gas absorbed at the surface, and then the substrate was flashed or annealed to ~1200°C several times to form a clean surface. After that, the substrate was cooled down to ~ 95 K by liquid nitrogen during growth to achieve atomically flat single crystal Pb thin films over a macroscopic area. By means of STM, scanning tunneling spectroscopy (STS), and transport measurements various intriguing properties of the 2D Pb systems have been discovered. These include "magic film thickness", "preferential island heights", "oscillating superconductivity transition temperatures" and other novel behaviors because of the strong quantum confinement effect (Bao et al., 2005; Eom et al., 2006; Guo, et al., 2004; Herzog et al., 1996; Markovic et al., 1998; Parendo et al., 2004; Xiong et al., 1997).

High quality Pb films are not only important for fundamental research, but also offer possibilities of electronic device fabrication. From the view point of applications, Pb films and related nanodevices have been predominantly prepared on semiconductor substrates. In recent years, J. Wang and his collaborators performed experiments on nanoscale systems based on Pb films grown on Si substrates (Wang et al., 2007, 2008a, 2008b, 2009b, 2009c, 2010b). In the Pb thin film with a unique fractal-like morphology grown on Si (111)-7×7 surface, the films exhibit distinctive negative magnetoresistance with wide magnetoresistance terrace under the perpendicular magnetic field. Analysis of the experimental results revealed that the observed effect originates from the coexistence of two superconducting phases in the system: the percolation structures with lower critical field (H_c) and the flat 2D Pb islands with higher H_c. In the low temperature regime, electron tunneling dominates the transport property of the percolation structures. When a magnetic field is applied, the superconducting gap is suppressed. Reducing the gap leads to a significant decrease of the resistance. The unusual magnetoresistance effect in the coexistence phase is of both fundamental interest and has possible applications in fabrication of hybrid devices based on the traditional microelectronics and the emerging superconducting quantum electronics (Leggett, 2002).

The high quality Pb films can also be etched to quasi-one dimensional (quasi-1D) single crystal Pb superconductor nanobridge (several nanometers thick and 100 to 500 nanometers wide) by using an FIB system (the measurement geometry are shown in figure 14). The Pb nanostructures thus formed were measured in a physical property measurement system (PPMS) with standard four-electrode method. Many interesting phenomena with the different widths and thicknesses of the Pb nanobridges were observed by studying the transport properties, such as the enhanced superconductivity (T_C) compared to the film with the same thickness (Wang et al., 2008a), magnetoresistance (MR) oscillations (Wang et al., 2008a, 2009c, 2010b), the broadening of the transition near T_C because of thermally activated phase-slip (TAPS), multiple voltage steps, and power-law Voltage vs. Current relationship (Wang et al., 2009b). The properties and the unique structures provide a new approach for integrating superconducting circuits on single Si chips, for example nanoscale SQUID-like devices (Johansson et al., 2005), "on-off" functional and logical nanodevices.

Furthermore, atomically uniform single-crystal epitaxial Pb films, several nanometers thick, were fabricated on a Si substrate to form a sharp superconductor–semiconductor heterojunction (SSH) by FIB etching (Wang et al., 2008b) as shown in figures 15 (a) and (b). The transport measurements of the junction showed an unusual magnetoresistance effect when the Pb film was superconducting. The resistance of the SSH decreased sharply with increasing magnetic field and was very sensitive to the temperature below T_C (figure 15 (c)). Such sensitivity and the SSH structure make the Pb–Si junction highly promising for mass production for use in developing a magnetic field controlled "on–off" device and new cryogenic temperature detectors on Si chips (Day et al., 2003; Irwin et al., 1996). The devices derived by superconductor films on semiconductor substrates have many promising applications in the future.

The giant magnetoresistance (GMR) effect was discovered in the late 1980s on Fe/Cr magnetic multilayer (Baibich et al., 1988) and Fe/Cr/Fe tri-layers (Binasch et al., 1989). It was seen that as the applied field was increased, the resistance of the multilayer structure decreased (6% for Fe/Cr/Fe tri-layers and 50% for Fe/Cr magnetic multilayer, respectively). This discovery was significant to the scientific community and caught widespread attention in the industry. The researchers of IBM recognized its potential and invented spin-valves (Tsang et al., 1994) and the commercial read head was created (Wolf et al., 2001). GMR's application in the read head of hard discs greatly increases the density of stored information. Since the introduction of GMR-type sensors as reading elements, storage capacities have increased approximately 100 times (Grunberg, 2008). Moreover, magnetic random access memory (MRAM), medical/biological sensors and various logic components have been invented by using the GMR effect. Meanwhile, GMR is often viewed as the founding stone of spintronics, where the electron spin is the key to physical functionality. Thus GMR of the magnetic multilayers opened the way to an efficient control of the motion of the electrons by acting on their spin through the orientation of a magnetization. In the SSH composed of Pb/Si system, described in the paragraph above, J. Wang and his collaborators observed a dramatic negative magnetoresistive effect (Wang et al., 2008b). Compared to the Fe/Cr magnetic multilayer in 1988 (Baibich et al., 1988), the resistance changes more with increasing the magnetic field, and the saturated critical field of the SSH is smaller than that of Fe/Cr multilayer by an order at the same temperature.

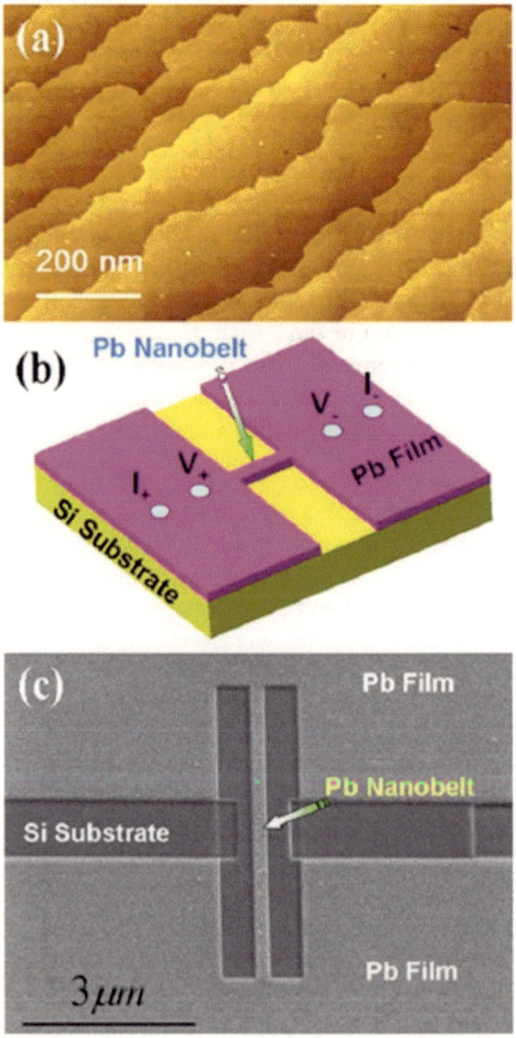

Fig. 14. (Wang et al., 2008b) (a) A scanning tunneling microscope image of the atomically smooth lead thin film on the Si(111) 7×7 substrate. (b) The schematic graph for the transport measurement across the lead nanobelt. (c) A SEM image of the lead nanobelt made by milling the lead film.

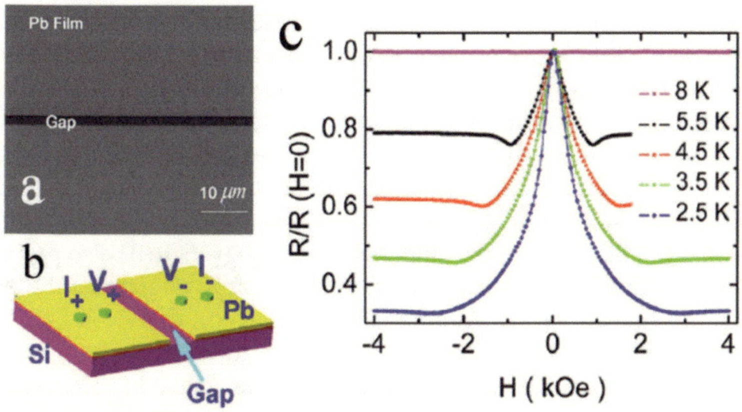

Fig. 15. (Wang et al., 2008b) (a) A scanning electron micrograph of the Pb film after a 2μm wide gap (the dark region) was fabricated. (b) The schematic graph for the transport measurement across the Pb/Si (111) heterojunctions. (c) Magnetoresistance of the heterojunctions with a magnetic field perpendicular to the film at different temperatures. The vertical scale is normalized to the resistance at zero magnetic field.

Although the physical mechanisms in traditional GMR and the giant negative magnetoresistive effect in SSH are completely different and the resistance changes in SSH only happens at low temperatures, it is reasonable to believe that the large magnetoresistance effect in SSH may also be extended and applied in technology and other fields such as magnetic switches, temperature or magnetic field sensors, high-density information storage and quantum computing circuits. It is to be remembered that Pb is a metal and is easily oxidized in atmosphere. Therefore, cover layers of Ge, Si, Au etc. need to be used to protect Pb based devices for applications.

3.7 Josephson effect and superconducting quantum interference device (SQUID)

A Josephson junction is composed of two superconducting layers coupled by a weak link. The weak link can consist of a thin insulating barrier (S-I-S), a short section of non-superconducting metal (S-N-S), or a physical constriction that weakens the superconductivity at the point of contact (S-c-S). The Josephson Effect is the phenomenon that the current can tunnel through the junction (or weak link) without any voltage applied (Josephson, 1962, 1974). It was observed by the British physicist B. D. Josephson, who was the first to predict the tunneling of superconducting Cooper pairs (before that, it was only known that normal electrons can tunnel through an insulating barrier). Josephson received the Nobel prize in 1973 for this finding. Josephson junctions have important applications in superconducting quantum interference devices (SQUID), single-electron transistors, superconducting qubits and rapid single flux quantum (RSFQ) digital electronics. SQUID is a very sensitive magnetometer used to measure extremely weak magnetic fields (Range, 2002), based on superconducting loops containing one or two Josephson junctions. For instance, the SQUID magnetometer reaches 7×10^{-11} Os/Hz$^{1/2}$ which is higher than that of the best traditional magnetometer by one order. SQUID is used as a sensitive sensor that can detect the tiny magnetic flux changes through measuring the other visible parameters. It can be used in many fields.

The most common use of SQUID is probably in magnetic property measurement systems (MPMS). The SQUID is being used as a detector to perform MRI because of its extreme sensitivity. It is an ideal device for studying biological systems. In the clinical environment, the applications of SQUID are wide. For example, Magnetoencephalography (MEG) (Cohen, 1972), which is an equipment that measures and traces the extremely weak biomagnetic signals emitted by the brain. The SQUID is also used in cardiology for magnetic field imaging (MFI) (Brockmeier et al., 1994), which detects the magnetic field of the heart for diagnosis and risk stratification. In the other fields of biology, SQUID also plays a significant role such as recording the weak magnetic fields of the stomach, tracing the path of orally applied drugs in human body. Additionally, in geological exploration, the SQUID-superconducting magnetic magnetometer can be used to measure the geomagnetic field accurately and analyze the distribution to observe the valuable mineral deposits. For earthquake prediction, it can be used to detect the gravity changes on a floating superconducting ball. The changes in gravity change the relative position of the ball and the superconducting ring and cause distortion of the magnetic field. The changing magnetic field can be detected by SQUID-Gravimeter (Verkin et al., 1976). Furthermore, it is also promising as a potential component in quantum computers (Bouchiat et al., 1998) and in precision sensors. The applications of SQUID are many.

With the development of semiconductor industry, electronic devices are rapidly becoming smaller. Although the sensitivity of the SQUID increases with the area of the superconducting ring, nano-SQUID is also needed for detecting the weak magnetic signals of small structures and integrating the logic components of computers. The British "New Scientist" magazine reported (Troeman et al., 2007) that researchers in Holland had fabricated the smallest SQUID in the world. It is 180 nm in diameter and was made by Nb. To maintain superconductivity of the device, the whole system needs to operate at low temperatures. Pb also has T_C higher than liquid helium temperature and is suitable for work in helium cooling conditions. In addition, the possibility of using focused ion beam and electron beam lithography techniques discussed in this chapter on silicon substrate offers the possibility of synthesizing millions or billions of SQUIDs devices in one chip.

4. Conclusion

Superconductivity and its applications in science and technology is currently one of the most important research fields. In recent decades, the developments in this field have been extremely rapid and superconductors are extensively used in many applications. With the dimensions of electronic circuits reduced to nanometer scale, it becomes important to study the physics of superconductivity on this scale for potential nano-superconducting circuit elements. Many interesting properties are exhibited in superconducting nanowires with various morphologies, which offer numerous opportunities for applications.

The interesting physics observed in nanowires and nanobelts discussed in the chapter proves that they are an ideal platform for helping in understanding the fundamental nature of superconductivity. Furthermore, the plethora of synthesis techniques available, offer flexibility in morphology and performance control. The very high quality, defect free nanostructures like Pb films discussed in the chapter are ideal candidates for studying low-dimensional superconductivity because they enable the separation of true low-dimensional physics from the effects caused due to defects and morphology. It is also possible to fabricate such devices for use in integrated circuits on Si substrates. The e-beam lithography technique used for gold, aluminum and lead nanobelts is especially suitable for this. Examples like nano-squid and single-electron transistors (Fulton & Dolan, 1987) enabling extremely sensitive magnetic field and charge measurements demonstrate the tremendous potential for applications of nanoscale superconducting devices. The factor limiting widespread use of these devices is still the low temperatures required and the prohibitive cost of the helium needed to reach those low temperatures. Much research has been done to achieve usable high T_c superconductors. However, till now, high temperature superconductors have realized applications only in very limited fields. Although their T_c is higher than conventional metallic superconductors, they are ceramic, making them inflexible and hard to machine compared to metallic superconductors. The low T_c of metallic superconductors, limiting their applications in devices, is not a problem in potential applications to space technology (ambient temperature in space ~ 3K and many metallic superconductors have T_c higher than this). Due to properties like high speed (response rate is 10-100 times faster than conventional devices), ultra-low noise (the thermal noise is much lower than conventional semiconductor devices) and low heat dissipation, superconductivity in aerospace applications such as space communications, radar guidance etc. is widely used.

In summary, the use of superconducting nanoscale components will produce denser and more rapid chips since the resistance is a major source of heat generation and charging time in integrated circuits. Studying and understanding superconductivity on this scale is of great interest and importance.

5. Acknowledgement

We are grateful to Professors Moses H. W. Chan, Mingliang Tian, Qi-Kun Xue, Jainendra K. Jain, Thomas E. Mallouk, Nitin Samarth and Xu-Cun Ma for the collaboration in this project. This work was financially supported by the National Natural Science Foundation of China (No. 11174007) and National Basic Research Program (NBRP) of China (No. 2012CB921300), the Penn State MRSEC under NSF grant DMR-0820404, and China Postdoctoral Science Foundation (No. 2011M500180).

6. References

Altomare, F., Chang, A. M., Melloch, M. R., Hong, Y. G., & Tu, C. W. (2006). Evidence for Macroscopic Quantum Tunneling of Phase Slips in Long One-dimensional Superconducting Al Wires. *Physical Review Letters,* Vol. 97, No. 1, (July 2006), pp. (017001), ISSN 0031-9007

Arthur, J. R. (2002). Molecular Beam Epitaxy. *Surface Science,* Vol. 500, No. 1-3, (March 2002), pp. (189-217), ISSN 0039-6028

Aumentado, J., & Chandrasekhar, V. (2001). Mesoscopic Ferromagnet-Superconductor Junctions and the Proximity Effect. *Physical Review B,* Vol. 64, No. 5, (August 2001), pp. (054505), ISSN 0163-1829

Baibich, M. N., Broto, J. M., Fert, A., Vandau, F. N., Petroff, F., Eitenne, P., Creuzet, G., Friederich, A., & Chazelas, J. (1988). Giang Magnetoresistance of (001)Fe/(001)Cr

Magnetic Superlattices. *Physical Review Letters,* Vol. 61, No. 21, (November 1988), pp. (2472-2475), ISSN 0031-9007

Bao, X. Y., Zhang, Y. F., Wang, Y. P., Jia, J. F., Xue, Q. K., Xie, X. C., & Zhao, Z. X. (2005). Quantum Size Effects on the Perpendicular Upper Critical Field in Ultrathin Lead Films. *Physical Review Letters,* Vol. 95, No. 24, (December, 2005), pp. (247005), ISSN 0031-9007

Bera, D., Kuiry, S. C., & Seal, S. (2004). Synthesis of Nanostructured Materials using Template-assisted Electrodeposition. *JOM,* Vol. 56, No. 1, (January 2004), pp. (49-53), ISSN 1047-4838

Bergeret, F. S., Volkov, A. F., & Efetov, K. B. (2005). Odd Triplet Superconductivity and Related Phenomena in Superconductor-Ferromagnet Structures. *Reviews of Modern Physics,* Vol. 77, No. 4, (October 2005), pp. (1321-1373), ISSN 0034-6861

Bezryadin, A., Lau, C. N., & Tinkham, M. (2000). Quantum Suppression of Superconductivity in Ultrathin Nanowires. *Nature,* Vol. 404, No. 6781, (April 2000), pp. (971-974), ISSN 0028-0836

Binasch, G., Grunberg, P., Saurenbach, F., & Zinn, W. (1989). Enhanced Magnetoresistance in Layered Magnetic-Structures wit Antiferromagnetic Interlayer Exchange. *Physical Review B ,* Vol. 39, No. 7, (March 1989), pp. (4828-4830), ISSN 1098-0121

Bouchiat, V., Vion, D., Joyez, P., Esteve, D., & Devoret, M. H. (1998). Quantum Coherence with Single Cooper Pair. *Physica Scripta ,* Vol. T76, (January 1998), pp. (165-170), ISSN 0281-1847

Brandt, N. B., & Ginzburg, N. I. (1969). Superconductivity at High Pressures. *Contemporary Physics,* Vol. 10, No. 4, (April 1969), pp. (355-384), ISSN 0010-7514

Brockmeier, K., Comani, S., Erne, S. N., Diluzio, S., Psquarelli, A., & Romani G. L. (1994). Magnetocardiography and Exercise Training. *Journal of Electrocardiology,* Vol. 27, No. 2, (April 1994), pp. (137-142), ISSN 0022-0736

Brugger, R. M., Bennion, R. B., & Worlton, T. G. (1967). Crystal Structure of Bismuth-2 at 26 Kbar. *Physcs Letters A,* Vol. A24, No. 13, (June 1967), pp. (714-717), ISSN 0375-9601

Chen, T. T., Chen, J. T., Leslie, J. D., & Smith, H. J. T. (1969). Phonon Spectrum of Superconducting Amorphous Bismuth and Gallium by Electron Tunneling. *Physical Review Letters,* Vol. 22, No. 11, (March 1969), pp. (526-530), ISSN 0031-9007

Chen, Y., Snyder, S. D., & Goldman A. M. (2009). Magnetic-field-induced Superconducting State in Zn Nanowires Driven in the Normal State by an Electric Current. *Physical Review Letters,* Vol. 103, No. 12, (September 2009), pp. (127002), ISSN 0031-9007

Chen, Y., Lin, Y. H., Snyder, S. D., & Goldman A. M. (2011). Stabilization of Superconductivity by Magnetic Field in Out-of-equilibrium Nanowires. *Physical Review B,* Vol. 83, No. 5, (February 2011), pp. (054505), ISSN 1098-0121

Chiang, Y. N., Shevchenko, O. G., & Kolenov, R. N. (2007). Manifestation of Coherent and Spin-dependent Effects in the Conductance of Ferromagnets Adjoining a Superconductor. *Low Temperature Physics,* Vol. 33, No. 4, (April 2007), pp. (314-320), ISSN 1063-777X

Cohen, D. (1972). Magnetoencephalography - Detection of Brains Electrical-activity with a Superconducting Magnetometer. *Science,* Vol. 175, No. 4022, (February 1972), pp. (664-666), ISSN 0036-8075

Day, P. K., LeDuc, H. G., Mazin, B. A., Vayonakis, A., & Zmuidzinas, J. (2003). A Broadband Superconducting Detector Suitable for Use in Large Arrays. *Nature,* Vol. 425, No. 6960, (October 2003), pp. (817-821), ISSN 0028-0836

Degtyareva, O., McMahon, M. I., & Nelmes, R. J. (2004). High-pressure Structural Studies of Group-15 Elements. *High Pressure Research,* Vol. 24, No. 3, (September 2004), pp. (319-356), ISSN 0895-7959

Eom, D., Qin, S., Chou, M. Y., & Shih, C. K. (2006). Persistent Superconductivity in Ultrathin Pb Films: A Scanning Tunneling Spectroscopy Study. *Physical Review Letters,* Vol. 96, No. 2, (January, 2006), pp. (027005), ISSN 0031-9007

Eschrig, M. (2011). Spin-polarized Supercurrents for Spintronics. *Physics Today,* Vol. 64, No. 1, (January 2011), pp. (43-49), ISSN 0031-9228

Fu, H. C., Seidel, A., Clarke, J., & Lee, D. (2006). Stabilizing Superconductivity in Nanowires by Coupling to Dissipative Environments. *Physical Review Letters,* Vol. 96, No. 15, (April 2006), pp. (157005), ISSN 0031-9007

Fu, L., Kane, C. L., & Mele, E. J. (2007). Topological Insulators in Three Dimensions. *Physical Review Letters,* Vol. 98, No. 10, (March, 2007), pp. (106803), ISSN 0031-9007

Fulton, T. A., & Dolan, G. J. (1987). Observation of single-electron charging effects in small tunnel junctions. *Physical Review Letters,* Vol. 59, No. 1, (July, 1987), pp. (109-112), ISSN 0031-9007

Giordano, N. (1988). Evidence for Macroscopic Quantum Tunneling in One-dimensional Superconductors. *Physical Review Letters,* Vol. 61, No. 18, (October 1988), pp. (2137-2140), ISSN 0031-9007

Giordano, N., & Schuler, E. R. (1990). Giant Conductance Fluctuations in Thin Metal Wires. *Physical Review B,* Vol. 41, No. 17, (June 1990), pp. (11822-11826), ISSN 0163-1829

Giordano, N. (1991). Superconductivity and Dissipation in Small-diameter Pb-In Wires. *Physical Review B,* Vol. 43, No. 1, (January 1991), pp. (160-174), ISSN 0163-1829

Giordano, N. (1994). Superconducting Fluctuations in One-dimension. *Physica B,* Vol. 203, No. 3-4, (December 1994), pp. (460-466), ISSN 0921-4526

Giroud, M., Courtois, H., Hasselbach, K., Mailly, D., & Pannetier, B. (1998). Superconducting Proximity Effect in a Mesoscopic Ferromagnetic Wire. *Physical Review B,* Vol. 58, No. 18, (November 1998), pp. (11872-11875), ISSN 1098-0121

Grunberg, P. A. (2008). Nobel Lecture: From Spin Waves to Giant Magnetoresistance and Beyond. *Reviews of Modern Physics* , Vol. 80, No. 4, (October - December 2008), pp. (1531-1540), ISSN 0034-6861

Guo, Y., Zhang, Y. F., Bao, X. Y., Han, T. Z., Tang, Z., Zhang, L. X., Zhu, W. G., Wang, E. G., Niu, Q., Qiu, Z. Q., Jia, J. F., Zhao, Z. X., & Xue, Q. K. (2004). Superconductivity Modulated by Quantum Size Effects. *Science,* Vol. 306, No. 5703, (December, 2004), pp. (1915-1917), ISSN 0036-8075

Gupta, A. K., Cretinon, L., Moussy, N., Pannetier, B., & Courtois, H. (2004). Anomalous Density of States in a Metallic Film in Proximity with a Superconductor. *Physical Review B,* Vol. 69, No. 10, (March, 2004), pp. (104514), ISSN 1098-0121

Hartman, R. (1969). Temperature Dependence of Low-field Galvanomagnetic Coefficients of Bismuth. *Physical Review,* Vol. 181, No. 3, (May, 1969), pp. (1070-1086), ISSN 0031-899X

Haussermann, U., Soderberg, K., & Norrestarrit, R. (2002). Comparitive Study of the High-pressure Behaviour of As, Sb and Bi. *Journal of the American Chemical Society,* Vol. 124, No. 51, (December, 2002), pp. (15359-15367), ISSN 0002-7863

He, L., & Wang, J. (2011). Periodic Magnetoresistance Oscillations Induced by Superconducting Vortices in Single Crysal Au Nanowires. *Nanotechnology,* Vol. 22, No. 44, (September, 2011), pp. (445704), ISSN 1361-6528

Herzog, A. V., Xiong, P., Sharifi, F., & Dynes R. C. (1996). Observation of a Discontinuous Transition from Strong to Weak Localization in 1D Granular Metal Wires. *Physical Review Letters,* Vol. 76, No. 4, (January, 1996), pp. (668-671), ISSN 0031-9007

Herzog, A. V., Xiong, P., & Dynes R. C. (1998). Magnetoresistance Oscillations in Granular Sn Wires Near the Superconductor-Insulator Transition. *Physical Review B,* Vol. 58, No. 21, (December 1998), pp. (14199-14202), ISSN 0163-1829

Hohenberg, P. C. (1967). Existence of Long-Range Order in 1 and 2 Dimensions. *Physical Review,* Vol. 158, No. 2, (June 1967), pp. (383-386), ISSN 0031-899X

Irwin, K. D., Hilton, G. C., Wollman, D. A., & Martinis, J. M. (1996). X-Ray Detection Using a Superconducting Transition-edge Sensor Microcalorimeter with Electrothermal Feedback. *Applied Physics Letters,* Vol. 69, No. 13, (September 1996), pp. (1945-1947), ISSN 0003-6951

Jalochowski, M., Hoffmann, M., & Bauer, E. (1995). Pb Layer-by-layer Growth at Very-low Temperatures. *Physical Review B,* Vol. 51, No. 11, (March 1995), pp. (723-7238), ISSN 1098-0121

Johansson, A., Sambandamurthy, G., Shahar, D., Jacobson, N., & Tenne, R. (2005). Nanowire Acting as a Superconducting Quantum Interference Device. *Physical Review Letters,* Vol. 95, No. 11, (September 2005), pp. (116805), ISSN 0031-9007

Josephson, B. D. (1962). Possible New Effects in Superconductive Tunneling. *Physics Letters,* Vol. 1, No. 7, (July 1962), pp. (251-253)

Josephson, B. D. (1974). Discovery of Tunneling Supercurrents. *Reviews of Modern Physics,* Vol. 46, No. 2, (April, 1974), pp. (251-254), ISSN 0034-6861

Keizer, R. S., Goennenwein, S. T. B., Klapwijk, T. M., Miao, G. X., Xiao, G., & Gupta, A. (2006). A Spin Triplet Supercurrent through a Half-metallic Ferromagnet CrO_2. *Nature,* Vol. 439, No. 7078, (February 2006), pp. (825-827), ISSN 0028-0836

Khaire, T. S., Khasawneh, M. A., Pratt, W. P., & Birge, N. O. (2010). Observation of Spin-Triplet Superconductivity in Co-Based Josephson Junctions. *Physical Review Letters,* Vol. 104, No. 13, (April 2010), pp. (137002), ISSN 0031-9007

Knoedler, C. M. (1990). Helium-ion Damage and Nanowire Fabrication in GaAs/AlGaAs Heterostructures. *Journal of Applied Physics,* Vol. 68, No. 3, (August 1990), pp. (1129-1137), ISSN 0021-8979

Konschelle, F., Cayssol, J., & Buzdin, A. (2010). Long-range Singlet Proximity Effect in Ferromagnetic Nanowires. *Physical Review B,* Vol. 82, No. 18, (November 2011), pp. (180509), ISSN 1098-0121

Langer, J. S., & Ambegaokar, V. (1967). Intrinsic Resistive Transition in Narrow Superconducting Channels. *Physical Review,* Vol. 164, No. 2, (December 1967), pp. (498-510), ISSN 0031-899X

Lau, C. N., Markovic, N., Bockrath, M., Bezryadin, A., & Tinkham, M. (2001). Quantum Phase Slips in Superconducting Nanowires. *Physical Review Letters,* Vol. 87, No. 21, (November 2001), pp. (217003), ISSN 0031-9007

Lee, W., Ji, R., Gosele, U., & Nielsch, K. (2006). Fast Fabrication of Long-range Ordered Porous Alumina Membranes by Hard Anodization. *Nature Materials,* Vol. 5, No. 9, (September 2006), pp. (741-747), ISSN 1476-1122

Leggett, A. J. (2002). Physics - Superconducting Qubits - A Major Roadblock Dissolved?. *Science,* Vol. 296, No. 5569, (May 2002), pp. (861-862), ISSN 0036-8075

Le Sueur, H., Joyez, P., Pothier, H., Urbina, C., & Esteve, D. (2008). Phase Controlled Superconducting Proximity Effect Probed by Tunneling Spectroscopy. *Physical Review Letters,* Vol. 100, No. 19, (May 2008), pp. (197002), ISSN 0031-9007

Little, W. A., & Parks, R. D. (1962). Observation of Quantum Periodicity in Transition Temperature of a Superconducting Cylinder. *Physical Review Letters,* Vol. 9, No. 1, (June 1962), pp. (9-12), ISSN 0031-9007

Liu, Y., & Allen, R. (1995). Electronic-structure of the Semimetals Bi and Sb. *Physical Review B,* Vol. 52, No. 3, (July, 1995), pp. (1566-1577), ISSN 0163-1829

Lucignano, P., Stornaiuolo, D., Tafuri, F., Altshuler, B. L., & Tagliacozzo, A. (2010). Evidence for a Minigap in BCO Grain Boundary Josephson Junctions. *Physical Review Letters,* Vol. 105, No. 14, (September 2010), pp. (147001), ISSN 0031-9007

Madou, M. J. (2002). *Fundamentals of Microfabrication: The Science of Miniaturization* (2nd edition), CRC Press, ISBN 0-8493-0826-7, USA

Markovic, N., Mack, A. M., Martinez-Arizala, G., Christiansen, C., & Goldman, A. M. (1998). Evidence of Vortices on the Insulating Side of the Superocnductor-Insulator Transition. *Physical Review Letters,* Vol. 81, No. 3, (July 1998), pp. (701-704), ISSN 0031-0097

McCumber, D. E., Halperin, B. I. (1970). Time Scale of Intrinsic Resistive Fluctuations in Thin Superconducting Wires. *Physical Review B,* Vol. 1, No. 3, (February, 1970), pp. (1054-1070)

Menon, V. P., & Martin, C. R. (1995). Fabrication and Evaluation of Nanoelectrode Ensembles. *Analytical Chemistry,* Vol. 67, No. 13, (July 1995), pp. (1920-1928), ISSN 0003-2700

Mermin, N. D., & Wagner, H. (1966). Absence of Ferromagnetism or Antiferromagnetism in one- or 2-dimensional Isotropic Heisenberg Models. *Physical Review Letters,* Vol. 17, No. 22, (November 1966), pp. (1133-1136), ISSN 0031-9007

Mooij, J. E., Harmans, C. J. P. (2005). Phase-slip Flux Qubits. *New Journal of Physics,* Vol. 7, (October 2005), pp. (219), ISSN 1367-2630

Nédellec, P., Dumoulin, L., & Noer, R. J. (1974). Electron Tunneling Study of Superconducting High-pressure Phase of Bi. *Journal of Phyics F,* Vol. 4, No. 6, (June, 1974), pp. (L145-L150), ISSN 0305-4608

Nielsch, K., Castano, F. J., Matthias, S., Lee, W., Ross, C. A. (2005). Synthesis of Cobalt/Polymer Multilayer Nanotubes. *Advanced Engineering Materials,* Vol. 7, No. 4, (April 2005), pp. (217-221), ISSN 1438-1656

Parendo, K. A., Hernandez, L. M., Bhattacharya, A., & Goldman, A. M. (2004). Anomalous Parallel-field Negative Magnetoresistance in Ultrathin Films Near the Superconductor-insulator Transition. *Physical Review B,* Vol. 70, No. 21, (December, 2004), pp. (212510), ISSN 1098-0121

Parks, R. D., & Little, W. A. (1964). Fluxoid Quantization in Multiply-connected Superconductor. *Physical Review,* Vol. 133, No. 1A, (January, 1964), pp. (A97-A103), ISSN 0031-899X

Pena, V., Sefrioui, Z., Arias, D., Leon, C., Santamaria, J., Varela, M., Pennycook, S. J., & Martinez, J. L. (2004). Coupling of Superconductors through a Half-metallic Ferromagnet: Evidence for a Long-range Proximity Effect. *Physical Review B,* Vol. 69, No. 22, (June 2004), pp. (224502), ISSN 1098-0121

Peng H. L., Kai, K. J., Kong, D. S., Meister, S., Chen, Y. L., Qi, X. L., Zhang, S. C., Shen, Z. X., & Cui, Y. (2010). Aharonov-Bohm Interference in Topological Insulator Nanoribbons. *Nature Materials,* Vol. 9, No. 3, (March 2010), pp. (225-229), ISSN 1476-1122

Petrashov, V. T., Sosnin, I. A., Cox, I., Parsons, A., & Troadec, C. (1999). Giant Mutual Proximity Effects in Ferromagnetic/Superconducting Nanostructures. *Physicsal Review Letters,* Vol. 83, No. 16, (October 1999), pp. (3281-3284), ISSN 0031-9007

Range, S. K. (2002). Balancing relevancy and accuracy: Presenting gravity probe B's science meaningfully, *Proceedings of NASA OSS Conference on Education and Public Outreach,* 1-58381-181-8, Chicago (USA), June 2002

Reguera, G., McCarthy, K. D., Mehta, T., Nicoll, J. S., Tuominen, M. T., & Lovley, D. R. (2005). Extracellular electron transfer via microbial nanowires. *Nature,* Vol. 435, No. 7045, (June 2005), pp. (1008-1101), ISSN 0028-0836

Rogachev, A., Wei, T. C., Pekker, D., Bollinger, A. T., Goldbart, P. M., & Bezryadin, A. (2006). Magnetic-field Enhancement of Superconductivity in Ultranarrow Wires. *Physicsal Review Letters,* Vol. 97, No. 13, (September 2006), pp. (137001), ISSN 0031-9007

Santhanam, P., Chi, C. C., Wind, S. J., Brady, M. J., & Bucchignano, J. J. (1991). Resistance Anomaly near the Superconducting Transition-temperature in Short Aluminum Wires. *Physical Review Letters,* Vol. 66, No. 17, (April 1991), pp. (2254-2257), ISSN 0031-9007

Sazio, P. J. A., Amezcua-Correa, A., Finlayson, C. E., Hayes, J. R., Scheidemantel, T. J., Baril, N. F., Jackson, B. R., Won, D. J., Zhang, F., Margine, E. R., Gopalan, V., Crespi, V. H., Badding, J. V. (2006). Microstructured Optical-fibers as High-pressure Microfluidic Reactors. *Science,* Vol. 311, No. 5767, (March 2006), pp. (1583-1586), ISSN 0036-8075

Singh, M., Wang, J., Tian, M. L., Zhang, Q., Pereira, A., Kumar, N., Mallouk, T. E., & Chan, M. H. W. (2009). *Chemistry of Materials,* Vol. 21, No. 23, (December, 2009), pp. (5557-5559), ISSN 0897-4756

Singh, M., Wang, J., Tian, M. L., Mallouk, T. E., & Chan, M. H. W. (2011). Antiproximity Effect in Aluminum Nanowires with no Applied Magnetic Field. *Physical Review B,* Vol. 83, No. 22, (June, 2011), pp. (220506), ISSN 1098-0121

Sosnin, I., Cho, H., Petrashov, V. T., & Volkov, A. F. (2006). Superconducting Phase Coherent Electron Transport in Proximity Conical Ferromagnets. *Physical Review Letters,* Vol. 96, No. 15, (April 2006), pp. (157002), ISSN 0031-9007

Stoltenberg, R. M., & Woolley, W. T. (2004). DNA-templated Nanowire Fabrication. *Biomedical Microdevices,* Vol. 6, No. 2, (June 2004), pp. (105 111), ISSN 1387 2176

Tian, M. L., Wang, J. G., Snyder, J., Kurtz, J., Liu, Y., Schiffer, P., Mallouk, T. E., & Chan, M. H. W. (2003). Synthesis and Characterization of Superconducting Single-crystal Sn Nanowires. *Applied Physics Letters,* Vol. 83, No. 8, (August 2003), pp. (1620-1622), ISSN 0003-6951

Tian, M. L., Kumar, N., Xu, S. Y., Wang, J. G., Kurtz, J. S., & Chan, M. H. W. (2005). Suppression of Superconductivity in Zinc Nanowires by Bulk Superconductors. *Physical Review Letters,* Vol. 95, No. 7, (August 2005), pp. (076802), ISSN 0031-9007

Tian, M. L., Kumar, N., Wang, J. G., Xu, S. Y., & Chan, M. H. W. (2006a). Influence of a Bulk Superconducting Environment on the Superconductivity of One-dimensional Zinc Nanowires. *Physical Review B* , Vol. 74, No. 1, (July 2006), pp. (014515), ISSN 1098-0121

Tian, M. L., Wang, J. G., Kumar, N., Tianheng, H., Kobayashi, Y., Liu, Y., Mallouk, T. E., & Chan, M. H. W. (2006b). Observation of Superconductivity in Granular Bi Nanowires Fabricated by Electrodeposition. *Nano Letters,* Vol. 6, No. 12, (November 2006), pp. (2773-2780), ISSN 1530-6984

Tian, M. L., Wang, J., Zhang, Q., Kumar, N., Mallouk, T. E., & Chan, M. H. W. (2009). Superconductivity and Quantum Oscillations in Crystalline Bi Nanowir. *Nano Letters,* Vol. 9, No. 9, (September 2009), pp. (3196-3202), ISSN 1530-6984

Troeman, A. G. P., Derking, H., Borger, B., Pleikies, J., Veldhuis, D., & Hilgenkamp, H. (2007). NanoSQUIDs Based on Niobium Constrictions. *Nano Letters* , Vol. 7, No. 7, (July 2007), pp. (2152-2156), ISSN 1530-6984

Tsang, C., Fontana, R. E., Lin, T., Heim, D. E., Speriosu, V. S., Gurney, B. A., & Mason, M. L. (1994). Design, Fabrication and Testing of Spin-valve Read Heads for High-density Recording. *IEEE Transactions on Magnetics* , Vol. 30, No. 6, (November 1994), pp. (3801-3806), ISSN 0018-9464

Upton, M. H., Wei, C. M., Chou, M. Y., Miller, T., & Chiang, T. C. (2004). Thermal Stability and Electronic Structure of Atomically Uniform Pb Films on Si(111). *Physical Review Letters,* Vol. 93, No. 2, (July 2004), pp. (026802), ISSN 0031-9007

Verkin, B. I., Mende, F. F., Trubitsin, A. V., Bondarenko, I. N., & Sinenko, V. D. (1976). Superconducting Resonance System in Precise Measuring Units. *Cryogenics,* Vol. 16, No. 9, (September 1976), pp. (519-520), ISSN 0011-2275

Vinet, M., Chapelier, C., & Lefloch, F. (2001). Spatially Resolved Spectroscopy on Superconducting Proximity Nanostructures. *Physical Review B,* Vol. 63, No. 16, (April, 2001), pp. (165420), ISSN 1098-0121

Vodolazov, D. Y. (2007). Negative Magnetoresistance and Phase Slip Process in Superconducting Nanowires. *Physical Review B,* Vol. 75, No. 18, (May, 2007), pp. (184517), ISSN 1098-0121

Wagner, R. S., & Ellis, W. C. (1964). Vapor-Liquid-Solid Mechanism of Single Crystal Growth. *Applied Physics Letters,* Vol. 4, No. 5, (March, 1964), pp. (89-90), ISSN 0003-6951

Wang, J., Ma, X. C., Qi, Y., Fu, Y. S., Ji, S. H., Lu, L., Jia, J. F. & Xue, Q. K. (2007). Negative Magnetoresistance in Fractal Pb Thin Films on Si(111). *Applied Physics Letters,* Vol. 80, No. 11, (March, 2007), pp. (113109), ISSN 0003-6951

Wang, J., Ma, X. C., Lu, L., Jin, A. Z., Gu, C. Z., Xie, X. C., Jia, J. F., Chen, X. & Xue, Q. K. (2008a). Anomalous Magnetoresistance Oscillations and Enhanced Superconductivity in Single-crystal Pb Nanobelts. *Applied Physics Letters,* Vol. 92, No. 23, (June, 2008), pp. (233119), ISSN 0003-6951

Wang, J., Ma, X. C., Qi, Y., Fu, Y. S., Ji, S. H., Lu, L., Xie, X. C., Jia, J. F., Chen, X., & Xue, Q. K. (2008b). An Unusual Magnetoresistance effect in the Heterojunction Structure of an Ultrathin Single-crystal Pb film on Silicon Substrate. *Nanotechnology,* Vol. 19, No. 47, (November, 2008), pp. (475708), ISSN 0957-4484

Wang, J., Shi, C. T., Tian, M. L., Zhang, Q., Kumar, N., Jain, J. K., Mallouk, T. E., & Chan, M. H. W. (2009a). Proximity-Induced Superconductivity in Nanowires: Minigap State and Differential Magnetoresistance Oscillations. *Physical Review Letters*, Vol. 102, No. 24, (June, 2009), pp. (247003), ISSN 0031-9007

Wang, J., Ma, X. C., Qi, Y., Ji, S. H., Fu, Y. S., Lu, L., Jin, A. Z., Gu, C. Z., Xie, X. C., Tian, M. L., Jia, J. F. & Xue, Q. K. (2009b). Dissipation in an Ultrathin Superconducting Single-crystal Pb Nanobridge. *Journal of Applied Physics*, Vol. 106, No. 3, (August 2009), pp. (034301), ISSN 0021-8979

Wang, J., Ma, X. C., Ji, S. H., Qi, Y., Fu, Y. S., Jin, A. Z., Lu, L., Gu, C. Z., Xie, X. C., Tian, M. L., Jia, J. F. & Xue, Q. K. (2009c). Magnetoresistance Oscillations of Ultrathin Pb Bridges. *Nano Research*, Vol. 2, No. 9, (September 2009), pp. (671-67), ISSN 1998-0124

Wang, J., Singh, M., Tian, M., Kumar, N., Liu, B. Z., Shi, C., Jain, J. K., Samarth, N., Mallouk, T. E., & Chan, M. H. W. (2010a). Interplay between Superconductivity and Ferromagnetism in Crystalline Nanowires. *Nature Physics*, Vol. 6, No. 5, (May 2010), pp. (389-394), ISSN 1745-2473

Wang, J., Jia, J. F., Ma, X. C., Shen, Q. T., Han, T. Z., Jin, A. Z., Lu, L., Gu, C. Z., Tian, M.L., Xie, X. C., & Xue, Q. K. (2010b). Semiconductor-superconductor Transition and Magnetoresistance Terraces in an Ultrathin Superconducting Pb Nanobridge. *Journal of Vacuum Science and Technology*, Vol. 28, No. 4, (July 2010), pp. (678-681), ISSN 1071-1023

Wolf, S. A., Awschalom, D. D., Buhrman, R. A., Daughton, J. M., von Molnar, S., Roukes, M. L., Chtchelkanova, A. Y., & Treger, D. M. (2001). Spintronics: A Spin-based Electronics Vision for the Future. *Science* , Vol. 294, No. 5546, (November 2001), pp. (1488-1495), ISSN 0036-8075

Xiong, P., Herzog, A. V., & Dynes, R. C. (1997). Negative Magnetoresistance in Homogeneous Amorphous Superconducting Pb wires. *Physical Review Letters*, Vol. 78, No. 5, (February 1997), pp. (927-930), ISSN 0031-9007

Yang, Y., Kung, S. C., Taggart, D. K., Xiang, C., Yang, F., Brown, M. A., Guell, A. G., Kruse, T. J., Hemminger, J. C., & Penner, R. M. (2008). Synthesis of PbTe Nanowire Arrays Using Lithographically Patterned Nanowire Electrodeposition. *Nano Letters*, Vol. 8, No. 8, (August, 2008), pp. (2447-2451), ISSN 1530-6984

Zhang, T., Cheng, P., Li, W. J., Sun, Y. J., Wang, G., Zhu, X. G., He, K., Wang, L. L., Ma, X. C., Chen, X., Wang, Y. Y., Liu, Y., Lin, H. Q., Jia, J. F., & Xue, Q. K. (2010). Superconductivity in One-atomic-layer Metal Films Grown on Si(111). *Nature Physics*, Vol. 6, No. 2, (February, 2010), pp. (104-108), ISSN 1745-2473

Zhang, Y. F., Jia, J. F., Han, T. Z., Tang, Z., Shen, Q. T., Guo, Y., Qiu, Z. Q., & Xue, Q. K. (2005). Band Structure and Oscillatory Electron-phonon Coupling of Pb Thin Films Determined by Atomic-layer-resolved Quantum-well States. *Physical Review Letters*, Vol. 95, No. 9, (August, 2005), pp. (096802), ISSN 0031-9007

Josephson Vortices in High T_c Superconductors

Kazuto Hirata
National Institute for Materials Science, Tsukuba
Japan

1. Introduction

When a magnetic field is applied to the type-II superconductors in a superconducting state, the magnetic field penetrates into the superconductors above the first critical field H_{c1}, and is quantized in a unit of ϕ_0 (=h/2e=2.07x 10^{-7}gauss·cm^2). This is a general view of magnetic flux (vortex) in the type-II superconductors as schematically drawn in Fig.1 (a). The wave function of superconductivity Φ extends to the coherence length ξ, and the applied magnetic field H is screened by the supercurrent in a range of the penetration depth of λ from the center of vortex core. In this case, vortex has a core of normal state. In a lower magnetic field, vortices behave as isolated vortices. When the magnetic field becomes higher, interaction between/among vortices, vortices form „Abrikosov" vortex lattice in triangular or square distribution, which has been observed by scanning tunneling microscope (Hess et al. 1994) and small angle neutron scattering (Yethiraj et al, 1993) mainly in mettalic and intermetallic superconductors.

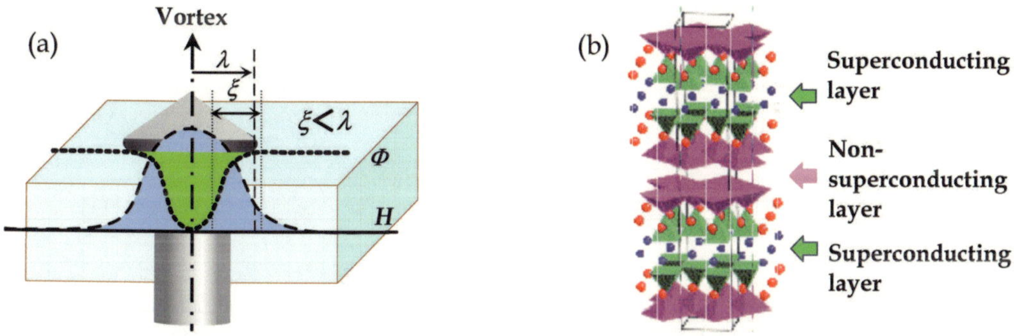

Fig. 1. Schematic drawing of a magnetic flux (vortex) in type-II superconductor (a) and the crystal structure of $Bi_2Sr_2CaCu_2O_{8+\delta}$ (b).

High-temperature superconductors (HTSCs) also belong to the type-II superconductors. However, vortex state is quite different from that in conventional superconductors. This has mainly its origin in the layered crystal structures in atomic scale. The crystal structure of HTSCs is consisted of the Cu-O superconducting layers and the nonsuperconducting ones iteratively as shown in Fig.1 (b). It causes a weak Josephson coupling between the Cu-O superconducting layers through the insulating layers in superconducting state. This leads intrinsically to the nanoscale Josephson junctions (JJs) in crystalline unit cells, and also induces a strong anisotropy of physical properties in HTSCs.

When the magnetic field is applied perpendicular to the superconducting layers of HTSCs, vortices in the superconducting layer distribute in a regular two-dimensional (2D) lattice of triangular or square as in the case of conventional superconductors. However, along the perpedicular direction to the layers, flux line is connected weakly to each other with pancake vortex in each layer. Vortices of HTSCs in the perpendicular field are called as pancake vortices (PVs). In the parallel field, based on the layered strucure of HTSCs, namely, intrinsic Josephson junctions (IJJs) (Kleiner et al., 1992; Rapp et al. 1996), vortices shown in Fig.2 (a) behave as those in the Josephson junctions artificially made of superconductor/insulator/superconductor structure. In the JJs, Josephson vortex (JV) is formed in a characteristic state of vortex without normal state core (Bulaevskii & Clem 1991; Koshelev 2000). JV is contracted along the direction perpendicular to the layers and extends to the direction parallel to the layers, depending on the anisotropy. Details are discribed in section 3.1.

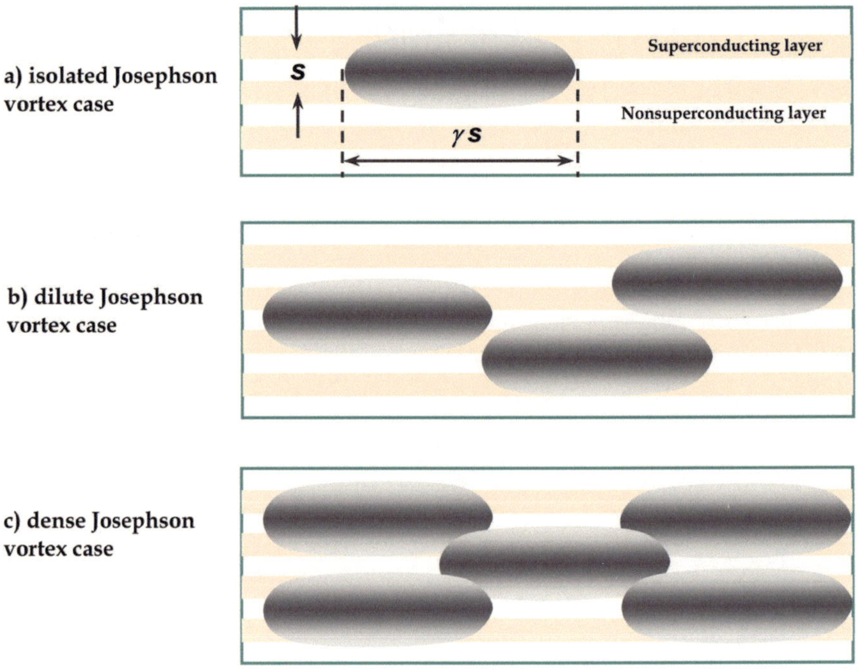

Fig. 2. Schematic drawing of distribution of Josephson vortices with a) isolated, b) dilute, and c) dense case.

Even in the perpendicular field to the superconducting layers, HTSCs show distinguished features of PVs in the vortex state. The first-order melting transition of PV-lattice (Zeldov et al. 1995), Bragg glass phase and vortex glass phase (Blatter et al. 1994) have been made clear experimentally and theoretically. In the parallel fields to the layers, magnetic field penetrates as JVs in a superconducting state of HTSCs. Theoretical studies have been made extensively from just after the discovery of HTSCs. Thermally-activated disordered vortex state, Kosterlitz-Thouless (K-T) transition in the smectic vortex state (Blatter et al. 1994; Balents & Nelson 1995), new vortex phases and a change of a tri-critical point in the dimension of the melting transition (Hu & Tachiki 2000)and the K-T type phase and the melting of 2D quasi-lattice have been predicted theoretically (Ivlev et al. 1990; Hu &

Tachiki 2004). However, JV system has not been understood well experimentally. The melting transition of the JV system (Kwok et al. 1994), the oscillatory melting temperature (Gordeev et al. 2000), and the vortex smectic phase have been found only in $YBa_2Cu_3O_{7-\delta}$ (YBCO). In strongly anisotropic HTSCs such as $Bi_2Sr_2CaCu_2O_{8+\delta}$ (Bi-2212), few experimental results have been reported on the magnetic phases of the JV system (Fuhrer et al. 1997; Mirkovic et al. 2001).

Most of the difficulties to study on JV system in experiments are considered due to a little change of physical parameters at the phase boundary. Shear modulus of JVs between the layers is very little and it is considered that a change in free energy is so small at the boundary to be observed by the experiments (Shilling et al. 1997). Recently we have found a new method to study the magnetic phases of JV system by using the periodic oscillations in flow resistance of JVs (Ooi et al. 2002), from which we can deduce the configurations of JVs. This phenomenon is caused by the intrinsic boundary effect of the sample, related to the configurations of JVs (Ooi et al. 2002; Koshelev 2002; Machida 2003). The magnetic phase, in which the periodic oscillations are observed, has been confirmed as a three-dimensional (3D) ordered vortex state. The periodic oscillations are observed independent to temperature and magnetic field, which is a universal nature of JV flow-resistance. Then, it is useful for determining an absolute magentic field. In this chapter, experimental methods for observing the periodic oscillations and for fabricating the sample for the measurements are described in section 2, general view of vortex flow in section 3, magnetic phases in section 4, application of the periodic oscillations in section 5, and finally conclusion in section 6.

2. Experimental method

Single crystals of Bi-2212 were grown with a travelling solvent floating zone method (Mochiku et al., 1997). The crystals have been cut and cleaved into the platelets with a thickness of about 20μm and an area of about 5mm x 3mm. Samples for the focused ion beam (FIB) structuring were obtained by dicing the platelets into small pieces of a bar-shape with an area of about 50μm x 3mm. The superconducting transition temperature (T_c) of the samples is about 85K. After four electrodes have been put on the bar by silver paste, IJJs have been fabricated at the center of the bar by using FIB as shown in Fig.3 (a). Detailed fabrication is described elsewhere (Kim et al. 1999). In the inset of Fig. 3 (a) and Fig.3 (b), a schematic drawing IJJ of the in-line symmetric type is shown used for four-probe measurements in the applied field *H* with ac (13 Hz:LR-700 AC resistance bridge) and dc current (Keithley 2400). Since the resistance of the IJJ part is 3 orders larger than that of the other part, we can measure the flow resistance of the IJJ part along the *c*-axis of the single crystal mainly as shown in Fig.3 (b).

The samples have been aligned to the parallel field by measuring the flow resistance at constant field and current as a function of relative angle. When the field is close to parallel to the layers, JVs are in a lock-in state (Hechtfischer et al. 1997), and they flow collectively parallel to the layers with a Lorentz force, which causes flow voltage and hence the flow resistance obtained. The mid point of the center of the plateaus in the flow-resistance obtained from both direction of scanning angle has been taken as a tentatively aligned position to the superconducting layers with a resolution of 0.005°. Refinement of the aligned position to the field has been made to show the maximum flow-resistance. Details of the JV flow measurements are also described in Ref. (Ooi et al. 2002). Figure 3 (c) shows a typical curve of the flow resistance at 70 K. The flow resistance begins to oscillate periodically from below 10 kOe, reaches to its maximum at 20 kOe, and suddenly drops to zero at around 23 kOe. This drop is caused by the misalignment of the sample. When the magnetic component H_c along the c-axis of the applied field H reaches to H_{c1}, then, PVs penetrate to the sample, are pinned at pinning centers (crystal defects), cause JVs pinning with cooperative phenomena, and reduce the JV flow-resistance. Refinement of the aligned position to the field has been made to show the maximum of the flow-resistance as large as possible in magnetic field. Then, the resistance reaches to saturate with a finite value after the maximum with increasing the parallel field. This angle is set as an optimum value for evaluating a magnetic phase of JVs.

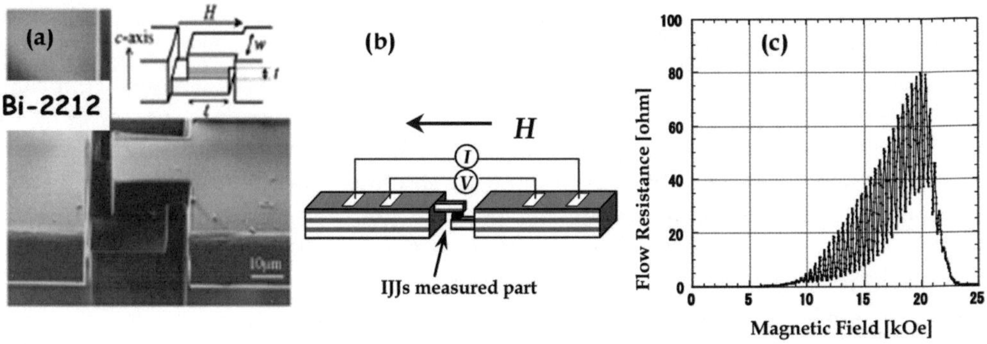

Fig. 3. (a) Secondary ion image of the measured sample after the FIB fabrication. Inset in (a) shows a schematic drawing of the sample geometry and applied magnetic field. (b) Schematic drawing of IJJs parts and electrodes configuration for the transport measurements. (c) Typical example of Josephson vortex flow-resistance against magnetic field.

In section 5, for the application of the periodic oscillations to measure a magnetic field precisely, pair of modulation coils has been set to the both side of the sample in Fig.3 (b) in addition to the main superconducting 8 Tesla split magnet. It produces about 200 Oe.

3. Flow resistance of Josephson vortices

When vortices in superconductors are driven by the applied current perpendicular to the direction of magnetic field, the vortices flow by the Lorentz force in the direction perpendicular to the field and the current. Movement of the vortex core, which is in normal state, causes a dissipation and hence flow-resistance. The flow velocity is determined by the vortex states and also the superconducting media. Vortex states are related to the glass state, vortex-lattice state, for example. Superconductors have many kinds of imperfections; inhomogenity of the materials, defects (stacking fault, dislocations, etc), and impurities. These imperfections sometimes become pinning centers for vortices and the vortex states cause a flow-resistance (Tinkham 1975; Blatter 1994). In HTSCs, PVs flow as to the vortex motion of vortices in conventional superconductors, although they have a variety of vortex phases. As HTSCs have a high T_c, the vortices are easily influenced by the thermal fluctuation. Furthermore, PVs are weakly connected between the superconducting layers through the insulating layer, which causes a complicated dissipation. On the other hand, the motion of JVs in HTSCs is quite different from that of PVs. JVs are located at the insulating layer as shown in Fig.2. They can easily move along the superconducting layers, but not along the perpendicular direction to the layers. This is so-called intrinsic pinning of JVs. For the applications of HTSCs to superconducting magnets, the intrinsic pinning mechanism is used to generate a magnetic field along the insulating layers.

3.1 Flow of Josephson vortices in HTSCs

Flow of JVs has been studied in artificially-made Josephson junctions. In principle, JV flows very fast close to the light velocity because of a very low dissipation without normal state core (Fujimaki 1987; Bulaevskii 1991). However, there are many imperfections in real materials, which lead to the larger dissipation and reduce the velocity. In HTSCs, Josephson junctions are formed intrinsically, consisted of a stack of atomic scale layers as shown in Fig.1 (b). In strongly anisotropic superconductors such as HTSCs, JV core is defined as Josephson length γs, where γ is the anisotropic parameter defined as λ_c/λ_{ab} (London penetration depth (λ_{ab} and λ_c)) and s is the interlayer spacing of superconducting layers, as shown in Fig.2 (a). JV flow accompanies a movement of JV core, which causes a dissipation of the *in-plane* resistivity ρ_{ab} and the c-axis quasi-particle tunneling resistivity ρ_c. Before discussing the flow resistivity ρ_{Jff} of JVs, we define the crossover magnetic field as $B_{cr}=\phi_0/\pi\gamma s^2$, following to the definition of Koshelev 2000, where JVs start to overlap and dense distribution of JVs form JV lattice as shown in Fig.2 (c). B_{cr} in Bi-2212 ($\gamma = 500$) is about 6 kOe with s = 1.5 nm. This value is relatively small compared to YBCO, which has a small value of γ about 10 and has a crossover magnetic field larger than a few Tesla.

JV flow-resistivity ρ_{Jff} is expressed as $\rho_{Jff} \sim \pi\gamma s^2 B/(\phi_0(\sigma_c+0.27\sigma_{ab}/\gamma^2))$ for the magnetic field $B< B_{cr}$ (Koshelev 2000), where σ_c and σ_{ab} are the c-axis quasi-particle tunneling conductivity and the *in-plane* conductivity, respectively. In this magnetic field range, ρ_{Jff} shows linear dependence to B and is proportional to the number of JVs. In the magnetic fields of $B>B_{cr}$, $\rho_{Jff} \sim \rho_c B^2/(B^2+B_\sigma^2)$, where $B_\sigma^2 = \sigma_{ab}\Phi_0^2/2\pi^2\sigma_c\gamma^4 s^4$ and ρ_c is the c-axis resistivity. With increasing magnetic fields, flow resistivity gradually deviates from linear dependency, and saturates to ρ_c at the field B_σ for strong in-plane dissipation. There are two kinds of dissipation dominated; the c-axis dissipation channel ($\sigma_{ab}/\sigma_c<<\gamma^2$) and in-plane dissipation channel ($\sigma_{ab}/\sigma_c>>\gamma^2$). B_σ is much larger than B_{cr}. In Bi-2212, B_σ is about 40 kOe for $\gamma = 500$, σ_{ab} = 50000 Ω^{-1}cm^{-1}, and σ_c = 0.002 Ω^{-1}cm^{-1}.

Figure 4 shows a magnetic field dependence of JV flow-resistance of Bi-2212 measured at 70 K with dc current density of 60 A/cm^2. Bi-2212 belongs to the resume of $\sigma_{ab}/\sigma_c<<\gamma^2$, estimated from the values in the last paragraph. Actually, the flow resistance increases linearly with magnetic field at lower magnetic fields, and shows a upward curvature at larger magnetic fields, which is the behaviour of the *in-plane* channel dominated flow resistivity (Koshelev 2000). At near the zero magnetic fields, the flow-resistance shows a downward curvature, which is considered as pinning effects of surface boundary of the junctions and pinning centres intrinsically existed in the sample.

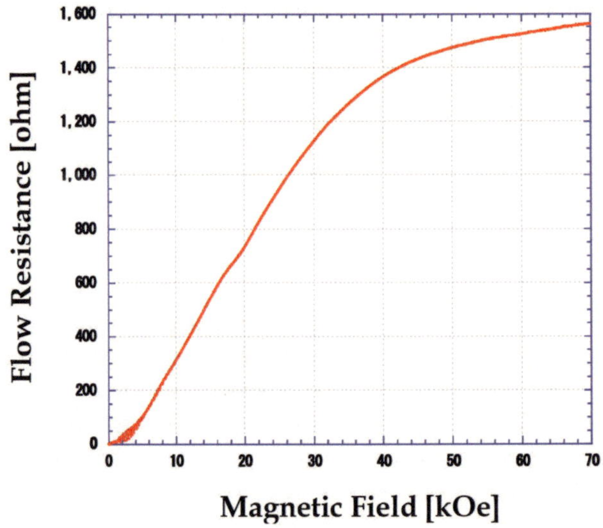

Fig. 4. Magnetic field dependence of the JV flow-resistance in Bi-2212 measured at 70K with dc current density of 60 A/cm². At lower magnetic fields, the flow-resistance shows linear magnetic field dependence, and saturates to a finite value at higher fields, which suggests the *in-plane* channel dominates in Bi-2212.

3.2 Periodic oscillations in flow resistance of Josephson vortices

In Fig.4, the JV flow-resistance of Bi-2212 was measured with dc current of 0.15 mA and the junction size of width 17.6 μm, length 14.0 μm, and thickness 1 μm. From the size of the sample, this current corresponds to the current density of about 60 A/cm². Figure 5 (a) shows current density dependence of the flow resistance of the sample at 70 K with the angle from the superconducting layers of 0.010 degree. Decreasing the current density from 60 A/cm², it can be clearly observed that the flow-resistance starts to oscillate at lower magnetic fields around 6-7 kOe below around 8 A/cm² (0.020 mA), and the oscillations disappear at higher magnetic fields. At lower current density than 8 A/cm², the oscillations become pronounced and the flow-resistance shows a maximum at around 40 kOe. The flow-resistance decreases from about 40 kOe, but still remains even at 70 kOe, which is the maximum magnetic field applied by the split superconducting magnet.

It is noted that the oscillations are observed only in sufficiently small current region. Figure 5 (b) shows an *I-V* characteristic of the junction size of width 18.0 μm, length 16.5 μm, and thickness 1 μm at 65K. *I-V* curves are plotted in magnetic fields from 14 kOe to 16 kOe for every 20 Oe. Each curve is shifted by a 1 mV step. At the magnetic fields where the flow-resistance shows minimum, the nonlinearity of *I-V* curves becomes large and a kink structure is seen at around 30-50 μA. The periodic nonlinearity corresponds to the periodic oscillation of the JV flow-resistance. In the current larger than 100 μA, the nonlinearity disappears. The current 30 μA corresponds to the current density of 9 A/cm². Appearance of the periodic oscillations at 30 μA (9 A/cm²) is quite well to coincide with the value of 8 A/cm² in the last paragraph.

Figure 6 (a) shows a part of Fig.3. (c), plotted from 15 to 18 kOe. The oscillations show quite constant periodicity H_p even in this magnetic field. The periodic oscillations are reproduceably observed in a wide temperature range from (T_c-3) to 4.2 K, and in a wide magnetic field range from B_{cr} to close to B_σ, which will be discussed in determining magnetic phases of JVs. We have investigated the origin of the periodicity by measuring the flow resistance on several samples with different width w. The width w is defined as the sample size in the perpendicular direction to the magnetic field sown in the inset of Fig.3(a).

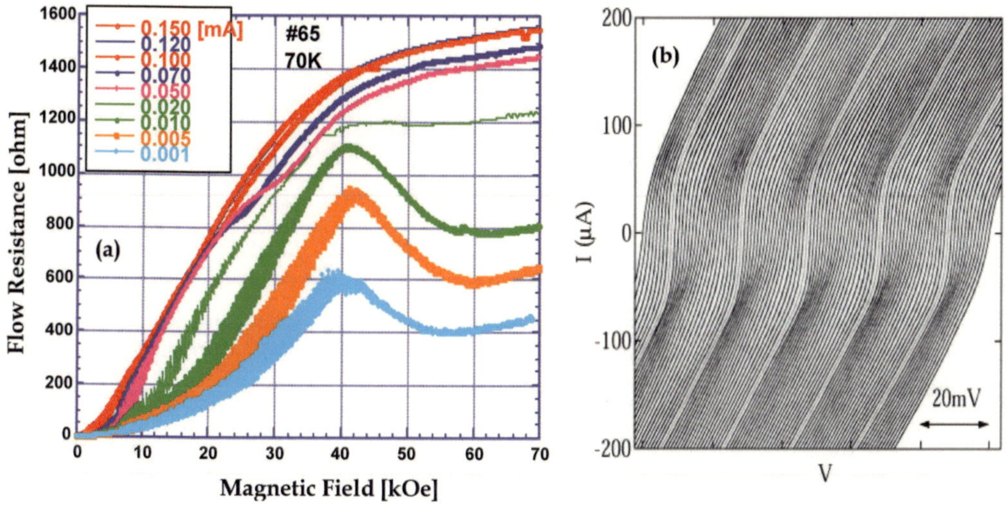

Fig. 5. (a) Current-dependency of the flow-resistance in a small current range. (b) I-V characteristic in magnetic fields from 14 kOe to 16 kOe for every 20 Oe at 65K. Each curve is shifted by a 1 mV step.

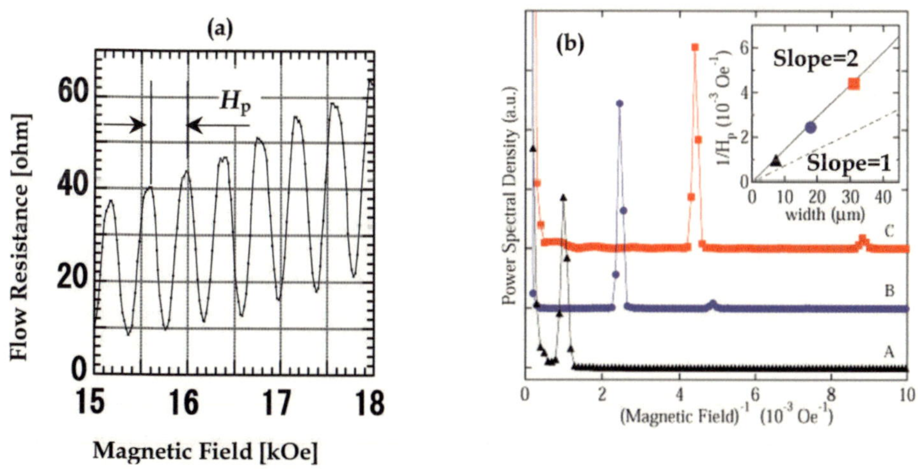

Fig. 6. (a) Periodic oscillations in JV flow-resistance; enlarged part of Fig.3. (b). (b) Power spectrum of the periodic oscillations in fast Fourier transform on samples A, B and C. Inset shows width (w) dependency of the inverse of H_p. The broken and solid lines represent $1/H_0(w)$ and its double, respectively.

Figure 6 (b) shows the fast Fourier transform of the periodic oscillations on samples A (w=7.3 μm), B (w=18.0 μm) and C (w=31.0 μm). The sharp fundamental peaks can be seen, which suggests that the periodicity H_p is quite constant in a wide range of fields and over the samples. The small peaks correspond to the second harmonic waves, which are caused by a little distortion from the sinusoidal curves. In the inset of Fig.6 (b), the position of the peaks is plotted against sample width (w). All the samples show a linear dependence to width (w) with a slope of 2, but not 1.

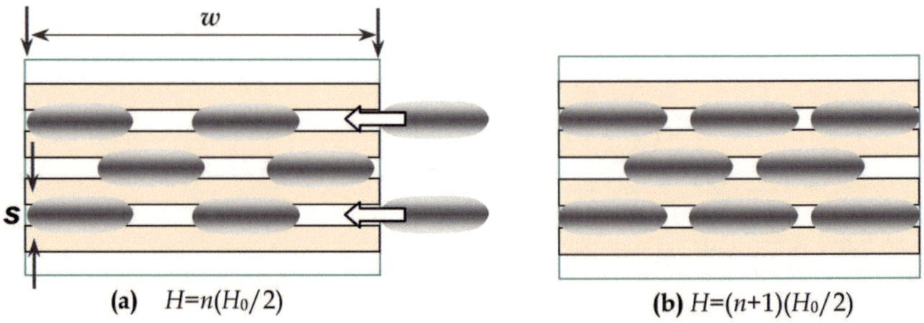

Fig. 7. Schematic drawing of Josephson vortex configuration to explain the magnetic period of the flow-resistance. The period indicates the magnetic increment of $\phi_0/2\,ws$, which means that the Josephson vortices form a triangular lattice above B_{cr}.

The value of the slope is very indicative for the interpretation of the periodic oscillations. When the JVs are supposed to be uniformly distributed in the junctions above B_{cr}, with increasing magnetic fields, JVs penetrate into each junction. Then, the increment of magnetic field ($H_0(w)$) to put one JV per one layer is ϕ_0/ws. Then, the slope $1/(\phi_0/ws)$ would be 1. However, the experiments show the value of 2. So, the slope is double of $1/(\phi_0/ws)$, and is rewritten as $2/(\phi_0/ws) = 1/(\phi_0/2ws)$. This means that ,in a increment of magnetic field H_p, one JV penetrates into every two-junctions.

The above experimental results can be explained by assuming that the lattice structure of JV system is triangular, and, at both sides of the sample, a surface barrier, Bean-Livingston barrier, for example, acts on the JVs' lattice. Figure 7 shows schematic drawings of the JV penetration into the junctions as (a) $H_p = n(H_0(w)/2)$ and (b) $H_p = (n+1)(H_0(w)/2)$, where n is the number of JV at certain magnetic field. Figures 7 (a) and (b) show the examples of JV system, in which the JVs are distributed in the ordered states, matched exactly with the size of the sample. In Fig.7. (a) of flowing JV-lattice system, JVs enter from the right side and exit from the left side simaultaneously. Then, the total potential in the JV-lattice becomes muximum. The average velosity of JV flow and hence the flow-resistance becomes minimum. With increasing the magnetic field from this sistuation, additional JVs are forced to enter into the ordered lattice by the applied current. The JV-lattice can then move more easily because the total potential decreases due to the mismatching between JV-lattice and the width of the sample. The flow resistance becomes larger than that of the matching situation. When the magnetic field exceeds $(n+1/2)(H_0(w)/2)$, the JV-lattice starts to form the next matching state as shown in Fig.7 (b). Therefore, it is plausible to consider that the oscillation period of the flow-resistance becomes half of $H_0(w)$. The effect of the boundary to the JV-lattice is crucial as shown in Fig.8. Difference in the two main effective width $\Delta w = |2l\sin\theta|$ shown in the inset of Fig.8 (b) acts to determine the beating period as a surface barrier to the flowing JV-lattice. The period H_p is determined only by the effective width, which means that JV-lattice flows is strongly affected by the boundary (Ooi et al. 2004).

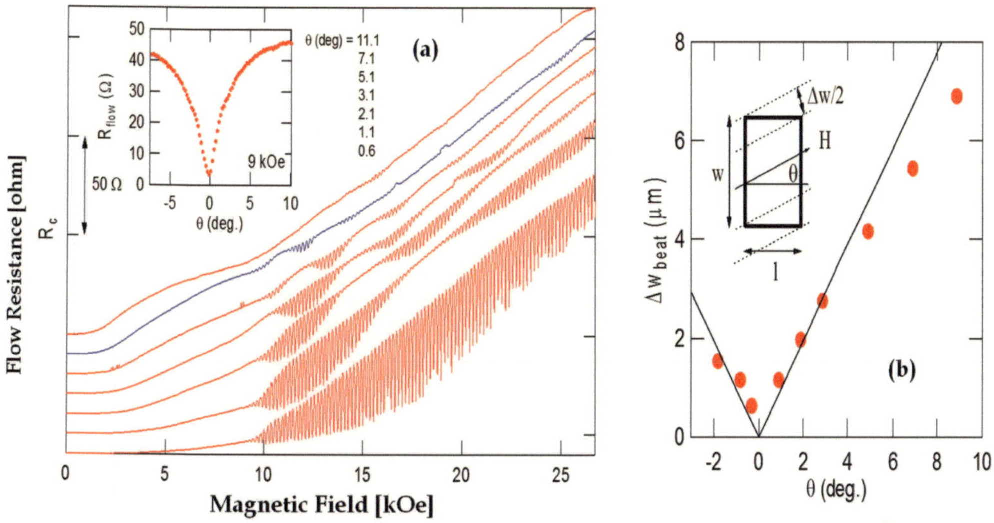

Fig. 8. (a) Beating in JV flow-resistance. The flow resistance was measured with changing the *in-plane* angle θ. With increasing angle θ, the beating effect becomes significant. Inset shows the angle dependence of the flow resistance at 9 kOe. (b) Difference of two effective widths as a function of θ estimated from the periods of beatings. Solid curve shows $\Delta w = |2l \sin\theta|$. A schematic drawing of the sample in an in-plane field is put in the inset to explain the effective width and definition of Δw.

The above mentioned interpretation can be applied to the *I-V* characteristics and the current density dependence of the oscillations in the magnetic fields qualitatively. When the surface potential is smaller than the energy of JV-system flowing with Lorentz force by the larger applied current, it is expected that the periodic oscillations would be smeared out. This is the speculation from the experimental results, but it was confirmed theoretically by analytical method (Koshelev 2002; Ikeda 2002) and numerical calculations (Machida 2003).

As the periodic oscillations can be observed in a lower current density, it is considered that as a ground state the JV-system is in a triangular lattice at intermediate magnetic fields. We also have investigated the JV flow-resistance with changing the anisotropy γ to confirm the effect of the anisotropy on forming a triangular lattice (Yu S. 2004 & 2007). As this is related to the magnetic phase diagram of Bi-2212 described in the next section, we only mentioned here that the starting magnetic field H_s of the periodic oscillations is inversely proportional to γ. This is also one evidence to prove the theories and the formation of triangular lattie. Furthermore, the triangular lattice of JV system changes to the rectangular lattice with changing the sample size smaller and increasing the magnetuc fields (Yu S. 2004 & Hirata K. 2003). The boundary magnetic field B_{TR} between triangular and rectangular lattice is defined as $B_{TR}=w\Phi_0/2\pi\gamma^2 s^3$ (Koshelev 2002), which is also well coincided with experiments (Yu S. 2004 & Hirata K. 2003). On contrary, it is useful to determine the anisotropic parameter from the JV flow resistance measurement directly. The periodic oscillations have a universal nature of JV-system in a strongly anisotropic superconductors, and hence can be applied to determine the magnetic phases of JV-system and the magnetic field precisely, which will be disccussed in section 4 and 5, respectively.

4. Magnetic phases of Josephson vortices

Magnetic phases of HTSCs have been well studied in the weakly anisotropic superconductors such as YBCO (Blatter et al. 1994). The first-order melting transition is one of the pronounced characters in HTSCs. In strongly anisoropic superconductors such as Bi-2212, theoretical approaches have been made extensively, but few experimental studies have been made due to the little change in free energy to confirm the phase boundary, as mentioned in the Introduction. One of the useful methods is to measure a flow resistance of JVs. This method is useful even in the vortex solid phase. In the solid phase, it is difficult to study the state because of that resistivity usually goes to zero due to the pinning effects from quenched disorder. However, a collective JV flow appears in the *c*-axis resistance even at low temperatures (Lee et al. 1995). The periodic oscillations mentionend above is an indirect experimental evidence of forming a triangular lattice in JV system of strongly anisotropic superconductors, but is considered to be a unique experimental probe to confirm the 3D-ordered state in JVs. Therefore we applied this method to determine magnetic phases in Bi-2212.

4.1 JV flow resistance at lower magnetic fields

To determine magnetic phases in Bi-2212, we apply the JV flow resistance method. The magnetic field region, in which we can observe the periodic oscillations, is assigned to the 3D-ordered state of JV system. In the measurements, the effect of the PVs is crucial, which are easily excited thermally in HTSCs with a strong anisotropy. As PVs are also incorporated with the misalignment of the parallel fields to the superconducting layers. However, the effect is much enhanced at higher magnetic fields, and there is no need to consider about the existence of PVs at lower fields. Figure 9 shows a JV flow resistance of the sample (width= 17 μm, length = 14 μm and thickness= 1 μm) as a function of the lower parallel field at the temperature close to T_c (85K) with an ac current of 1.0 μA. The periodic oscillations start at around 80.5 K, and are clearly seen from 80.0 K at 5.8 kOe, which is denoted as an oscillation-starting magnetic field of H_s. This magnetic field H_s corresponds to the boundary between a liquid state and a triangular lattice state. H_s has little temperature dependence seen from Fig.9. The boundary is almost independent of temperature. It is noted here that, at the temperatures without the oscillations in the flow resistance, the flow resistance increases proportional to the field and we can see a development of a hump shown as H_h in Fig.9. Just before the appearance of the oscillations, the hump H_h seemes to jump to the starting field H_s of the oscillations with decreasing temperature. H_h also appears continuously as a kink in the flow resistance at about 3 kOe independent of temperature at lower temperatures (not shown here). This value of H_h is almost half of the starting field H_s, and is considered to relate to the formation of the JV lattice with a half filling to the layers (Ikeda 2002). As the half filling state is one of ordered state (Ikeda 2002), the increment of the flow resistance will be stopped as indicated in the experiments, considering from the matching effect of the JV half filling lattice to the sample width.

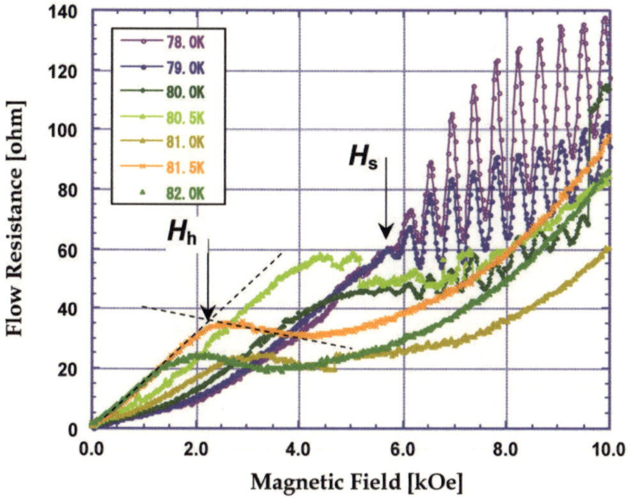

Fig. 9. Josephson-vortex flow resistance as a function of the parallel fields at the temperature close to T_c. We can find a hump, continuously connected to the beginning of the oscillations at 5.8 kOe with decreasing temperature.

4.2 JV flow resistance at higher magnetic fields

Figures 10 (a) and (b) show the flow resistance as a function of the parallel field at temperatures from 4.2 K to 80K on the same sample in Fig.9. Below 20 K, we can obseve a constant period H_p ($H_p = \phi_0/(2ws)$) of the oscillations till 70 kOe. The flow resistance still increases in this temperature range. At intermediate temperatures from 30 K to 50K, it shows a maximum and decreases and becomes constant at higher magnetic fields. But at higher temperatures it increases again. The decrease in flow resistance at the intermediate temperatures is not caused by the misalignment of the sample as following. In the case of the misalignment about 0.03°, on which the results are shown in Fig.3 (c), the resistance shows a maximum around 20 kOe and a sharp drop to zero at about 23 kOe. Incorporation of the PVs with the *c*-axis component of the field, JVs are easily pinned to the pinning sites, because of the strong pinning of the PVs at lower temperatures in Bi-2212 and of considering the free energy of the pinned JVs to the PVs (Koshelev 1999). Decrease of the flow resistance after the maximum may be caused by a disordered state of JVs magnetically and thermally from the 3D-ordered state. Above 60 K, the pinning strength of the PVs in Bi-2212 is decreased, which can explain the increment of the flow resistance. Then, it is considered that, when the sample is close to the aligned configuration like this experiment, the finite flow resistance in the intermediate temperatures at higher magnetic fields is an intrinsic one in JV system. With increasing temperature and magnetic field, it is suggested that JV system changes from a triangular-lattice state to another one. The phase boundaries are drawn in Fig. 11 as a magnetic phase diagram of JV system. This may be considered as a magnetically induced transition at a constant temperature, and as a thermally induced transition at a constant field. In this state, JV system is not in an ordered state in 3D, along the *c*-axis or along the *a*- or *b*-axis also. This state may be in a 2D quasi-long-range ordered state suggested in Ref. (Hu &Tachiki 2000). Further experiments should be made to make clear the phase diagram of JVs.

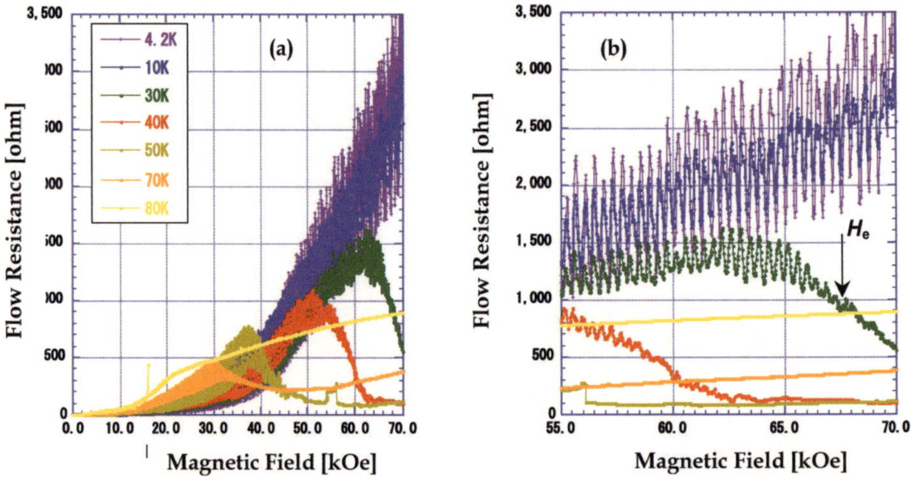

Fig. 10. Josephson-vortex flow resistance as a function of the parallel fields at various temperatures (a) and the enlarged part of (b) at higher magnetic fields (b). H_e denotes the end of the periodic oscillations.

The flow resistance increases monotonously with the periodic oscillations at lower temperatures with increasing magnetic field. With increasing temperature, the flow resistance always shows a maximum and decreases to a finite value in the flow resistance, in which the oscillations stop just before reaching to the finite flow resistance. To define the characteristic magnetic field H_e, a criterion was taken at some magnitude of the resistivity in the amplitude of the oscillations. Reproducibility was confirmed several times with changing the angle between the superconducting layers and the magnetic field, which showed a few kOe as in experimental error. Therefore, this characteristic feature of the flow resistance at higher temperature is considered as an intrinsic behaviour in JV flow.

In Fig.11, the phase boundaries obtained from the flow resistance measurements are shown. The magnetic field H_e in Fig.10 (b), indicated by the arrow, corresponds to the upper boundary of the 3D-ordered phase, which is defined as the magnetic field for the oscillations to be stopped. The boundary H_e decreases steeply at the temperatures lower than 40 K, is saturated at about 50kOe, and suddenly decreases at about 75 K towards 0 kOe at T_c. This means that the oscillations continue even in higher magnetic fields than 70kOe below 40 K, and JV system is in the 3D-ordered state below H_e and above H_s boundaries. It is noted here that H_s is almost constant with temperature because of depending only on the anisotropy γ and the samples width w.

Considering about the JV state in this magnetic field and temperature region, it is suggested that the JV state may really change from the 3D-ordered state to the other. In the flow resistance above H_e, the resistance always decreases to a finite value after the periodic oscillations in the region B of Fig.11, which may suggest that the JV state is in disordered along the *c*-axis, as the *in-plane* disorder causes an enhancement of the flow resistance with less effective potential barrier to the dynamical movement of JVs. With changing the angle between the layers and the magnetic field, little effect and the less systematic changes to the flow resistance were observed concerning on the decrement of the resistance. Then, we can get rid of the effect of the PVs caused by the misalignment in the higher magnetic fields. Then, the decrease of the resistance in this region may be caused by the excitation of abundant amount of pancake-vortex/anti-vortex pairs, thermally and magnetically, because the pinning force of the PVs is very weak in this magnetic field and temperature range far above the irreversibility region of the PVs.

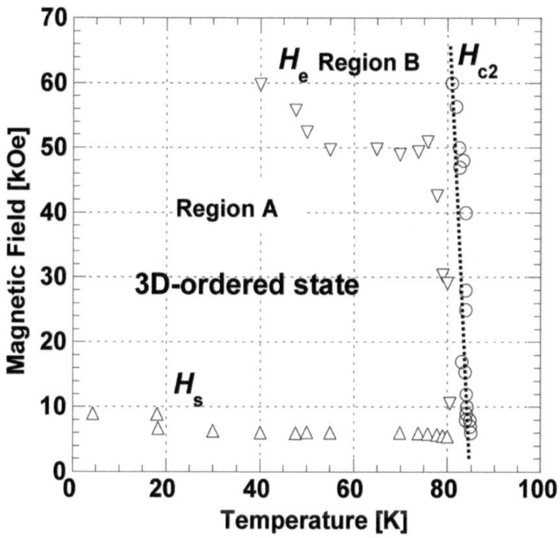

Fig. 11. Magnetic phase diagram of Josephson vortices in Bi-2212, obtained from the flow-resistance measurements. Open and solid symbols correspond to the boundaries defined on the sample in Fig. 10 and another sample with smaller γ, respectively.

To study the JV state above/below the upper boundary H_e, shown in Fig.11 as the region A/B, respectively, *I-V* measurements have been made. In Fig.12, the power-law dependence to the magnetic field in sample A is shown as the exponent α(T,H) of $V=AI^{\alpha}$, obtained from the *I-V* characteristics. In the figure, the data are plotted every 10 kOe, and are shifted in both sides at every magnetic field to see clearly. The measurements were made in the current range well below the critical current density of Bi-2212, in which the periodic oscillations can be observed in the region A. It shows a distinct change in α(T,H) to 1.6 and 2.0 above 50 kOe, compared with those at lower magnetic fields. The change in α is well coincided with the magnetic phase diagram shown in Fig.11.

Hu and Tachiki (Hu & Tachiki 2000) have studied on JV system, the first-order melting transition, phase transition between 3D ordered state and 2D QLRO (quasi-long-range ordered) state, and also the power-law dependence in K-T phase with Monte Carlo simulations. The region B in Fig.11 obtained from the present measurements is considered to correspond to the K-T phase in Fig.10 of Ref. (Hu & Tachiki 2000). Temperature dependence of $\alpha(T,H)$ is not explicitly shown in Eq.18; $\alpha=1+\Upsilon\pi/(k_BT)$ of Ref. (Hu & Tachiki 2004). However, the behavior of the exponent α(T,H) is inferred from the temperature dependence of the calculated helicity modulus Υ with $\gamma = 8$ and 20 in Fig.10 of Ref. (Hu & Tachiki 2000), which decreases with increasing temperature. Assuming that the behavior of Υ in temperature is the same as that in a larger anisotropic value of Bi-2212, $\alpha(T,H)$ decreases with increasing temperature. This shows a little different behavior from the experimental results. Furthermore, there have not been observed the oscillatory behavior of the melting transition and the first-order phase transition in the experiments on Bi-2212. It may need much more sensitive measurements than the present experimental set-up. Further experiments will be needed to make this JV state clear.

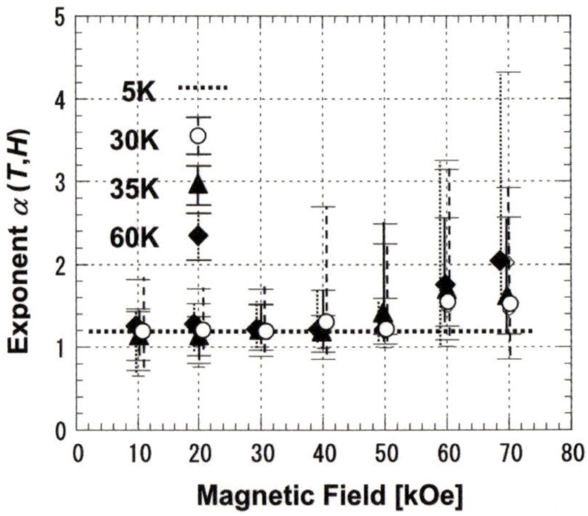

Fig. 12. Magnetic field dependence of the exponent $\alpha(T,H)$ in the I-V characteristics at 5, 30, 35, 60 K with $V=AI^{\alpha}$.

We have tried to confirm the effect of the anisotropy γ in the boundary with changing the anisotropy (Yu et al. 2005 & 2007). The tendency of the change of H_e is a good indication to increase the doping level of carriers into Bi-2212, namely, decreasing the anisotropic parameter γ, which can be related to the weaker anisotropic superconductors such as YBCO continuously.

5. Application of periodic oscillations for magnetic sensor

With magnetic-field cycles, the flow resistance reproduces quite well in a wide range of temperature, in which the oscillations can be observed. However, with heating/cooling cycles at a constant magnetic field, the flow resistance shows an irreversible feature as shown in Fig.13. After applying the magnetic field up to 20 kOe at 5 K with a zero-field cooling and confirming the periodic oscillations, for example, with increasing temperature, the flow resistance suddenly drops close to zero at about 50 K, even if a finite flow resistance can be observed at the same temperature and magnetic field with the zero-field cooling and magnetic-field cycles. This can be seen at every magnetic field where the oscillations can be observed. This may be closely related to the JV states, which have never been made clear before on the strongly anisotropic HTSCs experimentally. It has been theoretically discussed just after the discovery of HTSCs (Chakravaty et al. 1990; Blatter et al. 1994), and recently with Monte-Carlo simulations (Hu & Tachiki 2004). In the JV system, it is predicted that pairs of pancake vortex-anti-vortex are easily excited thermally. If these pancake pairs are excited in the JV system, they are easily pinned at the intrinsic pinning centers of Bi-2212, because this region belongs to the irreversible one in the magnetic phase diagram of Bi-2212 with the perpendicular field.

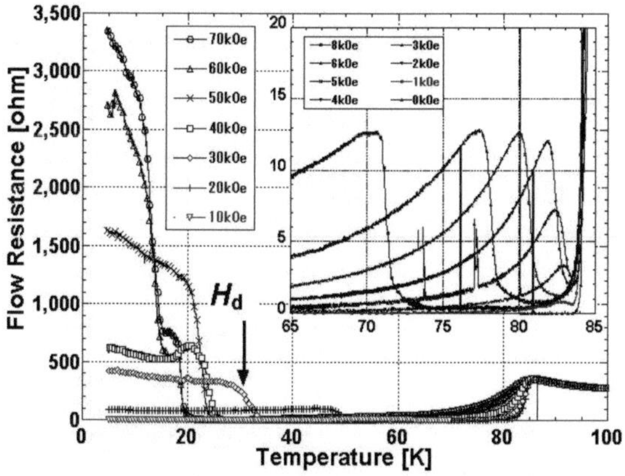

Fig. 13. Temperature dependence of the flow resistance at a constant magnetic field measured in a zero-field cooling&warming-up process. Inset shows the flow resistance at lower magnetic field region.

However, the magnetic field, at which the flow resistance suddenly drops, cannot be seen in the temperature dependence of magnetic-field cycles at every temperature. There exists only a boundary between the regions on the periodic oscillations and of the finite flow resistance without the oscillations at high temperatures and magnetic fields (Hirata et al. 2004). This boundary may correspond to the continuously melting transition (Hu & Tachiki 2004). In the field-cooling process, it is considered that these PV pairs are excited thermally at higher temperatures than the irreversibility temperature, they are frozen to the pinning centers with decreasing temperature, and the flow resistance does not show any sharp increment at low temperatures. Although there are still incomprehensive problems in the JV system, for the application, we have to avoid a temperature increase of the sensor higher than 50 K at 20 kOe, for example, as shown in Fig.13.

Using a pair of small coils, we have measured the JV flow resistance. Figure 14 shows the results of the JV flow resistance measured at the main magnetic field H_m with an interval of 20 Oe, an ac current of 1.0 μA, and the size of the junction t = 1.5 μm, w = 11.2 μm, l = 12.4 μm at 70.0 K as an example. The period H_p is obtained as about 900 Oe independent of temperature and magnetic field, which is well coincided with the calculated value H_p as 924 Oe, considering about Josephson penetration length and the damaged width by the FIB-fabrication of the sample (Hirata et al. 2004). Fixing the field H_m at 15.2 kOe, a modulation field H_i was applied to measure a resistance change by the JV flow. The signal was directly obtained from LR-700 as a raw output voltage without any filters, which is shown in Fig.14. To compare the magnitude of the flow resistance at 15.2 kOe of 70.0 K in Fig.13 with that in Fig.14, it is just a bit smaller than that about 30 ohm. Carrying out the measurements of the flow resistance, we have recognized that the c-axis resistance becomes very sensitive in superconducting state. By the extra set-up of the coils to the normal measurements, it may cause some noise in the circuit of LR-700.

In Fig.14, it is apparently seen that the signal is well scattered. This means that the JV flow velocity changes within less than 0.1 sec, because the measurements were made with the ac current of 13.6 Hz. This has been discussed by Machida (Machida 2003). The JV flow is a dynamical effect, interacting to each other. And JVs go over the potential barriers at the both edges of the sample, which is one of the principal reasons for the periodic oscillations, while inside the sample edges JVs move almost freely accompanied with some dissipation by the quasi-particles (Koshelev 2000). When we take a snapshot to see the behavior of the JV flow, there might be some distribution in velocity of JV-lattice flow. The velocity is determined mainly by the matching effect of JV lattice to the width of the sample (Machida 2003). Making these noise characteristics clear, it may be effective to take a noise spectrum of the flow voltage, which will be developed and obtained near future.

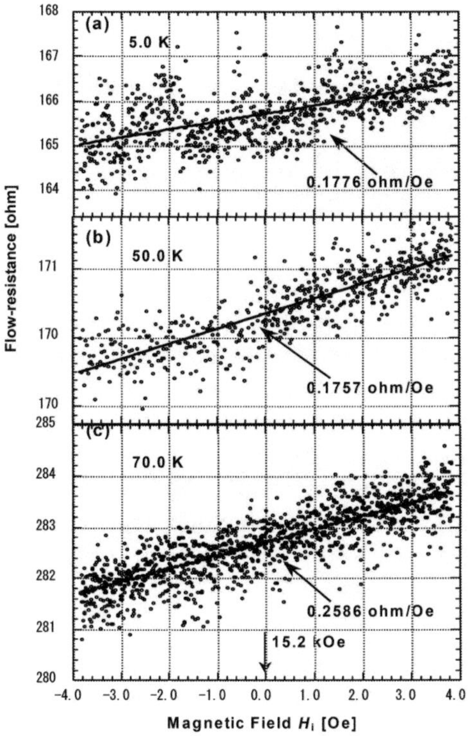

Fig. 14. Flow-resistance (a), (b) and (c) at the main fields H_m=15.2 kOe as a function of a modulation field H_i of small coils at 5, 50 and 70 K, respectively. The solid lines show a fitting line to the data, and show a good linearlity with a slope of 0.1776, 0.1757 and 0.2586 ohm/Oe, respectively.

It is worth for being noted that the linearity of the flow resistance to H_i is relatively good, which is shown as solid lines in Fig.14. The solid lines were obtained with a square-root minimum fitting method. The slope of the line is calculated as 0.2568 ohm/Oe at 70 K shown in Fig.14 (c), for example. The fitting line is considered to indicate average values in time at each field. And the measurement is easily improved to make averaging the voltage in the interval of a few second or longer. When we need to measure a high magnetic field precisely, we are usually using the magnetic field in a steady state; apparatus of nuclear magnetic resonance and magnetic resonance imaging, for example. It is also benefit for this magnetic-field sensor to put into the low T_c superconducting magnets, as they are using at a liquid He temperature, and the temperature is very stable. Once if we make a calibration of the sensors at 4.2 K and a fixed magnetic field, we can get a precise control of the magnetic field within a μT range.

6. Conclusion

In conclusion, we have measured JV flow-resistance on the in-line symmetric IJJs of Bi-2212 single crystals to study on the flow mechanism of JVs and on the dynamics of moving JVs. A pronounced feature has been found in the flow resistance. Periodic oscillations have been observed with sweeping the magnetic field parallel to the layers. The oscillations have been found in the wide range of the parallel fields larger than 6kOe and only in the smaller current region than about 10 A/cm^2. The period of the oscillations is inversely proportional to the width of the samples, and it corresponds to adding "one" flux quantum per "two" IJJs. These results suggest that the JVs form a triangular lattice as a ground state, which is related to the commensurability between the sample width and the lattice spacing of JVs along the layers in the present experiments using small IJJs.

We have studied the JV states with measuring the flow resistance. From the measurements, we can determine the 3D-ordered phase of JVs, in which JVs are well ordered. This has been never determined with other methods in strongly anisotropic superconductors. Furthermore, by the measurements of temperature dependence of the flow resistance, we find a sudden drop in the resistance in the zero-field cooling&warming-up process. The sudden drop is explained by the thermal and magnetical excitation of pancake vortex/anti-vortex pairs, which is an intrinsic property of Bi-2212. The boundary of the excitation exists between 20-30K above 15kOe and above 30K at lower magnetic fields. This boundary limits the application of HTSC materials in the magnetic fields.

We also have characterized the JV flow device as a magnetic sensor to measure the flow-resistance in Bi-2212 IJJs, using a remarkable feature of the periodic oscillations. The device has an irreversible character in temperature-cycles at high magnetic fields, which is caused by the thermal excitation of pancake vortex-anti-vortex pairs. Using a pair of small coils under high magnetic fields, good linearity of the flow-resistance was obtained to the internal field H_i. There is also a problem with noise, which may be caused by the intrinsic behavior of JV flow. However, it can be improved easily by the measurement method. The linearity of the flow-resistance to the small field H_i indicate that, once the device is calibrated, it has a high potential to measure and control high magnetic fields in Tesla range with a μT accuracy.

7. Acknowledgment

I acknowledged to Dr. S. Ooi for the fabrications of junctions and for useful discussions, and Dr. T. Mochiku for providing us single crystals of Bi-2212.

8. References

Bean C.P. & Livingstone J.D. (1964). Surface Barrier in Type-II Superconductors, *Phys. Rev. Letters* Vol.12, pp.14-16

Balents L. & Nelson D.R. (1995). Quantum Smectic and Supersolid order in Helium Films and Vortex Arrays, *Phys. Rev.* B Vol. 52, pp.12951-12968

Blatter G.; Ivlev B., & Ovchinnikov Y. (1990). Kosterlitz-Thouless transition in the smectic vortex state of a layered superconductor, *Phys. Rev. Letters* vol. 66, pp. 2392–2395

Blatter G.; Ivlev B. & Rhyner J. (1991). Kosterlitz-Thouless Transition in the Smectic Vortex State of a Layered Superconductor, *Phys. Rev. Letters* Vol.66, pp.2392-2395

Blatter G.; Feigel'man M.V., Geshkenbein V.B., Larkin A.I. & Vinokur V.M. (1994). Vortices in high-temperature superconductors, *Rev. Mod. Phys.* Vol.66, pp.1125-1388

Bulaevskii L. & Clem J.R. (1991). Vortex lattice of highly anisotropic layered superconductors in strong, parallel magnetic fields, *Phys. Rev.* B Vol. 44, pp.10234-10238

Chakravaty S.; Ivlev B. & Ovchinnikov Y. (1990), Resistivity of High-Temerature Superconductors: Is the Vortex State a Liquid?, *Phys. Rev. Letters* Vol.64, pp.3187-3190

Fuhrer M. S.; Ino K., Oka K., Nishihara Y. & Zettl A. (1997). Observation of Josephson Vortex Lattice Melting in a Highly Anisotropic Superconductor, *Solid State Commun.* Vol.101, pp. 841-845

Gordeev S.N.; Zhukov A.A., de Groot P.A.J., Jansen A.G.M., Gagnon R. & Taillefer L. (2000). Oscillatory Melting Temperature of the Vortex Smectic Phase in Layered Superconductors, *Phys. Rev. Letters* Vol.85 pp.4594-4597

Hechtfischer G.; Kleiner R., Schlenga K., Walkenhorst W. & Muller P. (1997). Collective motion of Josephson vortices in intrinsic Josephson junctions in $Bi_2Sr_2CaCu_2O_{8+y}$, *Phys. Rev.* B Vol.55, pp.14638-14644

Hess H.F.; Murray C.A. & Waszczak J.V. (1994). Flux Lattice and Vortex Structure in 2*H*-$NbSe_2$ in Inclined Fields, *Phys. Rev.* B Vol.50, pp.16528-16540

Hirata K.; Ooi S., Sadki E.H. & Mochiku T. (2003). Josephson Vortex flow in $Bi_2Sr_2CaCu_2O_{8+y}$, *Physica* B Vol.329-333 pp.1332-1333

Hirata K.; Ooi S., Yu S., Sadki E.H. & Mochiku T. (2004). Josephson Vortex States in $Bi_2Sr_2CaCu_2O_{8+\delta}$ Obtained from Vortex Flow Resistance measurements, *Physica* C Vol.412-414 pp.449-453

Hirata K.; Ooi S., Yu S., Sadki E.H. & Mochiku T. (2004). Novel Magnetic Sensors using High T_c Superconductors, *Supercon. Sci. Technol.*Vol.17 pp.S432-S435

Hu X. & Tachiki M. (2000). Possible Tricritical Point in Phase Diagrams of Interlayer Josephson-Vortex Systems in High-T_c Superconductors, *Phys. Rev. Letters* Vol.85 pp.2577-2580

Hu X. & Tachiki M. (2004). Decoupled two-dimensional superconductivity and continuous melting transition in layered superconductors immersed in a parallel magnetic field, *Phys. Rev. B*, vol. 70, pp. 064506-1- 13

Ikeda R.; (2002). Josephson Vortex State in Intermediate Fields, *J. Phys. Soc. Jpn.*, vol. 71, pp. 587–593

Ishida T.; Okuda K., Ryukov A.I., Tajima S. & Terasaki I. (1998). In-Plane Anisotropy of Vortex-Lattice Melting in Large $Yba_2Cu_3O_7$ Single Crystals, *Phys. Rev.* B Vol.58, pp.5222-5225.

Ivlev B.I.; Kopnin N.B. & Pokrovsky V.L. (1990). Shear Instabilities of a Vortex Lattice in Layered Superconductors, *J. Low Temp. Phys.* Vol.80, pp.187-195

Kim S. -J.; Latyshev Y.I. & Yamashita T. (1999). Submicron stacked-junction fabrication from whiskers by focused-ion-beam etching, *Appl. Phys. Letters* Vol.74 pp.1156-1158

Kleiner R.; Steinmeyer F., Kunkel G. & Muller P. (1992). Intrinsic Josephson Effects in $Bi_2Sr_2CaCu_2O_8$ Single Crystals, *Phys. Rev. Letters* Vol.68, pp.2394-2397

Koshelev A.; (1999). Crossing Lattices, Vortex Chains, and Angular Dependence of Melting Line in Layered Superconductors, *Phys. Rev. Letters* Vol.83 pp.187-190

Koshelev A.; (2000). Role of In-Plane Dissipation in Dynamics of a Josephson Vortex Lattice in High-Temperature Superconductors, *Phys. Rev.* B, vol. 62, pp. R3616–3619

Koshelev A.; (2002). Edge Critical Current of Dense Josephson Vortex Lattice in Layered Superconductors, *Phys. Rev.* B Vol.66 pp.224514-1-6

Kwok W.K.; Fendrich J., Welp U., Fleshler S., Downey J. & Crabtree G.W. (1994). Suppression of the First Order Vortex Melting Transition by Intrinsic Pinning in $YBa_2Cu_3O_{7-\delta}$, *Phys. Rev. Letters* Vol.72, pp.1088-1091

Lee J.U.; Nordman J.E. & Hohenwarter G. (1995). Josephson Vortex Flow in Superconducting Single-Crystal $Bi_2Sr_2CaCu_2O_x$, *Appl. Phys. Lett.* vol. 67, pp. 1471-1473

Machida M. (2003). Dynamical Matching of Josephson Vortex Lattice with Sample Edge in Layered High-T_c Superconductors: Origin of the Periodic Oscillations of Flow Resistance, *Phys. Rev. Letters* vol. 90, pp. 037001-1-4

Mirkovic J.; Savel'ev S.E., Sugawara E. & Kadowaki K. (2001). Stepwise Behavior of Vortex-Lattice Melting Transition in Tilted Magnetic Fields in Single Crystals of $Bi_2Sr_2CaCu_2O_{8+\delta}$, *Phys. Rev. Letters* vol. 85, pp. 886-889

Mochiku T.; Hirata K. & Kadowaki K. (1997). Crystallinity Improvement of $Bi_2Sr_2CaCu_2O_{8+\delta}$ Single Crystal by TSFZ Method, *Physica* C282-287 (1997) 475-476.

Fujimaki A.; Nakajima K. & Sawada Y. (1987). Spatiotemporal Observation of the Soliton-Antisoliton Collision in a Josephson Transmission Line, *Phys. Rev. Letters* vol. 59, pp. 2895-2899

Ooi S.; Mochiku T. & Hirata K. (2002). Periodic Oscillations of Josephson-Vortex Flow Resistance in $Bi_2Sr_2CaCu_2O_{8+y}$,*Phys. Rev. Letters* Vol. 89 247002-1-4

Ooi S.; Mochiku T., Yu S., Ishikawa H. & Hirata K. (2004). Beats of Josephson-Vortex Flow Oscillations, *Physica* C Vol.412-414 pp.454-457

Schilling A.; Fischer R.A., Phillips N.E., Welp U., Kwok W.K. & Crabtree G.W. (1997). Anisotropic Latent Heat of Vortex-Lattice Melting in Untwinned $YBa_2Cu_3O_{7-\delta}$,*Phys. Rev. Letters* Vol.78, pp.4833-4836.

Tinkham M.; (1975). *Introduction to Superconductivity*, McGraw-Hill, ISBN 0-07-064877-8, USA

Rapp M.; Murk A., Semerad R. & Prusseit W. (1996). c-Axis Conductivity and Intrinsic Josephson Effects in $YBa_2Cu_3O_{7-\delta}$, *Phys. Rev. Letters* Vol.77, pp.928-931

Zeldov E.; Majer D., Konczykowski M., Geshkenbein V.B., Vinokur V.M. & Shtrikman H. (1995). Thermodynamic Observation of First-Order Vortex-Lattice Melting Transition, *Nature* (London) Vol.375, pp.373-376

Yethiraj M.; Mook H.A., Wignall G.D., Cubitt R., Forgan E.M., Paul D.M. & Armstrong T. (1993). Small-Angle Neutron Scattering Study of Flux Line Lattices in Twinned $YBa_2Cu_3O_{7-\delta}$, *Phys. Rev. Letters* Vol.70, pp.857-860

Yu S.; Ooi S., Mochiku T. & Hirata K. (2005). Anisotropy Dependence of Josephson-Vortex flow Resistance, *Physica* C Vol.426-431 pp.51-55

Yu S.; Ooi S., Mochiku T. & Hirata K. (2007). Anisotropy Dependence of Traiangular lattice Formation of Josephson Vortices in $Bi_2Sr_2CaCu_2O_{8+y}$, *Phys. Rev.* B Vol. 76 pp. 092505-1-4

8

Shapiro Steps in BSCCO Intrinsic Josephson Junctions

Michihide Kitamura
Department of Electrical and Electronic System Engineering, Utsunomiya University, Yoto, Utsunomiya, Tochigi Japan

1. Introduction

In 1963, Shapiro did an experiment for the microwave irradiation effect on the *I-V* characteristics of a Josephson junction $Al/Al_2O_3/Sn$ and showed that when the microwave is impressed to the Josephson junction, clear current steps are observed on the quasiparticle (QP) branches of the *I-V* characteristics and those steps, which are now called "Shapiro steps", are observed on constant voltages given by $n\Phi_0 f_r$ $(n = \pm 1, \pm 2, ...)$ [1]. Here the Φ_0 is a flux quantum ($= 2.06785 \times 10^{-15}$ Wb) defined by $h/2e$ using Planck constant h and the f_r is the frequency of the impressed microwave. The interactions between the Josephson junction and the electromagnetic field are important and intersting problems on not only the fundamental but also the applicational points of view. The Shapiro steps are one of the important phenomena observed on a superconductor-insulator-superconductor (SIS)-type Josephson junction.

It is well known that the high-T_c superconductors such as $Bi_2Sr_2CaCu_2O_{8+\delta}$ (BSCCO) and $(Pb_yBi_{1-y})_2Sr_2CaCu_2O_{8+\delta}$ (PBSCCO) consist of a stack of atomic-scale Josephson junctions called intrinsic Josephson junctions (IJJs) and these IJJs are SIS-type Josephson junctions. [2–8] The stacking structure of IJJs along the crystallographic *c*-axis can be made *naturally*, so that these natural super-lattices can be regarded as a stack of good quality Josephson devices made under the atomic-scale. This is just a reason why the studies of the current flow along the *c*-axis have been done extensively for the high-T_c superconductors. It has been already established that the high-T_c superconductors such as BSCCO and PBSCCO are characterized by the $d_{x^2-y^2}$-symmetry superconducting gap rather than the *s*-one. [9–11] There are some excellent works for Shapiro steps on BSCCO single crystals. [12–18]

When a *large* SIS Josephson junction is embedded in the magnetic field applied parallel to the junction surface, the fluxon dynamics in the SIS Josephson junction can be understood by solving a following nonlinear partial differential equation. [19]

$$\frac{\partial^2 \varphi}{\partial x^2} - \frac{1}{\bar{c}^2}\frac{\partial^2 \varphi}{\partial t^2} - \frac{\beta}{\bar{c}^2}\frac{\partial \varphi}{\partial t} = \frac{1}{\lambda_J^2}\sin\varphi\,, \tag{1}$$

where $\bar{c}$ is Swihart velocity, λ_J is Josephson penetration depth and φ is the phase difference of a SIS Josephson junction as a function of spacial variable x and real time t. As already stated,

BSCCO consists of a stack of SIS-type IJJs, so it is expected that the BSCCO should be regarded as a good sample to study the fluxon dynamics in the SIS Josephson junctions. Actually, there are many excellent works for the effect of magnetic field on the BSCCO IJJs. [20–34] It is sure that a pioneering work done by Sakai, Bodin and Pedersen(SBP) [20] is the starting point to understand the fluxon dynamics in BSCCO IJJs with a stacking structure.

We adopt here a unified theory proposed by Machida and Sakai (MS) [29], which includes both the electric and magnetic field couplings between neighboring IJJs in multistacked structure. The MS-unified theory is a further extension of the model proposed by Sakai, Bodin and Pedersen (SBP). [20] Present paper is mainly based on my previous jobs. [35, 36]

2. Theory

Let us consider a stacked system consisting of N-identical SIS-Josephson junctions such as IJJs in BSCCO high-T_c superconductors, and set spacial coodinates x, y and z parallel to a, b and c axes of a BSCCO single crystal. If the external magnetic induction $\mathbf{B}_{ext}$ is selected parallel to y-axis, i.e., $\mathbf{B}_{ext} = (0, B_{ext}, 0)$, and the spacial variation of the gauge-invariant phase difference φ_ℓ for the ℓ-th SIS Josephson junction is assumed as a function of only x, then the $\varphi_\ell(x,t)$ as a function of x and real time t satisfies a following coupled sine-Gordon(CSG) equation with a matrix form. [29]

$$\mathbf{\Sigma_c}\lambda_J^2 \frac{\partial^2}{\partial x^2} \begin{pmatrix} \varphi_1(x,t) \\ \cdot \\ \varphi_\ell(x,t) \\ \cdot \\ \varphi_N(x,t) \end{pmatrix} = \mathbf{\Sigma_L} \begin{pmatrix} \mathcal{J}_1(x,t) \\ \cdot \\ \mathcal{J}_\ell(x,t) \\ \cdot \\ \mathcal{J}_N(x,t) \end{pmatrix}, \tag{2}$$

where λ_J is the Josephson penetration depth given by $\sqrt{\Phi_0/2\pi\mu_0 d'_L J_c}$ using a flux quantum Φ_0, the vacuum permeability μ_0 and the critical current density J_c. The $\mathbf{\Sigma_c}$ and $\mathbf{\Sigma_L}$ are matrices describing the electric and magnetic interactions between neighboring Josephson junctions and are written as

$$\mathbf{\Sigma_{c(L)}} = \begin{pmatrix} 1 & \Sigma_{c(L)} & 0 & \cdot & \cdot \\ \Sigma_{c(L)} & 1 & \Sigma_{c(L)} & 0 & \cdot \\ \cdot & \Sigma_{c(L)} & 1 & \Sigma_{c(L)} & \cdot \\ \cdot & \cdot & \cdot & \cdot & \cdot \\ \cdot & \cdot & 0 & \Sigma_{c(L)} & 1 \end{pmatrix}, \tag{3}$$

where the coupling constants Σ_c and Σ_L are given by [29]

$$\Sigma_{c(L)} = -\frac{\lambda_{e(L)}}{d'_{c(L)} \sinh(w/\lambda_{e(L)})}. \tag{4}$$

The d'_c and d'_L are the effective electric and magnetic thickness given by [29]

$$d'_{c(L)} = d + 2\lambda_{e(L)} \coth(w/\lambda_{e(L)}), \tag{5}$$

using electrode thickness w and barrier thickness d. The λ_e and λ_L are the Debye screening length and the London penetration depth, respectively. The $\mathcal{J}_\ell(x,t)$ in Eq.(2) is the current defined by $j_J^{(\ell)}(x,t) - j_{ext}(t)$. The $j_J^{(\ell)}(x,t)$ and $j_{ext}(t)$ are junction and external currents normalized to the critical current $I_c(=J_cS)$, therefore, the current $\mathcal{J}_\ell(x,t)$ can be written in SI units as

$$\mathcal{J}_\ell(x,t) = \left(\frac{\Phi_0}{2\pi}\right)\frac{C_u^*}{J_c}\frac{\partial^2\varphi_\ell(x,t)}{\partial t^2} + \left(\frac{\Phi_0}{2\pi}\right)\frac{G_u^*}{J_c}\frac{\partial\varphi_\ell(x,t)}{\partial t} + \sum_{\ell'=1}^{N}\{\boldsymbol{\Sigma}_c\}_{\ell,\ell'}\, i_{CP}^{(\ell')}(x,t) - j_{ext}(t)\,, \quad (6)$$

where C_u^* and G_u^* are the effective unit area capacitance $\varepsilon\varepsilon_0/d_c'$ and conductivity σ/d_c' per a junction, respectively. The current $i_{CP}^{(\ell)}(x,t)$ is the normalized Cooper-pair (CP) tunneling current of the ℓ-th SIS Josephson junction whose general form has been already presented in my previous paper. [35] Note that the simplest form of $i_{CP}^{(\ell)}(x,t)$ is $\sin\varphi_\ell(x,t)$. The external current $j_{ext}(t)$ is the same as the normalized external current $i_{ext}(t)$ presented in my previous paper, that is,

$$j_{ext}(t) = i_0 + i_r\sin\omega_r t + i_{noise}(t,T) \equiv j_{ext}(t,T)\,, \quad (7)$$

where i_0 is the normalized external dc, $i_r\sin\omega_r t$ is the normalized external ac modulation current with the frequency ω_r and $i_{noise}(t,T)$ is the normalized current due to thermal noise at a sample temperature T. As a noise, we consider here a "white noise" so that the $I_{noise}(t,T) = i_{noise}(t,T)I_c(T)$ satisfies following relations:

$$\left.\begin{array}{l} < I_{noise}(t,T) >_t = 0 \\ < I_{noise}(t,T)I_{noise}(t+t',T) >_t = \frac{2k_BT}{R_J}\delta(t') \end{array}\right\}, \quad (8)$$

where $< A(t) >_t$ means the time average of a time dependent function $A(t)$. The R_J is the resistance of a junction and is given as a function of $I_0 = i_0I_c(T)$;

$$\frac{1}{R_J} = \frac{1}{R_{QP}^{(eff)}(I_0)} + \frac{1}{R_{shunt}} \equiv \frac{1}{R_J(I_0)}\,, \quad (9)$$

where R_{shunt} is a shunt resistance added externally per a junction, and $R_{QP}^{(eff)}(I_0)$ is an effective resistance of a junction due to QP tunneling.

The white noises are made numerically by using random numbers. We write here the random numbers as $i_{Random}^{(Cal)}(t)$, which have been made by the normal random number generator in the library program. The normalized white noise $i_{noise}(t,T)$ defined by $I_{noise}(t,T)/I_c(T)$ must be proportional to $i_{Random}^{(Cal)}(t)$, that is,

$$i_{noise}(t,T) = \eta(T)i_{Random}^{(Cal)}(t)\,. \quad (10)$$

The autocorrelation $< i_{noise}(t,T)i_{noise}(t+t',T) >_t$ is equal to $\eta(T)^2 < i_{Random}^{(Cal)}(t)i_{Random}^{(Cal)}(t+t') >_t$. Therefore, by using relation (8) and doing an integral over the t', we get

$$\frac{2k_BT}{I_c(T)^2R_J} = \eta(T)^2\int_{-\infty}^{\infty} < i_{Random}^{(Cal)}(t)i_{Random}^{(Cal)}(t+t') >_t dt'\,. \quad (11)$$

Here we define an absolute time resolution Δt. By doing some numerical calculations for the kernel defined by Eqs.(27) and (28), I found that the time resolution Δt with the value of 1×10^{-14}s(=10fs) is small enough to resolve the structure of the kernel. [35] The autocorrelation $<$

$i^{(Cal)}_{Random}(t) i^{(Cal)}_{Random}(t+t') >_t$ is also proportional to the delta function $\delta(t')$, so that the integral in Eq.(11) can be well evaluated as $< i^{(Cal)}_{Random}(t)^2 >_t \Delta t$. Therefore, the $\eta(T)$ is given by

$$\eta(T) = \sqrt{\frac{2k_B T}{< i^{(Cal)}_{Random}(t)^2 >_t \Delta t \; I_c(T)^2 R_J}} , \tag{12}$$

so that from Eq.(10) we can get the normalized white noise $i_{noise}(t, T)$ at a finite temperature T as full numerical data with a function of time t defined by $n\Delta t$. All the experiments always include the effect of thermal noise. Therefore, all the calculations presented in the present paper also include the effect of the thermal noise, except for the special case that I notice. In the present study, the sample temperature T has been set to 4.2K.

By using the dimensionless time τ defined by $\omega_r t/2\pi (= f_r t)$ and the normalized distance ν defined by x/λ_J, equation (2) can be rewritten as

$$\begin{pmatrix} Q_1(\nu,\tau) \\ \cdot \\ Q_\ell(\nu,\tau) \\ \cdot \\ Q_N(\nu,\tau) \end{pmatrix} = \left(2\pi \frac{f_p^*}{f_r}\right)^2 \mathbf{\Sigma_L}^{-1} \mathbf{\Sigma_c} \frac{\partial^2}{\partial \nu^2} \begin{pmatrix} \varphi_1(\nu,\tau) \\ \cdot \\ \varphi_\ell(\nu,\tau) \\ \cdot \\ \varphi_N(\nu,\tau) \end{pmatrix} , \tag{13}$$

where f_p^* is the effective plasma frequency of a junction calculated as $\sqrt{J_c d_c'/2\pi\Phi_0 \varepsilon\varepsilon_0}$. In the MS-unified theory, the effect of the magnetic field is taken into account via the boundary condition at the junction edge. Therefore, we solve Eq.(13) under the boundary condition such that

$$\left.\frac{\partial \varphi_\ell(\nu,\tau)}{\partial \nu}\right|_{edge} = \frac{B^{(\ell)}_{eff}}{\mu_0 J_c \lambda_J}(1 + 2\Sigma_L) , \tag{14}$$

where $B^{(\ell)}_{eff}$ is an effective magnetic induction at the edge of the ℓ-th IJJ. By solving Eq.(13) full numerically, we can get the gauge-invariant phase differences $\varphi_\ell(\nu,\tau)$ of all the junctions $\ell = 1, 2, \cdot\cdot, N$ as a function of ν and τ. In the present paper, not only the case such as SIS-IJJs but also the case in which some insulating(I) layers in IJJs are replaced by ferromagnetic(F) layers is considered. The F-layer makes a magnetization M, therefore, if the ℓ-th I-layer is replaced by the F-layer with $M = M^{(\ell)}$, then the $B^{(\ell)}_{eff}$ in Eq.(14) is given in SI-units as

$$B^{(\ell)}_{eff} = B_{ext} + M^{(\ell)} . \tag{15}$$

For the SFS-junctions, there are many excellent works. [37–44] Especially, two review papers by Golubov et al. [39] and Buzdin [40] include many valuable information. Recently, the phase dynamics induced by spin waves in *a* SFS Josephson junction consisting of *s*-wave superconductors have been numerically studied within the framework of the resistively shunted junction (RSJ) model. [44] The calculations by Hikino et al. [44] have told us that the magnetization of F-layer plays an essential role to understand the nature of phase dynamics. In the SFS-junction, it is well known that the amplitude of the CP-tunneling critical current *oscillatory* varies as a function of the thickness d_F of the F-layer [40], which is called as "0-π

transition". The first crossover thickness $d_F^{(1)}$, in which the nature of junction changes from the 0 to the π state, is about 50Å for the SFS-junction using an NiCu-system ferromagnetic metal. [40, 43] As can be seen later, the barrier thickness d of BSCCO IJJs is 12Å. In the present paper, therefore, two cases are considered; one is the case in which the correction for the amplitude of the CP tunneling current is considered and other is not.

The $Q_\ell(\nu,\tau)$ in Eq.(13) is

$$Q_\ell(\nu,\tau) = \frac{\partial^2 \varphi_\ell(\nu,\tau)}{\partial \tau^2} + 2\pi \frac{f_p^*}{f_r} \frac{1}{\sqrt{\beta_c^*(i_0, J_c, SR_{shunt})}} \frac{\partial \varphi_\ell(\nu,\tau)}{\partial \tau}$$
$$+ \left(2\pi \frac{f_p^*}{f_r}\right)^2 \left\{ \sum_{\ell'=1}^{N} \{\mathbf{\Sigma_c}\}_{\ell,\ell'} \chi_{\ell'}(d_F) i_{CP}^{(\ell')}(\nu,\tau) - i_{ext}(\tau) \right\} . \tag{16}$$

Here note that we generally consider the case including the F-layer, so that the correction term $\chi_\ell(d_F)$ which has not been written in Eq.(6) has been added into Eq.(16). The $\chi(d_F)$ is defined by $I_c^{(F)}(d_F)/I_c^{(NM)}$ using the critical currents $I_c^{(F)}(d_F)$ and $I_c^{(NM)}$ of the SFS and SIS Josephson junctions, and the value of $\chi_\ell(d_F)$ oscillatory varies as a function of d_F bewteen 0 to 1 when the ℓ-th junction includes F-layer. Here it is clear that $\chi_\ell(d_F) = 1$ when the ℓ-th junction includes no F-layer. For example, $\chi(d_F)$ can be written as [40]

$$\chi(y) = 4y \left| \frac{\cos 2y \sinh 2y + \sin 2y \cosh 2y}{\cosh 4y - \cos 4y} \right| , \tag{17}$$

where $y \equiv d_F/\xi_F$ and ξ_F is the characteristic length of the superconducting correlation decay with oscillations in the F layer. [40] The $\chi(y)$ as a function of $2y$ is drawn in Fig.11 in the review paper by Buzdin. [40]

By using the McCumber parameter $\beta_c(i_0, J_c, SR_{shunt})$, [35] the $\beta_c^*(i_0, J_c, SR_{shunt})$ in Eq.(16) can be written as

$$\frac{1}{\sqrt{\beta_c^*(i_0, J_c, SR_{shunt})}} = \frac{\Phi_0 f_p^*}{J_c SR_{shunt}} + \frac{\Phi_0 f_p^*}{I_c R_{QP}^{(eff)}(I_0)}$$
$$= \gamma_p \left\{ \frac{\Phi_0 f_p}{J_c SR_{shunt}} + \mu(i_0) \right\}$$
$$= \frac{\gamma_p}{\sqrt{\beta_c(i_0, J_c, SR_{shunt})}} , \tag{18}$$

where

$$\gamma_p \equiv \frac{f_p^*}{f_p} = \sqrt{\frac{d_c'}{d}} . \tag{19}$$

It should be noted here that the $\mu(i_0)$ in Eq.(18) is a universal curve as a function of only the normalized external dc i_0 so that this curve is always valid for all the IJJs in BSCCO characterized by the $d_{x^2-y^2}$ symmetry superconducting gap. [35] Namely, if the no shunt case, i.e., the case of $R_{shunt} \to \infty$, is considered, then the McCumber parameter of BSCCO is given as a function of only the i_0 such as $\mu^{-2}(i_0)$. [35] Here I wish to say that when the effect of electric field coupling is not considered, the plasma frequencies ratio γ_p is equal to 1 and the matrix $\boldsymbol{\Sigma}_{\mathbf{c}}$ is equal to the unit matrix $\mathbf{1}$, therefore, the $Q_\ell(\nu,\tau)$ for this case is just equal to the left-hand side of Eq.(31) in my recent paper. [35] We will see later that $\gamma_p \simeq 1$ and $\boldsymbol{\Sigma}_{\mathbf{c}} \simeq \mathbf{1}$ even if the effect of the electric field coupling is considered for BSCCO IJJs.

Kautz and Monaco extensively studied the nature of Shapiro steps on the SIS Josephson junction [45]. As a result, they found that there is a chaotic region in which Shapiro steps cannot be observed. The chaotic region is found between f_{RC} and f_p, i.e., $f_{RC} \le f_r \le f_p$. In my previous paper, [35] I have already pointed out that the lower frequency f_{RC} of the chaotic region is given by

$$\left.\begin{aligned} f_{RC} &= f_{RC}^{(NS)}(i_0) + f_{RC}^{(Shunt)}(SR_{shunt}) \\ f_{RC}^{(Shunt)}(SR_{shunt}) &= \frac{1}{2\pi R_{shunt}C} \end{aligned}\right\} . \tag{20}$$

Moreover, I have mentioned that the frequency $f_{RC}^{(NS)}(i_0)$ for the case of no-shunt (NS) resistance is given by using a universal curve $\mu(i_0)$ as

$$f_{RC}^{(NS)}(i_0) = f_p \mu(i_0) . \tag{21}$$

Above two equations (20) and (21) tell us that the SR_{shunt}-product is an important parameter to understand the nature of Shapiro step observed on the BSCCO IJJ.

If the case such that $B_{ext} = 0$ and there are no F-layers in IJJs is considered, then the $\partial\varphi_\ell(\nu,\tau)/\partial\nu$ is zero for all junctions. Therefore, the equation (13) for the present case is equivalent to the equation $Q_\ell(\nu,\tau) = 0$. Equation (16) tells us that the equation $Q_\ell(\nu,\tau) = 0$ reflects *only* the electric interaction between neighboring Josephson junctions. It is clear that if $\boldsymbol{\Sigma}_{\mathbf{c}} = \mathbf{1}$ and $\chi_\ell(d_F) = 1$ for all ℓ, then the equation $Q_\ell(\nu,\tau) = 0$ is just equal to equation(31) in my recent paper. [35] In order to see the effects originated from both the electric and magnetic interactions, we must solve Eq.(13) with a finite value of B_{ext}, so that the Eq.(13) becomes a matrix form including not only $\boldsymbol{\Sigma}_{\mathbf{c}}$ but also $\boldsymbol{\Sigma}_{\mathbf{L}}$ describing the magnetic interaction between neighboring Josephson junctions.

The dc voltage $< V_\ell >$ observed on the ℓ-th IJJ satisfies

$$\varepsilon\varepsilon_0 \frac{L_x L_y}{d} < V_\ell >= \int dQ = \int_0^{L_x} \varepsilon\varepsilon_0 \frac{L_y}{d} < V_\ell(x,t) >_t dx , \tag{22}$$

where L_x and L_y are junction lengths along the x and y axes, i.e., $S = L_x L_y$, and $< V_\ell(x,t) >_t$ means the time average of $V_\ell(x,t)$ which is a voltage of the ℓ-th IJJ as a function of x and t. Therefore, the dc voltage $< V_\ell >$ for the ℓ-th IJJ is simply given by

$$< V_\ell >= \frac{1}{L_x} \int_0^{L_x} < V_\ell(x,t) >_t dx , \tag{23}$$

so that the total dc voltage $< V >_{total}$ observed experimentally is given by

$$< V >_{total}= \sum_{\ell=1}^{N} < V_\ell >\equiv N < V >_{reduce} , \tag{24}$$

where $< V >_{reduce}$ is the reduced voltage when $< V >_{total}$ is assumed as a simple sum of N-junctions system. The $< V_\ell(x,t) >_t$ is the same as Eq.(33) in my previous paper [35], that is,

$$< V_\ell(x,t) >_t = \frac{\Phi_0 f_r}{2\pi} \frac{\varphi_\ell(x,\tau_2) - \varphi_\ell(x,\tau_1)}{\tau_2 - \tau_1} \, . \tag{25}$$

Here, we must remember that τ_1 and τ_2 must be intergers in order to get a very well converged value for an averaging between τ_1 and τ_2. [35]

The general form of $i_{CP}^{(\ell)}(\nu,\tau)$ is given by [35]

$$i_{CP}^{(\ell)}(\nu,\tau) = \int_{-\infty}^{\tau} K_{CP}^{(\ell)}(\tau-\tau') \sin \frac{\varphi_\ell(\nu,\tau') + \varphi_\ell(\nu,\tau)}{2} d\tau' \, , \tag{26}$$

where the kernel $K_{CP}^{(\ell)}(\tau-\tau')$ can be written as

$$K_{CP}^{(\ell)}(\tau-\tau') = \frac{2\pi}{\hbar\omega_r} \kappa_{CP}^{(\ell)}(\tau-\tau') \, , \tag{27}$$

$$\kappa_{CP}^{(\ell)}(\tau-\tau') = \frac{\sum_{\mathbf{k}}^{BZ_1}\sum_{\mathbf{k}'}^{BZ_1} \left|H'^{(\ell)}_{\mathbf{k},\mathbf{k}'}\right|^2 \frac{|\Delta_{\mathbf{k}}\Delta_{\mathbf{k}'}|}{E_{\mathbf{k}}E_{\mathbf{k}'}} \left[\begin{array}{l} \{f(E_{\mathbf{k}}) - f(E_{\mathbf{k}'})\} \sin\left\{\frac{E_{\mathbf{k}}-E_{\mathbf{k}'}}{\hbar\omega_r} 2\pi(\tau-\tau')\right\} \\ + \{1 - f(E_{\mathbf{k}}) - f(E_{\mathbf{k}'})\} \sin\left\{\frac{E_{\mathbf{k}}+E_{\mathbf{k}'}}{\hbar\omega_r} 2\pi(\tau-\tau')\right\} \end{array} \right]}{\sum_{\mathbf{k}}^{BZ_1}\sum_{\mathbf{k}'}^{BZ_1} \left|H'^{(\ell)}_{\mathbf{k},\mathbf{k}'}\right|^2 \frac{|\Delta_{\mathbf{k}}\Delta_{\mathbf{k}'}|}{E_{\mathbf{k}}E_{\mathbf{k}'}} \left[\frac{f(E_{\mathbf{k}})-f(E_{\mathbf{k}'})}{E_{\mathbf{k}}-E_{\mathbf{k}'}} + \frac{1-f(E_{\mathbf{k}})-f(E_{\mathbf{k}'})}{E_{\mathbf{k}}+E_{\mathbf{k}'}} \right]} \, , \tag{28}$$

where BZ_1 means the 1st Brillouin zone, $f(E)$ is the Fermi-Dirac distribution function and $H'^{(\ell)}_{\mathbf{k},\mathbf{k}'}$ is the matrix element due to the $\mathbf{k} \to \mathbf{k}'$ tunneling in the ℓ-th IJJ. Here it should be noted that the $H'^{(\ell)}_{\mathbf{k},\mathbf{k}'}$ for the coherent tunneling is proportional to $\delta_{\mathbf{k},\mathbf{k}'}$ and that for the incoherent one is given by a constant irrespective of $\mathbf{k}$ and $\mathbf{k}'$. In the practical calculations of $\kappa_{CP}^{(\ell)}(\tau-\tau')$ defined by Eq.(28), therefore, $H'^{(\ell)}_{\mathbf{k},\mathbf{k}'}$ is set to $\delta_{\mathbf{k},\mathbf{k}'}$ for the coherent tunneling and to 1 for the incoherent one. The quasiparticle excitation energy $E_{\mathbf{k}}$ is given by $\sqrt{\xi_{\mathbf{k}}^2 + \Delta_{\mathbf{k}}^2}$. The $\Delta_{\mathbf{k}}$ is the superconducting energy gap and it is well known that the $\Delta_{\mathbf{k}}$ of d-wave symmetry superconductor is represented as $\Delta(T)\cos 2\phi_k$, where ϕ_k is the angle of $\mathbf{k}$-vector measured from x-axis. [10] The $\xi_{\mathbf{k}}$ is the one-electron energy relative to the Fermi level. By doing the band structure calculation, I got the values of $\xi_{\mathbf{k}}$ as full numerical data. [46]

In my previous paper, [35] I have carried out the sophisticated numerical calculations in which two types of CP tunneling currents such as coherent and incoherent ones have been correctly calculated within the framework of the $d_{x^2-y^2}$ symmetry superconducting gap, and found that the resultant I-V characteristics very well coincide with the result calculated in a usual way such as $i_{CP}(\tau) = \sin\varphi(\tau)$. [35] This finding is come from the fact that Shapiro steps are observed on the quasiparticle (QP) branch of the I-V characteristics so that the Shapiro steps are not so sensitive to the CP tunneling *mechanism*. [35] In the present paper, therefore, we evaluate $i_{CP}^{(\ell)}(\nu,\tau)$ as $\sin\varphi_\ell(\nu,\tau)$, in order to save the CPU times on the numerical calculations.

3. Calculation

We consider here the BSCCO high-T_c superconductors. The typical values of electrode thickness w, barrier thickness d and London penetration depth λ_L are 3, 12 and 1700Å, respectively [21], and that of the Debye screening length λ_e is 1Å. [29] Therefore, the effective electric and magnetic thickness d'_c and d'_L defined by Eq.(5) are calculated as 14.0 and 1.93×10^6Å , and the coupling constants Σ_c and Σ_L defined by Eq.(4) are calculated as -7.13×10^{-3} and -0.499996, respectively. Therefore, we can see that the plasma frequencies ratio γ_p calculated as $\sqrt{d'_c/d}$ is 1.08 and the matrix $\mathbf{\Sigma_c}$ describing the electric interaction between the neighboring Josephson junctions is very well approximated by the unit matrix **1**. This fact means that the effect of electric field coupling, the charge coupling in other words, could not be so crucial in BSCCO IJJs. In my previous paper, [36] I have checked this point *quantitatively*, and found that the effect of charge coupling is not crucial in the BSCCO IJJs. In the present calculations, therefore, γ_p is set to 1 and $\mathbf{\Sigma_c}$ is set to **1**. This reduction largely saves the CPU times, since the term $\sum_{\ell'=1}^{N} \{\mathbf{\Sigma_c}\}_{\ell,\ell'} \chi_{\ell'}(d_F) i_{CP}^{(\ell')}(\nu,\tau)$ in Eq.(16) is simply given by $\chi_\ell(d_F) i_{CP}^{(\ell)}(\nu,\tau)$ when $\mathbf{\Sigma_c} = \mathbf{1}$.

We adopt here 1000(A/cm^2) again as the value of critical current density J_c, therefore, the Josephson penetration depth λ_J and the plasma frequency f_p are calculated as 3687Å and 122GHz, respectively, using $\varepsilon = 7$. The f_p and J_c are given so that the SR_{shunt}-product becomes an essential value to decide the value of McCumber parameter $\beta_c(i_0, J_c, SR_{shunt})$. Recently, I have numerically studied the conditions for observing Shapiro steps in BSCCO IJJs and found that clear and stable Shapiro steps with good responses are obtained when the IJJ is operated under the condition such that the shunt resistance R_{shunt} is added and the external ac modulation frequency f_r is higher than the plasma frequency f_p. [35] In the present paper, therefore, the value of f_r is set to 200GHz again because of $f_p = 122$GHz.

The values of λ_J, $d'_{c(L)}$, $\Sigma_{c(L)}$, f_r, f_p and J_c are given, the effect of the normalized current $i_{noise}(\tau)$ due to the thermal noise is considered by using normal random numbers, and the normalized CP tunneling current $i_{CP}^{(\ell)}(\nu,\tau)$ is evaluated as $\chi_\ell(d_F)\sin\varphi_\ell(\nu,\tau)$, therefore, SR_{shunt}-product, i_0, i_r, B_{ext}, $M^{(\ell)}$, $\chi_\ell(d_F)$ and the number of junctions N become the important variables to be considered. Here, note that the i_0 is the basic variable in the *I-V* characteristics that I present here. The junction cross section $S(= L_x L_y)$ is set to 25μm^2 again and 1, 2, 3, 4 and 5Ω are selected as the values of shunt resistance R_{shunt} per a junction, i.e., the SR_{shunt}-products are set to 25, 50 75, 100 and 125μm^{2}Ω/junction. We write here the external magnetic induction B_{ext} as $\alpha B_0(L_x, d)$. Here the $B_0(L_x, d)$ is the magnetic induction needed to insert a flux quantum Φ_0 into a SIS-Josephson junction and is given by $\Phi_0/L_x d$ as a function of L_x and d. For example, the $B_0(L_x, d)$ is calculated as 0.345T for $L_x = 5\mu$m and $d = 12$Å. In the present paper, the value of barrier thickness d is fixed to 12Å, so that the $B_0(L_x, d)$ is simply written as $B_0(L_x)$.

In my previous paper, [36] I have studied the effect of N on the *I-V* characteristics very carefully. The calculated results have told us that (I) a stacking system consisting of 5 or 6-identical SIS Josephson junctions is enough to see the *I-V* characteristics of BSCCO IJJs with odd or even number of junctions, and (II) if the stacked system consists of 5 junctions, then *a* flux quantum enters an insulating layer of the 3@5 central junction, but if the system consists of 6 junctions, then *two* flux quanta must enter the insulating layers in such a way that one flux quantum enters the insulating layer of 3@6 junction and the other the 4@6 one, because of its symmetry. In the present paper, the number N of junction is fixed to 5. Therefore, if the case in which there is no ferromagnetic (F)-layer is considered, then the symmetry says that the 1@5 and 5@5 junctions, which are surface junctions, show the same characteristics and the 2@5 and 4@5 ones (intermediate junctions) also show the same nature.

In the following, previous results calculated for no F-layers, i.e., $M^{(\ell)} = 0$ for all the IJJs, are first presented, and next the results for $M^{(\ell)} \neq 0$ are given. Here note that in the present calculations, *an* ℓ-th insulating (I) layer in BSCCO IJJs is replaced by a F-layer which makes a magnetization with the value of $M^{(\ell)}$.

4. Results and discussion

4.1 Effect of L_x

It is important to check the effect of junction length L_x along the x-axis into the I-V characteristics. In my previous paper, [36] I have calculated the I-V characteristics for three shapes $(L_x, L_y) = (2.5, 10)$, $(5, 5)$ and $(10, 2.5)$ in μm with the same junction cross section $S(= L_x L_y) = 25\mu\text{m}^2$. The calculated results have told us that the overall profiles of the I-V characteristics calculated for the same external magnetic induction are fairly similar each other. This finding is reasonable since all the junction lengths L_x considered here satisfy the condition of $L_x > \lambda_J (= 0.3687\mu\text{m})$, that is, all the junctions adopted in the calculations belong to the category of so-called "large junction". In the following, therefore, only the square junction ($L_x = L_y = 5\mu$m) is considered for simplicity.

4.2 Effect of SR_{shunt}-product

The SR_{shunt}-product is an important parameter to understand the nature of Shapiro step. In my previous papers, [35, 36] I *extensively* studied the effect of the SR_{shunt}-product to the I-V characteristics. In the present paper, therefore, only the conclusions obtained are briefly presented.

First, from the calculations for a single junction, I have found that (I) the value of shunt resistance R_{shunt} added externally should be small such as 1Ω/junction, i.e., SR_{shunt}/junction$= 25\mu\text{m}^2\Omega$, if we wish to make a Shapiro step device which is *hard* for the external magnetic disturbance, and (II) in the case of the large R_{shunt}-value such as 5Ω/junction, not only the 1st-order Shapiro steps but also the Fiske steps [47, 48] are found, especially, in the case of the low magnetic field. Next, from the calculations for multi junctions such that $N = 5, 6$ and R_{shunt}/junction$= 1, 2, 3, 4$ and 5Ω, I have found that (III) the 1st-order Shapiro steps are found when R_{shunt}/junction$\leq 3\Omega$, but no detectable Shapiro steps are found when R_{shunt}/junction$= 4$ and 5Ω, and (IV) this result directly reflects the nature of the magnetic interactions between the neighboring Josephson junctions.

From the above results, I have concluded that it is essential to add the shunt resistance with a low-value into the BSCCO IJJs, if we wish to make a good quality Shapiro step device. In the present study, therefore, the value of SR_{shunt}/junction is fixed to $50\mu\text{m}^2\Omega$.

4.3 Effect of i_r

We consider here the effect of the normalized amplitude i_r of the external ac modulation current $i_r \sin\omega_r t$. In order to do so, first, I have calculated the I-V characteristics with no external magnetic field, i.e., $\alpha = 0$, under the condition such that $J_c = 1000\text{A/cm}^2$,

$f_p = 122\text{GHz}$, $f_r = 200\text{GHz}$, $SR_{shunt}/\text{junction} = 50\mu\text{m}^2\Omega$ and $N = 5$. Those are drawn as (i), (ii) and (iii) in Fig.1 for $i_r = 0, 1$ and 2 which are denoted as $(i_r, \alpha) = (0,0)$, $(1,0)$ and $(2,0)$, respectively. Next, I have calculated the *I*-*V* characteristics with finite external magnetic fields, i.e., $\alpha \neq 0$, under the same condition as in Fig.1. Here note again that the external magnetic induction B_{ext} is defined by $\alpha B_0(L_x)$ and the $B_0(L_x)$ is 0.345T for $L_x = 5\mu\text{m}$. The *I*-*V* characteristics calculated for $(i_r, \alpha) = (0,0.5)$, $(0,1)$, $(0,2)$, $(1,0.5)$, $(1,1)$, $(1,2)$, $(2,0.5)$, $(2,1)$ and $(2,2)$ are shown in Fig.2(a), (b), (c), (d), (e), (f), (g), (h) and (i), respectively.

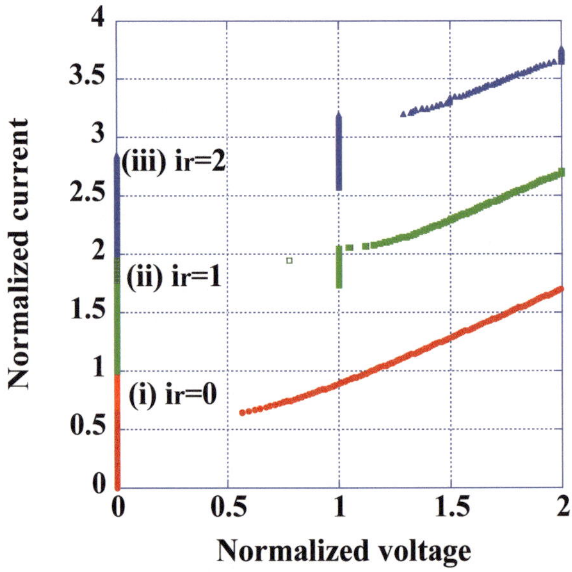

Fig. 1. The *I*-*V* characteristics with no external magnetic field, i.e., $\alpha = 0$, calculated for $N = 5$, $J_c = 1000\text{A/cm}^2$, $f_p = 122\text{GHz}$, $f_r = 200\text{GHz}$, $L_x = 5\mu\text{m}$, $S(= L_x L_y) = 25\mu\text{m}^2$, and $SR_{shunt}/\text{junction} = 50\mu\text{m}^2\Omega$. The effect of charge coupling is not taken into account. (i), (ii) and (iii) are those for $i_r = 0, 1$ and 2, and are denoted as $(i_r, \alpha) = (0,0)$, $(1,0)$ and $(2,0)$, respectively. Note that the *I*-*V* characteristics (ii) and (iii) have been shifted by 1 and 2 along the vertical axis, and that the (ii) is basically the same as the *I*-*V* characteristics (b) in Fig.10 in my previous paper. [35] It is also noted that no external magnetic field is applied, so that all the SIS-junctions show exactly the same characteristics because the no charge coupling case is considered. Namely, a relation "$1@N = 2@N = \cdots = N@N$" is satisfied for the *I*-*V* characteristics of respective junctions. The horizontal axis indicates the dc voltage normalized to $\Phi_0 f_r$, and the vertical one shows the normalized dc i_0. Shapiro steps are clearly found when the external ac modulation current is applied, i.e., $i_r \neq 0$.

First, let us focus our attention to the results calculated for $i_r = 0$, i.e., $(i_r, \alpha) = (0,0)$ in (i) in Fig.1 and $(0,0.5)$, $(0,1)$ and $(0,2)$ in Fig.2(a), (b) and (c), respectively. Since $i_r = 0$, these results directly show the change of the *I*-*V* characteristics due to the change of the external magnetic field. We can see that (I) the critical current decreases with the increasing the external magnetic induction B_{ext}, (II) this trend is remarkable in inner junctions 2@5, 3@5 and 4@5 rather than surface junctions 1@5 and 5@5, and (III) the central junction 3@5 shows a large change due to the change of the B_{ext}. Flux quanta enter the insulating layers when $B_{ext} \neq 0$. If we consider the case when *a* flux quantum just enters an insulating layer in the stacked junctions consisting

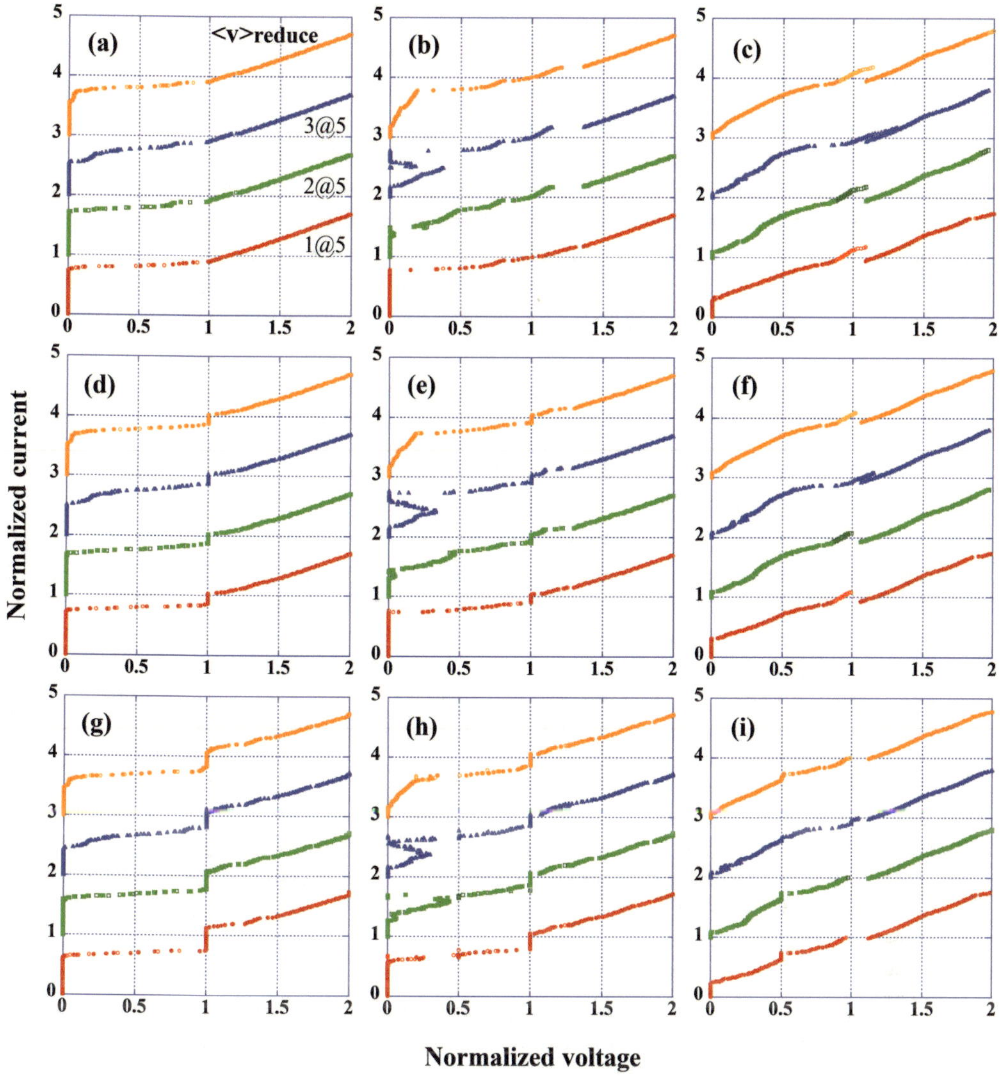

Fig. 2. The *I*-*V* characteristics of BSCCO IJJs calculated for the same condition as in Fig.1, except for the external magnetic induction B_{ext}. The B_{ext} is defined by $\alpha B_0(L_x)$ and the $B_0(L_x)$ is 0.345T because of $L_x = 5\mu$m. The (i_r, α) of (a), (b), (c), (d), (e), (f), (g), (h) and (i) are $(0, 0.5)$, $(0, 1)$, $(0, 2)$, $(1, 0.5)$, $(1, 1)$, $(1, 2)$, $(2, 0.5)$, $(2, 1)$ and $(2, 2)$, respectively. The reduced voltage $< V >_{reduce}$ defined by Eq.(24) is also shown. Shapiro steps are found when $i_r \neq 0$, but not so clear when $\alpha = 2$.

of the odd number of junctions, then it is sure that *its* flux quantum must enter the insulating layer of central SIS Josephson junction such as 3@5 junction because of the symmetry of the odd number of junctions. This is reasonable to the finding (III). Actually, in the case of Fig.2(a), we can see that only the 3@5 central junction just shows the voltage due to the flux flow when i_0 is about 0.6.

As can be seen in Fig.1, the height of Shapiro steps increases with the increasing the i_r. By comparing the calculated results for $i_r \neq 0$ shown in Fig.2(d), (e), (f), (g), (h) and (i) with those for $i_r = 0$ shown in Fig.2(a), (b) and (c), we can say that the *I*-*V* characteristics calculated for $(i_r, \alpha) = (1, 0.5), (1, 1), (1, 2), (2, 0.5), (2, 1)$ and $(2, 2)$ could be regarded as those for $(i_r, \alpha) = (0, 0.5), (0, 1)$ and $(0, 2)$ *added* Shapiro steps which have been calculated by taking into account the effect of the external magnetic field. The *I*-*V* characteristics for $i_r = 0$ shown in Fig.2(a), (b) and (c) *directly* reflect the nature of flux flow, and the Shapiro steps shown in Fig.2(d), (e), (f), (g), (h) and (i) are originated from the coupling between the ac Josephson effect and the external ac modulation current with $i_r \neq 0$. Therefore, the calculated results shown in Fig.2 seem to show that the Shapiro step and flux flow could be treated separately under the presence of the external magnetic field.

4.4 Effect of B_{ext}

We shall consider here the effect of the external magnetic field B_{ext} on the reduced voltage $< V >_{reduce}$ defined by Eq.(24) which is directly connected with the experimental *I*-*V* characteristics. By setting the calculation condition to such that $J_c = 1000\text{A/cm}^2$, $f_p = 122\text{GHz}$, $f_r = 200\text{GHz}$, $R_{shunt}/\text{junction} = 2\Omega$, $SR_{shunt}/\text{junction} = 50\mu\text{m}^2\Omega$, $i_r = 1$ and $L_x = 5\mu\text{m}$, i.e., $B_0(L_x) = 0.345\text{T}$, I have calculated the *I*-*V* characteristics as a function of α, which defines the external magnetic induction B_{ext} as $\alpha B_0(L_x)$, and the number N of stacked junctions. The values from 0 to 2 with 0.1-step have been selected as the α and two values 1 and 5 have been adopted as the N. The normalized height $\Delta i_1(\alpha, N)$ of the 1st-order Shapiro step and the normalized critical current $i_c(\alpha, N)$ evaluated from the reduced voltage $< V >_{reduce}$ are plotted in Fig.3 as a function of α.

Figure 3 shows that (I) for almost all the α-values, the values of $\Delta i_1(\alpha, 1)$ drawn by open circles are somewhat larger than those of $\Delta i_1(\alpha, 5)$ by solid circles, but its difference is not so large, i.e., the α-dependence of $\Delta i_1(\alpha, N)$ seems to be very similar even if the value of N differs each other such as $N = 1$ and 5. For the $i_c(\alpha, N)$, we can see that (II) the α-dependence of the $i_c(\alpha, 1)$ drawn by open squares largely differs from that of the $i_c(\alpha, 5)$ by solid squares. In the BSCCO IJJs with a stacking structure, there are considerable magnetic interactions between neighboring Josephson junctions when $\alpha \neq 0$. Therefore, the finding (II) clearly tells us that (III) the normalized critical current is strongly affected by the magnetic interaction between neighboring Josephson junctions.

4.5 Effect of ferromagnetic layer

As the case in which some insulating(I) layers in BSCCO-IJJs are replaced by ferromagnetic(F) ones, there are many cases. In this section, for the simplicity we consider the case such that only the ℓ-th I-layer in the BSCCO-IJJs is replaced by a F-layer whose value of the magnetization is $M(B_{ext})$ as a function of the external magnetic field B_{ext}. The magnetization curve $M(B_{ext})$ is characterized by three points; one is the saturation magnetization $M_{sat}(\neq 0)$ in which $dM/dB_{ext} = 0$, two is the residual magnetization $M_{res}(\neq 0)$ in which $B_{ext} = 0$, and three is the coercive force in which the magnetization is zero for the non-zero external field, i.e., $M(B_{ext} = B_{coe} \neq 0) = 0$. We consider here the non-zero magnetization case so that the case of coercive force is not considered here. The M_{sat} is obtained when a magnetic field B_L which is large enough to satisfy the condition of $dM/dB_{ext} = 0$ is applied to the F-layer, i.e.,

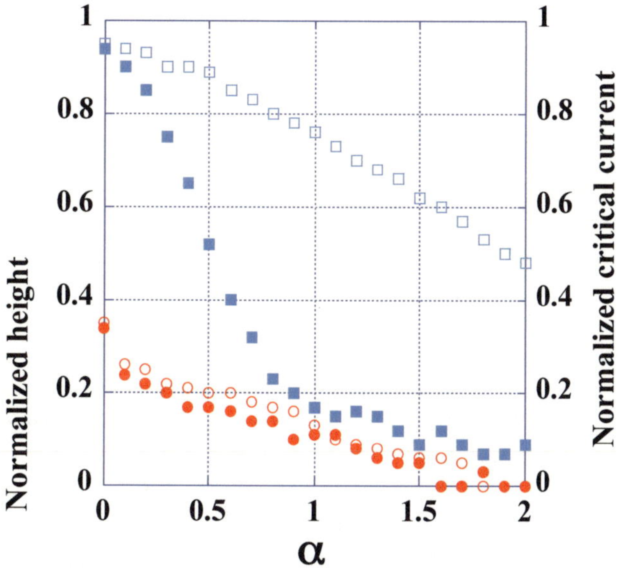

Fig. 3. Normalized height $\Delta i_1(\alpha, N)$ of the 1st-order Shapiro step and the normalized critical current $i_c(\alpha, N)$ as a function of α and N. In this figure, the values of α have been selected from 0 to 2 with 0.1-step, and two values 1 and 5 have been adopted as the number N of junctions. The calculation condition is that $J_c = 1000\text{A/cm}^2$, $f_p = 122\text{GHz}$, $f_r = 200\text{GHz}$, $L_x = 5\mu\text{m}$, R_{shunt}/junction$= 2\Omega$, and $i_r = 1$. As already stated, the external magnetic induction B_{ext} is defined by $\alpha B_0(L_x)$ and the $B_0(L_x)$ is 0.345T because of $L_x = 5\mu$m. The $\Delta i_1(\alpha, 1)$ and $\Delta i_1(\alpha, 5)$ are indicated by open and solid circles, respectively, and the $i_c(\alpha, 1)$ and $i_c(\alpha, 5)$ are drawn by open and solid squares, respectively. In the present paper, the effect of thermal noise has been always taken into account in all the calculations, so it must be stated that both the values of $i_c(0, 1)$ and $i_c(0, 5)$ are somewhat smaller than 1. The effect of charge coupling is not taken into account.

$M_{sat} = M(B_{ext} \geq B_L)$, but the strength of the B_L strongly depends on the magnetic property of the F-layer. On the contrary to the case of M_{sat}, the M_{res} is obtained when $B_{ext} = 0$, i.e., $M_{res} = M(0)$, so it is expected that it could be much easier to control the sample condition as compared with the case of M_{sat}.

The sizable effect for the magnetic field is mainly found in the 3@5 central junction as can be seen from Fig.2 for $N = 5$. In the following, therefore, first we focus our attention to the case such that the I-layer in the 3@5 central SIS-Josephson junction is replaced by the F-layer, hereafter we call it as "central case", and next the cases for 1@5 and 2@5 junctions are considered, called as "surface and intermediate cases". Equation (15) tells us that the effective magnetic induction $B_{eff}^{(j)}$ at the edge of the j-th IJJ is given by

$$B_{eff}^{(j)} = B_{ext} + M(B_{ext})\delta_{j,\ell} \,. \tag{29}$$

The external magnetic induction B_{ext} is written as $\alpha B_0(L_x)$ using a constant magnetic induction $B_0(L_x)$ so that we write the magnetization $M(B_{ext})$ as $m(\alpha)B_0(L_x)$ using a dimensionless parameter $m(\alpha)$ as a function of α. The ℓ in Eq.(29) is 3, 1 and 2 for the central, surface and intermediate cases, respectively. Therefore, the configuration $\mathbf{\Gamma}$ for the dimensionless effective magnetic induction defined by $B_{eff}^{(j)}/B_0(L_x)$ can be written as $(\alpha, \alpha, \alpha + m(\alpha), \alpha, \alpha)$, $(\alpha + m(\alpha), \alpha, \alpha, \alpha, \alpha)$ and $(\alpha, \alpha + m(\alpha), \alpha, \alpha, \alpha)$ for the central, surface and intermediate cases, respectively.

4.5.1 Case of $\chi_\ell(d_F) = 1$ for the ℓ-th SFS junction

First, we consider the case of the residual magnetization M_{res}, i.e., $\alpha = 0$. The configurations $\mathbf{\Gamma}$ for the central, surface and intermediate cases are therefore $(0, 0, m(0), 0, 0)$, $(m(0), 0, 0, 0, 0)$ and $(0, m(0), 0, 0, 0)$, respectively. By adopting the condition such that $N = 5$, $J_c = 1000\text{A}/\text{cm}^2$, $f_p = 122\text{GHz}$, $f_r = 200\text{GHz}$, R_{shunt}/junction$= 2\Omega$, SR_{shunt}/junction$= 50\mu\text{m}^2\Omega$, $i_r = 1$ and $L_x = 5\mu\text{m}$, i.e., $B_0(L_x) = 0.345\text{T}$, I have calculated the I-V characteristics for $m(0) \equiv m_{res} = 0.1$, 0.2, 0.5, 1 and 2. The I-V characteristics of the central case calculated for $m_{res} = 0$, 0.1, 0.2, 0.5, 1 and 2 are shown in Fig.4(a), (b), (c), (d), (e) and (f), respectively. Here note that only the I-V characteristics of 1@5 surface, 2@5 intermediate and 3@5 central junctions are shown from a symmetry consideration. Figure 4 shows that (I) Shapiro steps are clearly found except for the 3@5 central junction when $m_{res} = 2$, and due to the increasing the m_{res}, i.e., the residual magnetization M_{res} $(\equiv m_{res}B_0(L_x))$, (II) the I-V characteristics of 3@5 central junction are gradually changed, but (III) those of other junctions including no F-layer almost remain the same, that is, no remarkable change is found for the junctions without the F-layer. These findings are interesting. We are now solving a matrix equation (13) with $\mathbf{\Sigma_c} = \mathbf{1}$. The matrix $\mathbf{\Sigma_L}$ describing the magnetic interactions between neighboring junctions is given by Eq.(3) so that the inverse matrix $\mathbf{\Sigma_L}^{-1}$ is not a diagonal but a general form in which all the matrix elements are not equal to zero. Therefore, except for the case of $\mathbf{\Gamma} = (0, 0, 0, 0, 0)$, it seems to be reasonable to suppose that the effect of ferromagnetic layer should be observed on not only the 3@5 central junction but also the others. The calculated results are far from such a conjecture. The results calculated for the surface and intermediate cases, namely, $\mathbf{\Gamma} = (m_{res}, 0, 0, 0, 0)$ and $(0, m_{res}, 0, 0, 0)$, are shown in Figs.5 and 6, respectively. From these figures, we can see a similar result as in Fig.4 such that (I) Shapiro steps are clearly found except for the junction including F-layer when $m_{res} = 2$, (II) the I-V characteristics of the junction including F-layer are gradually changed due to the increasing the m_{res}, but (III) no remarkable change is found for others.

From the above, we can conclude that if the case such that an I-layer in the BSCCO-IJJs is replaced by a F-layer is considered, then the effect of the F-layer to the I-V characteristics is restricted when $B_{ext} = 0$, namely, no remarkable change is found for the junctions without the F-layer.

Finally, let us consider the case of $B_{ext} \neq 0$, i.e., $\alpha \neq 0$. The configurations $\mathbf{\Gamma}$ to be considered are $(\alpha, \alpha, \alpha + m(\alpha), \alpha, \alpha)$, $(\alpha + m(\alpha), \alpha, \alpha, \alpha, \alpha)$ and $(\alpha, \alpha + m(\alpha), \alpha, \alpha, \alpha)$ for the central, surface and intermediate cases. As an example, in the following the α is set to 1 tentatively, and 0.1, 0.2, 0.5 and 1 are selected as the value of $m(1)$.

The I-V characteristics calculated for the central cases $(1, 1, 1 + m(1), 1, 1)$ with $m(1) = 0$, 0.1, 0.2, 0.5 and 1 are shown in Fig.7(a), (b), (c), (d) and (e), respectively, together with those for $\mathbf{\Gamma} = (2, 2, 2, 2, 2)$ shown in (f). Figure 7 shows that (I) Shapiro steps are clearly found

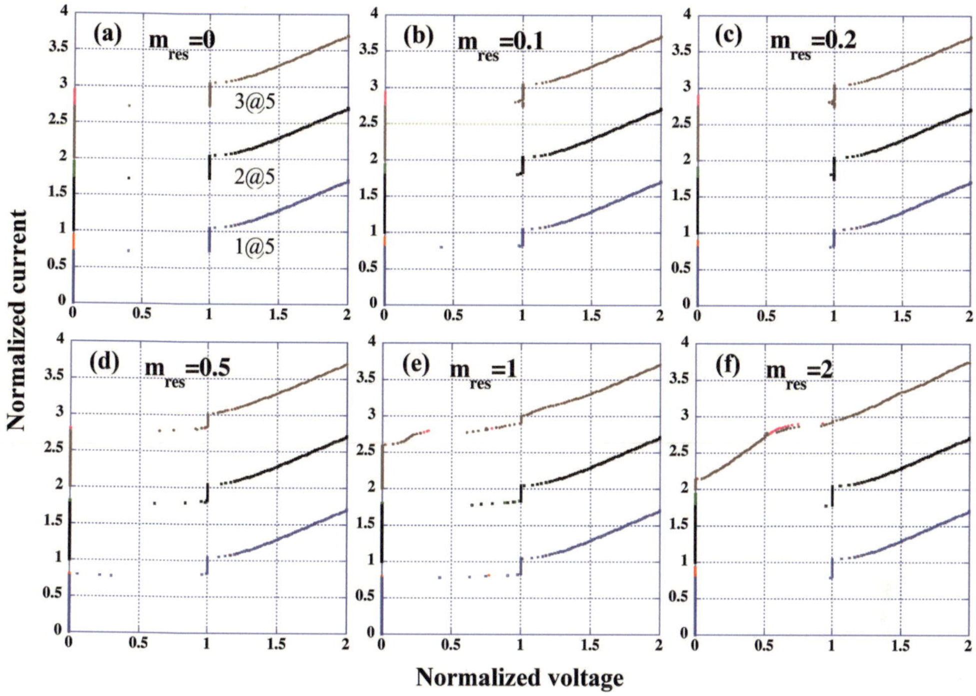

Fig. 4. The I-V characteristics of BSCCO IJJs calculated for $N = 5$, $J_c = 1000\text{A}/\text{cm}^2$, $f_p = 122\text{GHz}$, $f_r = 200\text{GHz}$, R_{shunt}/junction= 2Ω, SR_{shunt}/junction= $50\mu\text{m}^2\Omega$, $i_r = 1$, $L_x = 5\mu\text{m}$, i.e., $B_0(L_x) = 0.345\text{T}$ and $\chi_\ell(d_F) = 1$ for all junctions. The case such that there is no external magnetic field, i.e., $B_{ext} = 0$, and the I-layer in the 3@5 central SIS-Josephson junction is replaced by a F-layer is considered, so that the configuration $\boldsymbol{\Gamma}$ for the dimensionless effective magnetic induction defined by $B^{(j)}_{eff}/B_0(L_x)$ is $(0, 0, m_{res}, 0, 0)$. (a), (b), (c), (d), (e) and (f) are for $m_{res} = 0$, 0.1, 0.2, 0.5, 1 and 2, respectively. From the symmetry, only the I-V characteristics of 1@5 surface, 2@5 intermediate and 3@5 central junctions are shown. The horizontal axis indicates the dc voltage normalized to $\Phi_0 f_r$, and the vertical one shows the normalized dc i_0. Shapiro steps are clearly found except for the 3@5 central junction when $m_{res} = 2$ shown in (f).

in not only the junctions without F-layer but also the junction including F-layer, (II) the I-V characteristics of the 3@5 central junction including F-layer are surely changed due to the increasing the value of $m(1)$, but (III) those of other junctions without F-layer almost remain the same for the change of $m(1)$. Findings (II) and (III) mentioned now are fairly similar to those for the case of no external magnetic field, however, we wish to say that Shapiro step is found even in the 3@5 junction shown in (e) whose the $\boldsymbol{\Gamma}$ is $(1, 1, 2, 1, 1)$, in spite of the fact that clear Shapiro step cannot be found in (f) whose the $\boldsymbol{\Gamma}$ is $(2, 2, 2, 2, 2)$.

The I-V characteristics calculated for the surface case $(1 + m(1), 1, 1, 1, 1)$ are shown in Fig.8 and those for the intermediate one $(1, 1 + m(1), 1, 1, 1)$ are in Fig.9, where (a), (b) and (c) are for $m(1) = 0.2$, 0.5 and 1, respectively. Figure 8 shows that (I) Shapiro steps are clearly found in all the junctions even in the case of $m(1) = 1$, and (II) the I-V characteristics of 1@5 surface

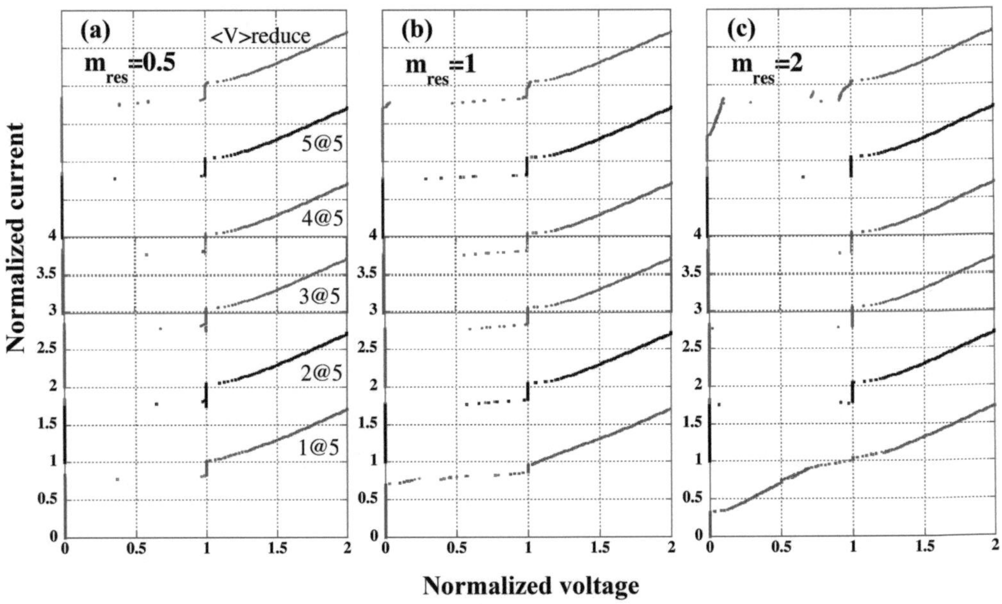

Fig. 5. The I-V characteristics of BSCCO IJJs calculated for $N = 5$, $J_c = 1000\text{A/cm}^2$, $f_p = 122\text{GHz}$, $f_r = 200\text{GHz}$, R_{shunt}/junction= 2Ω, SR_{shunt}/junction= $50\mu\text{m}^2\Omega$, $i_r = 1$, $L_x = 5\mu\text{m}$, i.e., $B_0(L_x) = 0.345\text{T}$, $B_{ext} = 0$ and $\chi_\ell(d_F) = 1$ for all junctions. The I-layer in 1@5 surface SIS junction is replaced by a F-layer so that the corresponding configuration $\mathbf{\Gamma}$ is $(m_{res}, 0, 0, 0, 0)$. Therefore, the I-V characteristics of all junctions are shown together with the reduced voltage $< V >_{reduce}$. (a), (b) and (c) are for $m_{res} = 0.5$, 1 and 2, respectively. The horizontal axis indicates the dc voltage normalized to $\Phi_0 f_r$, and the vertical one shows the normalized dc i_0. Shapiro steps are clearly found except for the 1@5 surface junction when $m_{res} = 2$ shown in (c).

junction including F-layer are gradually changed due to the increasing the value of $m(1)$. These findings are basically the same as the finding obtained in the central case. For the 2@5, 3@5, 4@5 and 5@5 junctions without F-layer, we can see that (III) the I-V characteristics of 2@5 and 3@5 junctions, which are close to the 1@5 junction including the F-layer, are fairly changed due to the increasing the $m(1)$, but (IV) those of 4@5 and 5@5 junctions, which are far from the 1@5 junction, almost remain the same as well as the central case. Finding (III) clearly tells us that the magnetic interaction between neighboring junctions is not negligible in BSCCO IJJs when $B_{ext} \neq 0$. From Fig.9, we can see a similar result as in the surface case, excepting 1@5 and 5@5 surface junctions in which the I-V characteristics are almost the same for the change of the value of $m(1)$.

4.5.2 Case of $\chi_\ell(d_F) < 1$ for the ℓ-th SFS junction

First, we consider the case of no external magnetic field, i.e., $\alpha = 0$. The calculation condition is the same as above, that is, $N = 5$, $J_c = 1000\text{A/cm}^2$, $f_p = 122\text{GHz}$, $f_r = 200\text{GHz}$, R_{shunt}/junction= 2Ω, SR_{shunt}/junction= $50\mu\text{m}^2\Omega$, $i_r = 1$ and $L_x = 5\mu\text{m}$, i.e., $B_0(L_x) = 0.345\text{T}$. There are three cases as the configurations that we should study, but in the following,

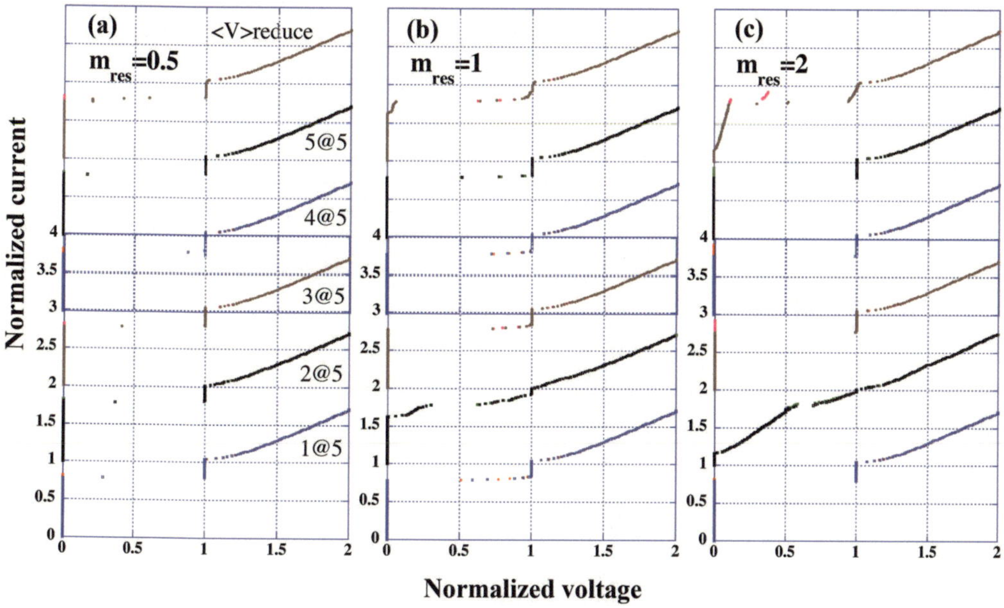

Fig. 6. The same as in Fig.5 but for $\mathbf{\Gamma} = (0, m_{res}, 0, 0, 0)$, that is, the I-layer in 2@5 intermediate SIS junction is replaced by a F-layer. Shapiro steps are clearly found except for the 2@5 intermediate junction when $m_{res} = 2$ shown in (c).

we consider only the central case with $m(0) = 0.5$. Namely, only an I-layer in the central SIS Josephson junction is replaced by a F-layer, therefore, the corresponding configuration $\mathbf{\Gamma}$ is $(0, 0, 0.5, 0, 0)$. The aim of the present section is to see the effect of $\chi_\ell(d_F)$ defined in Eq.(16) to the *I*-*V* characteristics.

The central case for $N = 5$ is considered so that the value of ℓ is 3. By using 0, 0.2, 0.4, 0.6 and 0.8 as a value of $\chi_3(d_F)$, I have calculated the *I*-*V* characteristics. The *I*-*V* characteristics calculated for $\chi_3(d_F) = 1.0$, 0.8, 0.6, 0.4, 0.2 and 0.0 are shown in Fig.10(a), (b), (c), (d), (e) and (f), respectively. Here note that from a symmetry consideration, only the *I*-*V* characteristics of 1@5-surface, 2@5-intermediate and 3@5-central junctions are shown in the respective figure. Figure 10 shows that (I) as is expected, the *I*-*V* characteristic of 3@5-central junction including F-layer surely changes due to the change of the value of $\chi_3(d_F)$, especially, the height of the Shapiro step decreases with the decreasing the $\chi_3(d_F)$ and reaches to zero when $\chi_3(d_F) = 0$, however, (II) those of 1@5-surface and 2@5-intermediate junctions without the F-layer fairly well remain the same for the changing the $\chi_3(d_F)$, and (III) clear Shapiro steps are found in 1@5-surface and 2@5-intermediate junctions irrespective of the value of $\chi_3(d_F)$. Equation (16) tells us that when $\chi_3(d_F) = 0$, the 3@5-central junction can be regarded as a simple non-superconducting RC-junction. This fact is clearly found in the *I*-*V* characteristic of 3@5-junction shown in (f), that is, in which a purely ohmic characterictic is found.

Next, let us consider the case of $\alpha \neq 0$ for the same calculation condition mentioned above. We consider again the central case with $m(\alpha) = 0.5$. As an example, we set the α to 1 so that the corresponding configuration $\mathbf{\Gamma}$ is $(1, 1, 1.5, 1, 1)$. The *I*-*V* characteristics calculated for $\chi_3(d_F) = 1.0$, 0.8, 0.6, 0.4, 0.2 and 0.0 are shown in Fig.11(a), (b), (c), (d), (e) and (f),

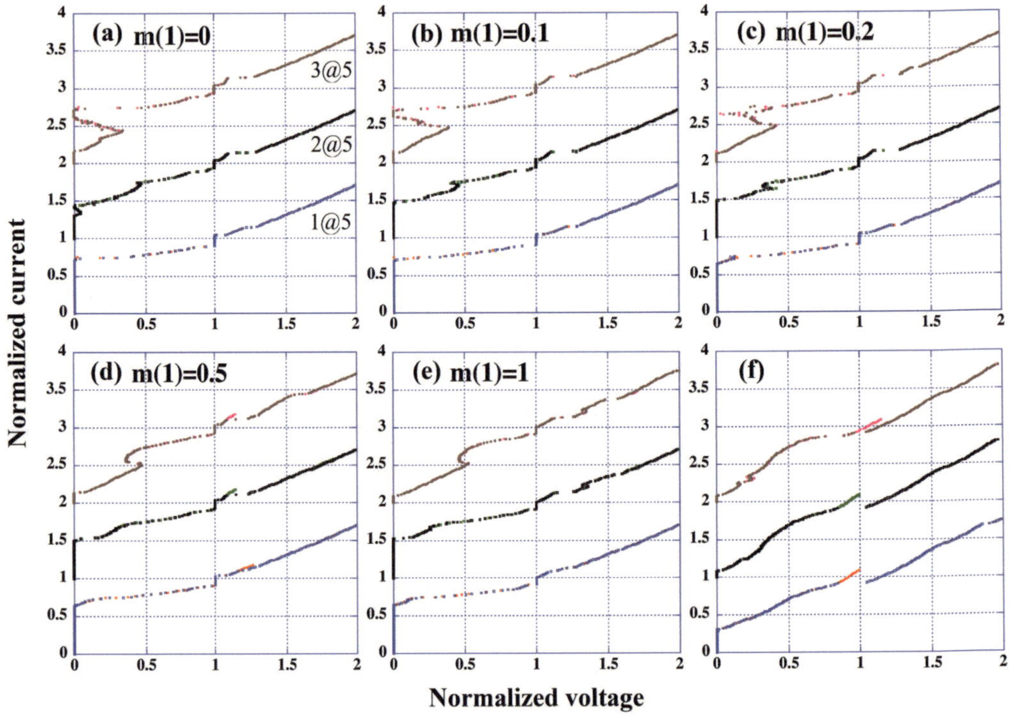

Fig. 7. The *I*-*V* characteristics of BSCCO IJJs calculated for $N = 5$, $J_c = 1000\text{A/cm}^2$, $f_p = 122\text{GHz}$, $f_r = 200\text{GHz}$, R_{shunt}/junction= 2Ω, SR_{shunt}/junction= $50\mu\text{m}^2\Omega$, $i_r = 1$, $L_x = 5\mu\text{m}$, i.e., $B_0(L_x) = 0.345\text{T}$ and $\chi_\ell(d_F) = 1$ for all junctions. The external magnetic induction B_{ext} is equal to $B_0(L_x)$ because of $\alpha = 1$. The I-layer in the 3@5 central SIS-Josephson junction is replaced by a F-layer so that the corresponding configuration $\mathbf{\Gamma}$ is $(1, 1, 1 + m(1), 1, 1)$. (a), (b), (c), (d) and (e) are for $m(1) = 0, 0.1, 0.2, 0.5$ and 1, respectively. For the comparison, the *I*-*V* characteristics calculated for $\mathbf{\Gamma} = (2, 2, 2, 2, 2)$ are also shown in (f). From the symmetry, only the *I*-*V* characteristics of 1@5 surface, 2@5 intermediate and 3@5 central junctions are shown. The horizontal axis indicates the dc voltage normalized to $\Phi_0 f_r$, and the vertical one shows the normalized dc i_0. (a) and (f) are equal to Fig.2(e) and (f), respectively. Shapiro steps are clearly found in all the *I*-*V* characteristics drawn in (a) to (e).

respectively. Figure 11 leads us a same conclusion mentioned above such that due to the decreasing the value of $\chi_3(d_F)$, (I) the *I*-*V* characteristic of 3@5 junction including the F-layer gradually changes, but those of 1@5 and 2@5 junctions without the F-layer well remain the same, and (II) Shapiro steps are clearly found in 1@5 and 2@5 junctions, but the height of the Shapiro step found in 3@5 junction decreases and reaches to zero when $\chi_3(d_F) = 0$.

From the above, we can say that the effect of $\chi_3(d_F)$ defined in the 3@5 central junction to the other junctions is negligible small irrespective of the value of the external magnetic field H_{ext}. This finding tells us that even if the BSCCO junction array with a multi-stacking structure is considered, the influence of the F-layer is restricted within a SFS-junction *itself*, i.e., that is negligible for others.

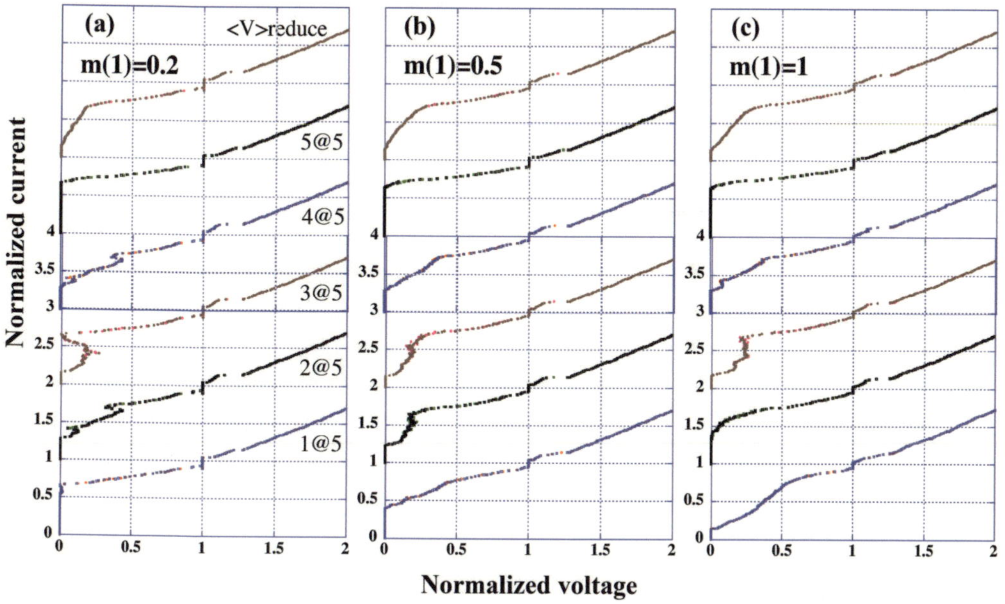

Fig. 8. The same as in Fig.7 but for the surface case, i.e., $\mathbf{\Gamma} = (1 + m(1), 1, 1, 1, 1)$. The I-layer in 1@5 surface SIS junction is replaced by a F-layer so that the *I*-*V* characteristics of all junctions are shown together with the reduced voltage $< V >_{reduce}$. (a), (b) and (c) are for $m(1) = 0.2$, 0.5 and 1, respectively. The horizontal axis indicates the dc voltage normalized to $\Phi_0 f_r$, and the vertical one shows the normalized dc i_0. Shapiro steps are clearly found in all the *I*-*V* characteristics.

5. Summary

Shapiro steps in BSCCO IJJs have been studied full numerically. In the study, the *I*-*V* characteristics of not only the case of no external magnetic field but also the case in which an external magnetic field is applied parallel to the junction surface have been calculated by using the MS-unified theory including both the electric and magnetic field couplings between neighboring Josephson junctions.

For the effect of the junction length L_x along the x-axis, we have found that the overall profiles of the *I*-*V* characteristics including Shapiro steps calculated for the same external magnetic induction are fairly similar each other even if the junction length L_x differs each other.

For the effect of SR_{shunt}-product, we have found that it is essential to add the shunt resistance with a low-value into the BSCCO IJJs, if we wish to make a good quality Shapiro step device which is *hard* for the external magnetic disturbance. In the present study, therefore, the value of SR_{shunt}-product has been fixed to $50\mu\text{m}^2\Omega/\text{junction}$.

In order to see the effect of i_r to the *I*-*V* characteristics including Shapiro steps, I have calculated the *I*-*V* characteristics of nine cases such that $(i_r, \alpha) = (0, 0.5)$, $(0, 1)$, $(0, 2)$, $(1, 0.5)$, $(1, 1)$, $(1, 2)$, $(2, 0.5)$, $(2, 1)$ and $(2, 2)$. Here note that the i_r is the normalized amplitude of the external ac modulation current with the frequency ω_r and the α defines the external magnetic induction B_{ext} as $\alpha B_0(L_x)$ using a constant magnetic induction $B_0(L_x)$. We have found that the

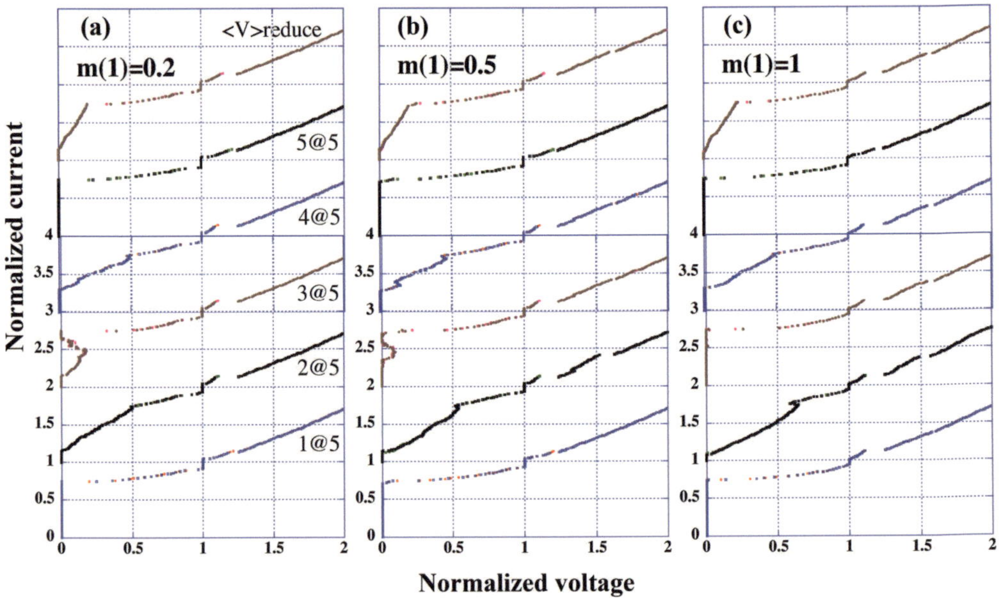

Fig. 9. The same as in Fig.7 but for the intermediate case, i.e., $\mathbf{\Gamma} = (1, 1 + m(1), 1, 1, 1)$. The I-layer in 2@5 intermediate SIS junction is replaced by a F-layer so that the I-V characteristics of all junctions are shown together with the reduced voltage $< V >_{reduce}$. (a), (b) and (c) are for $m(1) = 0.2$, 0.5 and 1, respectively. The horizontal axis indicates the dc voltage normalized to $\Phi_0 f_r$, and the vertical one shows the normalized dc i_0. Shapiro steps are clearly found in all the I-V characteristics.

Shapiro step and the flux flow could be treated separately under the presence of the external magnetic field.

From the I-V characteristics as a function of α, we have found that (I) the α-dependence of the normalized height $\Delta i_1(\alpha, N)$ of the 1st-order Shapiro step is very similar even if the N differs each other, however, (II) the α-dependence of the normalized critical current $i_c(\alpha, N)$ largely differs for $N = 1$ and 5, that is, the α-dependence of the $i_c(\alpha, N)$ is strongly affected by the magnetic interaction between neighboring Josephson junctions.

Within the framework of the MS-unified theory, I have studied the effect of ferromagnetic(F) layers into the I-V characteristics of BSCCO IJJs. In the present paper, we have studied the simplest case such that only the ℓ-th insulating(I)-layer in the BSCCO-IJJs is replaced by a F-layer whose the value of the magnetization is $M(B_{ext})$ as a function of the external magnetic induction B_{ext}. The effect of the magnetic field is taken into account via the boundary condition at the junction edge, so that the effective magnetic induction $B_{eff}^{(j)}$ at the edge of the j-th IJJ is given by $B_{ext} + M(B_{ext})\delta_{j,\ell}$. We have found that when $B_{ext} = 0$, (I) Shapiro steps are clearly found except for the junction including F-layer with a large residual magnetization $M_{res} \equiv m_{res} B_0(L_x)$ such as $m_{res} = 2$, (II) the I-V characteristics of the junction including F-layer are gradually changed due to the increasing the value of m_{res}, but (III) no remarkable change is found for the other junctions without F-layer. For $B_{ext} \neq 0$, i.e., $\alpha \neq 0$, in which the magnetization $M(B_{ext})$ is written as $m(\alpha)B_0(L_x)$, we have found that (IV) even in the

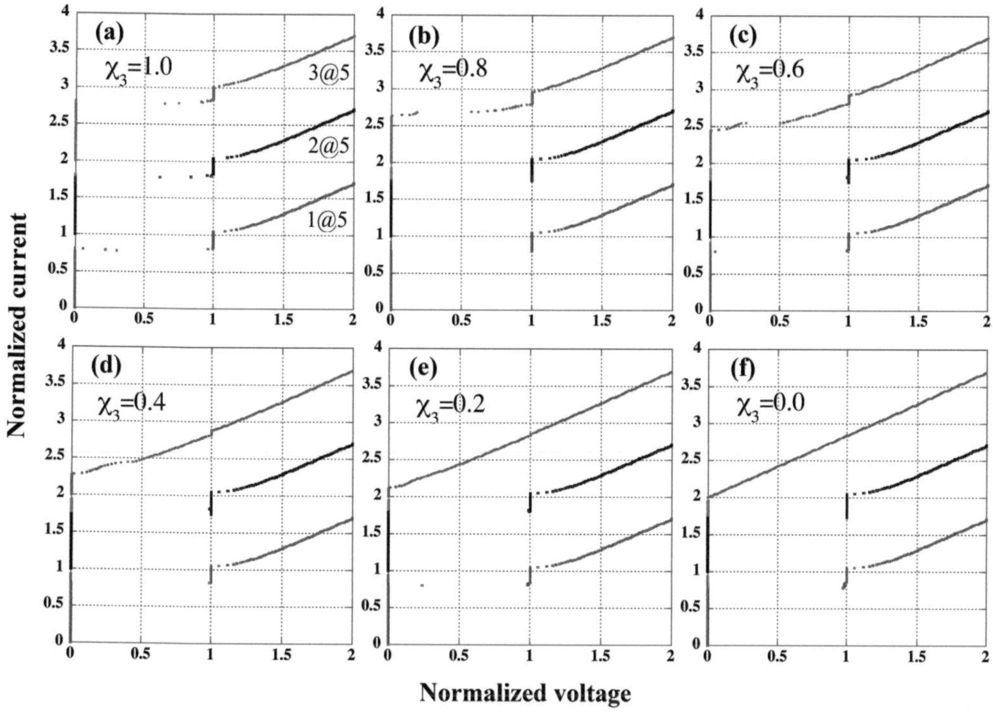

Fig. 10. The *I-V* characteristics of BSCCO IJJs calculated for $N = 5$, $J_c = 1000\text{A/cm}^2$, $f_p = 122$GHz, $f_r = 200$GHz, R_{shunt}/junction= 2Ω, SR_{shunt}/junction= 50μm$^2\Omega$, $i_r = 1$, $L_x = 5\mu$m, i.e., $B_0(L_x) - 0.345$T. In this figure, the central case has been considered and the configuration Γ has been tentatively selected as $(0,0,0.5,0,0)$. For the $\chi_\ell(d_F)$ defined in Eq.(16), only the value for $\ell = 3$ has been corrected, i.e., that of others has been held to 1. The value of $\chi_3(d_F)$ is 1.0, 0.8, 0.6, 0.4, 0.2 and 0.0 for (a), (b), (c), (d), (e) and (f), respectively. From the symmetry, only the *I-V* characteristics of 1@5 surface, 2@5 intermediate and 3@5 central junctions are shown. The horizontal axis indicates the dc voltage normalized to $\Phi_0 f_r$, and the vertical one shows the normalized dc i_0. Note that (a) is equal to Fig.4(d).

case of $\alpha \neq 0$, the findings (II) and (III) mentioned above are basically held again, but for the tinging (I) we have found that (V) clear Shapiro steps are found in not only the junction without F-layer but also the junction including F layer. In addition to the above calculations, I have further calculated the *I-V* characteristics for the surface and intermediate cases when $\alpha = 1$, whose the configurations are $(1 + m(1), 1, 1, 1, 1)$ and $(1, 1 + m(1), 1, 1, 1)$, respectively. We have concluded that even in the case in which a junction in BSCCO IJJs includes a F-layer with a finite magnetization, (I) the effect of the magnetic interaction between neighboring junctions is not so clearly found when $B_{ext} = 0$, but (II) when $B_{ext} \neq 0$, that is found clearly. Conclusion (II) obtained here is the same as a previous statement such that in the BSCCO IJJs with a stacking structure, there are considerable magnetic interactions between neighboring Josephson junctions when $\alpha \neq 0$. [36]

In order to see the effect of the F-layer in the SFS-junction to the *I-V* characteristics of BSCCO IJJs with a stacking structure, I have calculated the *I-V* characteristics as a function of $\chi_\ell(d_F)$ defined in Eq.(16). From the results calculated for the no external magnetic field and the

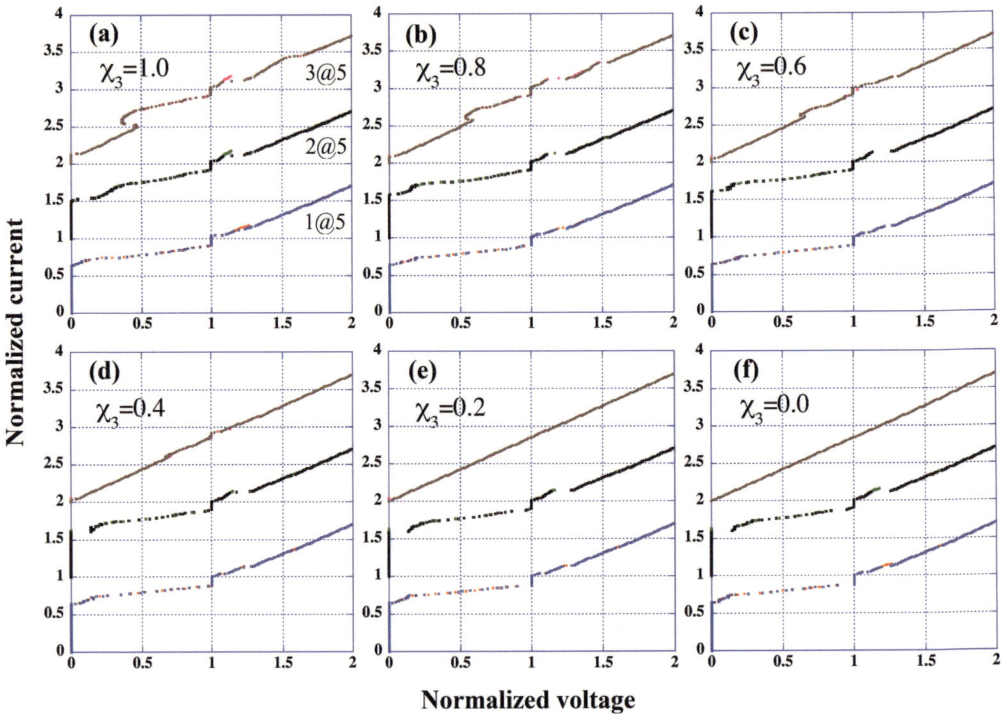

Fig. 11. The same as in Fig.10 but for the case of $\mathbf{\Gamma} = (1, 1, 1.5, 1, 1)$. Note that (a) is equal to Fig.7(d).

nonzero field such as $\alpha = 0$ and 1, we have found that the effect of $\chi_3(d_F)$ defined in the 3@5-central junction to the other junctions is negligible small irrespective of the value of the external magnetic field H_{ext}, namely, the influence of the F-layer is restricted within a SFS-junction *itself*, i.e., that is negligible for others.

6. Acknowledgments

The author would like to thank Professors G. Oya and A. Irie, Utsunomiya University, for their valuable comments and discussion.

7. References

[1] S. Shapiro, Phys. Rev. Lett. 11, 80 (1963).

[2] R. Kleiner, F. Steinmeyer, G. Kunkel, and P. Müller, Phys. Rev. Lett. 68, 2394 (1992).

[3] G. Oya, N. Aoyama, A. Irie, S. Kishida, and H. Tokutaka, Jpn. J. Appl. Phys. 31, L829 (1992).

[4] R. Kleiner and P. Müller, Phys. Rev. B 49, 1327 (1994).

[5] A. Yurgens, D. Winkler, N. V. Zavaritsky, and T. Claeson, Phys. Rev. B 53, R8887 (1996).

[6] Yu. I. Latyshev, J. E. Nevelskaya and P. Monceau, Phys. Rev. Lett. 77, 932 (1996).

[7] K. Schlenga, R. Kleiner, G. Hechtfischer, M. Mößle, S. Schmitt, P. Müller, Ch. Helm, Ch. Preis, F. Forsthofer, J. Keller, H. L. Johnson, M. Veith, and E. Steinbeiß, Phys. Rev. B 57, 14518 (1998).

[8] A. Irie, Y. Hirai, and G. Oya, Appl. Phys. Lett. 72, 2159 (1998).
[9] D. J. Van. Harlingen, Rev. Mod. Phys. 67, 515 (1995).
[10] C. C. Tsuei and J. R. Kirtley, Rev. Mod. Phys. 72, 969 (2000).
[11] A. Damascelli, Z. Hussain, and Z. -X. Shen, Rev. Mod. Phys. 75, 473 (2003).
[12] Y. J. Doh, J. Kim, K. T. Kim, and H. J. Lee, Phys. Rev. B 61, R3834 (2000).
[13] H. B. Wang, P. H. Wu and T. Yamashita, Phys. Rev. Lett. 87, 107002 (2001).
[14] Yu. I. Latyshev, M. B. Gaifullin, T. Yamashita, M. Machida, and Y. Matsuda, Phys. Rev. Lett. 87, 247007 (2001).
[15] H. B. Wang, P. H. Wu, J. Chen, K. Maeda, and T. Yamashita, Appl. Phys. Lett. 80, 1604 (2002).
[16] G. Oya, A. Terada, N. Takahashi, A. Irie, and T. Hashimoto, Jpn. J. Appl. Phys. 44, L491 (2005).
[17] G. Oya, T. Hashimoto, and A. Irie, Supercond. Sci. Technol. 19, S191 (2006).
[18] H. B. Wang, S. M. Kim, S. Urayama, M. Nagao, T. Hatano, S. Arisawa, T. Yamashita, and P. H. Wu, Appl. Phys. Lett. 88, 063503 (2006).
[19] A. Barone and G. Paterno, *Physics and Applications of the Josephson Effect* (Wiley, New York, 1982).
[20] S. Sakai, P. Bodin, and N. F. Pedersen, J. Appl. Phys. 73, 2411 (1993).
[21] R. Kleiner, P. Müller, H. Kohlstedt, N. F. Pedersen, and S. Sakai, Phys. Rev. B 50, 3942 (1994).
[22] R. Kleiner, Phys. Rev. B 50, 6919 (1994).
[23] S. Sakai, A. V. Ustinov, H. Kohlstedt, A. Petraglia, and N. F. Pedersen, Phys. Rev. B 50, 12905 (1994).
[24] T. Koyama and M. Tachiki, Phys. Rev. B 54, 16183 (1996).
[25] V. M. Krasnov and D. Winkler, Phys. Rev. B 56, 9106 (1997).
[26] M. Machida, T. Koyama and M. Tachiki, Physica C 362, 16 (2001).
[27] A. E. Koshelev, Phys. Rev. B 66, 224514 (2002).
[28] M. Machida, Phys. Rev. Lett. 90, 037001 (2003).
[29] M. Machida and S. Sakai, Phys. Rev. B 70, 144520 (2004).
[30] A. V. Ustinov and N. F. Pedersen, Phys. Rev. B 72, 052502 (2005).
[31] S. Madsen and N. F. Pedersen, Phys. Rev. B 72, 134523 (2005).
[32] B. Y. Zhu, H. B. Wang, S. M. Kim, S. Urayama, T. Hatano, and X. Hu, Phys. Rev. B 72, 174514 (2005).
[33] M. Machida, Phys. Rev. Lett. 96, 097002 (2006).
[34] A. Irie and G. Oya, Supercond. Sci. Technol. 20, S18 (2007).
[35] M. Kitamura, A. Irie, and G. Oya, Phys. Rev. B 76, 064518 (2007).
[36] M. Kitamura, A. Irie, and G. Oya, J. Appl. Phys. 104, 063905 (2008).
[37] V. V. Ryazanov, V. A. Oboznov, A. Yu. Rusanov, A. V. Veretennikov, A. A. Golubov, and J. Aarts, Phys. Rev. Lett. 86, 2427 (2001).
[38] T. Kontos, M. Aprili, J. Lesueur, F. Genêt, B. Stephanidis, and R. Boursier, Phys. Rev. Lett. 89, 137007 (2002).
[39] A. A. Golubov, M. Yu. Kupriyanov, and E. Il'ichev, Rev. Mod. Phys. 76, 411 (2004).
[40] A. I. Buzdin, Rev. Mod. Phys. 77, 935 (2005).
[41] V. A. Oboznov, V. V. Bol'ginov, A. K. Feofanov, V. V. Ryazanov, and A. I. Buzdin, Phys. Rev. Lett. 96, 197003 (2006).

[42] J. W. A. Robinson, S. Piano, G. Burnell, C. Bell, and M. G. Blamire, Phys. Rev. Lett. 97, 177003 (2006).
[43] M. Weides, M. Kemmler, H. Kohlstedt, R. Waser, D. Koelle, R. Kleiner, and E. Goldobin, Phys. Rev. Lett. 97, 247001 (2006).
[44] S. Hikino, M. Mori, S. Takahashi, and S. Maekawa, J. Phys. Soc. Jpn. 77, 053707 (2008).
[45] R. L. Kautz and R. Monaco, J. Appl. Phys. 57, 875 (1985).
[46] M. Kitamura, A. Irie, and G. Oya, Phys. Rev. B 66, 054519 (2002).
[47] M. D. Fiske, Rev. Mod. Phys. 36, 221 (1964).
[48] V. M. Krasnov, N. Mros, A. Yurgens, and D. Winkler, Phys. Rev. B 59, 8463 (1999).

9

Magnetical Response and Mechanical Properties of High Temperature Superconductors, $YBaCu_3O_{7-x}$ Materials

J.J. Roa[1,*], F.T. Dias[2] and M. Segarra[3]
[1]*Institut P' (UPR 3346), Département de Physique et Mécanique des Matériaux, CNRS - Université de Poitiers – ENSMA, SP2MI - Téléport 2 - BP 30179, F-86962 Futuroscope, Chasseneuil Cedex*
[2]*Federal University of Pelotas, Physics and Mathematics Institute, Department of Physics,*
[3]*DIOPMA Centre. Universitat de Barcelona, Facultat de Quimica, Departament de Ciència dels Materials i Enginyeria Metal·lúrgica, Barcelona*
[1]*France*
[2]*Brazil*
[3]*Spain*

1. Introduction

Since the discovery of the second generation High Superconductor Materials (2G-HTSC) in 1986, their magnetic properties have been widely studied by different research groups around the world. During the last years, the mechanical properties at micro-/ and nanometric scale are starting to be studied in order to know and improve the durability of conventional devices.

In this book chapter, we would like to focus our attention on the magnetical response and also the mechanical properties of 2G-HTSC.

In relation to the magnetic response, the Meissner effect is one important signature of the superconductivity. In this case, a diamagnetic response is observed due to exclusion of the magnetic flux of the interior of the superconducting material when the temperature is below the critical temperature. This important property allows to distinguish a superconducting material from a conducting one. But in several cases the superconducting materials exhibit a paramagnetic response instead of the conventional diamagnetism. This effect is frequently called *paramagnetic Meissner effect (PME)*. In this case, the magnetic flux is not expelled, and a paramagnetic state can be originated. This effect is observed in several magnetic field regimes, and in some cases the paramagnetic response increases with the applied magnetic field, but in others the paramagnetic response decreases when the magnetic field increases.

The paramagnetic Meissner effect (PME) was first observed in polycrystalline samples of high temperature superconductors. The discovery of this effect has originated an enormous effort of investigation by several groups around the world and several theories and models have been proposed to explain this interesting and controversial behavior, but without a conclusive understanding yet. One of the first interpretations is based on the occurrence of Josephson π junctions between superconducting grains into the sample, and in this case the superconducting samples showing the PME can be modeled as a Josephson medium where the π junctions are randomly distributed.

Some experimental results have motivated the proposal of mechanisms based on flux trapping and flux compression into the samples, and a non-equilibrium compressed flux states can be stabilized by inhomogeneously cooling the sample, thus yielding to a paramagnetic response. The formation of a giant vortex state on the sample, or the flux capture by samples with a particular shape, may produce a strong compression of the trapped flux in their interior. Several groups have used models based on these ideas to explain their results mainly when high magnetic fields are applied.

Other models based on arrays of multiply-connected superconductors have been proposed to explain the PME. Josephson junctions are taken into account and a paramagnetic response appears when inner currents may flow through these junctions under special conditions. Many groups have pointed that the PME observed in granular superconductors can be described as an intrinsic effect of multiply-connected superconductors, since the granularity emulates a network of Josephson junctions.

An important result observed in superconducting samples that exhibit the PME is the anomalous time-dependence on the magnetization. In some cases, the paramagnetic moments obtained after field cooling process relax monotonically to increasingly positive values. Some results show that this tendency remains unaltered for large time intervals. In other cases, the relaxation may show a change of sign in the magnetization from negative to positive. This behavior is opposite to expectations based on a flux-creep scenario, and some models have been proposed in order to explain this anomalous effect.

One of the sections of this chapter is dedicated to a review of the experimental results and main models and ideas about the paramagnetic Meissner effect in high-Tc materials. A special attention to results observed in $YBa_2Cu_3O_{7-x}$ superconductor prepared by melt-textured techniques will be paid. These materials are very important to technological applications, and in some cases the observed PME is very strong. In relation to melt-textured samples, the importance of the Y_2BaCuO_5 (Y-211) phase added during the growth is also discussed in order to connect with the PME causes.

Concerning the mechanical properties, attention will be focused on the following: hardness, Elastic modulus, toughness fracture and the stress-strain curve, taking into account the different works published since the discovery of this material. However, their determination at nanometric scale is highlighted, as this technique is very functional and fast, and can give us a lot of information about the mechanical behavior of each phase present in this kind of ceramic materials. The principle of the technique is the evaluation of the response of a material to an applied load, it allows to extract the mechanical properties of an heterogeneous material at very low applied loads. Depending on the tip employed, different equations should be applied to study the response of the material (Oliver and Pharr or Hertz equations for elastic-to-plastic and elastic regimen respectively), and thus calculate the stress-strain curves in order to obtain the yield strength of the material, and conventional parameters like hardness, Elastic modulus and toughness. In this chapter we would like to make a state of the art of the different mechanical properties obtained using hardness tests up to now of a ceramic superconductor material ($YBa_2Cu_3O_{7-x}$ or Y-123) textured by Bridgman and Top-Seeding Melt Growth techniques. This material presents a phase transition from tetragonal to orthorhombic that promotes a change in its electrical properties, from insulating to superconductor, and that can be achieved by partially oxygenating the material. On the other hand, the structure of the texture material presents a heterogeneous distribution of two different phases: Y-123) as a matrix and Y-211 as quasi spherical inclusions. So, a review and a discussion about the different equations and method employed to isolate the mechanical response for each phase and for the whole material is presented. Finally, the different fracture mechanisms activated during the indentation process such as chipping, or radial cracks at the corners of the imprints will be presented.

1.1 Magnetic response: The paramagnetic Meissner effect

This section is dedicated to the description of the magnetic response exhibited by some superconducting materials, which is known as the paramagnetic Meissner effect.

1.1.1 The Meissner effect

The Meissner effect is one important signature of the superconductivity. In this case, a diamagnetic response is observed due to exclusion of the magnetic flux from inside the superconducting material when the temperature is below its critical temperature. This important property allows us to distinguish a superconducting material from a perfectly conducting material.

Type-II superconductors present a perfect diamagnetism (Meissner effect) just when the magnetic applied field is lower than $H_{C1}(T)$, known as the lower critical field, as showed by **Fig. 1**. Consequently this region is known as Meissner state. For magnetic applied fields higher than $H_{C1}(T)$ but lower than $H_{C2}(T)$, or upper critical field, the magnetic field can penetrate inside the superconductor in the form of vortices. This state is known as mixed state, and the $H_{C1}(T)$ value determines the penetration of the first magnetic flux-line inside the superconductor.

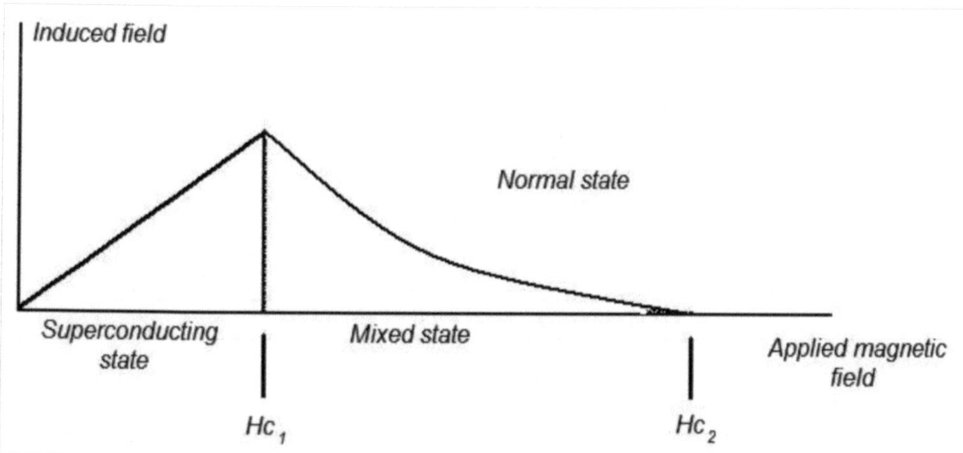

Fig. 1. Magnetic behaviour of a type-II superconductor.

The Meissner effect can be observed by the field-cooled warming (FCW) or field-cooled cooling (FCC) prescriptions. The FCW prescription consists of a first cooling down of the superconducting material to temperatures well below the critical temperature (T_C) in magnetic applied field; then the magnetization can be measured under constant magnetic field while the sample warms up to temperatures above the critical temperature. The FCC prescription is similar, but in this case the magnetization is measured while cooling the sample back to temperatures well below the critical temperature in magnetic applied field. In these conditions the magnetic flux uniformly distributes into the sample, and when the sample is cooled below T_C a diamagnetic moment appears as a consequence of the expulsion of magnetic flux from within it. Due to vortex pinning effects, in general the field-cooled (FC) response is lower than the zero-field cooling (ZFC - diamagnetic shielding) response, as showed by **Fig. 2**.

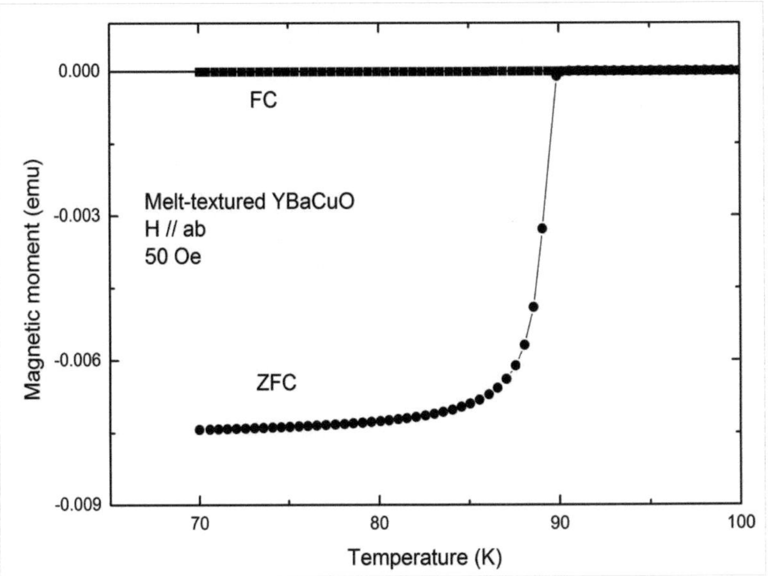

Fig. 2. ZFC and FC curves for a melt-textured $YBa_2Cu_3O_{7-x}$ sample.

1.1.2 Paramagnetic Meissner effect

The Meissner effect is characterized by a diamagnetic response of the superconducting material when a magnetic field is applied, but in several cases this magnetic response in FC experiments can be paramagnetic, challenging the conventional Meissner effect. This effect is known as paramagnetic Meissner effect (PME), but sometimes is called Wohlleben effect in honor of some of its discoverers (Bräuchle et al., 1994; Khomskii, 1994).

The first observations of the PME were reported in a pioneer work by P. Svedlindh et al. (Svedlindh et al., 1989) in the BiSrCaCuO superconductor. The discovery of the PME has prompted an enormous effort of investigation due to the striking contrast of this effect with the expected diamagnetic response. Initially, the effect was credited as an artifact of the measurement procedure (Blunt et al., 1991), but several further results showed that the PME is a genuine response of the superconducting material.

The PME has been observed in several superconducting systems, since low-T_C materials such as Nb (Thompson et al., 1996; Minhaj et al., 1994), high-Tc materials such as $Bi_2Sr_2CaCu_2O_8$ and $YBa_2Cu_3O_{7-x}$ (Braunisch et al., 1992; Dias et al., 2000), exotic systems such as $Ba_{1-x}K_xBiO_3$ (Kim et al., 1996), multiphase In-Sn alloys (Schwartz et al., 2006), $YBa_2Cu_3O_{7-x}$ /$La_{0.7}Ca_{0.3}MnO_3$ superlattices (Arias et al., 2006), organic compounds (Lebed, 2008) and many others. Recently, the PME has been also observed in Pb films by Brandt et al. (Brandt et al., 2011). The effect can be observed in polycrystalline sintered samples (Svendlindh et al., 1989), melt-textured samples (Pureur et al., 2001), single-crystals (Kusmartsev et al., 1997; Dias et al., 2010), alloys (Schwartz et al., 2006) and thin films (Pan et al., 2006; Terentiev et al., 1999) prepared by several techniques.

The PME is qualitatively distinct in relation to the applied magnetic field, and can be observed from few Oe up to several kOe. In some cases the effect decreases when the magnetic field increases, as can be seen in **Fig. 3** for an $YBa_2Cu_3O_{7-x}$ single-crystal. However, in other cases the effect increases when the magnetic field increases, as showed by **Fig. 4** for another $YBa_2Cu_3O_{7-x}$ single-crystal.

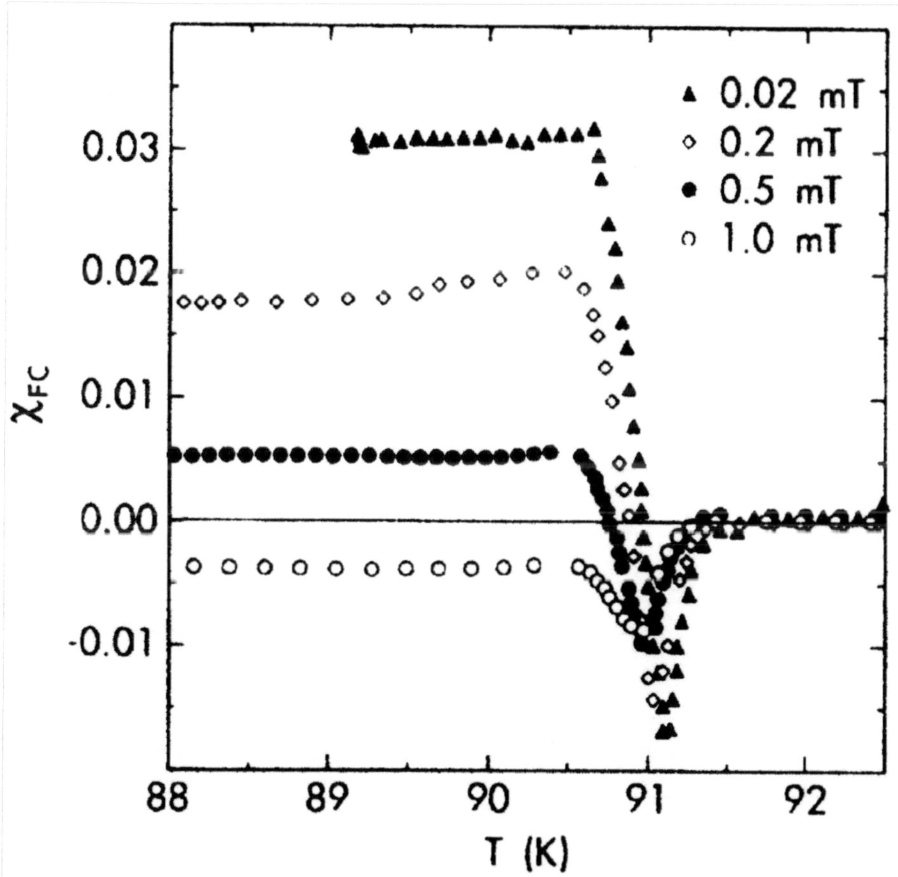

Fig. 3. Magnetic susceptibility in an $YBa_2Cu_3O_{7-x}$ single crystal. The magnetic susceptibility decreases when the magnetic field increases. From Bräuchle et al., 1994.

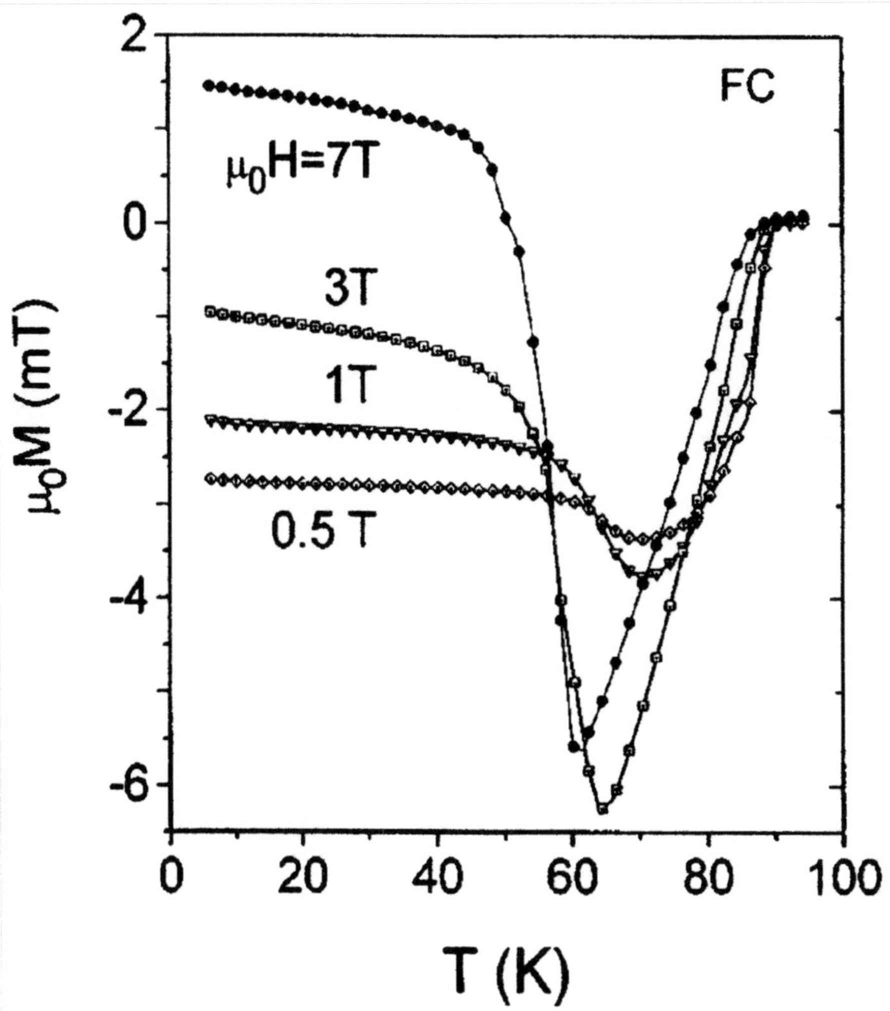

Fig. 4. FC magnetizations in an overdoped $YBa_2Cu_3O_{7-x}$ single crystal. In this case the magnetization increases when the magnetic field increases. From Kusmartsev et al., 1992.

Sometimes the PME presents a curious time dependence of the FC magnetization, as showed by **Fig. 5** for a melt-textured $YBa_2Cu_3O_{7-x}$ sample. In **Fig. 5** a paramagnetic signal was observed when the FC moment relaxes at constant magnetic field and temperature. The inset of this figure shows the sign reversal obtained in a relaxation experiment in the same sample, but cooled under another rate and magnetic field configuration.

It is possible to observe that an originally negative FC moment relaxes towards positive values and reverses sign after a wait time of about 3000 s. The time dependence of the PME has been related by other authors in different superconducting materials (Pan et al., 2006; Terentiev et al., 1999; Dias et al., 2004).

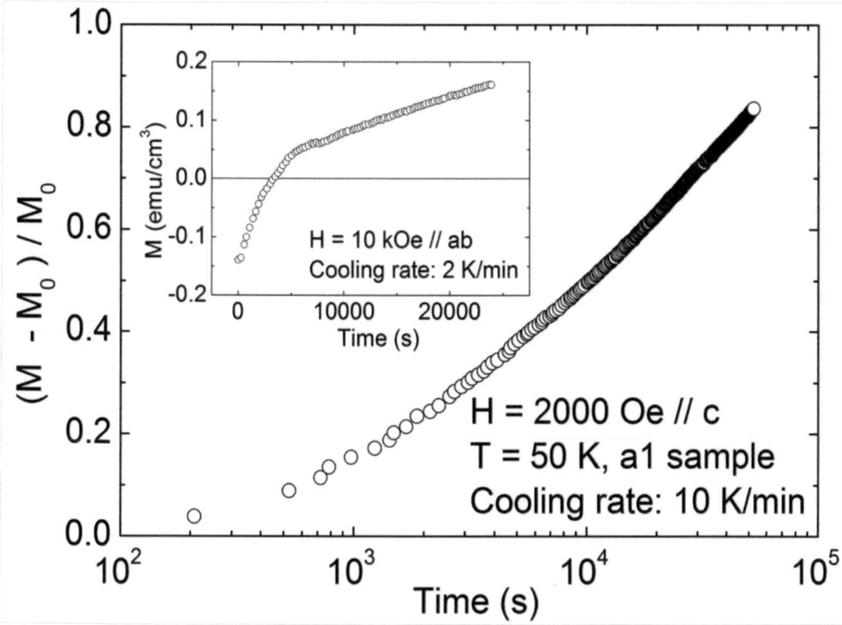

Fig. 5. Time dependence of normalized FC magnetization in a melt-textured $YBa_2Cu_3O_{7-x}$ sample exhibiting the PME. The inset shows the sign reversal obtained in a relaxation experiment in the same sample. From Pureur et al., 2001.

1.1.3 Models and experimental results

The paramagnetic Meissner effect is a controversial and challenging effect that can be observed in a variety of superconducting systems. However, in several cases there is a strong dependence on the applied magnetic field (magnitude and configuration), cooling rate during measurements, time effects, pinning effects, and others circumstances, and hence there is no a definitive model that can explain the paramagnetic moments observed during the FC experiments.

After the first observations of the PME, an enormous effort of investigation has emerged in order to explain this anomalous effect. The initial motivation was given by its interpretation based on the occurrence of π junctions between the superconducting grains in the material (Kusmartsev et al., 1992).

According to this, the sample showing PME may be modeled as a Josephson medium where the π junctions are randomly distributed. Assuming that closed superconducting loops containing an odd number of these anomalous junctions are likely to occur in these networks, spontaneous and polarizable orbital currents may be generated, giving rise to the PME when very low magnetic fields are applied. The required π-phase shifts in the Josephson links were proposed to arise from *d*-wave pairing symmetry in the high-*Tc* superconducting materials (Sigrist & Rice, 1995).

The **Fig. 6** shows PME results obtained in a polycrystalline $Bi_2Sr_2CaCu_2O_8$ sample by Braunisch et al. (Braunisch et al., 1992) and explained on the basis of this model.

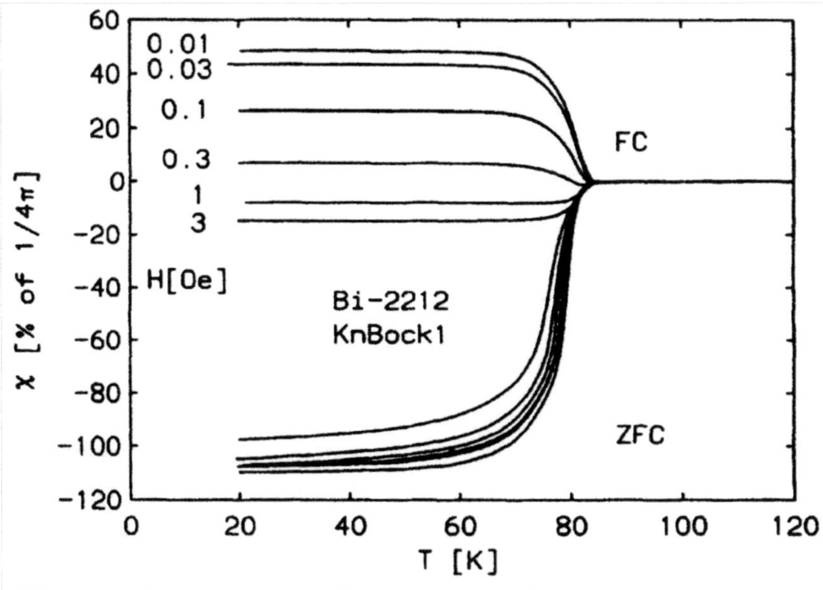

Fig. 6. ZFC and FC signals in a polycrystalline $Bi_2Sr_2CaCu_2O_8$ sample exhibiting the PME. From Braunisch et al., 1992.

Several results can be explained based on this model, specially when low magnetic fields are applied. The concept of polarized orbital currents to explain the low field PME was further developed by studies of the characteristic microstructure of samples showing this effect (Freitag et al., 1999), and by investigations on the time dependence of the paramagnetic moment (Magnusson et al., 1995), ac susceptibility (Nordblad et al., 1998) and magnetic aging (Svedlindh et al., 1999).

A controversial point to this model arises from similar results obtained in Nb samples, because some authors (Sigrist & Rice, 1995) attribute that π junctions are a consequence of the *d*-wave pairing symmetry, which is nonexistent in this system. The PME observed in Nb (Thompson et al., 1996; Minhaj et al., 1994; Kostic et al., 1996) and Nb thin films (Ortiz et al., 2001; Terentiev et al., 1999) was demonstrated to be strongly influenced by the sample geometry and surface. In some cases, the polishing of the surfaces affects the PME significantly, leading to its disappearance. **Fig. 7** shows the magnetization results of a Nb disk before and after polishing the sample surface, where it is possible to see that the PME disappears after the sample is polished. A surface dependent PME in $YBa_2Cu_3O_{7-x}$ single crystals was also found (Lucht et al., 1995).

These results motivated the proposal of mechanisms to explain the PME without taking into account effects from intrinsically nonconventional superconductivity, and models based on flux trapping and flux compression effects were proposed. According to these models, a nonequilibrium compressed flux state may be generated and stabilized by inhomogeneous cooling of the sample, so that its surface becomes superconducting prior to the bulk (Koshelev & Larkin, 1995). Then, the magnetic flux may be pushed into the sample, creating a flux compressed state upon further cooling.

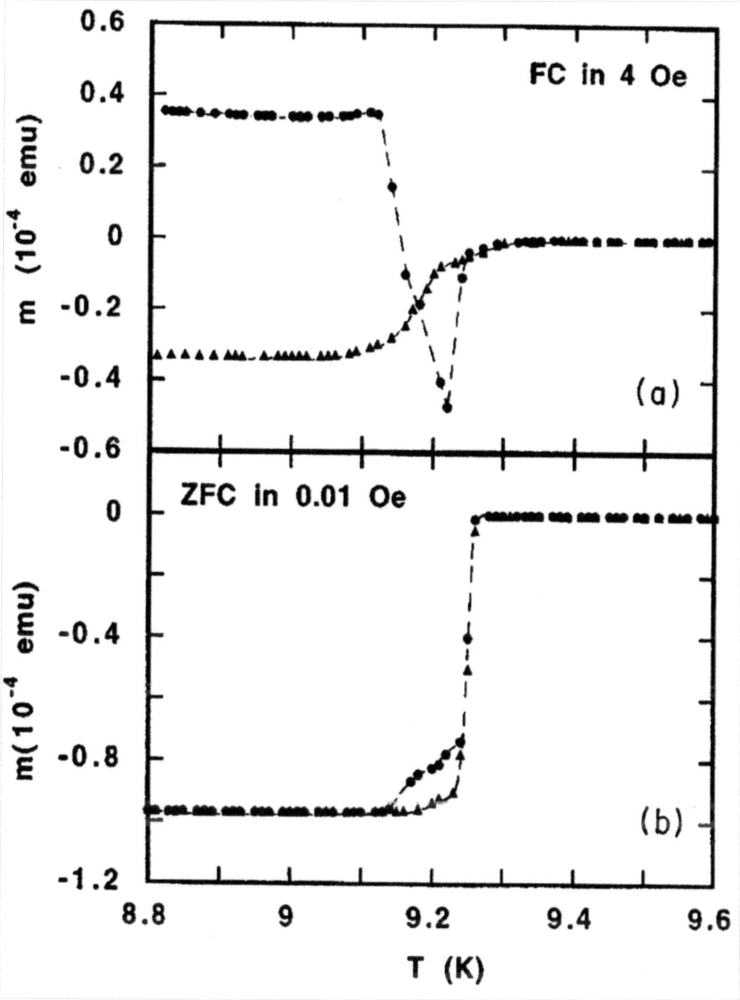

Fig. 7. FC (a) and ZFC (b) experiments in a Nb disk. The circles were taken before polishing the sample surface, and the triangles were taken after polishing the sample surface. In this case, the polishing of the surface affects the PME, leading to its disappearance. From Kostic´ et al., 1996.

Similarly, the formation of a giant vortex state on the sample surface proposed by Moshchalkov et al. (Moshchalkov et al., 1997) may produce a compression of the trapped flux. According to the authors, the PME observed in high-T_C and Nb superconducting samples can be explained by the persistence of a giant vortex state with a fixed orbital quantum number *L*, formed during the FC regime at the third critical field. However, the amplitude of the PME is suppressed by applying a magnetic field. The giant vortex state was also shown to be responsible for PME in small disks of Al samples (Geim et al., 1998), where transitions between metastable giant vortex states having different angular momenta (*L*) have been observed. The transitions compress the giant vortex into smaller volumes, allowing extra flux to penetrate in the sample surface and producing the paramagnetic moments.

The flux capture by samples with a special shape proposed by Obukhov (Obukhov et al., 1998), may also produce compression of the trapped flux. According to the author, an unusual behavior of the long-time relaxation of the magnetization after a FC procedure may show not only an increase in the magnetization over the time but a change of sign from negative to positive. This interesting result was observed in melt-textured $YBa_2Cu_3O_{7-x}$ samples (Pureur et al., 2001), as shown by the inset in the **Fig. 5**.

A model to explain the PME observed in Josephson-junctions arrays, such as Nb-Al_2O_3-Nb (Nielsen et al., 2000), was proposed to account for the paramagnetism without invoking the role of π junctions or *d*-wave superconductivity (Araujo-Moreira, 1997; Barbara, 1999; Nielsen et al., 2000). According to this model, the screening of the array of these multiply-connected superconductors is provided by the loops at its boundaries, and the diamagnetic currents running at the external perimeter induce currents in the opposite sense inside the array. These inner currents generate a paramagnetic scenario under certain conditions. This mechanism is shown in **Fig. 8**.

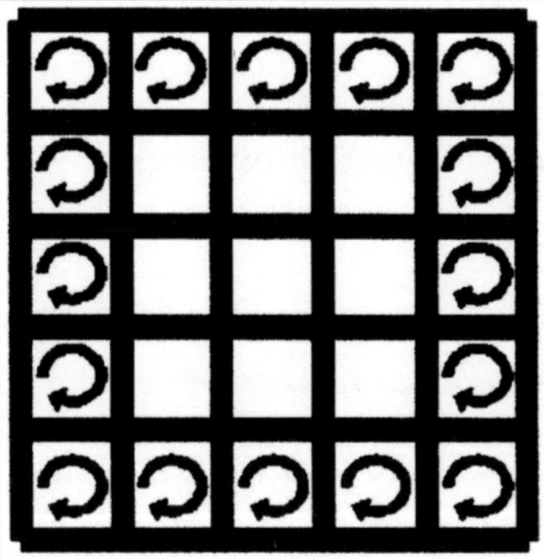

Fig. 8. Mechanism for generation of a paramagnetic current inside a multiply-connected superconducting sample. From Nielsen et al., 2000.

According to this model, only the exterior plaquettes create a diamagnetic screening current on the outside of the sample, and as a consequence a paramagnetic current of the same magnitude is created inside the sample. Simulations taking into account the mutual-inductance interactions between loops in the array confirm that PME might be the dominant response, depending on the field strength and parameters as the self-inductance and capacitance of the Josephson junctions (Rotoli et al., 2001).

The granularity in superconductors emulates a disordered network of Josephson junctions, and the paramagnetic moments observed in granular thin films of Nb and $YBa_2Cu_3O_{7-x}$ has also been described as an intrinsic effect of multiply-connected superconductors.

Important results were obtained by Kusmartsev et al. (Kusmartsev et al., 1997) after cooling large $YBa_2Cu_3O_{7-x}$ single crystals in strong applied fields. The authors called this phenomenon the high-field paramagnetic effect (HFPME) so that it may be distinguished from the low-field PME. The HFPME was found to depend on the cooling rate and the sample size, which suggests that the paramagnetic moment in this case might be induced by a flux compression mechanism. The **Fig. 4** shows representative results obtained by Kusmartsev et al.

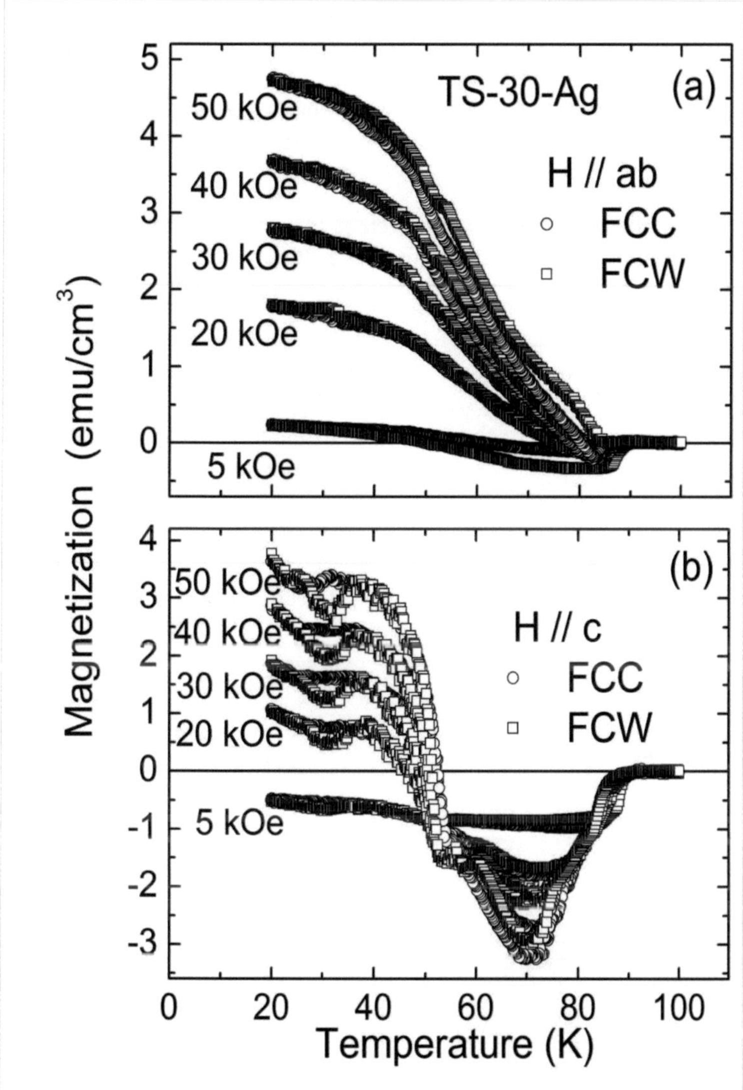

Fig. 9. FCC and FCW magnetizations in a melt-textured $YBa_2Cu_3O_{7-x}$ sample exhibiting a strong PME. From Dias et al., 2004.

The PME at high magnetic fields was also observed by Dias et al. (Dias et al., 2004) in different melt-textured $YBa_2Cu_3O_{7-x}$ samples containing large amounts of the Y_2BaCuO_5 (Y-211) phase. Magnetic fields up to 50 kOe were applied either parallel or perpendicular to the *ab* planes and the magnitude of the high-field paramagnetic moment increases when the field is augmented. The **Fig. 9** shows FCC and FCW magnetizations as a function of the temperature in a melt-textured $YBa_2Cu_3O_{7-x}$ sample measured in the indicated fields applied parallel and perpendicular to the *ab* planes. In the **Fig. 9a** the data are obtained when the field is applied parallel to the *ab* plane. These results present the general trends of the PME shown by the melt-textured YBaCuO samples. In the **Fig. 9b** are depicted the peculiar results obtained when the field is applied along the *c*-axis. In this case the positive contribution to the moment rises abruptly at an almost field-independent temperature around 55 K, then shows a tendency to saturate at low temperatures at values varying roughly proportional to the applied field. When the magnetic field increases the magnetization also increases and for magnetic fields below 5 kOe the magnetization is negative, as shown in the **Fig 9b**.

The effect shows a strong and anomalous relaxation, such that the paramagnetic moment increases as a function of the time, as shown in **Fig. 10**.

According to the authors, the pinning by the Y-211 particles is relevant and the formation of a flux-compressed scenario modulated by pinning and a strong flux-creep is a possible origin of the observed PME. Recently, the high-field PME was also observed in an $Y_{0.98}Ca_{0.02}Ba_2Cu_3O_{7-x}$ single crystal (Dias et al., 2010), and the results were attributed to strong flux compression state into the sample modulated by pinning. An important review of the paramagnetic Meissner effect and related dynamical phenomena was presented by M.S. Li (Li, 2003), with emphasis on the theoretical description of the π junctions.

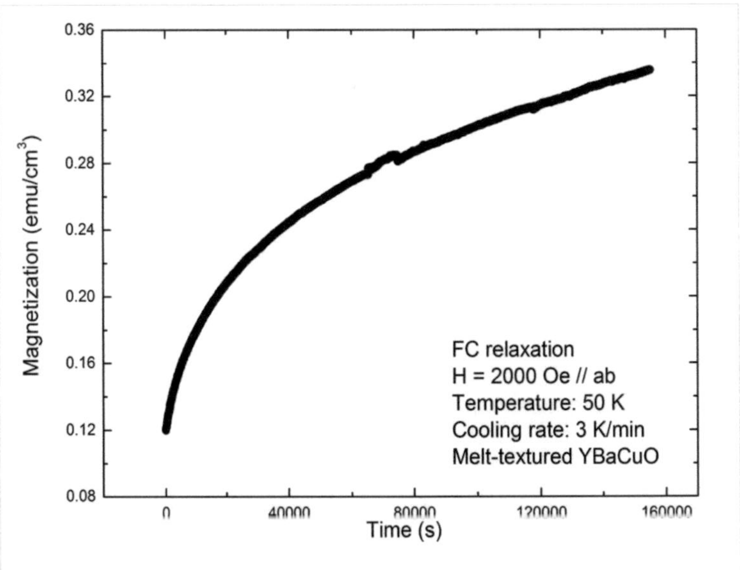

Fig. 10. Strong time dependence of the FCC magnetization in a melt-textured $YBa_2Cu_3O_{7-x}$ sample exhibiting the PME.

1.2 Mechanical properties of YBCO-bulk materials

1.2.1 Theoretical fundamentals

Indentation testing is a simple method that consists essentially of touching the material of interest, whose mechanical properties such as elastic modulus and hardness are unknown, with another material whose properties are known (Fischer-Cripps et al., 2004). The technique has its origins in Mohs' hardness scale of 1822, in which materials that are able to leave a permanent scratch in another were ranked as harder materials, with diamond assigned the maximum value of 10 on the scale. The establishment of the Brinell, Knoop, Vickers and Rockwell tests all follow from a refinement of the method of indentation test, in which the length scale of the penetration is measured in nanometers rather than microns or millimeters, the latter being common in conventional hardness tests. Apart from the displacement scale involved, the distinguishing feature of most nanoindentation testing is the indirect measurement of the contact area.

In conventional hardness tests, the area of contact is calculated from direct measurements of the dimensions of the residual impression in the specimen surface upon the removal of load. In nanoindentation tests, the size of the residual impression is of the order of microns and too small to be conveniently directly measured. Thus, it is usual to determine the area of contact by measuring the depth of indentation of the indenter into the specimen of interest. This, together with the known geometry of the indenter, provides an indirect measurement of the contact area at full load. For this reason, nanoindentation can be considered a special case of depth-sensing indentation or instrumented indentation testing.

In this section, we would like to do a brief explanation of the different hardness tests employed to characterize the $YBa_2Cu_3O_{7-x}$ samples textured by Bridgman and TSMG technique. Concretely, we will explain in detail the micro-, nano- and even picoindentation, according to the characteristic indentation sizes that fall into a particular size range and the applied load employed.

1.2.1.1 Microindentation

The purpose of microindentation hardness testing is to study fine scale changes in hardness. This technique is also commonly called microhardness testing, although this term is misleading because it implies that the hardness is extremely low. The Vickers and the Knoop tests, which are the two most common microindentation tests, will be described in more detail in the present section.

The term microhardness test usually refers to static indentations made with loads not exceeding 1kgf. The indenter is either the Vickers diamond pyramid or the Knoop elongated diamond pyramid. The surface being tested generally requires a metallographic finishing; the smaller the load used, the higher the surface finish required.

The Knoop indenter is a diamond ground to pyramidal form that produces a diamond shaped indentation, having an approximate ratio between long and short diagonals of 7:1. The depth of indentation is about 1/30 of its length. The Knoop hardness number (*KHN*) is the ratio of the load applied to the indenter to the unrecovered projected area as:

$$KHN = \frac{F}{A} = \frac{P}{CL^2} \tag{1}$$

Where F or P is the applied load in *kgf*, A is the unrecovered projected area of the indentation in mm^2, L is the measured length of the long diagonal of indentation in *mm*, and C is a constant of the indenter relating the projected area of the indentation to the square of the length of the long diagonal, and equal to 0.07028.

The Vickers diamond pyramidal indenter is ground in the form of a square pyramid with an angle of 136° between faces. The depth of indentation is about 1/7 of the diagonal length. The Vickers diamond pyramid hardness number (HV) can be obtained by using the following equation:

$$HV = \frac{2F\sin\frac{136°}{2}}{d^2} \quad or \quad HV \approx 1.854\frac{F}{d^2} \tag{2}$$

Where F is the load in *kgf*, d is the arithmetic mean of the two diagonals, d_1 and d_2 in *mm* and HV is the Vickers hardness.

Comparing the indentations made with Knoop and Vickers diamond pyramid indenters for a given load and test material we can conclude:

- Vickers indenter penetrates about twice as deep as Knoop indenter
- Vickers indentation diagonal about 1/3 of the length of Knoop major diagonal
- Vickers test is less sensitive to measurement errors than Knoop test
- Vickers test is more sensitive to measurement errors than Knoop test
- Vickers test is best for small rounded areas
- Knoop test is best for small elongated areas
- Knoop test is good for very hard brittle materials and very thin sections

1.2.1.2 Nanoindentation

During the last years, Instrumented Indentation Testing has been widely employed. Also known as depth-sensing indentation, continuous-recording indentation, ultra-low load indentation or nanoindentation, ITT is a relatively new form of mechanical testing, which significantly expands the capabilities of traditional hardness tests at macro- and microscale. It employs high-resolution instrumentation to continuously control and monitor the loads and displacements of an indenter (Golovin et al., 2008; Pharr et al., 1992). This technique presents several advantages compared to the microindentation technique:

- It is a local probe that can evaluate the properties of a material in different areas.
- It is depth-sensing, so it can characterize a material at different depths, unlike most scanning probe techniques, which usually only work at a given indentation depth.
- It can also measure the Elastic modulus (E) of the material.

Moreover, this technique is especially well suited for the characterization of small volumes of material, such as single grains or phases in a composite (Roa et al., 2007), dislocation dynamics (Gaillard et al., 2006), small structures (Choi et al., 2003) or thin films and coatings (Gaillard et al., 2008; Beegan et al., 2005; Roa et al., 2009a, 2011a; Rayon et al., 2011).

Finally, this technique eliminates the need to visualize the imprints produced during the test for homogeneous materials, which makes the extraction of mechanical properties easier. Due to all these advantages, nanoindentation could be a good and versatile technique for characterizing the mechanical properties of a given component.

The nanoindenter is an equipment which measures load and maximum indentation depth as a function of time during loading and unloading, and displays load-displacement data. Depending on the details of the specific testing systems, loads as small as 1 nN can be applied, and displacements of 0.1 nm can be measured. Stress-strain curves (σ–ε) and mechanical properties such as hardness (*H*), elastic modulus (*E*), fracture toughness (K_{IC}), yield strength (σ_{ys}), or shear stress (τ), can be obtained from the P-h data and with the corresponding tip indenter. The different indenters can be classified in four different groups:

i. *Pyramidal:* the most used sharp indenter is the Berkovich. It is a three-sided pyramid with the same depth-to-area relation as the four sided Vickers pyramid commonly used in microhardness. This indenter allows obtaining the *H* and *E*. The K_{IC} of brittle materials can also be determined with this indenter without high accuracy.
ii. *Spherical:* for this kind of indenters, the contact stress is initially small and produces only elastic deformation. As the spherical indenter is driven into the surface, a transition from elastic-to-plastic deformation occurs, which can theoretically be used to examine the σ_{ys} and obtain the σ–ε curve (Field et al., 1993).
iii. *Cube-corner:* a three-sided pyramid with mutually perpendicular faces arranged in geometry like the corner of a cube. The center-line-to-face angle for this indenter is 34.3°. The sharper cube-corner indenter produces much higher stress and strain in the vicinity of the contact, which is useful, for example, in producing very small and well defined cracks at the corners of the residual imprints in brittle materials.
iv. *Conical indenters:* this type of indenter is interesting due to the existence of stress concentrations at the sharp edges of the indenter. It is difficult to manufacture conical diamonds with sharp tips, making them of little use in nanoindentation technique (Tsui et al., 1997).

Moreover, in order to obtain reproducible results, several factors should be controlled during the indentation tests:

- *Choosing an appropriate indenter:* this is one of the most important steps and requires the consideration of a number of factors. One of them is the strain the tip imposes on the test materials. Although the indentation process produces a complex strain field beneath the indenter, it has been proven to be useful to quantify this field with a single quantity, often termed as the characteristic strain. There are problems, however, in obtaining accurate measurements of *H* and *E* with cube-corner indenters (Hay et al., 1999). Although not entirely understood, the problems appear to have two separate origins: first, as the angle of the indenter decreases, friction in the specimen-indenter interphase and its influence on the contact mechanics becomes increasingly important; on the other hand, to obtain the relation between the contact stiffness (*S*), contact area ($A(h_c)$) and effective elastic modulus (E_{eff}), corrections are required, and the magnitude of the correction factor depends on the angle of the indenter.
- *Environmental control:* to take full advantage of the fine displacement resolution available in most ITT systems, several precautions must be taken when choosing and

preparing the testing environment. Uncertainties and mistakes in measured displacement arise from two separate environmental sources: vibration and temperature. To reduce vibration, the nanoindenter should be located on quite, solid foundation and mounted on vibration-isolation system. Thermal stability can be provided by closing the apparatus in a thermally buffered cabin and controlling the room temperature to within ± 0.5°C.

- *Surface preparation:* one of the most important factors is the surface roughness of the samples, which is extremely important during the nanoindentation test because the mechanical properties are obtained from contact areas, which are calculated from the contact depth and area function. The main problems are found when the characteristic wavelength of the roughness is comparable to the contact diameter.
- *Testing procedure:* to avoid interference from previous assays, and ensure the independence of the measurement, successive indentations should be separated up to 20-30 times the maximum indentation depth (h_{max}) when Berkovich or Vickers indenters are employed. However, for other geometries, the rule is to perform indentations separated from 7 to 10 times the maximum contact radius.
- *Detecting the surface:* the most important step of any good nanoindentation test procedure is to accurate identificate the location of the specimen's surface. For hard and stiff materials, such as hardened metals and ceramics, the load and/or contact stiffness, both of which increase upon contact, are often employed. However, for soft compliant materials (like polymers and biological tissues), the rate of increase in load and contact stiffness is often too small to allow for accurate surface identification. In these situations, a better method is sometimes offered by dynamic stiffness measurement (Pethica et al., 1989; Lucas et al., 1998).

a. Hardness and Elastic modulus determination

In this section we will explain the method developed by Oliver and Pharr (Pharr et al., 1992; Oliver et al., 1992) to determine the hardness (*H*) and the elastic modulus (*E*) from the unloading part of recorded indentation load/unload (or *P-h* curves, such as **Fig. 11**).

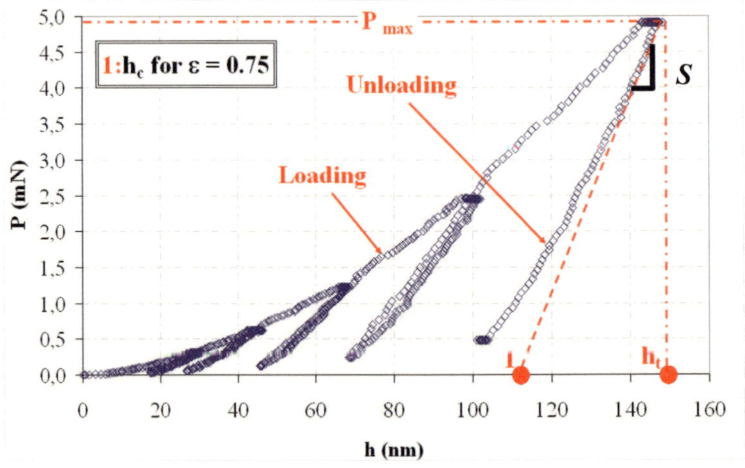

Fig. 11. Schematic representation of indentation P-h curve showing important measured parameters (From Roa et al., 2007).

The H is calculated by dividing the load by the projected contact area ($A(h_c)$) at maximum applied load, P_{max}:

$$H = \frac{P_{\max}}{A(h_c)} \quad (3)$$

where the contact depth (h_c) is determined as:

$$h_c = h_{\max} - \varepsilon \frac{P_{\max}}{S} \quad (4)$$

being h_{max} the maximum indentation depth, ε a parameter approximately equal to 0.75 for a Berkovich tip indenter, and S the unloading contact stiffness at maximum indentation depth (dP/dh, see **Fig. 11**). In the case of a perfect Berkovich tip, the $A(h_c)$ is determined as:

$$A(h_c) = 24.56 h_c^2 \quad (5)$$

However, in practice the projected contact area differs from this value when the tip is worn out, because the different angles of the faces may be somehow different from the nominal. Hence, the projected contact area must be regularly calibrated. When catastrophic fracture mechanisms occur, the $A(h_c)$ obtained by this theoretical method is overestimated, and results in an H underestimation. To avoid this problem, the real contact area should be determined using a technique of microscopy (as Scanning Electron Microscopy or Atomic Force Microscopy) and the contact area must be recalculated in order to avoid the sink-in or pile-up effects for brittle and/or ductile materials, respectively (Roa et al., 2009d, 2010c, 2011b).

The effective elastic modulus (E_{eff}) has been expressed in terms of S and $A(h_c)$ as:

$$E_{eff} = \frac{1}{\beta} \frac{\sqrt{\pi}}{2} \frac{S}{\sqrt{A(h_c)}} \quad (6)$$

where β is a geometrical factor (1.034). This parameter takes into account the lack of symmetry of the Berkovich indenter. Then, according to the Oliver and Pharr approach (Oliver et al., 1992), the Elastic modulus can be obtained as:

$$\frac{1}{E_{eff}} = \frac{1 - \nu^2}{E} + \frac{1 - \nu_i^2}{E_i} \quad (7)$$

where E and υ are the Elastic modulus and the Poisson's ratio ($\nu_{YBa2Cu3O7-x}$ = 0.3 (Roa et al, 2007, 2009d, 2010a, 2010c, 2011b, 2011d), respectively. The subindex i denotes the indenter properties (E_i = 1141GPa and $\upsilon_i = 0.07$) (Oliver et al., 1992; Pharr et al., 1992).

b. Fracture toughness

Moreover, nanoindentation provides information about contact mechanisms taking place in the material. In brittle materials indented with a Berkovich tip, surface examination of residual imprints has revealed the appearance of several typical features: surface deformation effects such as *sink-in* (Johnson et al., 1985; Alcala et al., 2000), microcracks or damage inside the imprints (Burghard et al., 2004); and fracture mechanisms such as *radial cracks* emanating from the imprint corners (Roa et al., 2008a, 2008b), and *chipping* (Lawn et al., 1975).

The resistance of the material against crack propagation, characterised by the fracture toughness K_{IC} (Anderson et al., 2005), is a critical mechanical parameter in brittle materials. Fracture toughness estimation by microfracture Vickers indentation is a well-known and broadly employed technique in ceramic materials. Usually, a Vickers diamond pyramid indenter at loads greater than 1 N is used to nucleate radial cracks at the imprint corners. The cracks emanating from the corners of a Vickers or Berkovich indenter are arrested when the residual stress driving force at the crack tip is in equilibrium with the K_{IC}. Several expressions are widely employed to extract the fracture toughness, depending on the indenter geometry and crack morphology. In this experimental work, the method we are pursuing is based on the radial cracking which occurs when brittle materials are indented by sharp indenter such as Vickers or Berkovich diamond (Casellas et al., 2007; Jang et al., 2008). K_{IC} is then evaluated using the equation developed by Lawn, Evans, and Marshall (Lawn et al., 1980):

$$K_{IC} = a\left(\frac{E}{H}\right)^{1/2}\left(\frac{P}{c^{32}}\right) \quad (8)$$

where α is an empirical constant depending on the indenter geometry (α=0.016 for pyramidal tips), P is the peak indentation load, c is the length of the radial cracks, E is the elastic modulus, and H is the hardness value. The fracture toughness of $YBa_2Cu_3O_{7-x}$ melt-textured samples has been determined by Vickers indentation (Leblond-Harnois et al., 2000; Li et al., 1997; Yoshino et al., 2001; Leenders et al., 1999; Cook et al., 1987; Fujimoto et al., 1992), and also by the single-edge notch beam (Joo et al., 1998) and bending (Giese et al., 1990) techniques. Values in the range of K_{IC}=0.4-2.8 MPa m$^{1/2}$ have been found, with a scattering attributed to differences between the techniques employed to obtain this parameter. In addition, K_{IC} is influenced by exact matrix composition. For instance, Joo *et al.* (Joo et al., 1998), reported that the fracture toughness of $YBa_2Cu_3O_{7-x}$ increased from K_{IC}=1.60 MPa m$^{1/2}$ to 2.80 MPa m$^{1/2}$ by adding a 15% in volume of silver to the matrix.

c. *Yield strength and stress-strain curve*

The stress and deflection arising from the contact between two elastic solids are of particular interest to understand the first steps of indentation testing. The contact depth between a rigid sphere and a flat surface for small indentation depths was described by Hertz (Hertz et al., 1881). This phenomenon is shown in **Fig. 12**. The radius of the circle of contact is a, the total indentation depth is h_{max}, the depth of the circle of contact from the specimen free surface is h_a, and h_p is the distance from the bottom of the contact to the contact circle.

When using spherical tips, the contact point is more difficult to be determined, due to the moderate increase in stiffness during the initial contact, and the interaction between the tip and samples before contacting. The indentation load-displacement data were analyzed based on the Hertz equation in the elastic region as follows (Oliver et al., 2004; Field et al., 1993; Field et al., 1995):

$$P = \frac{3}{4}E_{eff}\sqrt{R}^{3}\!/\!\sqrt[2]{h_e} \quad (9)$$

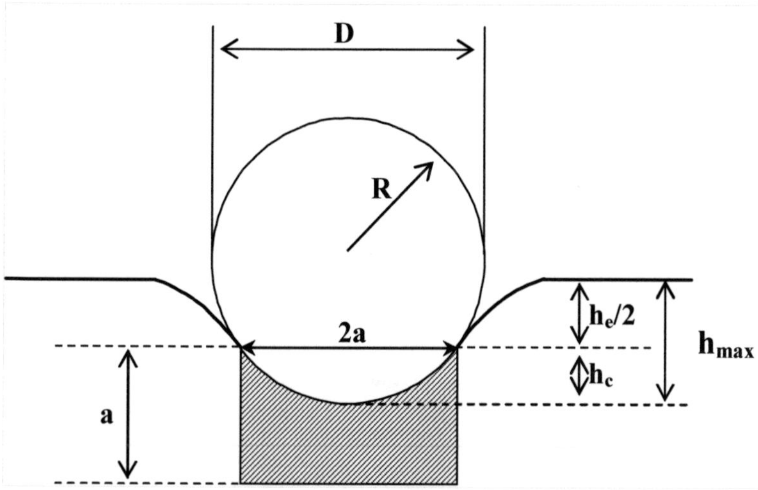

Fig. 12. Schematic representation of spherical indentation (From Roa et al., 2010b).

where h_e is the elastic indentation depth. Recently, Barsoum *et al.* (Moseson et al., 2008) proposed the fitting of the stiffness vs. indentation depth data, and extrapolating to zero, as predicated by a relation between E_{eff} and a, as follows (Roa et al., 2010b, 2010c, 2011b):

$$S = 2E_{eff}a \tag{10}$$

Hertz found that the radius of the circle of contact, a, is related to the indentation load, P, the indenter radius, R, and the elastic properties for small indentation depths of the material by:

$$j = \sqrt[3]{\frac{3}{4}\frac{PR}{E_{eff}}} = \sqrt[3]{\frac{3}{4}\frac{PR}{\frac{1}{\beta}\frac{\sqrt{\pi}}{2}\frac{S}{\sqrt{A(h_c)}}}} \tag{11}$$

The mean contact pressure, p_m, between the indenter and the material is the applied load divided by the contact area. For small indentation depths, it can be obtained from the Hertzian equations as:

$$p_m = \frac{P}{\pi a^2} = \frac{4E_{eff}}{3\pi}\left(\frac{a}{R}\right) \tag{12}$$

The mean contact pressure is referred to as the "*indentation stress*" and the quantity a/R as the "*indentation strain*". This equation allows us to plot the stress-strain curves when we use a spherical tip. The left side of this equation represents the indentation stress or mean contact pressure, also referred to as the Meyer hardness (Basu et al., 2006). The expression in parentheses on the right side of this equation represents the indentation strain (Rayon et al., 2011).

The maximum tensile stress occurs at the edge of the contact circle in the surface and can be expressed as (Fischer-Cripps, 1999):

$$\sigma_{\max} = \frac{1}{2}(1-2\nu)p_m \tag{13}$$

The yield stress of the material, σ_{ys}, can be obtained using the following expression:

$$p_m = 1.1\sigma_{ys} \tag{14}$$

This parameter will give information about the transition from elastic to elastic-to-plastic response (Jiménez-Piqué et al., 2007). The maximum shear stress (τ_{max}) is produced beneath the indentation axis at a depth close to $0.5a$, and equal to:

$$\tau_{\max} = 0.46p_m \tag{15}$$

The maximum shear stress is also known as the Tresca criterion. This assumes that yield occurs when the shear stress τ exceeds the shear yield strength, τ_{ys} (Tsui et al., 1997; Roa et al., 2008a):

$$\tau = \frac{\sigma_1 - \sigma_3}{2} \le \tau_{ys} \tag{16}$$

Where σ_1 and σ_3 are the principal stresses (also, can be re-written as σ_x and σ_z, respectively).

1.2.1.3 Picoindentation

The applied load in this case is much lower than in the previous one. Normally, this technique produces some indentations of few nanometers (generally fully elastic) obtaining the Elastic modulus. Picoindentation by means of atomic force microscopy-force spectroscopy (AFM-FS) presents several advantages over standard methods with regard to the quantification of *E* values:

- First, the measurements are extremely local (small contact areas between the sample and the probe, of few nm^2) and can be performed on different areas so as to average the mechanical properties of the samples.
- Second, it is a form of depth-sensing, so it allows characterization of a material at different indentation depths without surpassing the yield strength (fully elastic regimen).
- Finally, it is not necessary to visualize and characterize the imprint produced during the indentation test, which simplifies the calculation in order to extract the *E* values.

These advantages mean that picoindentation is a suitable and reliable technique for measuring E values while applying loads ranging from a few nN down to the pN level. The study of force versus displacement (F-z) curves can shed new light on the elastic properties of hard materials, as these curves are similar to those obtained with classic microscopic or nanoscopic indentation tests.

The different steps of the method to extract the elastic modulus value using the AFM-FS and Hertz equations are described below (Roa et al., 2011e, 2011f):

1. **Cleaning the tip:** it is important to remove contaminants covering the tip. First, rinse the tip with miliQ water, acetone and ethanol and dry it under a nitrogen flow. After that, put the tip in an ozone cleaning chamber during 30-35 minutes, and ultraviolet light.
2. **Measuring the tip radius by means of a NiO Pattern:** A NiO sample that features extremely sharp crystallographic ridges is used for measuring the tip radius. Then, as the sample ridges are nominally sharper than the tip apex, it is possible to obtain a "tip image" using blond reconstruction software. AFM images of the pattern should be collected in tapping mode with scan of 0.5μm by 0.5μm, scan rates of 1 Hz and 512 by 512 pixels. The blind reconstruction software employed is Scanning Probe Image Processor, see **Fig. 13** (Roa et al., 2011e).

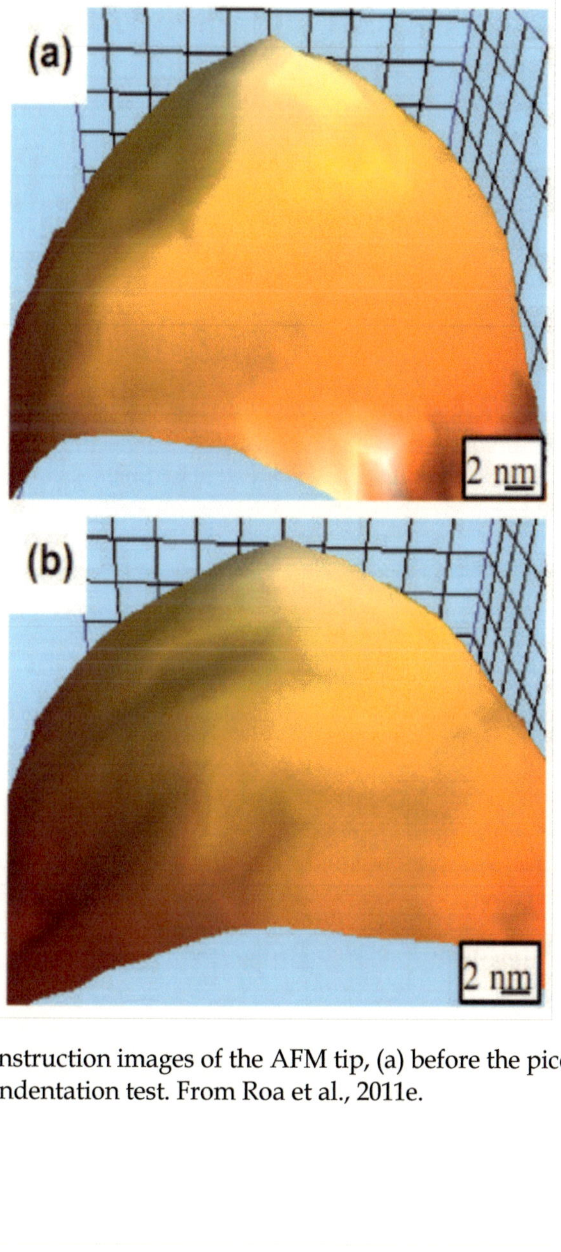

Fig. 13. Software reconstruction images of the AFM tip, (a) before the picoindentation test, and (b) after the picoindentation test. From Roa et al., 2011e.

3. **Indentation process:** after that, an image of the sample should be taken, and several spots (at least 40 spots) should be chosen in order to perform one indentation per spot, see **Fig. 14a** (from Roa et al., 2011e).

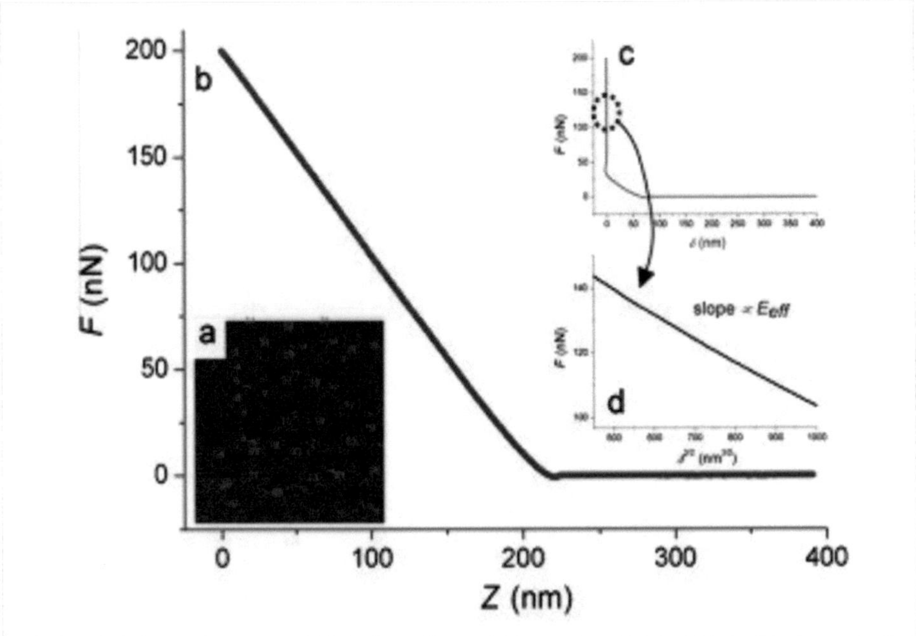

Fig. 14. Images of the different steps taking place during the picoindentation process. (a) AFM image with 40 different spots. (b) *F* vs. *z* cantilever deflection. (c) *F* vs δ, and (d) *F* vs. $\delta^{3/2}$ from the slope of Elastic modulus, which is calculated using Hertzian equations. From Roa et al., 2011e.

F is calculated as (see **Fig. 14b**):

$$F = k_v \Delta z \tag{17}$$

where Δz is the cantilever deflection, defined as:

$$\Delta z = \frac{\Delta V}{S} \tag{18}$$

where ΔV is the increment in photodetector vertical signal, as the tip comes into contact with the sample, and *S* is the sensitivity, which is the slope of the contact region of a force curve performed on a rigid sample.

The indentation depth (δ) due to F is calculated as (see **Fig. 14c**):

$$\delta = z - \Delta z \tag{19}$$

where *z* represents the piezo-scanner displacement along the axis perpendicular to the sample plane. After puncturing the sample, a topographic image should be captured in order to ensure that the sample had not undergone any plastic deformation. Only *F-z* curves that presented negligible lateral deflection should be considered to extract the mechanical response. *F-z* curves obtained at a certain *F* value were analyzed using the Hertz model in the elastic region (Oliver et al., 2004; Field et al., 1993, Field et al., 1995) by means of equation 20 (see **Fig. 14d**):

$$F = \left(\frac{3}{4} E_{eff} \sqrt{R}\right) \delta^{3/2} \tag{20}$$

where E_{eff} can be obtained using the equation 7. In this case the subindex *i* corresponds to the mechanical properties of the AFM probe (SiO_2 with E = 76 GPa, and υ_i = 0.17 (Oliver et al., 1992)). The quality of *S* value relies on the absence of elastic deformation of the reference sample.

1.2.2 State of the art

The mechanical properties of $YBa_2Cu_3O_{7-x}$ (Y-123 or YBCO) samples have been studied during the last years. The most important properties studied have been the elastic modulus, *E*, the hardness, *H* (*at micro-/ nanometric scale*), the toughness fracture, K_{IC}, and the yield stress value, σ_{ys}. Some authors studied the mechanical properties at room or at cryogenic temperatures, also known as work temperature (T_w). Lots of different techniques could be used to determine these parameters. The reported values of *E*, *H*, K_{IC} and σ_{ys} of $YBa_2Cu_3O_{7-x}$ samples (bulk materials) obtained using indentation techniques (macro-, micro-, nano- and/or picoindentation techniques) are summarized in *table 1, 2, 3 and 4*, respectively.

Author	Material	E (GPa)	Method	Year
Lucas et al.	$YBa_2Cu_3O_{7-x}$	154.30 ± 16.34	Indentation	1996
Güçlü et al.	Polycrystalline, 50K	47.20	Vickers indentation	2005
	Polycrystalline, 160K	29.63		
	Polycrystalline, 180K	28.47		
	Polycrystalline, 293K	9.39		
Goyal et al.	Y_2BaCuO_5 (Y 211)	213	Nanoindentation	1991
	Texturized	182	Nanoindentation	1991
Roa et al.	Y-123[1] at 5mN	197 ± 7	Nanoindentation	2007
	Y-211[1] at 5mN	194 ± 9		
	Y-123/Y-211[1] at 5mN	201 ± 7		
	Y-123[1] at 10mN	189 ± 4		
	Y-211[1] at 10 mN	206 ± 4		
	Y-123/Y-211[1] at 10mN	204 ± 6		
	Y-123/Y-211[1] at 30mN	180 ± 5		
	Y-123/Y-211[1] at 100mN	173 ± 3		

Table 1. Literature values of *E* for $YBa_2Cu_3O_{7-x}$ obtained by indentation techniques.

[1] *Texture process:* Bridgman technique

Author	Material	E (GPa)	Method	Year
Foerster et al.	YBCO-TSMG (*ab*-plane)	177 ± 10[2]	Nanoindentation	2008
	YBCO-TSMG + 5% Ag_2O (*ab*-plane)	220		
	YBCO-TSMG + 10% Ag_2O (*ab*-plane)	149 ± 4		
	YBCO-TSMG + 15% Ag_2O (*ab*-plane)	≅ 150		
	YBCO-TSMG (*ca(b)*-plane)	177 ± 10[2]		
	YBCO-TSMG + 5% Ag_2O (*ca(b)*-plane)	≅ 175		
	YBCO-TSMG + 10% Ag_2O (*ca(b)*-plane)	≅ 175		
	YBCO-TSMG + 15% Ag_2O (*ca(b)*-plane)	≅ 150		
Roa et al.	Y-123[3] at 5mN	176 ± 15	Nanoindentation	2009d
	Y-211[3] at 5mN	224 ± 18		
	Y-123/Y-211[3] at 5mN	208 ± 21		
	Y-123[3] at 10mN	174 ± 17		
	Y-211[3] at 10 mN	207 ± 11		
	Y-123/Y-211[3] at 10mN	190 ± 16		
	Y-123/Y-211[3] at 30mN	140 ± 14		
	Y-123/Y-211[3] at 100mN	129 ± 6		
Roa et al.	$YBa_{1.75}Sr_{0.25}Cu_3O_{7-\square}$	106 ± 4	FS[4]-AFM	2009b
Roa et al	Y-123/Y-211-Bridgman	123.5 ± 3.4	Nanoindentation	2010b,c
Bartolomé et al.	YBCO-TSMG	120 ± 5	Nanoindentation	2010a
Roa et al.	YBCO-Bridgman	128 ± 5	Nanoindentation	2011b,d
Roa et al.	Y-123 at 300K (tetragonal, T) [3]	185 ± 7	FS*-AFM	2011c
	Y-211 at 300K (T) [3]	202 ± 5		
	Y-123 at 300K (orthorhombic, O) [3]	189 ± 4		
	Y-211 at 300K (O) [3]	201 ± 6		
	Y-123 at 300K (T) [3]	192 ± 3		
	Y-211 at 300K (T) [3]	208 ± 7		
	Y-123 at 300K (O) [3]	193 ± 4		
	Y-211 at 300K (O) [3]	205 ± 5		
	Y-123 at 300K (T)[5]	179 ± 4		
	Y-123 at 300K (O)[5]	182 ± 5		
	$YBa_{1.75}Sr_{0.25}Cu_3O_{7-x}$[5]	106 ± 4		
	$Y_{0.98}Ca_{0.02}Ba_2Cu_3O_{7-x}$[5]	100 ± 4		

Table 1. Literature values of E for $YBa_2Cu_3O_{7-x}$ obtained by indentation techniques (continuation).

[2]At maximum indentation depth (around *1300nm*)

[3] *Texture process:* TSMG technique

[4] *FS:* Force Spectroscopy

[5] *Texture process:* Self-flux technique

Author	Material	H (GPa)	Method	Year
Cook et al.	YBCO	8.7	Vickers Indentation	1987
Goyal et al.	Textures	6.7		1992
		10.8	Nanoindentation	
Goyal et al.	Y-211	14.0	Nanoindentation	1992
Lucas et al.	Y-123	10.28 ± 1.67	Indentation	1996
Li et al.	MTG-1100°C, 5min	5.4	Vickers indentation	1997
	MTG-1100°C, 10min	11.0		
	MTG-1100°C, 15min	10.3		
	Solid State reaction	10.3		
Lim et al.	Single crystal	7.81 ± 0.23	Nanoindentation	2001
Yoshimo et al.	YBCO 40K	18 ± 2.5	Vickers Indentation	2001
	YBCO 293K	5.2 ± 0.5		
Güçlü et al.	Polycrystalline, 50K	3.58	Vickers Indentation	2005
	Polycrystalline, 160K	1.03		
	Polycrystalline, 180K	0.95		
	Polycrystalline, 293K	0.53		
Roa et al.	Y-123[6] at 5mN (O)	10.4 ± 0.5	Nanoindentation (Berkovich)	2007
	Y-211[6] at 5mN (O)	19.1 ± 0.8		
	Y-123/Y-211[6] at 5mN (O)	14.8 ± 0.5		
	Y-123[6] at 10mN (O)	11.2 ± 0.4		
	Y-211[6] at 10 mN (O)	16.9 ± 0.6		
	Y-123/Y-211[6] at 10mN (O)	15.1 ± 0.3		
	Y-123/Y-211[6] at 30mN (O)	11.1 ± 0.3		
Roa et al.	Y-123/Y-211[6] at 100mN (O)	8.9 ± 0.2	Nanoindentation (Berkovich)	2007
Roa et al.	Y-123[6] at 5mN (T)	11.9 ± 0.7		2008a, b
	Y-211[6] at 5mN (T)	17.9 ± 1.1		
	Y-123/Y-211[6] at 5mN (T)	15.4 ± 0.9		
	Y-123[6] at 10mN (T)	10.2 ± 0.7		
	Y-211[6] at 10 mN (T)	17.0 ± 0.9		
	Y-123/Y-211[6] at 10 mN (T)	14.8 ± 0.4		
	Y-123/Y-211[6] at 30 mN (T)	8.3 ± 0.2		
	Y-123/Y-211[6] at 100 mN (T)	8.1 ± 0.5		
Foerster et al.	YBCO-TSMG (*ab*-plane)	8	Nanoindentation	2008
	YBCO-TSMG (*ca(b)* plane)	7		
	YBCO-TSMG + 5% Ag_2O (*ab*-plane)	≅ 9		
	YBCO-TSMG + 5% Ag_2O (*ca(b)*-plane)	≅ 8		
	YBCO-TSMG + 10% Ag_2O (*ab*-plane)	≅ 6.5		
	YBCO-TSMG + 10% Ag_2O (*ca(b)*-plane)	≅ 7		

Table 2. Literature values of *H* for $YBa_2Cu_3O_{7-x}$ obtained by indentation techniques.

[6] *Texture process:* Bridgman technique

Author	Material	H (GPa)	Method	Year
	YBCO-TSMG + 15% Ag_2O (*ab*-plane)	≅ 6.5		
	YBCO-TSMG + 15% Ag_2O (*ca(b)*-plane)	≅ 6		
	YBCO-TSMG + 5% Ag_2O	5.2 ± 0.2	Microindentation (Vickers)	
	YBCO-TSMG + 10% Ag_2O	4.1 ± 0.4		
Bartolomé et al.	YBCO[7] (T)	10.6 ± 0.4	Nanoindentation (Berkovich)	2010a
	YBCO[7] (O)	7.6 ± 0.4		
	YBCO weld (0° misorientation) (T)[8]	9.5 ± 0.3		
	YBCO weld (14° misorientation) (T) [8]	6.5 ± 0.4		
	YBCO weld (0° misorientation) (O) [8]	6.8 ± 0.5		
	YBCO weld (14° misorientation) (O) [8]	4.4 ± 0.5		
	YBCO weld (15° misorientation) (O) [8]	5.2 ± 0.3		
Roa et al.	YBCO[7] (O)	8.8 ± 0.3	Nanoindentation	2011b
Roa et al.	YBCO-TSMG (T)	10.5 ± 0.1	Nanoindentation (Berkovich)	2011c
	YBCO-TSMG (O)	7.8 ± 0.1		

Table 2. Literature values of *H* for YBCO obtained by indentation techniques. (continuation)

Author	Material	K_{IC} (MPa $m^{1/2}$)	Method	Year
Lenblond-Harnois et al	YBCO	1.53	Vickers Indentation	2000
	Y-123 + 5% wt Ag	1.88		
Li et al.	MTG-1100°C, 5min	1.9		1997
	MTG-1100°C, 10min	1.7		
	MTG-1100°C, 15min	1.7		
	Solid State Reaction	1.3		
Yoshino et al.	YBCO 40K	0.4		2001
	YBCO 293K	1.3		
Lenders et al.	Y-123 + 30 mol% Y-211	1.01	Vickers Indentation	1999
	Y-123 + 60 mol % Y-211	1.44		
Cook et al.	YBCO	1.10		1998
Foerster et al.	YBCO-TSMG pure	1.43 ± 0.02	Vickers Indentation	2008
	YBCO-TSMG + 10% Ag_2O	1.6 ± 0.1		
Fujimoto et al.	Textured	0.99-1.20	Vickers Indentation	1992
	Y-123 with Ag	1.60-2.10		

Table 3. Literature values of toughness fracture, K_{IC}, for $YBa_2Cu_3O_{7-x}$ obtained by indentation techniques.

[7] *Texture process:* TSMG technique

[8] Hardness value in the welding region (YBCO/Ag/YBCO)

Author	Material	σ_{ys} (GPa)	Method	Year
Roa et al.	YBCO-Bridgman	3.2	Nanoindentation	2010b
Roa et al.	YBCO-TSMG	3.58 ± 0.10	(Spherical)	2009c

Table 4. Literature values of yield stress, σ_{ys}, for $YBa_2Cu_3O_{7-x}$ obtained by indentation techniques.

1.2.3 Mechanical properties

The elastic modulus values reported by Roa et al. (Roa et al., 2007, 2009d, 2011b) presented in **Table 1** for Bridgman and TSMG texture processes present a dependency with the applied load. This phenomenon is far from what are expected, as elastic modulus is an intrinsic property of each material and only depends on its composition. Then, this difference of elastic modulus is able to be attributed to one of the following factors:

- Firstly, the contact area value, which is strongly affected by the presence of macro-/microcracks, porosity, roughness and a wide variety of defects. In order to avoid this effect, the polishing process before the indentation should be carefully done.
- Secondly, the elastic modulus is obtained using the unloading slope of the loading-unloading curve, also denoted as *S*. If the contact point (indenter-sample) is not well estimated, the *S* value will be underestimated, yielding to a lower E_{eff}.
- Finally, the β factor could not modify the elastic modulus because it remains equal to 1.034 for each test performed (Oliver et al., 1992).

After considering all the different parameters that can modify the elastic modulus, we believe that the main contribution to change it is the contact area, and then the scattered values presented in the **Table 1** and reported by Roa et al. (Roa et al., 2007, 2008b) are attributed to this effect. Moreover, the Bridgman samples present elastic modulus values higher than for TSMG, as this last presents a higher density of macro- and microcracks, which can produce a relaxation of the elastic deformation range, see **Fig. 15.**

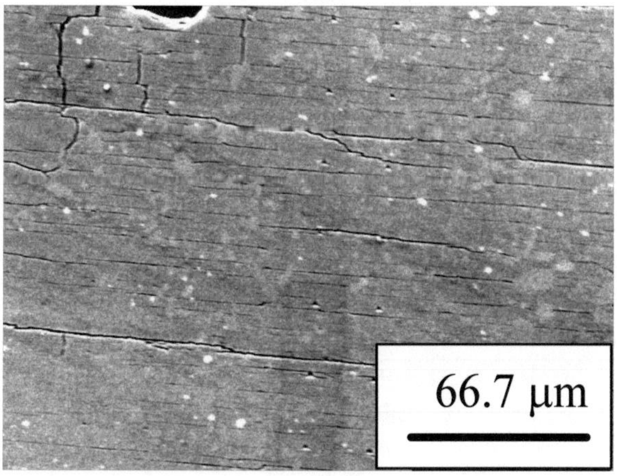

Fig. 15. FE-SEM image of the *ab*-plane of TSMG samples, which present a high microcracks density performed during the oxygenation process due to the tetragonal to orthorhombic transition (from Roa et al., 2011b).

Moreover, the **Table 1** presents several measurements of the elastic modulus for Y-123, which resulted in values scattered within the range between 40 - 200 GPa, caused by residual porosity and bad contact between the grains, as reported Johansen et al. (Johansen et al., 2001). If we compare all the elastic modulus reported in **Table 1**, we can conclude that the broad distribution observed can be attributed to the different measuring techniques (nanoindentation, AFM-FS) and to the different quality of the studied $YBa_2Cu_3O_{7-x}$ samples (grain structure, texture, and others).

Table 2 exhibits the hardness values for $YBa_2Cu_3O_{7-x}$ samples using different techniques. It is well known that the $YBa_2Cu_3O_{7-x}$ samples textured by Bridgman and TSMG techniques present a heterogeneous microstructure with Y-123 matrix and a minor phase (Y-211 or inclusions). Using the conventional micro hardness technique it is not possible to isolate the hardness value for each phase due to that the size of the residual imprint is higher than the size of the different inclusions. However, in 2007, Roa et al. (Roa et al., 2007) using the Nanoindentation technique obtained the hardness value for each phase when the applied load was lower or equal to 10mN. With this applied load, the residual contact area is lower than the size of the inclusions. Then, the mechanical property can be obtained, see **Table 2**. We can observe that the hardness value for Y-211 is around twice times that for Y-123. This fact could be due to different reasons (Roa et al., 2007):

- ionic bond of Y-211 is stronger than Y-123 (related to the different melting point of the two phases, $T_{Y\text{-}123}$ = 1010°C, and $T_{Y\text{-}211}$ = 1200°C (Aselage et al., 1988)),
- high anisotropy of dislocations confined onto a (001) plane (Sandiumenge et al., 2000), and/or
- the melt-processed ceramic composites contain a dense population of fine peritectic particles, which drastically affects the microstructure acting as nucleation sites for dislocations (Sandiumenge et al., 2000).

If the different residual imprints have been performed near the grain boundary, the plastic deformation mechanisms under the imprint could be affected by the Y-123 matrix, and the hardness value would be modified, thus giving us an average of the studied property.

At very small applied loads (lower than 10 mN), the hardness value can be strongly affected by the presence of defects and impurities that can cause almost no change in dislocation movement. In other words, hardness does not remain constant at different applied loads as corroborate Roa *et al.* in 2007 and 2008 (Roa et al., 2007, 2008b). Using Nanoindentation technique this effect can be related to one physical phenomenon known as *Indentation Size Effect*. Normally, hardness presents a maximum value for low loads. However, when the applied load increases this property is widely modified by the superficial defects, such as roughness, macro-cracks, and others. When this phenomenon takes place, the contact area could be over- or underestimated yielding to a false hardness value.

When the applied load using Nanoindentation technique is high (100 mN), overall nanohardness is very similar for both texturing techniques (8.9 GPa and 7.9 GPa for Bridgman and TSMG techniques, respectively). However, these values are lower than the nanohardness value of the Y-123 phase. The different hardness values reported in **Table 2** by Micro or Nanohardness tests at room temperature present the same value at high applied loads. However, when the $YBa_2Cu_3O_{7-x}$ samples are doped with a ductile material as silver oxide, the mechanical response tends to decrease, as reported by Foester et al. (Foester et al.,

2008). Moreover, several studies performed at cryogenic temperature (Güçlü et al., 2005) exhibit that the mechanical properties tend to be higher due to the nitrogen going inside the pre-existing cracks, yielding a homogeneous material. As we can observe in **Table 2,** the different $YBa_2Cu_3O_{7-x}$ samples studied at room temperature are much brittle than at cryogenic temperature due to the effect explained above.

Table 3 shows the different toughness fracture values reported in the literature and obtained using Vickers and Nanoindentation tests. The different results reported in this table present a high scattered due to the $YBa_2Cu_3O_{7-x}$ samples are strongly heterogeneous due to the presence of internal cracks, secondary phases and grain boundaries. Due to all this defects the different toughness values do not present any tendency. Lenblond-Harnois et al. (Lenblond-Harnois et al., 2000) and Foester et al. (Foester et al., 2008) obtained a toughness fracture of 1.88 and 1.6 $MPam^{1/2}$ for $YBa_2Cu_3O_{7-x}$ with 5 and 10% of Ag and Ag_2O, respectively. In that case, this value is higher than the others reported in this table due to the silver introduced inside of the $YBa_2Cu_3O_{7-x}$ samples reduce the fragility of these materials.

Finally, another mechanical parameter to take into account is the yield stress, see **Table 4.** As can be observed in this table, the intrinsic properties obtained using the stress-strain plots present similar values for both samples. The TSMG samples, present a higher density of macro-/microcracks and pore density (see **Fig. 15**) along the *c*-axis than Bridgman samples. Then, during the first contact between the indenter and the samples, the energy applied by the tip onto the surface is employed to close the different cracks produced during the oxygenation process. For this reason, TSMG samples present a higher contact radius value than Bridgman samples.

1.2.4 Fracture mechanisms

In this section we have analysed the different brittle effects appearing in as-grown samples textured by Bridgman and TSMG techniques. **Fig. 16** and **Fig. 17**, show Field Emission Scanning Electron Microscopy images of some residual imprints performed at 10 mN of applied load for Bridgman and TSMG samples, respectively (from Roa et al., 2011b).

All the different imprints in **Fig. 16** present the *sink-in* effect, typically found in elastic, brittle materials (Oliver et al., 1992). Sink-in is an elastic displacement of the surface at the contact perimeter. This effect can lead to an overestimation of the contact area, and thus an underestimation of the hardness value obtained by nanoindentation. As we can observe in this **Figure**, we noted a high density of micrometric pores around the residual imprint. Presumably, this micrometric porosity is created during the recrystalization process. The nanoindentation process would render the pores exposed to the surface, an effect that has been previously described (Gaillard et al., 2008).

In addition, all these images show the presence of microcracks inside the imprints, generated during the indentation process. This kind of damage is found in brittle materials containing inclusions harder than the matrix, in this case Y-211 (see **table 2**). Upon indentation, microcracks tend to be created at inclusions, and are recognised as microflaws within the imprint when observed from the surface.

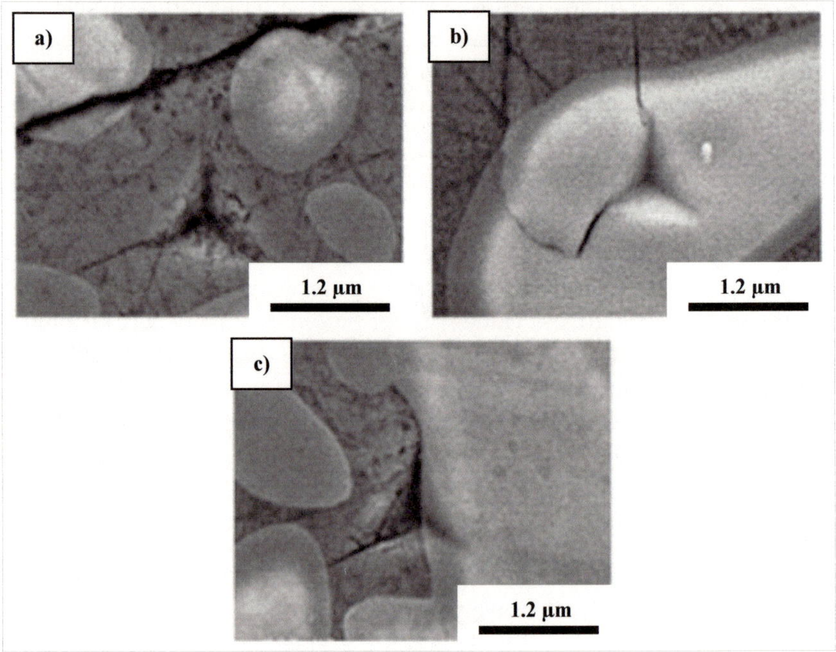

Fig. 16. Micrographs of nanoindentation imprints obtained by FE-SEM when the applied load was 10 mN for Bridgman samples, a) Y-123, b) Y-211 and c) Y-123/Y-211 (from Roa et al., 2011b).

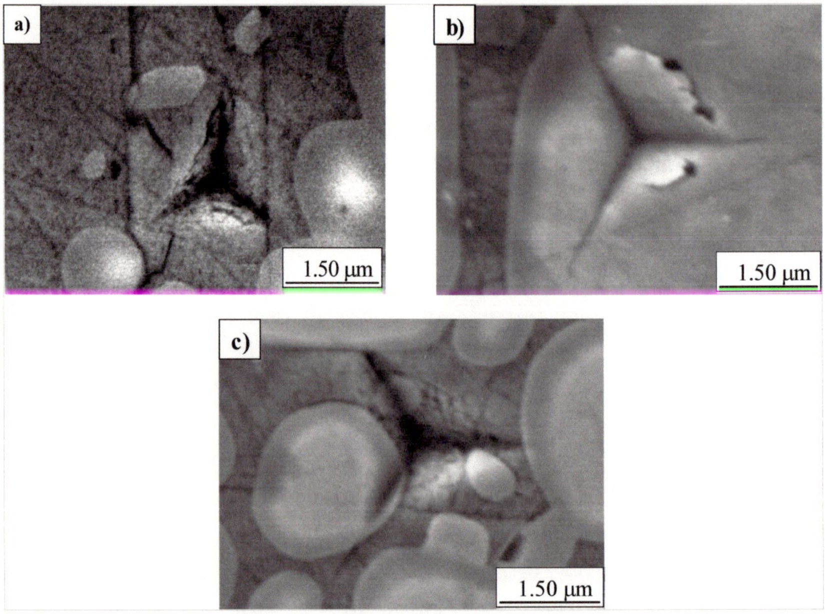

Fig. 17. Micrographs of nanoindentation imprints obtained by FE-SEM when the applied load was 10 mN for TSMG samples, a) Y-123, b) Y-211 and c) Y-123/Y-211 (from Roa et al., 2011b).

Another common feature in all imprints is the presence of radial cracks emanating from the corners of the imprints. This fracture mechanism, typical for ceramic materials, is nucleated due to the tensile strength originated during the indentation process.

Another common fracture mechanism is the *chipping*, i.e. a partial removal of surface material around the imprint. This phenomenon occurs as a result of Palqmvist cracks nucleating at the lateral sides of the imprint, beneath the surface.

In order to understand the fracture mechanisms that take place under the residual imprint, one nanoindentation performed at 10mN has been visualized by FIB, see **Fig. 18** (from Roa et al., 2011b, 2011d).

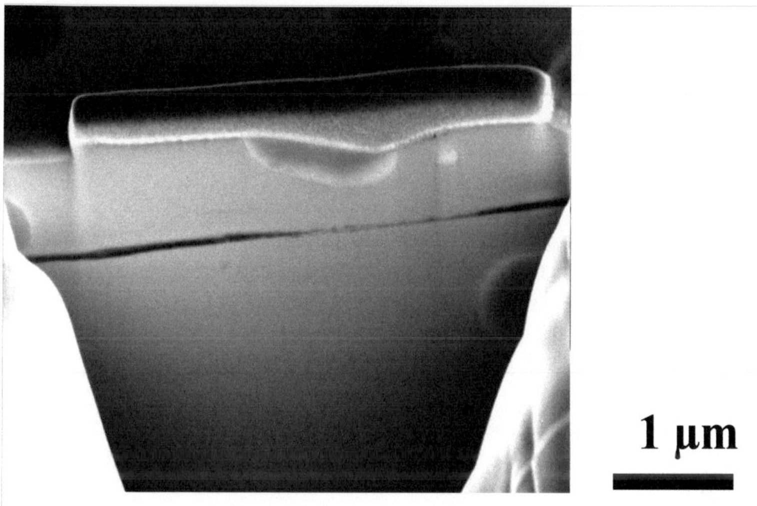

Fig. 18. Cross-sectioning and imaging of damage under an indentation imprint performed at 10mN of applied load in Bridgman samples using FIB-SEM. Cross-sectioning in the middle of the imprint (from Roa et al., 2011d).

Fig. 18 shows a heterogeneous distribution of Y-211 particles around the matrix, with different sizes from 1 to 5 µm. Moreover, no cracks or failure events can be appreciated under the nanoindentation imprint, thus implying that the deformation can be attributed to the closing of the oxygenation micro-cracks. The TSMG samples present the same effect –not shown in this work.

Due to all of these fracture mechanisms, the different values reported in the literature and summarized in the **table 3**, are not reproducible.

2. Conclusions

The paramagnetic Meissner effect has been observed in several superconducting materials grown by different techniques since their discovery, and different models and explanations have been proposed in order to explain their origin. Apparently, there is no a definitive model neither a tendency to saturation and the paramagnetic moments persist up to very high magnetic fields.

The mechanical properties (hardness, elastic modulus, toughness fracture and yield stress) for YBCO samples can be determined using the nanoindentation technique. This technique allows us to isolate the mechanical behavior of each phase, independently of the growing technique used. The mechanical properties for Bridgman samples are higher than that for TSMG samples. This phenomenon can be attributed to the texturing process, as TSMG process creates a high density of macro- and microcracks during the texture and oxygenation steps, respectively.

The values for yield stress for both texturing methods (Bridgman and TSMG), cannot be compared to previous works, as scare information in the literature is available.

With a cross-section, and using FIB technique, we can observe that the samples present a heterogenous distribution of Y-211 particles imbedded in the Y-123 matrix. No generation of porosity or microcracking have been observed, so the deformation mechanisms can be attributed to dislocations movement and twin generation in the *ab*-plane and *c*-axis, respectively.

3. Acknowledgments

The author thanks the Brazilian Ministry of Science and Technology and the State of Rio Grande do Sul, under the Grant PRONEX FAPERGS/CNPq 10/0009-2, and Ministry of Science and Technology, under the Grant CAPES/Procad 059/2007.

4. References

Alcalá, J., Barone, A. C., Anglada, M. (2000). The influence of plastic hardening on surface deformation modes around Vickers and spherical idnents. *Acta Materialia*, Vol. 48, No. 13 pp. 3451-3464

Anderson T. L. (2005). Fracture Mechanics: Fundamentals and Applications. *Taylor & Francis*, ISBN: 978-0-8493-1656-2

Araujo-Moreira, F., Barbara, P., Cawthorne, A., & Lobb, C. (1997). Reentrant ac Magnetic Susceptibility in Josephson-Junction Arrays. *Physical Review Letters*, Vol.78, No.24, (June 1997), pp. 4625-4628, ISSN 0031-9007

Arias, D., Peña, V., Sefrioui, Z., Leon, C., Santamaria, J., Martinez, J., & de la Torre, M. (2006). Paramagnetic Meissner effect in $YBa_2Cu_3O_7/La_{0.7}Ca_{0.3}MnO_3$ superlattices. *Physical Review B*, Vol.73, No.5, (February 2006), pp. 052503-1-4, ISSN 1098-0121

Aselage, T., Keefer, K. (1988). Liquidus relations in Y-Ba-Cu oxides. *Journal of Materials Research*, Vol. 3, pp. 1279-1291

Barbara, P., Araujo-Moreira, F., Cawthorne, A., & Lobb, C. (1999). Reentrant ac magnetic susceptibility in Josephson-junction arrays: An alternative explanation for the paramagnetic Meissner effect. *Physical Review B*, Vol.60, No.10, (September 1999), pp. 7489-7495, ISSN 1098-0121

Bartolomé, E., Roa, J. J., Bozzo, B., Segarra, M., Granados, X. (2010). Effective silver-assisted welding of YBCO blocks : mechanical versus electrical properties. *Superconductor Science and Technology*, Vol. 23, pp. 045013/1-045013/6

Basu, S., Moseson, M. W., Barsoum, M. W. (2006). On the determination of spherical nanoindentation stress-strain curves. *Journal Materials Research*, Vol. 21, Issue 10, pp. 2628-2637

Beegan, D., Maugier, M. T. (2005). Application of composite hardness models to copper thin hardness measurements. *Surface & Coatings Tehcnology*, Vol. 199, pp. 32-37

Blunt, F., Perry, A., Campbell, A., & Siu, R. (1991). An investigation of the appearance of positive magnetic moments on field cooling some superconductors. *Physica C*, Vol.175, No.5/6, (May 1991), pp. 539-544, ISSN 0921-4534

Brandt, D., Binns, C., Gurman, S., Torricelli, G., & Gray, D. (2011). Paramagnetic Meissner transitions in Pb films and the vortex compression model. *Journal of Low Temperature Physics*, Vol.163, No.3-4, (January 2011), pp. 170-175, ISSN 0022-2291

Bräuchle, G., Riedling, S., Lucht, R., Rohberg, K., Lohneysen, H. v., Claus, H., Erb, A., & Müller-Vogt, G. (1994). Observation of the Wohlleben effect in $YBa_2Cu_3O_{7-\delta}$ single crystals. *Physical Review B*, Vol.49, No.18, (May 1994), pp. 13283-13286, ISSN 1098-0121

Braunisch, W., Knauf, N., Kataev, V., Neuhausen, S., Grütz, A., Kock, A., Roden, B., Khomskii, D., & Wohlleben, D. (1992). Paramagnetic Meissner effect in Bi high-temperature superconductors. *Physical Review Letters*, Vol.68, No.12, (March 1992), pp. 1908-1911, ISSN 0031-9007

Burghard, Z., Zimmermann, A., Rödel, J., Aldinger, F., Lawn, B. R. (2004). Crack opening profiles of indentation cracks in normal and anomalous glasses. *Acta Materialia*, Vol. 52, pp. 293-297

Casellas, D., Caro, J., Molas, S., Prado, J. M., Valls, I. (2007). Fracture toughness of carbides in tool steels evaluated by nanoindentation. *Acta Materialia*. Vol. 55, pp. 4277-4286

Choi, Y., Suresh, S. (2003). Nanoindentation of patterned metal lines on a Si substrate. *Scripta Materialia*, Vol. 48, Issue: 3, pp. 249-254

Cook, R. F., Timothy, R. D., Clarke, D. R. (1987). Fracture toughness measurements of $YBa_2Cu_3O_x$ single crystals. *Applied Physics Letters*, Vol. 51, pp. 454-456

Dias, F., Pureur, P., Rodrigues Jr., P., & Obradors, X. (2000). Paramagnetic Meissner effect in melt-textured $YBa_2Cu_3O_{7-\delta}$. *Physica C*, Vol.341-348, No.2, (November 2000), pp. 1377-1378, ISSN 0921-4534

Dias, F., Pureur, P., Rodrigues Jr., P., & Obradors, X. (2004). Paramagnetic effect at low and high magnetic fields in melt-textured $YBa_2Cu_3O_{7-\square}$. *Physical Review B*, Vol.70, No.22, (December 2004), pp. 224519-1-9, ISSN 1098-0121

Dias, F., Vieira, V., de Almeida, M., Falck, A., Pureur, P., Pimentel Jr., J., & Obradors, X. (2010). Paramagnetic Meissner effect at high fields in YCaBaCuO single crystal and melt-textured YBaCuO. *Physica C*, Vol.470, No.1, (December 2010), pp. S111-S112, ISSN 0921-4534

Field, J. S., Swain, M. V. (1993). A simple predictive model for spherical indentation. *Journal of Materials Research*, Vol. 8, Issue: 2, pp. 297-306

Field, J. S., Swain, M. V. (1995). Determining the Mechanical Properties of Small Volumes of Material from Submicron Spherical Indentations. *Journal Materials Research*, Vol. 10, Num. 1 pp. 101-112

Fisher-Cripps, A. C. (1999). The Hertzian contact surface. *Journal of Materials Science*, Vol. 34, No. 1, pp. 129-137

Fischer-Cripps, A. C. (2004). Nanoindentation. *Mechanical Engineering Series. 2nd Edition. Springer.* ISBN 0-387-22045-3

Foester, C. E., Lima, E., Rodriguez, P., Serbena, F. C., Lepienski, C. M., Cantao, M. P., Jurelo, A. R., Obradors, X. (2008). Mechanical properties of Ag-doped top-seeded melt-grown YBCO pellets. *Brazilian Journal of Physics*, Vol. 38, No. 3, pp. 341

Freitag, B., Büchner, B., Knauf, N., Roden, B., Micklitz, H., Freimuth, A., & Kataev, V. (1999). Characteristic microstructure in small Bi-2212 grains showing the Wohlleben effect as revealed by High-Resolution Electron Microscopy. *Europhysics Letters*, Vol.45, No.3, (February 1999), pp. 393-398, ISSN 0295-5075

Fujimoto, H., Murakami, M., Koshizuka, N. (1992). Effect of Y_2BaCuO_5 on fracture toughness of YBCO prepared by a MPMG process. *Physica C*, Vol. 203, Issues 1-2, pp. 103-110

Gaillard, Y., Tromas, C., Woirgard, W. (2006). Quantitative analysis of dislocation pile-ups nucleated during nanoindentation in MgO. *Acta Materialia*, Vol. 54, Issue: 5, pp. 1409-1417

Gaillard, Y., Jiménez-Piqué, E., Soldera, F., Mücklich, F., Anglada, M. (2008). Quantification of hydrothermal degradation in zirconia by nanoindentation. *Acta Materialia*, Vol. 56, Num: 16, pp. 4206-4216

Geim, A., Dubonos, S., Lok, J., Henini, M., & Maan, J. (1998). Paramagnetic Meissner effect in small superconductors. *Nature*, Vol.396, No.6707, (November 1998), pp. 144-146, ISSN 0028-0836

Giese, R. F., Sheahen, T. P., Wolsky, A. M., Sharma, D. K. (1990). High-temperature superconductors; Their potential role in electric utility applications and what must be done to use them. *Proceeding American Conference*, Vol. 52, pp. 776-78

Golovin, Y. I. (2008). Nanoindentation and Mechanical properties of solids in submicrovolumes, thin near-surface layers, and films: a review. *Physics of the Solid State*, Vol. 50, No. 12, (December 2008), pp. 2205-2236, ISSN 1063-7834

Goyal, A., Oliver, W. C., Funkenbusch, P. D., Kroeger, C. M., Burns, S. J. (1991). Mechanical properties of highly aligned $YBa_2Cu_3O_{7-□}$ effect of Y_2BaCuO_x particles. *Physica C*, Vol. 183, No. 4-6, pp. 221-233

Güçlü, N., Kölemen, U., Uzum, O., Selebi, S. (2005). Work of indentation approach for investigation of mechanical properties of YBCO bulk superconductor at cryogenic temperatures. *Physica C*, Vol. 433, pp. 115-122

Hay, J. C., Bolshakov, A., Pharr, G. M. (1999). Critical Examination of the Fundamental Relations in the Analysis of Nanoindentation Data. *Journal of Materials Research*, Vol. 14, No. 6, pp. 2296-2305

Hertz, H., Reina, Angew, J. Math 92 (1982) 156. (Translated and reprinted in English in Hertz's Miscellaneous Papers, MacMillan and Co., London (1986) Ch. 5)

Jang, J., Pharr, G. M. (2008). Influence of indenter angle on cracking in Si and Ge during nanoindentation. *Acta Materialia*, Vol. 56, pp. 4458-4469

Jiménez-Piqué, E., Gaillard, Y., Anglada, M. (2007). Instrumented Indentation of Layered Ceramic Materials. *Key Enginnering Materials*, Vol. 333, pp. 107-116

Johansen, T. H. (2000). Flux-pinning-induces stress and magnetostriction in bulk superconductors. *Superconductor Science and Technology*, Vol. 13, No. 10, pp. R121

Johnson, K. L. (1985). Contact Mechanics. *Cambridge University Press*, Vol. 176, p. 94

Joo, J., Kim, J. G., Nah, W. (1998). Improvement of mechanical properties of YBCO-Ag composite superconductors made by mixing with metallic Ag powder and $AgNO_3$ solution. *Superconductors Science and Technology*, Vol. 11, No. 7, pp. 645

Khomskii, D. (1994). Wohlleben effect (paramagnetic Meissner effect) in high-temperature superconductors. *Journal of Low Temperature Physics,* Vol.95, No.1/2, (April 1994), pp. 205-223, ISSN 0022-2291

Kim, H., Minami, H., Schmidbauer, W., Hodby, J., Lyo, A., Iga, F. & Uwe, H. (1996). Paramagnetic Meissner effect in superconducting single-crystals of $Ba_{1-x}K_xBiO_3$. *Journal of Low Temperature Physics,* Vol.105, No.3/4, (November 1996), pp. 557-562, ISSN 0022-2291

Koshelev, A., & Larkin, A. (1995). Paramagnetic moment in field-cooled superconducting plates: paramagnetic Meissner effect. *Physical Review B,* Vol.52, No.18, (November 1995), pp. 13559-13562, ISSN 1098-0121

Kostic´, P., Veal, B., Paulikas, A., Welp, U., Todt, V., Gu, C., Geiser, U., Williams, J., Carlson, K., & Klemm, R. (1996). Paramagnetic Meissner effect in Nb. *Physical Review B,* Vol.53, No.2, (January 1996), pp. 791-801, ISSN 1098-0121

Kusmartsev, F. (1992). Destruction of the Meissner Effect in Granular High-Temperature Superconductors. *Physical Review Letters,* Vol.69, No.15, (October 1992), pp. 2268-2271, ISSN 0031-9007

Kusmartsev, F., Rykov, A., & Tajima, S. (1997). High-field paramagnetic effect in large crystals of $YBa_2Cu_3O_{7-□}$. *Physical Review B,* Vol.55, No.13, (April 1997), pp. 8557-8563, ISSN 1098-0121

Lawn, B. R., Evans, A. G., Marshall, D. B. (1980). Elastic/plastic indentation damage in ceramics: The median/radial crack system. *Journal of the American Ceramic Society*. Vol. 25, pp. 1053-1067

Lawn, B. R., Swain, M. V., Philips, K. (1975). On the mode of chipping fracture in brittle solids. *Journal of Materials Science*. Vol. 10, pp. 1236-1239

Lebed, A. (2008). Paramagnetic intrinsic Meissner effect in layered superconductors. *Physical Review B,* Vol.78, No.1, (July 2008), pp. 012506-1-4, ISSN 1098-0121

Leblond-Harnois, C., Caillard, R., Monot-Laffez, I., Desgardin, G., Raveau, B. (2000). Texturing process, superconducting and mechanical properties of Ag-doped top-seeded melt-grown YBCO pellets. *Physica C,* Vol. 341-348, No. 4, pp. 2439-2440

Leenders, A., Ullrich, M., Freyhardt, H. C., Kesten, M., Fieseler, H., Canders, W. R., May, H., Bock, J. (1999). Fabrication of HTS monoliths for a bearing system in a cryogenic vessel. *IEEE Transactions on Applied Superconductivity*, Vol. 9, Issue 2, pp. 992-995

Leenders, A., Ullrich, M., Freyhardt, H. C. (1999). Mechanical properties of TSMG-YBCO. *IEEE Transaction on Applied Superconductivity*, Vol. 9, Issue 2, pp. 2074-2077

Li, M. (2003). Paramagnetic Meissner effect and related dynamical phenomena. *Physics Reports,* Vol.376, No.3, (March 2003), pp. 133-223, ISSN 0370-1573

Li, L. F., Zhang, Z., Jin, D., Li, Y. Y., Meriani, S. (1997). Microstructure and mechanical properties of MTG YBCO. *Physica C,* Vol. 282-287, No. 4, pp. 2599-2600

Lim, Y. Y., Chaudhri, M. M. (2001). Do residual nanoindentations in metals and ceramics relax with time?. *Journal of Physics D: Applied Physics*, Vol. 34, L70-L78

Lucas, B. N., Oliver, W. C., William, R. K., Brynestad, J., O'Hern, M. E. (1991). The Hardness and Young's modulus of Bulk $YBa_2Cu_3O_{7-x}$ (1:2:3) and $YBa_2Cu_4O_8$ (1:2:4) as Determined by Ultra Low Load Indentation. *Journal Materials Research*, Vol. 6, pp. 2519-2522

Lucas, B. N., Oliver, W. C., Swindeman, J. E. (1998). The Dynamics of Frequency-Specific, Depth-Sensing Indentation Testing, Fundamentals of Nanoindentation and Nanotribology. *Materials Research Society Symposium Proc.* Vol. 522, pp. 3-14

Lucht, R., Löhneysen, H., Claus H., Kläser, M., & Müller-Vogt, G. (1995). Surface-sensitive paramagnetic Meissner effect in $YBa_2Cu_3O_x$ single crystals. *Physical Review B,* Vol.52, No.13, (October 1995), pp. 9724-9726, ISSN 1098-0121

Magnusson, J., Anderson, J., Björnander, M., Nordblad, P., & Svedlindh, P. (1995). Time dependence of the paramagnetic Meissner effect: Comparison between model calculations and experiments. *Physical Review B,* Vol.51, No.18, (May 1995), pp. 12776-12781, ISSN 1098-0121

Minhaj, M., Thompson, D., Wenger, L., & Chen, J. (1994). Paramagnetic Meissner effect in a niobium disk. *Physica C,* Vol.235/240, No.4, (December 1994), pp. 2519-2520, ISSN 0921-4534

Moseson, A. J., Basu, S., Barsoum, M. W. (2008). Determination of the effective zero point of contact for spherical nanoindentation. *Journal Materials Research,* Vol. 23, No. 1, pp. 204-209

Moshchalkov, V., Qiu, X., & Bruyndoncx, V. (1997). Paramagnetic Meissner effect from the self-consistent solution of the Ginzburg-Landau equations. *Physical Review B,* Vol.55, No.17, (May 1997), pp. 11793-11801, ISSN 1098-0121

Nielsen, A., Cawthorne, A., Barbara, P., Wellstood, F., Lobb, C., Newrock, R., & Forrester, M. (2000). Paramagnetic Meissner effect in multiply-connected superconductors. *Physical Review B,* Vol.62, No.21, (December 2000), pp. 14380-14383, ISSN 1098-0121

Nordblad, P., Magnusson, J., Papadopoulou, E., & Svedlindh, P. (1998). ac susceptibility of a paramagnetic Meissner effect sample. *Physica C,* Vol.297, No.3-4, (March 1998), pp. 317-325, ISSN 0921-4534

Obukhov, Y. (1998). The "Paramagnetic" Meissner Effect in Superconductors. *Journal of Superconductivity,* Vol.11, No.6, (July 1998), pp. 733-736, ISSN 0896-1107

Oliver, W. C., Pharr, G. M. (1992). An improved Technique for Determining Hardness and Elastic Modulus Using load and Displacement Sensing Indentation Experiments. *Journal of Materials Research.* Vol. 7, No. 6, pp. 1564-1583

Oliver, W. C., Pharr, G. M. (2004). Measurement of hardness and elastic modulus by instrumented indentation: Advances in understanding and refinements to methodology. *Journal Materials Research,* Vol. 19, Issue 1, pp. 3-20

Pan, A., Luzhbin, D., Komashko, V., Flis, V., Pan, V., Dou, S., & Esquinnazi, P. (2004). Origin of paramagnetic magnetization in field-cooled $YBa_2Cu_3O_{7-\square}$ films. *Physical Review B,* Vol.69, No.2, (January 2004), pp. 024506-1-7, ISSN 1098-0121

Pethica, J. B., Oliver, W. C. (1989). Mechanical Properties of Nanometer Volumes of Material: Use of the Elastic Response of Small Area Indentations, in Thin Films-Stresses and Mechanical Properties. *Materials Research Society Symposium Proc.* Vol. 130, pp. 13-23

Pharr, G. M.., Oliver, W. C., Brotzen, F. R. (1992). On the generality of the relationship among contact stiffness, contact area and elastic modulus during indentation. *Journal of Materials Research,* Vol. 7, pp. 613-617

Pureur, P., Dias, F., Rodrigues Jr., P., & Obradors, X. (2001). Paramagnetic Meissner effect at high fields in melt-textured $YBa_2Cu_3O_{7-\delta}$. *Physica C,* Vol.354, No.1-4, (May 2001), pp. 219-222, ISSN 0921-4534

Rayon, E., Bonache, V., Salvador, M. D., Roa, J. J., Sanchez, E. (2011). Hardness and Young's modulus distributions in atmospheric plasma sprayed WC-Co coatings using nanoindentation. *Surface & Coatings Tehcnology*, Vol. 205, pp. 4192-4197

Roa, J. J., Capdevila, X. G., Martinez, M., Espiell, F., Segarra, M. (2007). Nanohardness and Young's modulus of YBCO samples textured by the Bridgman technique. *Nanotechnology*, Vol. 18, pp. 385701/1-385701/6

Roa, J. J., Jiménez-Piqué, E., Capdevila, X. G., Martinez, M., Segarra, M. (2008a). Yield strength, shear stress and toughness of YBCO samples textured by Bridgman technique. *Journal of Physics: Conference Series*, Vol. 97, pp. 012116/1-012116/6

Roa, J. J., Capdevila, X. G., Segarra, M. (2008b). Estudio de los Mecanismos de Fractura durante la oxigenacion del YBCO. *Anales de Mecanica de la fractura 25*, Vol. 1, pp. 55-60

Roa, J. J., Gilioli, E., Bissoli, F., Pattini, F., Rampino, S., Capdevila, X. G., Segarra, M. (2009a). Study of the mechanical properties of CeO_2 layers with the nanoindentation technique. *Thin Solid Films*, Vol. 518, pp. 227-232

Roa, J. J., Dias, F. T., Vieira, V. N., Oncins, G., Diaz, J., Schaf, J., Capdevila, X. G., Segarra, M. (2009b). Caracterizacion mecanica a escala picometrica de $YBa_{1.75}Sr_{0.25}Cu_3O_{7-\delta}$ monocristlaino mediante la tecnica de autoflujo. *Anales de Mecanica de la Fractura* 26, Vol. 2, pp. 489-494

Roa, J. J., Morales, M., Martinez, M., Capdevila, X. G., Segarra, M. (2009c). Curva tension-deformacion mediante indentacion esferica para materiales ceramicos superconductores de YBCO-TSMG. *Anales de Mecanica de la Fractura 26*, Vol. 2, pp. 495-500

Roa, J. J., Capdevila, X. G., Segarra, M. (2009d). Mechanical characterization at Nanometric Scale of Cermic Superconductor Composites. *Superconducting Magnets and Superconductivity: Research, Technology and Applications; Publisher: Nova Science Publishers, Chapter 3.* pp. 77-168. ISBN: 978-1-60741-017-1

Roa, J. J., Bartolomé, E., Bozzo, B., Capdevila, X. G., Granados, X., Segarra, M. (2010a). Nano-mechanical properties of silver-welded YBCO bulks. *Journal of Physics: Conference Series*, Vol. 234, pp. 012034/1-012034/6

Roa, J. J., Jiménez-Piqué, E., Capdevila, X. G., Segarra, M. (2010b). Nanoindentation with spherical tips of single crystals of YBCO textured by the Bridgman technique: Determination of indentation stress-strain curves. *Journal of the European Ceramic Society*, Vol. 30, pp. 1477-1481

Roa, J. J., Capdevila, X. G., Segarra, M. (2010c). Mechanical characterization at Nanometric Scale of Cermic Superconductor Composites. *Nanotechnology: Nanofabrication, Patterning and Self Assembly; Publisher: Nova Science Publishers, Chapter 5.* pp. 151-236. ISBN: 978-1-60692-162-3

Roa, J. J., Jiménez-Piqué, Puig, T., Obradors, X., Segarra, M. (2011a). Nanoindentation of multilatered epitaxial $YBa_2Cu_3O_{7-\delta}$ thin films and coated conductors. *Thin Solid Films*, Vol. 519, pp. 2470-.2476

Roa, J. J., Segarra., M. (2011b). Mechanical properties of HTSC. *Publisher: Lambert Academic Publishing*. ISBN: 978-3-8465-0957-9

Roa, J. J., Dias, F. T., Martinez, M., Padilla, J. A., Segarra, M. (2011c). Oxygenation Kinetics of YBCO-TSMG samples using the nanoindentation technique. *Journal of the European Ceramic Society*. DOI: 10.1016/j.jeurceramsoc.2011.09.013

Roa, J. J., Kosntantopoulou, K., Jiménez-Piqué, E., Martin, V., Segarra, M., Pastor, J. Y. (2011d). Nanoindentation of Bridgman YBCO samples. *Ceramics International.* DOI: 10.1016/j.ceramint.2011.10.039

Roa, J. J., Oncins, G., Dias, F. T., Vieira, V. N., Schaf, J., Segarra, M. (2011e). AFM as an alternative for Young's modulus determination in ceramic materials in elastic deformation range. *Physica C,* Vol. 471, pp. 544-548

Roa, J. J., Oncins, G., Diaz, J., Capdevila, X. G., Sanz, F., Segarra, M. (2011f). Study of the friction, adhesion and mechanical properties of single crystals, ceramics and ceramic coatings. *Journal of the European Ceramic Society,* Vol. 31, pp. 429-449

Roa, J. J., Oncins, G., Diaz, J., Sanz, F., Segarra, M. (2011g). Calculation of Young's modulus Value by Means of AFM. *Recent Patents on Nanotechnology,* Vol. 5, pp. 27-36

Rotoli, G., De Leo, C., Barbara, P., Nielsen, A., & Lobb, C. (2001). Mutual-inductance route to the paramagnetic Meissner effect in two-dimensional Josephson-junction arrays. *Physical Review B,* Vol.64, No.14, (September 2001), pp. 144518-1-5, ISSN 1098-0121

Sandiumenge, F., Puig, T., Rabier, J., Plain, J., Obradors, X. (2000). Optimization of Flux Pinning in Bulk Melt Textured 1-2-3 Superconductors: Bringing Dislocations under Control. *Advanced Materials,* Vol. 12, Issue 5, pp. 375-381

Schwartz, A., Chu, S., Massalski, T., & Laughlin, D. (2006). Extrinsic paramagnetic Meissner effect in multiphase indium-tin alloys. *Applied Physics Letters,* Vol.89, No.11, (September 2006), pp. 111903-1-3, ISSN 1077-3118

Sigrist, M., & Rice, T. (1995). Unusual paramagnetic phenomena in granular high-temperature superconductors - A consequence of *d*-wave pairing? *Review of Modern Physics,* Vol.67, No.2, (April 1995), pp. 503-513, ISSN 0034-6861

Svedlindh, P., Niskanen, K., Norling, P., Nordblad, P., Lundgren, L. Lönnberg, B., & Lundström, T. (1989). Anti-Meissner effect in the BiSrCaCuO-system. *Physica C,* Vol.162/164, No.2, (December 1989), pp. 1365-1366, ISSN 0921-4534

Svedlindh, P., Papadopoulou, E., Nordblad, P., Schöneberger, R., & Gross R. (1999). Magnetic Aging in $Bi_2Sr_2CaCu_2O_8$ Displaying the Paramagnetic Meissner Effect. *Physical Review Letters,* Vol.82, No.1, (January 1999), pp. 173-176, ISSN 0031-9007

Terentiev, A., Watkins, D., De Long, L., Morgan, D., & Ketterson, J. (1999). Paramagnetic relaxation and Wohlleben effect in field-cooled Nb thin films. *Physical Review B,* Vol.60, No.2, (July 1999), pp. R761-R764, ISSN 1098-0121

Thompson, D., Wenger, L., & Chen, J. (1996). Paramagnetic Meissner effect in conventional Nb superconductors. *Journal of Low Temperature Physics,* Vol.105, No.3/4, (November 1996), pp. 509-514, ISSN 0022-2291

Tsui, T. Y., Oliver, W. C., Pharr, G. M. (1997). Indenter Geometry Effects on the Measurement of Mechanical Properties by Nanoindentation with Sharp Indenters, in Thin Films-stresses and Mechanical Properties VI. *Materials Research Society Symposium Proc.,* Vol. 436, pp. 147-152

Wortiz, W., Lisboa-Filho, P., Passos, W., & Araújo-Moreira, F. (2001). Field-induced networks of weak-links: an experimental demonstration that the paramagnetic Meissner effect is inherent to granularity. *Physica C,* Vol.361, No.4, pp. 267-273, ISSN 0921-4534

Yoshino, Y., Iwabuchi, A., Noto, K., Sakai, N., Murakami, M. (2001). Vickers hardness properties of YBCO bulk superconductor at cryogenic temperatures. *Physica C,* Vol. 357-360, pp. 796-798

10

Magnetic Irreversibility and Resistive Transition in YBaCuO Superconductors: Interpretations and Possible Correlations

Fábio Teixeira Dias
Universidade Federal de Pelotas
Brazil

1. Introduction

The second generation of high-temperature superconductors (HTSC) have been extensively studied since the discovery. Because of the potential for technological applications several research groups around the world have studied the electric and magnetic properties of HTSC that can provide important information about the performance of these materials when, e.g., magnetic fields are applied in many different applications.

This chapter will be dedicated to describe the main results obtained with YBaCuO superconductors and the interpretations on the resistive transition and magnetic irreversibility limit with special attention to their correlations. It will be presented experimental results obtained by magnetization and magnetotransport measurements in several magnetic field regimes with the purpose of disclosing the connection between the magnetic irreversibility limit and the zero resistance temperature as a function of applied magnetic field.

The discussion will be centered in experimental evidences obtained not only in pure YBaCuO materials but also in doped materials. It will be presented results obtained with single crystals, poly-crystals (sintered and melt-textured) and thin film materials grown by different techniques aiming to discuss the influence of the microstructure in the electric and magnetic properties.

Initially it will be done a rapid review about the YBaCuO superconductor and its basic properties, with special attention to magnetic and electric properties, such as magnetic irreversibility and resistive transition.

The next topic to be approached is the central theme proposed in this chapter i.e. the correlation between the magnetic irreversibility and the zero resistance in YBaCuO superconductors and the subsequent fundamental interpretations. Several experimental results will be presented and a special attention will be done in relation to results obtained in grain-oriented samples, which allow us to investigate anisotropic effects of the zero resistance line and the magnetic irreversibility line.

The idea that the magnetic irreversibility line is a boundary which an electric current may flow without electric resistance comes from the conventional homogeneous metallic superconductors. On the another hand, the high-temperature superconductors such as YBaCuO and other cuprates are usually granular superconductors and in this case the onset of magnetic irreversibility takes place when the first loop of coupled grains is formed. An array of coupled grains is responsible to trap the Josephson flux and establish an irreversible behavior. However in this case the electric transport remains resistive because the well-coupled grain clusters (which are responsible for the magnetic irreversibility) still remain disconnected from each other. It's important to observe that the resistivity vanishes only when the grain coupling strength overcomes the phase entropy and leads to a long-range coherence of the order parameter. As a consequence the electrical resistance in granular superconductors is expected to persist down to temperatures below the irreversibility line and to vanish only after the phase coherence percolates through the whole sample. In materials who exhibit a weak superconducting granularity the zero resistance data are expected to fall closely below the irreversibility line. Although in some cases the zero resistance line falls below the magnetic irreversibility line, some authors admitted that in high magnetic fields, where the flux dynamics is dominated by the intragrain Abrikosov flux, the zero resistance line again follows the irreversibility line. But apparently some results obtained in grain-oriented materials exclude this possibility when high magnetic fields are applied and in this case the expected behavior is not observed.

These important aspects of the correlation between magnetic irreversibility and zero resistance line will be discussed based on the main models and ideas currently accepted.

2. YBaCuO superconductor

It will be described in this section the structural properties, the main growth techniques and some potential technological applications of the YBaCuO superconductor.

2.1 Structural properties

The discovery of superconducting oxides of high critical temperature occurred in 1986 with the pioneer work by Müller and Bednorz (Müller & Bednorz, 1986), who reported the observation of the superconductivity around 30 K in a compound containing La, Ba, Cu and O. The intense interest in these systems occurred a year later, after the work of Wu et al. (Wu et al., 1987) who substituted lanthanum by yttrium in the original formula of Müller and Bednorz. The critical temperature increased to 93 K, well above the boiling point of liquid nitrogen. This material, $YBa_2Cu_3O_{7-\delta}$, or simply YBaCuO, was the responsible to inaugurate a new era in the materials research. The relative simplicity in growth of ceramic samples was crucial to mobilize several research groups around the world in a deep study for its physical properties, as well as possible technological applications.

Unlike conventional metallic superconductors, the crystal structure of the HTSC is complex. One of the most important characteristic in high-temperature superconductors is the planar anisotropy. Figure 1 shows the crystal structure of $YBa_2Cu_3O_{7-\delta}$ superconductor.

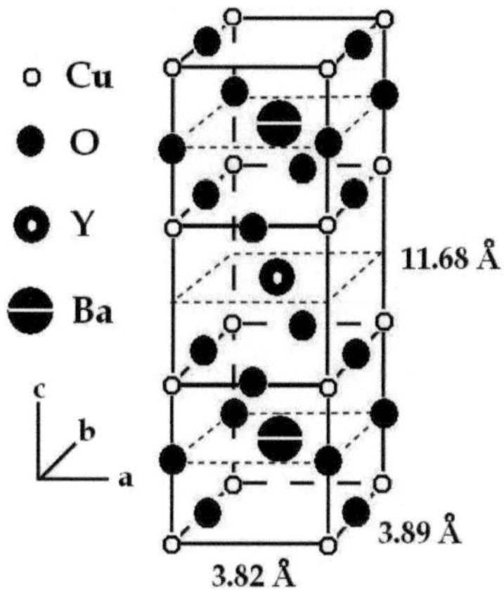

Fig. 1. The crystal structure of $YBa_2Cu_3O_{7-\delta}$ superconductor.

The typical lattice parameters are a = 3.82 Å, b = 3.89 Å and c = 11.68 Å. As shown in Figure 1, the $YBa_2Cu_3O_{7-\delta}$ superconductor is a perovskite-type cuprate with orthorhombic symmetry and the main structural characteristic is the presence of Cu-O_2 double planes separated by an yttrium atomic layer. The Cooper pairs, responsible to superconducting properties, are placed in the CuO_2 double planes. The atomic structures between the conducting planes are denominated charge reservoir. The oxygen content can vary between 6 and 7 atoms by unit cell. Deoxygenated systems (δ = 1 or simply 6 oxygen atoms per unit cell) has tetragonal structure and insulating anti-ferromagnet behaviour. For the optimum oxygen content of $\delta \approx 0.07$ the $YBa_2Cu_3O_{7-\delta}$ system has the biggest superconducting temperature at around 93 K.

2.2 Growth techniques

Several techniques have been employed to grow superconducting samples of $YBa_2Cu_3O_{7-\delta}$. It's possible to prepare samples as bulk (polycrystalline sintered or melt-textured), thin films and single crystals, among others. Initially the sintering procedure was the most promising technique because of its simplicity and possibility to obtain samples in a specific desired shape. The samples obtained by sintering techniques in general present a microstructure characterized by reduced grain size (few μm), porosity, cracks or microcracks and an aleatory grain alignment. This microstructure is responsible for the low critical current density values observed in sintered samples as compared to single-crystalline samples.

The melt-textured techniques have arisen as an alternative way to produce samples with high critical current density values needed for some technological applications. The pioneer work by Jin et al. (Jin et al., 1988a) set the basis to the melt-textured growth of $YBa_2Cu_3O_{7-\delta}$ samples. Melt-textured samples exhibit a microstructure characterized by large grains, low-angle grain boundaries, directional alignment (c-axis oriented), reduced porosity and dense structure. This microstructure plays an important role at the high critical density current values observed in melt-textured samples. Figure 2 shows a comparative image of (a) sintered and (b) melt-textured $YBa_2Cu_3O_{7-\delta}$ samples (Jin et al., 1988b).

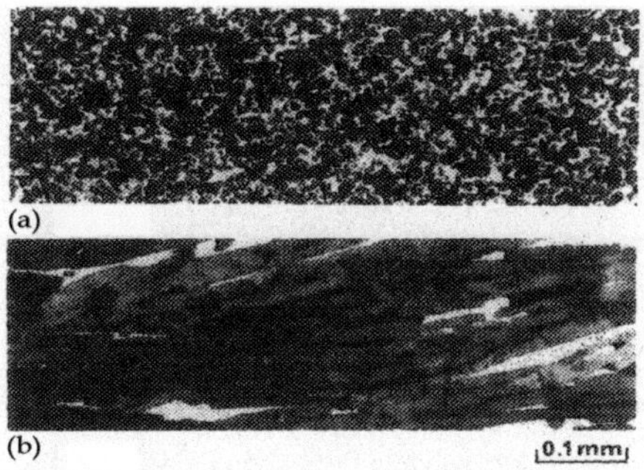

Fig. 2. Comparative image of (a) sintered and (b) melt-textured YBaCuO samples. From Jin et al.

Due to the large size of the oriented single domains obtained by melt-textured processes is possible to access directly the physics properties along the c-axis, such as electrical transport. Some works have been done using this important characteristic to study, for example, conductivity fluctuations in melt-textured $YBa_2Cu_3O_{7-\delta}$ samples (Jurelo et al., 2008; Dias et al., 2009).

$YBa_2Cu_3O_{7-\delta}$ samples can be grown as thin films and several techniques have been employed. Recently the chemical solution deposition (CSD technique) has been the more attractive technique due to its low cost and high performance of the samples obtained by this method. The growth of $YBa_2Cu_3O_{7-\delta}$ samples by metal-organic decomposition (MOD) using trifluoroacetate precursor solutions is used to produce high-performance superconductor thin films, as reviewed by some authors (Sandiumenge et al., 2006; Araki & I. Hirabayashi, 2003; Obradors et al., 2004).

2.3 Technological applications

The $YBa_2Cu_3O_{7-\delta}$ superconductor is the more used high-T_C system for technological applications due to its high critical current density and high-performance in applications involving high-magnetic fields. The relative facility in the growth of samples with specific shape permits their use in large scale. Due to the high performance in magnetic and electrical properties the materials prepared by melt-textured and thin films techniques have contributed decisively in this direction.

Several applications involving $YBa_2Cu_3O_{7-\delta}$ superconductor have been made since its discovery in 1987. Masato Murakami (Murakami, 2000) proposed two definitions for classify the applications of bulk high-temperature superconductors, named passive applications and magnet applications. In passive applications the high-temperature superconductors can be used in superconducting bearing systems, flywheels, cryogenic pumps, hysteresis motors (as rotor), etc. Examples for magnet applications include quasi-permanent magnets, maglev trains, magnetic clamps, motors (as stator), magnetic bumpers, in high-field physics, magnetization devices.

An important review by David Larbalestier et al. (Larbalestier et al., 2001) on power applications of high-T_C superconductors points out the need to low-cost fabrication associated to high-performance in electric current transport. This would allow producing large-scale superconducting electric devices for power industry, such as power cables, magnetic energy-storage devices, transformers, fault current limiters, etc. $YBa_2Cu_3O_{7-\delta}$ superconductor is pointed as one of the best candidates for these applications.

The importance of high-performance in current transport associated with a low-cost process to the fabrication of single grain bulk superconductors is also pointed out as one of main factors to effective applications in engineering devices by Hari Babu et al. (Hari Babu et al., 2011).

3. Fundamental properties

This section is dedicated to a review about electrical and magnetic properties of the high-temperature superconductors, with special attention to the $YBa_2Cu_3O_{7-\delta}$ system.

3.1 Electrical transport properties

In this sub-section it will be discussed some fundamental aspects of the electric properties of the high-temperature superconductors, with emphasis in the anisotropy of the normal state and in the granularity and disorder effects.

3.1.1 Anisotropy of the normal state

The transport properties of the high-temperature superconductors are highly anisotropic due to their crystalline structure. The CuO_2 atomic planes (ab planes) are good conductors but separated by highly resistive sheets, resulting in a strong planar anisotropy, with a low conductivity along the c-axis. The electrical current transport in $YBa_2Cu_3O_{7-\delta}$ superconductor is metallic-type along the ab plane. The same transport behaviour along the c-axis is also achieved in good quality samples. Figure 3 shows the electrical resistivity behaviour along the three crystallographic axes for an untwinned $YBa_2Cu_3O_{7-\delta}$ single crystal. The strong anisotropic character can be observed with a ratio at room temperatures of $\rho_a/\rho_b \approx 2.5$ and $\rho_c/\rho_a \approx 30$.

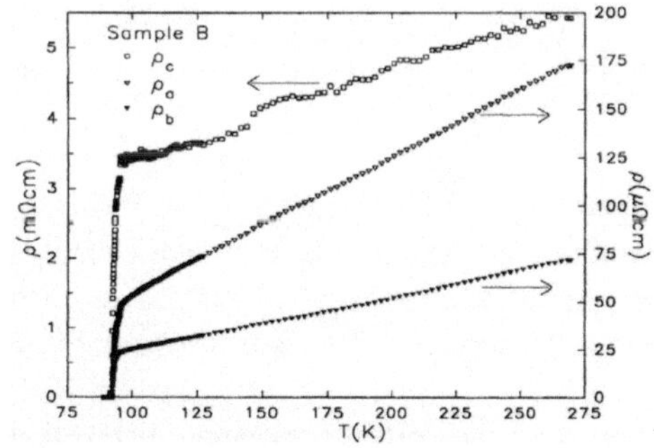

Fig. 3. Electrical resistivity behaviour along the three crystallographic axes for an untwinned $YBa_2Cu_3O_{7-\delta}$ single crystal. From Friedmann et al.

3.1.2 Granularity and disorder

The coherence length (T = 0 K) of an $YBa_2Cu_3O_{7-\delta}$ superconductor is approximately 12-20 Å along the ab plane and 1-3 Å along the c-axis, while the penetration length is about 1000 Å.

As a consequence the $YBa_2Cu_3O_{7-\delta}$ system is classified as an extreme type-II superconductor. Therefore defects of the same (or higher) order of magnitude are very important in these materials. The high-temperature superconductors present several intrinsic defects in different size scales: macroscopic (grain-boundaries, porosity, non-superconducting phases, such as the $Y_2Ba_1Cu_1O_5$ phase in some melt-textured $YBa_2Cu_3O_{7-\delta}$ samples), mesoscopic (twin-planes, dislocations, stacking-faults, columnar defects) and microscopic (oxygen vacancy, substituting atoms). The existence of these defect levels is responsible for the complex topology of the order parameter. Samples with higher disorder degree present an intrinsic and complex granularity, and this inhomogeneous character reflects in the magnetic and transport properties. Disorder is relevant since discovery of the HTSC's, when its presence was associated to the superconducting-glass state (Ebner & Stroud, 1985).

The resistive transition to the superconducting state in granular systems occurs in two steps-like transition, defining two important temperatures, namely T_C and T_{C0}, as presented in Figure 4.

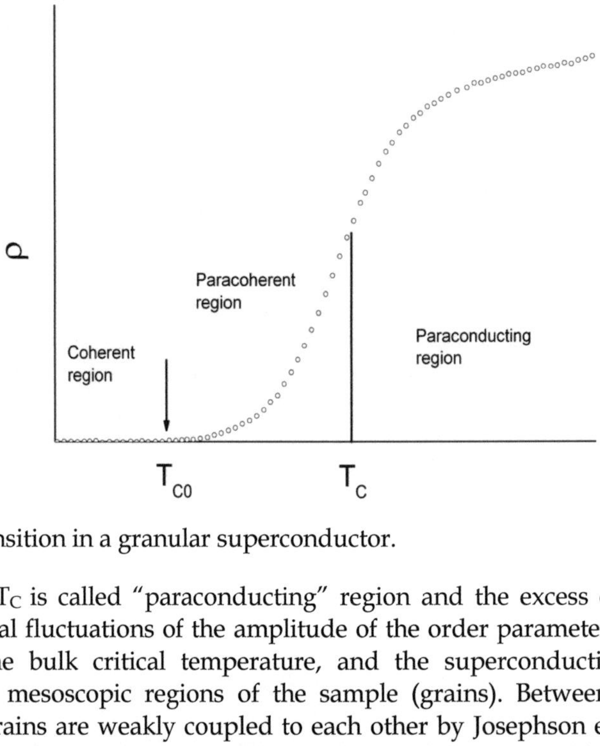

Fig. 4. Resistive transition in a granular superconductor.

The region above T_C is called "paraconducting" region and the excess of conductivity is attributed to thermal fluctuations of the amplitude of the order parameter. T_C is practically coincident with the bulk critical temperature, and the superconductivity stabilizes in homogeneous and mesoscopic regions of the sample (grains). Between T_C and T_{C0} the superconducting grains are weakly coupled to each other by Josephson effect and thermal fluctuations occur in the phase of the order parameter of the grains. This region is called "paracoherent" region and the electrical resistivity is not zero. The long-range superconducting state is achieved at T_{C0} through a percolation process at the grains. This process controls the activation of weak links between the superconducting grains and is called coherence transition. A detailed investigation of this process is given by Rosenblatt et al. (Rosenblatt et al., 1988). At the critical temperature (T_{C0}) where the coherence transition sets in, the fluctuating phases of the order parameter in each grain couple to each other into a long-range ordered state and a zero resistivity state is established.

3.2 Magnetic properties

In this section it will be discussed some fundamental aspects of the magnetic properties of the high-temperature superconductors, focusing in the irreversibility behaviour of the magnetization.

3.2.1 Phase diagram

The zero resistivity and the perfect diamagnetism (Meissner effect) are the main characteristics to any material be considered a superconductor. The Figure 5 shows a mean-field type phase diagram for a type-II superconductor, such as $YBa_2Cu_3O_{7-\delta}$. Type-II superconductors present a perfect diamagnetism (Meissner effect) just when the applied magnetic field is lower than $H_{C1}(T)$, called as lower critical field. Consequently this region is known as Meissner state. For applied magnetic fields in the region between $H_{C1}(T)$ and $H_{C2}(T)$, the magnetic flux can penetrate inside the superconductor in form of filamentary structures known as vortices. This state is titled as mixed state and the $H_{C1}(T)$ value marks the penetration of the first magnetic flux-line inside the superconductor. When the external magnetic field is higher than $H_{C2}(T)$ (also known as upper critical field), the superconductivity is suppressed and the material passes to the normal state.

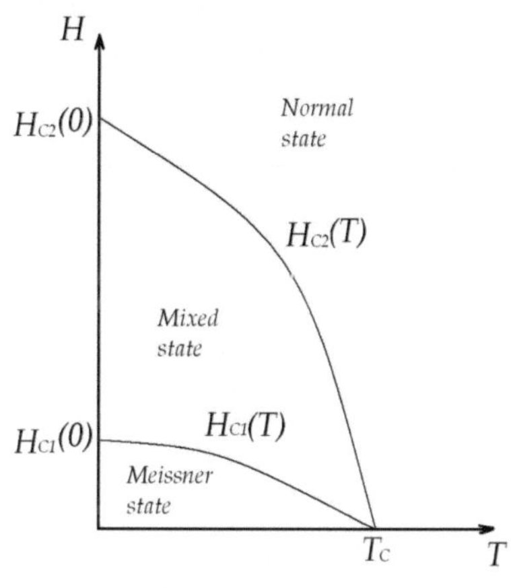

Fig. 5. Mean-field type phase diagram for a type-II superconductor.

3.2.2 Mixed state and vortex structure

The Figure 6 shows the mixed state structure in a superconducting sample where H_a is the applied magnetic field ($H_a > H_{C1}$). The Figure 6(a) shows a lattice of normal cores and the respective associated vortices, while the Figure 6(b) shows the variation with position of superparticles density (superelectrons density or Cooper pairs density).

The Figure 6(c) shows the variation of flux density with position. The superparticles density, designed by n_S, falls to zero at the centre of each isolated vortex, and the dip in the n_S is about two coherence lengths wide. The flux density in not cancelled inside the normal cores, but decreases into a small value in a distance about λ (London penetration length) away from the normal cores, as shown in Figure 6(c). Each individual vortex is surrounding by the shielding currents and the total flux generated is just one fluxon Φ_0, given by $2.067\text{x}10^{-15}$ Weber.

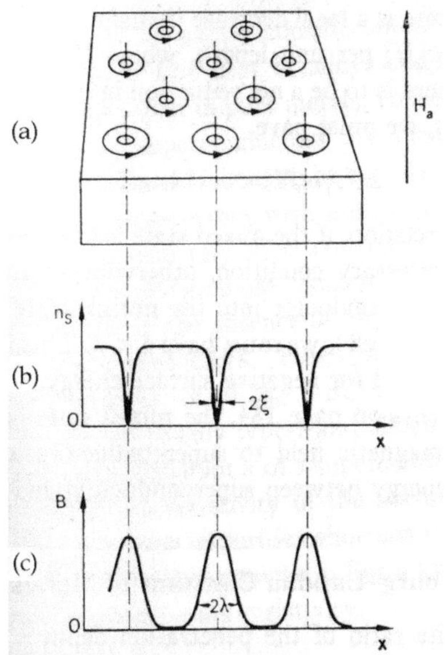

Fig. 6. The mixed state structure in a type-II superconductor. The Figure (a) shows a lattice of normal cores and the respective vortices, while the Figure (b) shows the variation with position of superparticles density and the Figure (c) shows the variation of flux density with position. From Rose-Innes & Rhoderick.

The interaction between vortices is repulsive and mediated by the Lorenz force. In a homogeneous superconductor the cores are arranged in a hexagonal (triangular) periodic array due to lower energy state. This array is known as Abrikosov lattice and can be observed in the Figure 6(a). In granular superconductors the vortices placed inside the superconducting grains are denominated intragranular vortices and are arranged in this lattice type.

The vortices can be pinned by defects of the crystal lattice. This effect is known as flux pinning and the mechanism can be improved by introducing microstructural or compositional changes like irradiation, chemical doping or precipitates.

In an isotropic and perfect superconductor (without the presence of defects) the vortices can be arranged in an Abrikosov lattice (hexagonal lattice). In a granular material, the superconducting grains can be coupled by Josephson effect, and based on this idea John R. Clem proposed a theoretical model to describe a granular superconductor using the Josephson effect as starting point (Clem, 1988). In this model a granular superconductor can be represented as anisotropic grains coupled by Josephson junctions (junctions based on Josephson effect).

The magnetic field can penetrate into a granular superconductor in the region between the grains, in the form of intergranular vortices, known as Josephson vortices. An intergranular Josephson vortex transports just one fluxon, similar to intragranular Abrikosov vortex and the magnetic structures are similar. The main difference is the region where both vortices are located. The Figure 7 shows a scheme of both vortices in a superconducting granular sample.

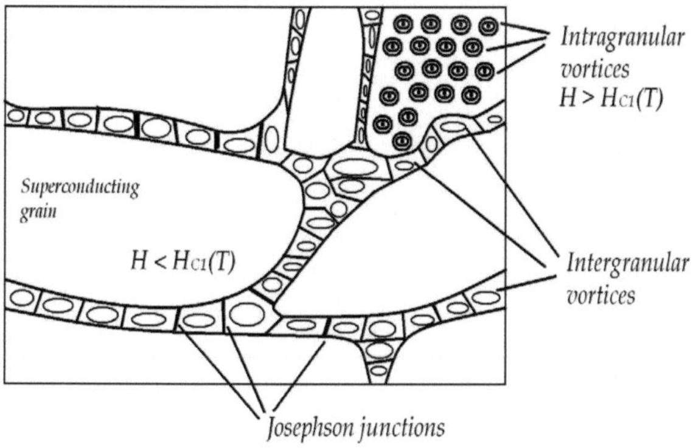

Fig. 7. Representative scheme showing intragranular and intergranular vortices in a type-II superconductor.

3.2.3 Magnetic irreversibilities

The magnetization of high-T_C superconductors shows an irreversible behaviour due to flux pinning combined with thermal activation of the vortices. In the next sections will be discussed the origin, the characteristics and some theoretical models used to explain this behaviour.

3.2.3.1 Magnetic irreversibility line

Through the realization of experiments based on ZFC (zero-field cooling) and FC (field-cooling) procedures in a superconducting sample it's possible to observe strong magnetic irreversibility effects, such as presented by Figure 8, due to flux pinning into the sample.

Below a temperature called irreversibility temperature (T_{irr}), the ZFC and FC curves present distinct behaviours. Below T_{irr} the vortices have its mobility strongly decreased due to pinning centres. As T_{irr} is dependent on the magnetic applied field, it´s possible to obtain a H versus T phase diagram similar to the one presented in Figure 9, where the line separate the diagram in two distinct regions: a region of high temperatures ($T > T_{irr}$) where the magnetization is reversible, and a region of low temperatures ($T < T_{irr}$) where the magnetization is irreversible. This line is known as magnetic irreversibility line or simply irreversibility line (IL). The study of the irreversibility line was originated with the pioneer work by Müller et al. in granular samples of $La_{2-x}Ba_xCuO_{4-\delta}$ superconductor (Müller et al., 1987).

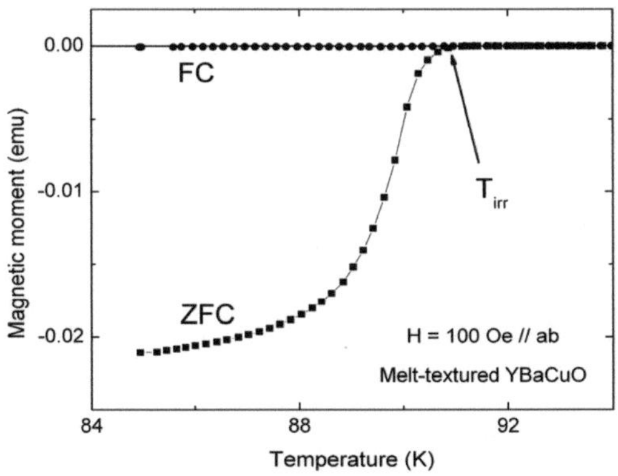

Fig. 8. Magnetic irreversibility effects in a superconducting sample obtained through the realization of ZFC (zero-field cooling) and FC (field-cooling) procedures.

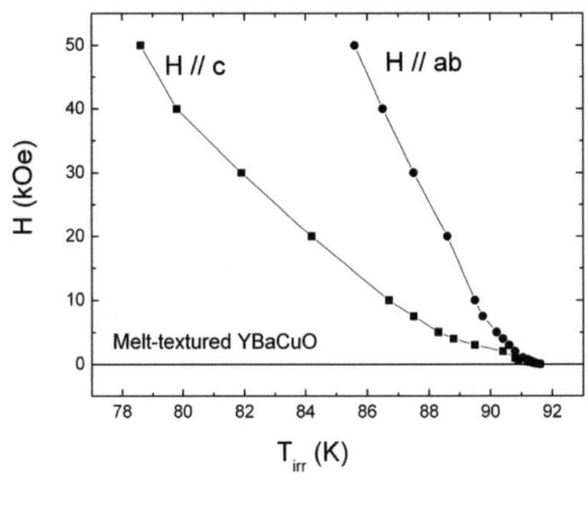

Fig. 9. The anisotropic character of the magnetic irreversibility line in a melt-textured sample.

The Figure 9 shows the anisotropic character of the magnetic irreversibility, attributed to intrinsic planar structure of the high-temperature superconductors. When H // ab the intrinsic pinning mechanism is evidenced and the vortices have their mobility restricted to the ab plane, reinforcing the effective pinning. By another hand, when H // c the lattice take the form of a stacked two-dimensional pancake-type vortices confined to the ab planes weakly coupled. At low temperatures the vortex lattice can be stabilized in a three-dimensional arrangement, but at high temperatures the thermal depinning effects passes to be relevant and a crossover from 3D to 2D occurs, resulting in a two-dimensional lattice of pancake vortices (Vinokur et al., 1990).

3.2.3.2 Theoretical models

This section is dedicated to describe some of the main models used to explain the magnetic irreversibility line (IL).

3.2.3.2.1 Giant flux creep

The model was proposed by Yeshurun and Malozemoff (Yeshurun & Malozemoff, 1988) based on the flux creep model initially developed by Anderson and Kim (Anderson & Kim, 1964). This model is based on the thermal activation effects of the vortex lattice. The model assumes that a vortex can be thermally activated over a pinning barrier even if the Lorentz force exceeds the pinning force. The main differences between both models are the high temperature and the low energy of the vortex pinning presented by the high-temperature superconductors.

According to the authors the experimental behaviour presented by the IL in a H versus T diagram obeys the power-law

$$H^{2/3} \propto (1-t) \ , \tag{1}$$

where $t = T_{irr}/T_C$ is the reduced temperature and T_{irr} is the irreversibility temperature.

According to the giant flux creep model the IL is essentially a depinning line in a H versus T phase diagram. Above the IL the vortices can move freely resulting in a zero critical current density while below the IL a finite value occurs once the vortices are pinned.

3.2.3.2.2 Vortex melting

According to this model the thermal fluctuations are responsible by the melting of the Abrikosov lattice, and the IL behaves as a boundary between two regions, namely vortex solid and vortex liquid. The instability produced by thermal fluctuations results in a second order phase transition of the vortex lattice at a temperature called vortex lattice melting temperature, or T_M. The aleatory dislocations produced by instabilities can promote collisions and loss of correlations between vortices. According to Nelson and Seung (Nelson & Seung, 1989), when the lattice melts two new regimes can arise, namely disentangled flux liquid and entangled flux liquid. The criterion used to define the vortex melting is known as Lindemann criterion (Lindemann, 1910).

The Figure 10 shows a representation of the vortex lattice melting phases. The disentangled flux liquid is characterized by high temperatures and low flux-line densities, as represented in Figure 10(b). The entangled flux liquid is reached when the flux-line density is high, as presented in Fig. 10(c). According to Houghton et al. (Houghton et al., 1989) the melting temperature (T_M) is slightly lower than the critical temperature (T_C) and the behaviour follow the power law

$$(1-t)^2 \approx H \ , \tag{2}$$

where $t = T/T_C$ is the reduced temperature.

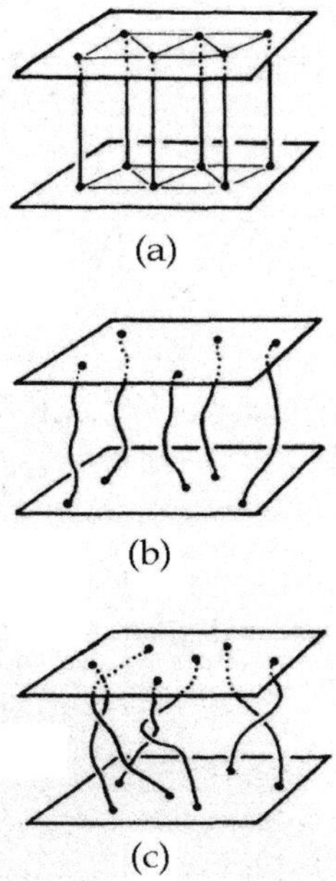

Fig. 10. Representation of the vortex lattice melting phases, showing (a) the Abrikosov lattice, (b) the disentangled and (c) the entangled flux liquid. From Nelson & Seung.

The Figure 11 shows experimental results obtained in an untwinned single crystal of $YBa_2Cu_3O_{7-\delta}$. The solid curve represents a fitting using equation (2).

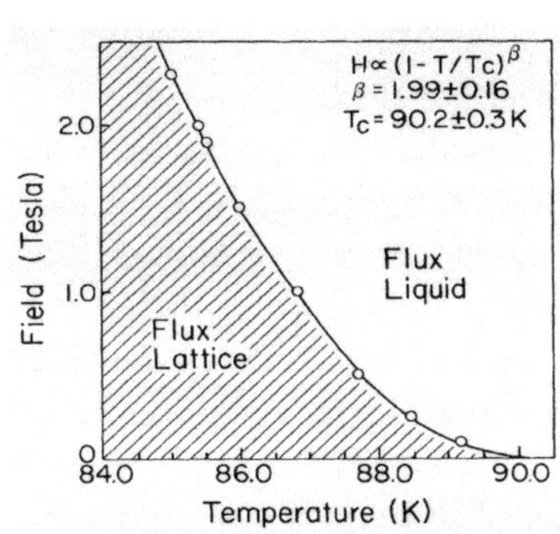

Fig. 11. H versus T phase diagram for the flux melting regime obtained in an untwinned single crystal of $YBa_2Cu_3O_{7-\delta}$. The solid curve represents a fitting using equation (2). From Farrell et al.

3.2.3.2.3 *Superconducting glass*

This model was based on the propositions of Ebner and Stroud (Ebner & Stroud, 1985) to study the diamagnetic susceptibility of superconducting clusters. The superconducting glass model proposes the existence of a disordered array of superconducting grains weakly coupled by Josephson effect or proximity effect. The Hamiltonian used to describe a system of weakly coupled grains is given by

$$H = -\sum_{i,j} J_{ij}\cos(\theta_i - \theta_j - A_{ij}), \tag{3}$$

where J_{ij} is the phase coupling energies between neighboring grains i and j and $\theta_i - \theta_j$ is the phase difference of the Ginzburg-Landau parameter on the grains i and j. The phase displacements A_{ij}, caused by the applied magnetic field, are given by

$$A_{ij} = \frac{2\pi}{\varphi_0}\int_i^j \vec{A}.d\vec{l}, \tag{4}$$

where ϕ_0 is the elementary flux quantum (fluxon), and the line integral is evaluated between the centres of grains i and j.

This model predicts the existence of a low temperature phase called superconducting glass, where disorder and frustration are dominants. The irreversibility line (IL) can be interpreted as a separation between two phases: namely superconducting glass and a system of decoupled grains. According to Müller et al. (Müller et al., 1987) it's possible to correlate the functionality of the IL with the same functionality used in spin glass systems. Using this analogy Müller et al. proposed that the behaviour of the IL at low applied magnetic fields in high-T_C superconductors can be described by de Almeida-Thouless type power law (de Almeida & Thouless, 1978), and it's given by

$$H = H_0\left(1 - \frac{T_g(H)}{T_g(0)}\right)^{\gamma}, \tag{5}$$

where $\gamma = 3/2$ and $T_g(H)$ can be interpret as $T_{irr}(H)$ in superconducting materials. The Figure 12 shows experimental results obtained in a $La_2BaCuO_{4-\delta}$ sample. The solid curve represents a numerical fit using equation (5). Another similarity occurs when higher magnetic fields are applied. In this case it's possible to observe a crossover from de Almeida-Thouless to Gabay-Toulouse (Gabay & Toulouse, 1981), where the behaviour can be described by the power law

$$H = \beta\left(1 - \frac{T_{irr}(H)}{T_{irr}(0)}\right)^{1/2}, \tag{6}$$

where β is a constant and $T_{irr}(H)$ and $T_{irr}(0)$ are the irreversibility temperatures obtained when $H \neq 0$ and $H = 0$.

3.2.3.2.4 Vortex glass

The vortex glass model was proposed by Fisher (Fisher, 1989) to describe the crossover from reversible to irreversible magnetization in high-T_C superconducting oxides using disorder effects. This model predicts the existence of a vortex glass phase between the Meissner state and the line defined by the temperature $T_G(H)$, as showed in Figure 13. The Abrikosov lattice in the vortex glass state has no long-range symmetry and the vortices freeze in a 2D lattice exhibiting a low-range order. The disorder and frustration are related to aleatory pinning centres distribution. As the temperature increases the vortex glass can be destabilized by thermal fluctuations and it melts above the $T_G(H)$ line, defining a state known as vortex liquid, as showed in Figure 13. The $T_G(H)$ line can be identified as the irreversibility line (IL).

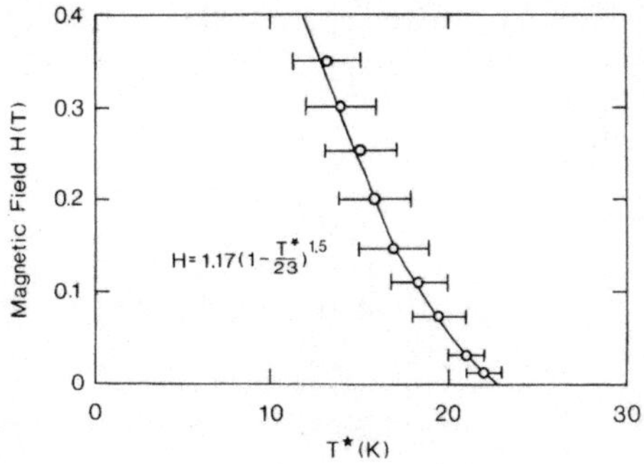

Fig. 12. Irreversibility line obtained in a $La_2BaCuO_{4-\delta}$ sample. The solid curve represents a numerical fit using equation (5). From Müller et al.

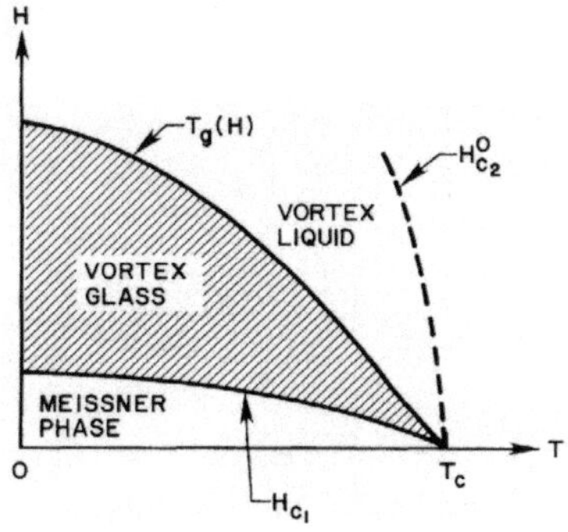

Fig. 13. Phase diagram for the vortex glass model. The $T_G(H)$ line can be identified as the irreversibility line (IL). From Fisher.

3.2.3.2.5 Bose glass

The Bose glass theory was developed by Nelson and Vinokur (Nelson & Vinokur, 1992) to treat the effects due to flux pinning by correlated defects (columnar defects, twin planes, grain boundaries, etc.). For a high-temperature superconductor with strongly correlated disorder the model predicts a H versus T diagram divided into three phases: superfluid, Bose glass and Mott insulator.

The superfluid phase is characterized by high temperature and viscous motion of vortices, which can jump across the correlated defects trough the sample. The Bose glass phase is essentially characterized by the strongly anisotropic pinning due to extended and correlated defects in the sample. This behaviour is strongly dependent on the alignment between the applied magnetic field and the direction of the defects dispersed in the superconducting sample. Misalignments greater than 5° may remove this phase in accordance with experimental results obtained by Safar et al. (Safar et al., 1996) in twinned $YBa_2Cu_3O_{7-\delta}$ single crystals. The Mott insulator phase occurs at low temperatures and can be characterized by a highly non-linear vortex dynamics governed by similar properties to the Meissner state.

4. Correlations between the magnetic irreversibility and the resistive transition

This section will be dedicated to present and to discuss experimental results on the magnetic irreversibility line, on the zero resistance line and on the respective correlations in high-temperature superconductors. The results will be focused in the $YBa_2Cu_3O_{7-\delta}$ superconductor.

4.1 Brief considerations

The magnetic irreversibility line in homogeneous superconductors defines a boundary in the H versus T plane below which the magnetization is irreversible and up to which a nonzero electric current can flow without any electrical resistance. Above this line the magnetization is reversible and all electric transport is resistive due to dissipation by flux dynamic effects. Very clean and well-oxygenated $YBa_2Cu_3O_{7-\delta}$ single crystals can be considered homogeneous superconductors and in this case the electrical resistivity is expected to vanish at the irreversibility line and even to comply with the planar anisotropy of the magnetic irreversibility. However high-temperature superconducting cuprates are generally inhomogeneous superconductors (Vieira & Schaf, 2002) or intrinsically inhomogeneous (Dagotto, 2005).

The magnetic irreversibility and the electrical resistivity in granular superconductors do not depend on the same parts of the superconducting sample. While the irreversibility depends on well-coupled grain clusters, the electrical resistance depends on grain arrays traversing the whole sample. Along such long-range paths, zero electrical resistance can be attained only at some temperature below the irreversibility temperature (Rosenblatt et al., 1988). On the other hand, at magnetic fields above several kOe the magnetic field penetrates the grains and the magnetic irreversibility is dominated by the Abrikosov flux dynamics (intragrain flux) while the electrical resistivity is still ruled by the grain junctions and can vanish only after the achievement of a long-range coherence state.

Magnetic irreversibility studies on pure and doped $YBa_2Cu_3O_{7-\delta}$ single crystals (Schaf et al., 2001, 2002) show the signature of superconducting granularity in the low-field region, dominated by Josephson flux (intergrain flux). Although in some cases the data for zero resistance falls below the magnetic irreversibility line, the authors admitted that in high magnetic fields, where the flux dynamics is dominated by the Abrikosov flux, the zero resistance line again follows the irreversibility line (Pureur et al., 2000, 2001).

High-quality melt-textured $YBa_2Cu_3O_{7-\delta}$ materials normally do not exhibit the signature of superconducting granularity, in spite of the polycrystalline structure. The linking between the crystallites is so strong that in some cases the coupling between superconducting grains occurs almost simultaneously with the superconducting transition.

4.2 Experimental results in YBaCuO superconductors

The results in this section are presented for several $YBa_2Cu_3O_{7-\delta}$ samples grown by different techniques, and some of experimental results and discussions were previously published by Schaf et al. (Schaf et al., 2008) and Dias et al. (Dias et al., 2008).

The Figure 14 shows results in an $YBa_2Cu_{2.97}Zn_{0.03}O_{7-\delta}$ single crystal grown by self-flux method. The continuous lines represent the irreversibility lines, $T_{irr}(H)$, for H // ab and

H // c and are fits to the power law predicted by the giant-flux-creep theory, while the circles represent the zero resistance data, $T_{C0}(H)$, for H // ab and H // c. The inset in the Figure 14 highlights the data at low magnetic fields where the continuous lines are fits to the de Almeida-Thouless and Gabay-Toulouse power laws. The $T_{irr}(H)$ data show a weak superconducting granularity and consequently the zero resistance data are expected to fall closely below the irreversibility line, as showed in the Figure 14. While at high magnetic fields the $T_{C0}(H)$ data fall about the irreversibility lines, in the low-field region they fall systematically underneath the irreversibility line for both field configurations (see inset).

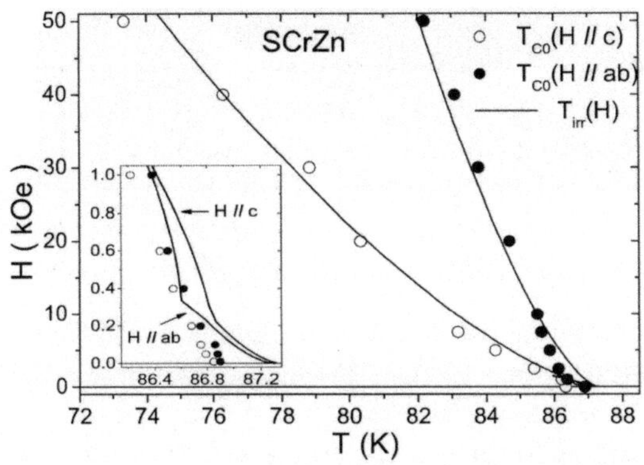

Fig. 14. Irreversibility lines in an $YBa_2Cu_{2.97}Zn_{0.03}O_{7-\delta}$ single crystal, together with the zero resistance data T_{C0}. The closed circles are for fields applied parallel to the ab plane and the open circles are for fields along the c-axis. The inset highlights the data in low fields. From Schaf et al.

The data in the Figure 15 illustrates the case of a melt-textured $YBa_2Cu_3O_{7-\delta}$ sample grown by top-seeding technique and containing 30 wt% of Y_2BaCuO_5 (Y211 phase) and 1 wt% of CeO_2. The purpose of the Y211 phase is to introduce pinning centers and the CeO_2 physically stabilizes the melt at high temperatures during the growth, to limit the size of the Y211 particles and to improve the distribution of the Y211 phase. This procedure strongly enhances the flux pinning potential in melt-textured $YBa_2Cu_3O_{7-\delta}$ samples. On the other hand, high-quality melt-textured samples in general do not display superconducting granularity. However, the high concentration of Y211 phase causes considerable misalignment of the c-axis, thereby encumbers the grain coupling and reduces the planar anisotropy (Carrillo et al., 2000). The weak superconducting granularity presented by the melt-textured sample, whose results are presented in the Figure 15, arises mainly from misalignment of the c-axis, resulting in a reduced planar anisotropy.

In the Figure 15 the irreversibility lines (H // ab and H // c) are represented by the experimental data and the continuous lines are fits to the power law predicted by the giant-flux-creep theory. The measuring current in the zero resistance data flows along the same ab planes for both field orientations (H // ab and H // c). While for H // ab plane and H // J the zero resistance data (closed circles) fall closely underneath the corresponding irreversibility line, they split strongly away from the irreversibility line toward lower temperatures for H // c and H $\perp$ J (open circles).

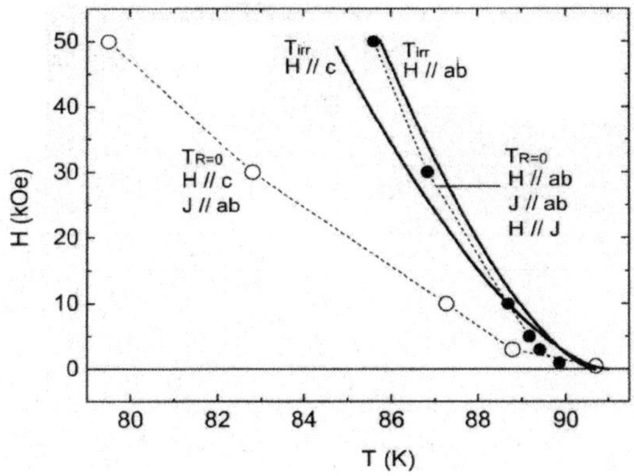

Fig. 15. Irreversibility lines obtained in a melt-textured $YBa_2Cu_3O_{7-\delta}$, together with the zero resistance data T_{C0}. The closed circles are for fields applied parallel to the ab plane and the open circles are for fields along the c-axis. From Schaf et al.

The twinning domains in the investigated melt-textured sample are nearly random in the ab plane and their direction also must be scattered somewhat about the average c-axis, due to the misalignment of the c-axis. Therefore, while flux pinning by the twinning planes certainly contributes somewhat for fields applied along the c-axis, its contribution for fields along the ab plane must be irrelevant. In the Figure 15 the irreversibility data show a much reduced anisotropy, due to the spreading out of the c-axes of the different crystallites and the consequent noncoplanarity of the ab planes.

Similar results (not published) are showed in the Figure 16, for a $YBa_2Cu_3O_{7-\delta}$ thin film grown by chemical solution deposition. It can be observed that the zero resistance data for H // c and H $\perp$J split away from the irreversibility line (H // ab) towards lower temperatures.

The data in the Figure 17 are representative for a polycrystalline $YBa_{1.75}Sr_{0.25}Cu_3O_{7-\delta}$ sample grown by conventional sintered technique using the standard solid state reaction method (Schaf et al., 2008). This polycrystalline sample exhibits a strong superconducting granularity, as shown by the presence of the de Almeida-Thouless and Gabay-Toulouse regimes in the low-field irreversibility data. The asterisks in the Figure 17 denote the approximate superconducting temperature transition (T_C); the open circles are the irreversibility data; the closed circles are the zero resistance data (T_{C0}) and the continuous lines are fits to de Almeida-Thouless, Gabay-Toulouse and giant-flux creep power laws.

In spite of the magnetic field being applied parallel to the measuring current, H // J, the zero resistance data of this polycrystalline sample fall considerably below the irreversibility line already in low magnetic fields and split away from this line steeply when the applied magnetic field increases. In polycrystalline samples the current flow through the grain aggregate is highly dispersive and therefore the H // J configuration is only a microscopic approximation. The data in the Figure 17 also show no indication the zero resistance data might come back and meet the irreversibility line somewhere in higher magnetic fields.

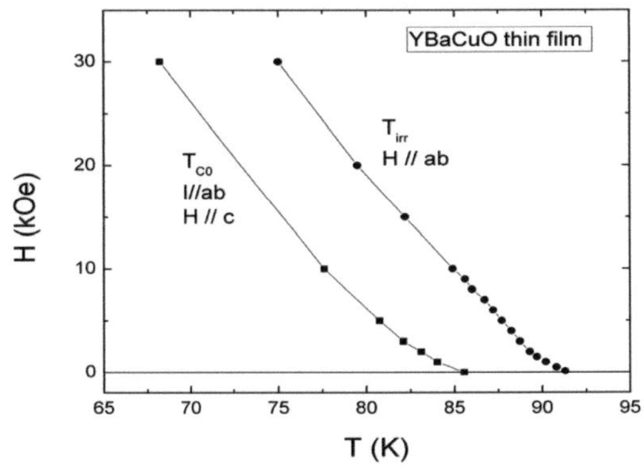

Fig. 16. Irreversibility line (T_{irr}) and zero resistance data (T_{C0}) obtained in a $YBa_2Cu_3O_{7-\delta}$ thin film. The closed circles are for fields applied parallel to the ab plane and the closed squares are for fields along the c-axis. Not published.

An applied magnetic field to a granular superconductor increases the phase entropy (Stroud et al., 1984). This magnetic field, while randomly produces a distortion of the phase of the order parameter, weakens and frustrates the grain coupling and favors the phase fluctuations that can result in the increase of the magnetoresistance. The random phase distortions also weakens the ability of the grain aggregate to pin the intergrain Josephson flux that dominates in the low-field regime. Therefore, in the low-field regime the irreversibility line exhibits the de Almeida-Thouless and Gabay-Toulouse power laws that are the signature of frustrated systems.

The response of a granular superconductor to a magnetic field is usually described in terms of the Josephson coupling Hamiltonian given by the equation (3) together with the equation (4). The equation (4) shows that a magnetic field causes phase displacements of the Ginzburg-Landau order parameter along those weak links which extend along the vector potential A , i.e. the A_{ij} are large along weak links that lie transversely to the magnetic field but may vanish along the weak links oriented parallel to the magnetic field.

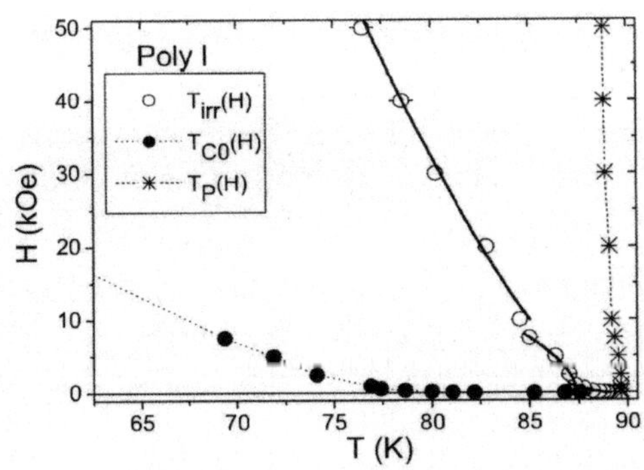

Fig. 17. Irreversibility line (T_{irr}) and zero resistance data (T_{C0}) in a polycrystalline $YBa_{1.75}Sr_{0.25}Cu_3O_{7-\delta}$ sample. The asterisks indicate the approximate superconducting transition temperatures. From Schaf et al.

Due to the dependence of the phase displacements A_{ij} and the associated grain decoupling effect on the direction of the applied field with respect to the weak links, the magnetic irreversibility line and the zero resistance of granular superconductors depend strongly on the field-current configuration. For instance, if the magnetic field is applied along the c-axis and the current flows in the ab plane, the degradation of the conducting junctions is expected to be strong and leads to a considerable increase of the resistivity and a large displacement of the zero resistance temperature towards lower temperatures. A representative result can be observed in the Figure 15. On the other hand, if the magnetic field orientation is parallel to the current in the ab plane, the effect of the field on the weak links carrying the current is expected to be much smaller, and this can be seen also in the Figure 15.

In the case of the polycrystalline sintered sample, in which the links between the grains are much weaker, the effects of the applied field on the grain coupling is drastic even for fields applied parallel to the intended measuring current, as can be seen in the Figure 17. In these granular superconductors the direction of the current is disperse due to the coupled grain arrays that percolate through the whole sample in randomic loops.

5. Conclusion

The connection between the magnetic irreversibility limit and the zero resistance in high-T_C superconductors is very important in order to understand the mechanisms of the electric current transport in these systems. In granular superconductors the zero resistance state is ruled by grain couplings and the effect of an applied magnetic field on the electrical resistivity depends strongly on the field-current configuration. In a superconducting sample which has a weak superconducting granularity, the electrical resistance vanishes very close to and underneath the magnetic irreversibility limit. On the other hand, for samples with a strong superconducting granularity, the magnetic field severely degrades the grain couplings. The electrical resistance, for current flowing along the weak links affected by the magnetic field, vanishes far below the irreversibility line.

Because of the potential for technological applications, the correlation between electric and magnetic properties in HTSC systems can provide important information on the performance of these materials when magnetic fields are applied in different configurations.

6. Acknowledgment

I'm very grateful to Dr. Frederik Wolff-Fabris for a careful reading of the text and to Dr. Jacob Schaf, Dr. Paulo Pureur and Dr. Valdemar Vieira for important discussions. This work was partially supported by the Brazilian Ministry of Science and Technology and the State of Rio Grande do Sul, under the Grant PRONEX FAPERGS/CNPq 10/0009-2, and Ministry of Science and Technology, under the Grant CAPES/Procad 059/2007.

7. References

Almeida, J., & Thouless, D. (1978). Stability of the Scherrington-Kirkpatrick solution of a spin glass model. *Journal of Physics A*, Vol.11, No.5, (May 1978), pp. 983-990, ISSN 1751-8113

Anderson, P., & Kim, Y. (1964). Hard superconductivity: theory of the motion of Abrikosov flux lines. *Review of Modern Physics,* Vol.36, No.1, (January 1964), pp. 39-43, ISSN 0034-6861

Araki, T., & Hirabayashi, I. (2003). Review of a chemical approach to $YBa_2Cu_3O_{7-x}$-coated superconductors - metalorganic deposition using trifluoroacetates. *Superconductor Science & Technology,* Vol.16, No.11, (September 2003), pp. R71-R94, ISSN 0953-2048

Bednorz, J., & Müller, K. (1986). Possible high Tc superconductivity in the Ba-La-Cu-O system. *Zeitschrift für Physik B,* Vol. 64, No. 2, (September 1986), pp. 189-193, 0722-3277

Carrillo, A., Puig, T., Plain, J., Figueras, J., & Obradors, X. (2000). Y_2BaCuO_5-free melt textured $YBa_2Cu_3O_7$: a search for the reference sample. *Physica C,* Vol.336, No.3-4, (July 2000), pp. 213-226, ISSN 0921-4534

Chu, C., Wu, M., Ashburn, J., Torng, C., Hor, P., Meng, R., Gao, L., Huang, Z., & Wang, Y. (1987). Superconductivity at 93 K in a new mixed-phase Y-Ba-Cu-O compound system at ambient pressure. *Physical Review Letters,* Vol.58, No.9, (March 1987), pp. 908-910, ISSN 0031-9007

Clem, J. (1988). Granular and superconducting-glass properties of the high-temperature superconductors. *Physica C,* Vol.153-155, No.1, (June 1988), pp. 50-55, ISSN 0921-4534

Dagotto, E. (2005). Complexity in Strongly Correlated Electronic Systems. *Science,* Vol.309, No.5732, (July 2005), pp. 257-262, ISSN 0036-8075

Dias, F., Vieira, V., Rodrigues Jr., P., Obradors, X., Pimentel Jr., J., Pureur, P., & Schaf, J. (2008). Magnetic irreversibility and zero resistance in melt-textured YBaCuO. *Journal of Magnetism and Magnetic Materials,* Vol.320, No.14, (July 2008), pp. e481-e483, ISSN 0304-8853

Dias, F., Vieira, V., Pureur, P., Rodrigues Jr., P., & Obradors, X. (2009). Fluctuation conductivity along the c-axis and parallel to the ab-planes in melt-textured $YBa_2Cu_3O_{7-\delta}$ samples doped with Y211 phase. *Physica B,* Vol.404, No.19, (October 2009), pp. 3106-3108, ISSN 0921-4526

Ebner, C., & Stroud, D. (1985). Diamagnetic susceptibility of superconducting clusters: spin-glass behavior. *Physical Review B,* Vol.31, No.1, (January 1985), pp. 165-171, ISSN 1098-0121

Farrell, D., Rice, J., & Ginsberg, D. (1991). Experimental evidence for flux-lattice melting. *Physical Review Letters,* Vol.67, No.9, (August 1991), pp. 1165-1168, ISSN 0031-9007

Fisher, M. (1989). Vortex-glass superconductivity: a possible new phase in bulk high-T_c oxides. *Physical Review Letters,* Vol.62, No.12, (March 1989), pp. 1415-1418, ISSN 0031-9007

Friedmann, T., Rabin, M., Giapintzakis, J., Rice, J., & Ginsberg, D. (1990). Direct measurement of the anisotropy of the resistivity in the a-b plane of twin-free, single-crystal, superconducting $YBa_2Cu_3O_{7-\delta}$. *Physical Review B,* Vol.42, No.10, (October 1990), pp. 6217-6221, ISSN 1098-0121

Gabay, M., & Toulouse, G. (1981). Coexistence of spin-glass and ferromagnetic orderings. *Physical Review Letters,* Vol.47, No.3, (July 1981), pp. 201-204, ISSN 0031-9007

Hari Babu, N., Shi, Y., Pathak, S., Dennis, A., & Cardwell, D. (2011). Developments in the processing of bulk (RE)BCO superconductors. *Physica C,* Vol.471, No.5-6, (December 2010), pp. 169-178, ISSN 0921-4534

Houghton, R., Pelcovits, R., & Sudbo, A. (1989). Flux lattice melting in high-Tc superconductors. *Physical Review B,* Vol.40, No.10, (October 1989), pp. 6763-6770, ISSN 1098-0121

Jin, S., Tiefel, T., Sherwood, R., van Dover, R., Davis, M., Kammlott, G., & Fastnacht, R. (1988a). Melt-textured growth of polycrystalline $YBa_2Cu_3O_{7-\delta}$™ with high transport JC at 77K. *Physical Review B,* Vol.37, No.13, (May 1988), pp. 7850-7853, ISSN 1098-0121

Jin, S., Tiefel, T., Sherwood, R., van Dover, R., Davis, M., Kammlott, G., Fastnacht, R., & Keith. H. (1988b). High critical currents in Y-Ba-Cu-O superconductors. *Applied Physics Letters,* Vol.52, No.24, (June 1988), pp. 2074-2076, ISSN 1077-3118

Jurelo, A., Rodrigues Jr., P., de Azambuja, P., Dias, F., & Costa, R. (2008). Fluctuation conductivity in melt-textured $YBa_2Cu_3O_{7-\delta}$. *Modern Physics Letters B,* Vol.22, No.18, (March 2008), pp. 1717-1725, ISSN 0217-9849

Larbalestier, D. Gurevich, A., Feldmann, D. & Polyanskii, A. (2001). High-T_C superconducting materials for electric power applications. *Nature,* Vol.414, (November 2001), pp. 368-377, ISSN 0028-0836

Lindemann, F. (1989). The calculation of molecular vibration frequencies. *Physik Zeitschrift,* Vol.11, No.14, (July 1910), pp. 609-612

Müller, K., Takashige, M., & Bednorz, J. (1987). Flux trapping and superconductive glass state in La_2CuO_{4-y}:Ba. *Physical Review Letters,* Vol.58, No.11, (March 1987), pp. 1143-1146, ISSN 0031-9007

Murakami, M. (2000). Progress in applications of bulk high temperature superconductors. *Superconductor Science & Technology,* Vol.13, No.5, (May 2000), pp. 448-450, ISSN 0953-2048

Nelson, D., & Seung, H. (1989). Theory of melted flux liquids. *Physical Review B,* Vol.39, No.13, (May 1989), pp. 9153-9174, ISSN 1098-0121

Nelson, D., & Vinokur, V. (1992). Boson localization and pinning correlated disorder in high-temperature superconductors. *Physical Review Letters,* Vol.68, No.15, (April 1992), pp. 2398-2401, ISSN 0031-9007

Obradors, X., Puig, T., Pomar, A., Sandiumenge, F., Piñol, S., Mestres, N., Castaño, O., Coll, M., Cavallaro, A., Palau, A., Gázquez, J., González, J., Gutiérrez, J., Romà, N., Ricart, S., Moretó, J., Rossell, M., & van Tendeloo, G. (2004) Chemical solution deposition: a path towards low cost coated conductors. *Superconductor Science & Technology,* Vol.17, No.8, (June 2004), pp. 1055-1064, ISSN 0953-2048

Pureur, P., Costa, R., Roa-Rojas, J., & Prieto, P. (2000). Pairing transition, coherence transition, and the irreversibility line in granular $GdBa_2Cu_3O_{7-\delta}$. *Physical Review B,* Vol.61, No.18, (May 2000), pp. 12457-12462, ISSN 1098-0121

Pureur, P., Roa-Rojas, J., & Fabris, F. (2001). Magnetotransport properties and the irreversibility line in ceramic $DyBa_2Cu_3O_{7-\delta}$. *Physica C,* Vol.354, No.1-4, (May 2001), pp. 304-308, ISSN 0921-4534

Rose-Innes, A., & Rhoderick, E. (1994). *Introduction to Superconductivity* (Revised second edition). Pergamon, ISSN 0-08-021652-8, Great Britain

Rosenblatt, J., Peyral, P., Raboutou, A., & Lebeau, C. (1988). Coherence in 3D networks: Application to high-T_c superconductors. *Physica B,* Vol.152, No.1-2, (August 1988), pp. 95-99, ISSN 0921-4526

Safar, H., Foltyn, S., Jia, Q., & Maley, M. (1996). Bose glass vortex phase transition in twinned $YBa_2Cu_3O_{7-\delta}$ superconductors. *Philosophical Magazine B,* Vol.74, No.5, (November 1996), pp. 647-654, ISSN 1364-2812

Sandiumenge, F., Gázquez, J., Coll, M., Pomar, A., Mestres, N., Puig, T., Obradors, X., Kihn, Y., Casanove, M., & Ballesteros, C. (2009). Precursor Evolution and Nucleation Mechanism of $YBa_2Cu_3O_x$ Films by TFA Metal-Organic Decomposition. *Chemistry of Materials,* Vol.18, No.26, (December 2006), pp. 6211-6219, ISSN 0897-4756

Schaf, J., Vieira, V., & Silva, J. (2001). Irreversibility limits of the Abrikosov and Josephson flux dynamics in homogeneous and granular high-T_C superconductors. *Physical Review B,* Vol.64, No.9, (August 2001), pp. 094516-1-7, ISSN 1098-0121

Schaf, J., & Vieira, V. (2002). Anisotropic irreversibility of the Abrikosov and Josephson flux dynamics in $YBa_{2-x}Sr_xCu_3O_{7-\delta}$ single crystals: Bose-glass and vortex-glass features. *Physical Review B,* Vol.65, No.14, (April 2002), pp. 144531-1-9, ISSN 1098-0121

Schaf, J., Pureur, P., & Vieira, V. (2002). Effects of Zn and Mg in Cu sites of $YBa_2Cu_3O_{7-\delta}$ single crystals on the resistive transition, fluctuation conductivity, and magnetic irreversibilities. *Physical Review B,* Vol.66, No.22, (December 2002), pp. 224506-1-11, ISSN 1098-0121

Schaf, J., Pureur, P., Dias, F., Vieira, V., Rodrigues Jr., P., & Obradors, X. (2008). Correlation between the magnetic irreversibility limit and the zero resistance point in different granular $YBa_2Cu_3O_{7-\delta}$ superconductors. *Physical Review B,* Vol.77, No.13, (April 2008), pp. 134503-1-8, ISSN 1098-0121

Stroud, D., Ebner, C., & Shih, W. (1984). Frustration and disorder in granular superconductors. *Physical Review B,* Vol.30, No.1, (July 1984), pp. 134-144, ISSN 1098-0121

Vinokur, V., Kes, P., & Koshelev, A. (1990). Flux pinning and creep in the very anisotropic high temperature superconductors. *Physica C,* Vol.168, No.1-2, (June 1990), pp. 29-39, ISSN 0921-4534

Yeshurun, Y., & Malozemoff, A. (1988). Giant flux creep and irreversibility in an Y-Ba-Cu-O crystal: an alternative to the superconducting-glass model. *Physical Review Letters,* Vol.60, No.21, (May 1988), pp. 2202-2205, ISSN 0031-9007

11

Manufacturing Technology of the $(Bi,Pb)_2Sr_2Ca_2Cu_3O_{10}$ High-Temperature Superconductor

Yury Evgenevich Grigorashvili
National Research University of Electronic Technology
Russia

1. Introduction

Practical use of high-temperature superconductors is considerably limited by the available technology to fabricate superconductors with characteristics, suitable for use in electrical products and electronics applications. All known high-temperature superconductors, which have critical temperature above the boiling point of nitrogen (77 K in normal conditions), consist of complex elemental composition and crystal lattice. All types of high-temperature superconductors are fragile.

Nowadays the most advanced technology in manufacturing the high-temperature superconductors and their products is done for the $Y_1Ba_2Cu_3O_7$ (Y-123) system. These are conductors of the second generation, multi-layer electronic structures, and devices with SQUIDs base. However, the temperature change of this system into the superconductive state is about 90 K. If affordable nitrogen is used to cool down the system, the working temperature difference would be just 14 percent shy of the critical one. Also, operational reliability is insufficient.

The high-temperature superconductor of the following composition - $Bi_2Sr_2Ca_2Cu_3O_{10}$ (Bi-2223) is the only one having the temperature in superconductive state above 100 K. Experimental data and research allow us to consider the creation of commercially-viable technologies based on the use of this superconductor. This superconductor has 30 percent operational reliability in temperature. It is also degradation-resistant under normal weather conditions. Yet, very limited research records exist in forming this high-temperature superconductor.

This chapter contains research results that focus on commercial implementation of technology that is based on manufacturing structures with the Bi-2223 superconductor, which are used in electronic industry. In this chapter we report the study results of the volume formation of the 110 K superconducting phase, including influences of lead, silver, and fluorine.

We show the process of obtaining thick films on inert substrate. Major attention is given to a manufacturing technology of epitaxial, thin, single-phased and defect-free layers situated on a monocrystal lattice. Also, additional information is included on its electrical and magnetic properties, as well as the analysis of possible implementation into electronic technology.

2. The manufacturing technology of superconductive structures with $(Bi,Pb)_2Sr_2Ca_2Cu_3O_{10}$ composition

In order to make high-quality superconductive structures on a monocrystal substrate we need to collect the information on processes happening during the crystal structure formation, the influences of material composition and dopants, manufacturing methods and technological conditions, as well as obtaining of electro-physical and magnetic properties in samples. In this chapter we describe the manufacturing technology of Bi - 2223 high-temperature superconductor and its structures that can be used in electronic industry (Grigorashvili, Y.E., Volkov, S.I., 1999; Grigorashvili, Y.E., 2005; Grigorashvily, Yu.E., Ichkitidze, L.P. & Volik, N.N., 2006).

2.1 The formation of superconductive phases in bulk material

Bulk material synthesis allows us gather information easily on the formation of high-temperature superconductors of various compositions and on optimal technological modes of thermal treatment. During the manufacturing process bulk material presence neutralizes changes in material surface structure. Prepared samples have dominant bulk properties as opposed to boundary properties.

This section describes the technology of making bulk superconductors of Bi-2223 composition, using methods of solid-phase synthesis. The original research results show the impact of developed technology on the formation of different superconducting phases in the BSCCO system. The conducted research determined the effect of lead and silver doping onto electrical characteristics of samples, including critical temperatures in the beginning and end of the superconductor's transition and critical currents in magnetic fields.

The first stage in bulk superconductors manufacturing is the production of micro dispersive charge of stoichiometric composition. In our research the charge was obtained from nitrates in metals. First, the manufacturing of nitrates mixture was done in proportions stated in 2223 formula for metals. In order to get rid of water and partial nitrates' decomposition the mixture has been annealed at 400 ^{0}C for two hours.

Formed spec grinded and the mixture stirred. The average size of grains in powdered mixture measured 4.5 µm. The second thermal treatment has been completed in oxygen atmosphere in 16 hours at 810-850 ^{0}C temperature range. The manufactured charge had superconductive properties. It was registered with diamagnetic response. The method of such diagnostics is presented in (Afanasev & Chaplygin, 1992).

Pellets were pressed from the charge into shapes sized 12-20 mm in diameter and 3-5 mm in thickness. The pellets were annealed in a gas mixture of nitrogen and oxygen in 10 to 50 hours under 800-850 ^{0}C. It was experimentally proven that the superconductive phases appear after a 20-hour annealing in temperatures registered above 820^{0}C. It is proven by temperature dependence on diamagnetic response and direct current resistance. The content of superconductive phases rises in relation to time of thermal treatment. Yet, their amount is small in comparison to the volume of charge. Based on measured results of diamagnetic response, their bulk makes about 15 percent at 849 degree C thermal treatment in 50 hours.

Right after the discovery of Bismuth-based superconductor it was determined that the lead presence considerably speeded up the process of superconductive phase formation (Sunshine S., et al., 1988). In our experiments lead was injected either as an additional element or as partial substitute for Bismuth. The amount of lead and Bismuth varied between 2/0.4 and 1.4/0.7.

Just like in previous case the pellets were pressed from powder, which were annealed in argon, oxygen, and air atmospheres. It was shown that the argon atmosphere annealing produced slight melting of pellets. Yet, under the oxygen atmosphere annealing the pellets' sides remained sharp and even. However, the amount of superconductive phases was too small in both cases. The best results were obtained during the annealing of argon and oxygen mixture with a partial pressure of 0.8 and 0.2 respectively. Similar results were received after annealing in air atmosphere.

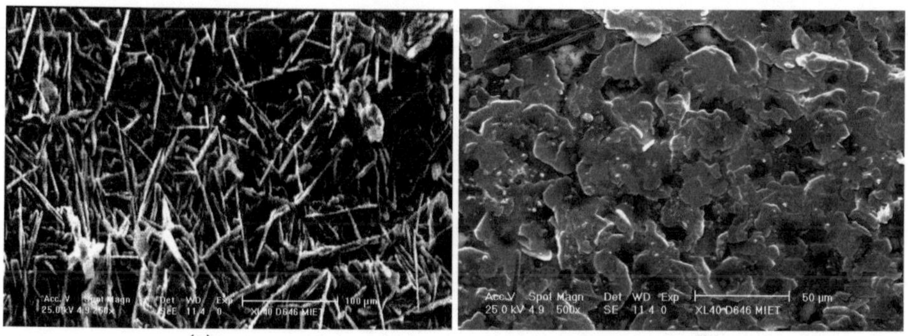

(a) (b)

a – $Bi_2Sr_2Ca_1Cu_2O_8$ Phase. Crystals are in a shape of needle like crystallites.
b – $Bi_2Sr_2Ca_2Cu_3O_{10}$ Phase. Crystals are in a shape of platelet crystallites.

Fig. 1. The superconductor's surface morphology of Bismuth system.

It's important to notice that the formation of Bi- 2212 phase happens when the temperature range of annealing is between 820 - 840 ^{0}C. The surface morphology of such samples has needle-like crystallite appearance (Fig. 1a). Needles' diameter is 2 to 4 µm and their length is 10 to 15 µm. The resistance dependence on temperature is shown in a spike that starts at 90 K temperature and ends at around 80 K (Fig. 2). The increased time of annealing does not lead to a temperature increase in the beginning of transition into a superconductive state.

The process of annealing at 840-860 ^{0}C temperature range forms a two-phase system of Bi-2212 and Bi- 2223. It is shown with two little steps situated on curves, explaining resistance dependence on temperature (Fig. 2). There is a formation of the through channel at a superconducting temperature reaching near 110 K during the 30-hour annealing at 860 ^{0} C. Critical current density is above 10^4 A/cm^2. It's typical for these samples to have crystallite formation in a shape of flat disks (platelet crystallites) that are 15-20 µm in diameter and 1-2µm thickness (Fig. 1b).

The results of experimental research show that it's difficult to make the Bi-2223 mono-phased material. Diamagnetic response measurements prove existence of two phases with critical temperatures measured at 110 K and 85 K. The synthesis of superconductive phase is very long and demands thermal treatment maintenance locked in a very narrow range.

With a temperature range at +/-0.5 at 860 ⁰C and the process of annealing around 100 hours the samples can be made with the superconducting state transition beginning at 115 K and completing at 108 K respectively. General findings in literature mention the development of bismuth-system phase with a critical temperature transition at only 110 K.

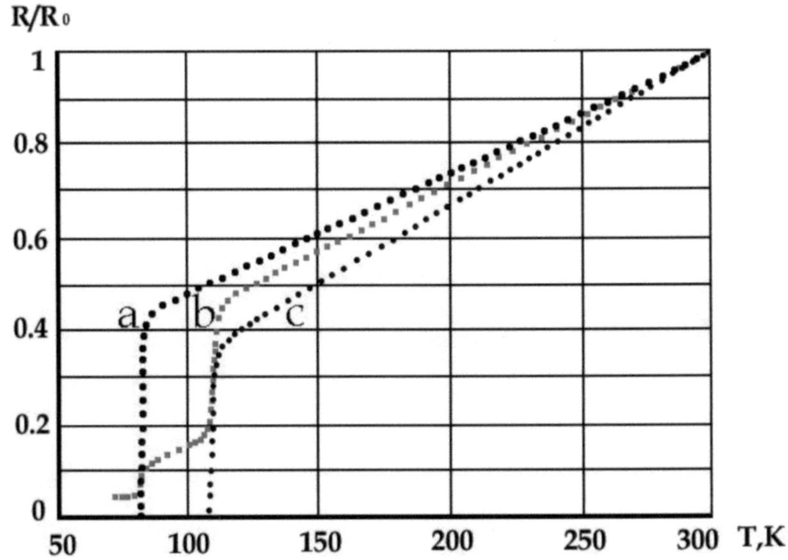

a – $Bi_2Sr_2Ca_1Cu_2O_8$ phase; b– mixture of two phases c - $Bi_2Sr_2Ca_2Cu_3O_{10}$ phase.

Fig. 2. Ceramic pellets resistivity dependence on temperature at various modes of annealing.

The author researched the opportunity of getting new characteristics of Bismuth system where the oxygen was replaced by fluorine. The samples were manufactured using the same technology mentioned above. Considerable superconductor volumes were not present. Alloying of silver powder with an atomic fraction of up to 0.5 increased the critical current density over 10 times.

2.2 Formation of bulk superconductors on a solid substrate

Next step in technological development of Bi-2223 formation is the creation of a bulk superconductor on a solid substrate. In this case, the processes described above should remain the same but the new sample would have a double-layer structure. To meet the described conditions, it's necessary to make an inert substrate relative to metals and their oxidations, entering the superconductor.

Bulk-layered Bi-2223 superconductor with a solid substrate was manufactured using a two-step system. First, the mixture was placed on a substrate, and then, a high-temperature annealing was carried out. This approach allowed us to divide the problem into two parts and research the conditions of each part separately.

The thickness of a deposited layer was several hundred microns. Such layer may be considered as a bulk one with the condition that there is neither interaction nor interfusion of the mixture at the superconductor-substrate border.

Widely accepted substrates of monocrystal silicon, oxidized silicon, and sapphire get mixed with superconductor's metals and its oxidizers at temperatures above 500 ^{0}C. Chemically inert substrates are MgO and $SrTiO_3$ monocrystal substrates.

Aerosol method is used to spray the oxides of metals onto the substrate. It is achieved by sputtering the microdrops of the solution with metal salts and their oxidation onto a hot substrate. The composition of the solution was chosen in such a way that when the microdrops hit the hot substrate, evaporation of a solvent, decomposition of metals' salts and the formation of oxide mixture took place.

The diagram of a unit, illustrating the manufacturing of films using aerosol method is shown at Fig. 3. The source is 0.05M water solution of lead nitrate, strontium, calcium, bismuth, and copper with added nitric acid and acetic acid. Ultrasonic disperser was used to make the aerosol. The aerosol mixture was delivered to the hot substrate via inert gas. In experimental research the substrate was heat-treated at 820-830 ^{0}C temperature. It required use of a silicon carbide heater that was able to work for a long time in aggressive atmosphere. The evaporation of water occurred at high temperature when water-drops reached the substrate. After that the nitrates decomposed and oxides developed. To increase the size of area of homogeneous deposition and maintain high temperature of a substrate, the aerosol nozzle scans the surface in two coordinates.

In experimental research the power of ultrasonic transmitter, speed of gas current, temperature of a substrate, frequency and range of scanning, and composition of a solution were varied. The composition of obtained films was controlled by chemical methods and the X-ray specter microanalysis method.

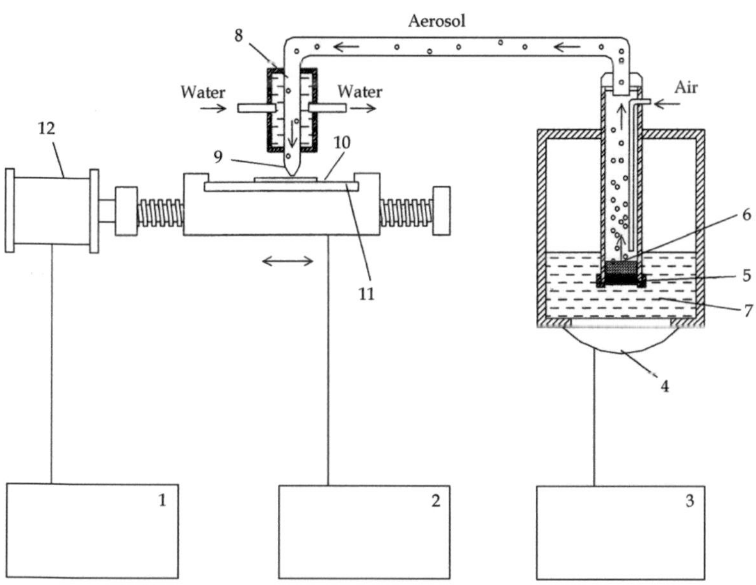

1- Control unit of the mechanical scanning system. 2- The temperature controlling unit. 3- High-frequency generator 12.6 MHz, 4- piezoceramic transmitter, 5- membrane, 6- the salt solution of basic metals. 7- water. 8- water refrigerator. 9- nozzle. 10- substrate. 11- Lamellar heater. 12- step motor.

Fig. 3. The diagram of a device showing the deposition of a bulk-layered superconductor on a substrate.

Main advantage of the described method is the simplicity of making changes in film composition. It is achieved by a varied concentration of basic metals' nitrate salts in a solution. The second important advantage of this method is the ability to make compositionally homogeneous layers in a big area.

Manufactured layers had amorphous structure of oxide mixture of basic metals. To form the crystal structure of Bi-2223 superconductive phase we used annealing in a mixture of either oxygen and argon, or 0.2/0.4 nitrogen. During the process of annealing there is a partial loss of bismuth, lead, and copper. That is why the aerosol mixture had enrichment with salts of listed metals at the depositional stage.

Film resistivity dependence on temperature with annealing modes that result in Bi-2223 phase formation is shown on Fig. 4. As seen on this picture the formation of superconductive phase is noticed after 10 hours of annealing. It is accompanied by appearance of graph zones with sharp decline of resistance at 110 K. There is a proportional increase of a superconductive phase during further annealing. Fair superconductive through channel forms after 50 hours of annealing.

Major disadvantage of this method is the difficulty of forming films with even relief. Superconductive 1-2 μm-thick layer had a bumpy surface of about the same height. The reduction in thickness by up to 0.5 μm decreased the height of unevenness. However, some zones appeared free of any superconductive phase on a substrate's surface.

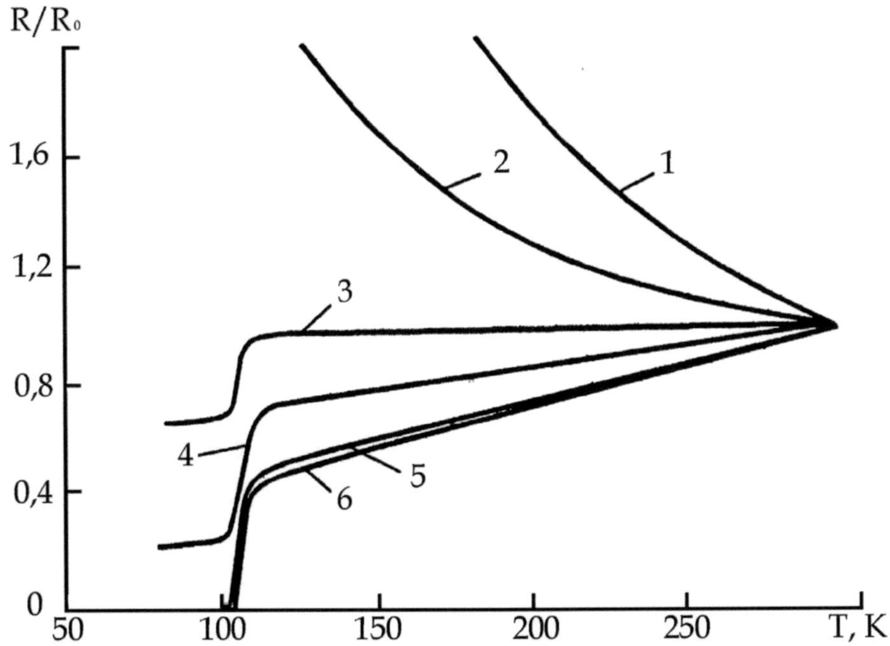

Period of annealing is 3, 5, 10, 20, 50, 100 hours. Curves are 1….6 respectively.

Fig. 4. Sample resistivity dependence on temperature at various annealing periods. Annealing temperature is 860 ^{0}C. R_0 – is the resistance at room temperature.

2.3 The technology of manufacturing thin film superconductors Bi-2223

During our preliminary research for this chapter we studied the available publications on manufacturing technology of Bi-2223 thin films. Publications were studied for 1990-2011 period in the following journals: Supercond. Sci. Technol., J. Appl. Phys., Appl. Phys. Lett., IEEE Transactions on Magnetics, IEEE Transactions on Applied Superconductivity, Physica C: Superconductivity and its Applications, Journal of Alloys and Compounds, Journal of Crystal Growth, Thin Solid Films, Journal of Vacuum Science & Technology A, Semiconductors, Technical Physics. Besides journal publications, we also studied the U.S., European and Russian patents.

It's worth mentioning that almost all publications on thin film manufacturing are for the Bi-2212 system. There are just a few publications available that have printed the results on thin film manufacturing with thickness less than 100 nm and critical temperature above 100 K. The problem of the formation of a superconductor Bi-2223 with a superconducting transition temperature above 100 K on single-crystal substrate remains topical today.

2.3.1 Basic methods of thin film manufacturing

Method of molecular-beam epitaxy (MBE) was developed in the beginning of the 70-s for manufacturing of high-quality, very thin films (Cho, A. Y., Arthur, J. R., 1975). This technology was used by a number of authors for BiSrCaCuO superconductive film manufacturing (Steinbeck, J., 1989; Yoon, D.H., Agung, I., Saito, M., Yoshizawa, M., 1997; Zakharov, N.D., Hoffschulz, H. et al., 1997). In this technology, the evaporation of individual components of the superconductor material is made from separate the Effusion (Knudsen cells). The composition and structure get formed by continuous opening of these cells for a required amount of time. (Varilci, A., Altunbas, M. et al., 2002) Working in MBE method BiPbSrCaCuO films were manufactured on MgO substrates (001). There were received critical temperatures of 105 K and current of $6x10^4$ A/cm^2. However, MBE method uses very expensive and complex equipment. Thus it's practically important to develop the alternative technologies of manufacturing superconductive thin film structures.

The technology of laser deposition (PLD) of BiSrCaCuO system films was used in this research (Ivanov, Z. et al., 1989). The process of film deposition consists of target material evaporation with a laser-radiation at energy density above 1 J/cm^2. There are excimer lasers - ArF ($\lambda = 193$ nm), KrF ($\lambda = 248$ nm), and XeCl ($\lambda = 308$nm) used for film deposition as well as infrared CO_2 lasers ($\lambda = 193$ nm). HTSC-film growth occurs at molecular oxygen pressures around 10 Pa.

This method has a number of important advantages. Thanks to having a laser radiation source outside the working chamber, the deposition process occurs in "clean" conditions. Laser beam with optical system is focused onto a small area using small-diameter targets that increase the process of efficiency in superconducting film manufacturing. Laser evaporation lets us manufacture multi-layer structures easily that consist of multi-component material (Jannah, A.N., 2009). Pulse-controlled laser radiation lets us determine precisely the thickness of the growing film.

In a study (Pilosyan, S.H., 1990) the manufacturing of $(BiPb)_2Sr_2Ca_2Cu_3O_{10}$ films is on MgO substrates (100) with Tco=110 K and Tce= 105 K parameters. Working pressure in a chamber during the deposition is 10^{-1} Torr. The impulse energy is 0.05-0.11 J at the frequency of 15 Hz pulse repetition rate. Target-substrate distance is 20-25mm. Thermal treatment of films occurs at 850 ^{0}C temperature. It is 3 hours long. It's necessary to point out that in this process the target of $Bi_{2}Pb_{0.5}Sr_{1.4}Ca_{2}Cu_{3.6}O_{x}$ composition was used.

Main disadvantage of the PLD method is the existence of clusters (drops) found in evaporated material. It happens because of high density of laser- radiation energy. Attempts to reduce the amount of microdrops via the reduction of energetic density current, lead to an increase in a non-stoichiometric component in a depositional condensate on a substrate. It limits dimensions of substrates that can be used for high-quality film manufacturing. Usually they don't exceed 10 mm in diameter.

Among chemical methods used for Bi-2223 film manufacturing we can highlight Liquid epitaxy (Pandey, et al., 1994), Spray Pyrolysis and the deposition from gas-vapor mixture (Kimura, T., Nakao, H. et al., 1991), including the MOCVD metalorganic compounds (Stejskal, J., Leitner, J.J. et al., 2000). In the MOCVD method, the components of superconductive film are transported into a reactor in a shape of gases of metalorganic volatile compounds. They are mixed with gaseous oxidizer resulting in gases' decay either inside the reactor with hot walls or on a heated substrate and the HTSC-film formation.

The advantages of this method include: the ability to make deposition of material onto substrates of large scale, production of homogenous layers in thickness and composition, and the ease of equipment use. At the same time, data found in publications shows that manufactured films are of low-quality. They contain non-superconductive inclusions, the surface has considerable unevenness, and there is no data available on the Bi-2223 phase formation.

Method of thermal deposition of material is the most simple and inexpensive resource for BiPbSrCaCuO film manufacturing. There are several types of used sources, including resistive evaporators, flash evaporation, ion beam (electron beam) evaporators and crucibles with radioactive and high-frequency inductive heating.

In a study (Azoulay, J., 1989) the researchers made BiSrCaCuO films using resistive heating of tungsten boat, with a source crucible of $Bi_{2.3}Sr_{1.5}Ca_{1}Cu_{3}$ material composition. After the deposition process, films were thermally processed in an oven at 725 ^{0}C for 15 minutes and 840 ^{0}C for 5 minutes. Typical superconductive temperature reaching the superconductive state T_{ce} was around 78-K for ZrO_2 substrates and 88-K for $SrTiO_3$ substrates.

Similar film parameters (Tce = 78K) are listed in (Patil, J.M., Bhangale, A.R. et al., 1993) that were manufactured with thermal deposition onto MgO substrate (100). In a study (Silver, R.M., Ogawa, E.T., Pan. S. de Lozanne, A.L., 1991) the BiSrCaCuO manufacturing in situ was completed using thermal evaporation method with high-frequency plasma generator. The device was constructed to make plasma at a distance from a substrate in order to stop recombination of oxygen atoms. After the deposition at 600-660 ^{0}C temperatures, films were annealed at 850 ^{0}C. The dominant phase was 2212, while the 2223 phase was present in small quantities. Similar critical temperatures were produced in studies (Basturk, N., 2005).

The analysis of conducted studies shows that high-quality HTSC-film with Bi-2223 phase and T_c around 110 K remains unresolved. To find a compromise solving a problem of Bi-2223 superconductor manufacturing might be the use of magnetron sputtering method.

2.3.2 The magnetron sputtering method – The computer model for the sputtering process

Magnetron sputtering for superconductive film manufacturing is used in modes of constant current (Schultz et al., 2001) and alternating current (Grigorashvily, Y.E., Volkov, S.I., Sotnikov, I.L. & Mingazin, V.T., 1999; Grigorashvili Y.E., Bukhlin, A.V. & Veryuzhskii, I.V., 2010) with the use of one or several sources (Kuroda, K., Kojima, K. et al., 1991). In this research the following strategy of forming Bi-2223 thin films with superconductive temperature around 110 K on a 40mm-radius substrate was used:

1. Film manufacturing was complete in two stages. First stage included the deposition onto a substrate's surface of a mechanical mixture of metal atoms and their oxides of stechometric composition. The second stage consisted of the crystal lattice formation of superconductive phase.
2. Method of magnetron sputtering is used for deposition of metals onto a substrate.
3. Target composition corresponds to $Bi_2Sr_2Ca_2Cu_3O_{10}$ superconductor composition.
4. Annealing is used to form structure. During this process the stimulation of epitaxial expansion of superconductor's thin layer takes place on a substrate.

The construction and technology of magnetron sputtering unit should take into account features of a problem, such as the formation of a five-component mixture of oxide metals on a substrate with various physical and chemical characteristics. To justify the construction of the sputtering system via experimental research is extremely expensive. Thus this research process consisted of experiments with the use of a mathematical model of transporting metal atoms from a target onto a substrate.

The most advanced mathematical model of the process of solution transportation with magnetron sputtering belongs to Volpyas (Volpyas, V.A. & Kozirev, A.B., 1997). It was supplemented in accordance with this research problem-solving.

The computer model of the sputtering process used in this research had the following algorithm:

- Calculation of magnetostatic field vector in magnetron source with a complex configuration of a magnetic conductor.
- The calculation of evaporation speed from the target's surface of various atoms' types depending on the coordinates with axisymmetric system.
- The calculation of atoms' movement from the target to the substrate. In this section the computer model was used to imitate the movement of each atom from the moment of leaving the target, trajectory change during the collision with gas atoms, conditions of return to the target or adhesion to a substrate, transition to evacuation system. Atom's type for each of five components was considered as well as speed distribution and the angle of departure from a target. To determine the angle of scattering we used the quasi-hard sphere potential (Volpyas, 2000).

The model allows to determine the thickness of sputtered layer precisely and the stoichiometry of composition depending on coordinates on a substrate. Partial pressures of argon and oxygen can vary in range widely, depending on the conditions of collision-free sputtered atoms until their complete thermalisation.

Computer model was used to calculate the construction of magnetron source, substrate's position, and the pressure range of gas mixture. The magnetron source type, magnetic fields' configuration, and calculated profiles of sputtered film in a multi-component system are shown at Fig. 5.

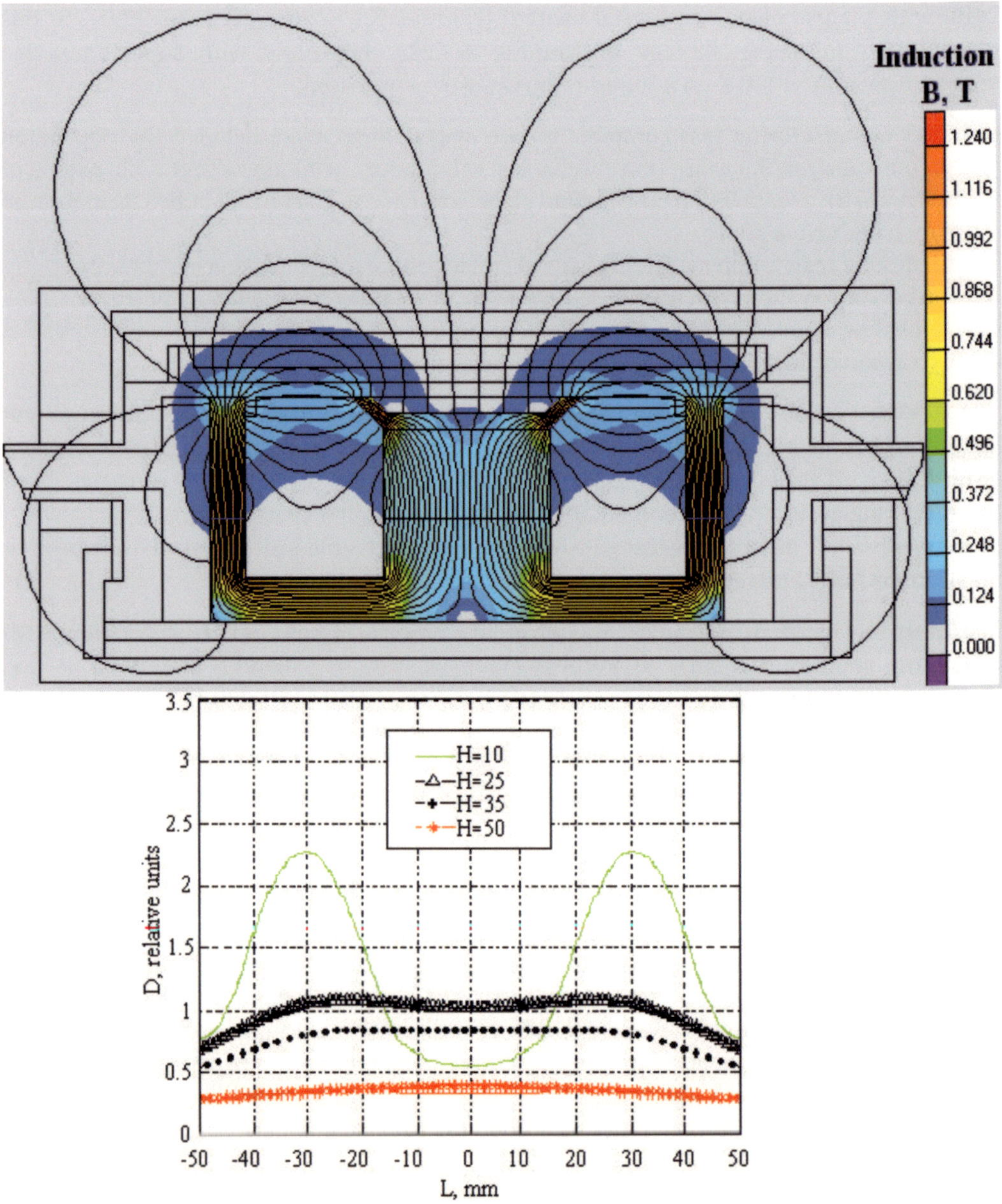

Fig. 5. The construction of magnetron sputtering system. The magnetic field induction is shown in magnetic conductor and inter-electrode space. The graphs illustrate the thickness of sputtered layer dependence on the target-substrate distance.

2.3.3 The influence of technological regimes onto composition of the multi-component mixture

In order to shorten time and reduce material expenses while searching for optimal gas pressures, we used method of planned experiment (Adler, Yu.P., Markova, E.V. & Granovskiy, Yu.V., 1976). The following arguments were chosen: the relationship of partial pressures of argon and oxygen - $P(O_2/Ar)$ and the overall pressure of gas mixture (Psum). The response function consisted of atomic concentration of each component of sputtered layer.

Method of planned experiment assumes to find the analytical dependence between function of response and varied arguments (Mongomery, D.C., 1980). The input information for obtaining such correlations comes from experimental results, received in limited quantities of points of argument values. In our research we needed to find concentration correlations of each component (5 in total) in relationship to partial pressures of oxygen/argon and the summarized pressure of gas mixture. Next, we used the optimal plan D that works well is conditions under considerable influence of the noise factor (Mongomery, D.C., 1980).

During the development of matrix of planed experiment it was important to consider that the expected effect of one factor depends on the level at which the other factor is located. Thus we have the effect of reciprocity of two factors. To quantitatively determine the effects of reciprocity is to use a full-factor experiment. The planning matrix is shown in a table 1. It consists of four real experiments.

Experiment's number	X_0	Psum	$P(O_2/Ar)$	Psum *$P(O_2/Ar)$	C_i
1	+1	+1	+1	+1	C_i (1)
2	+1	-1	+1	-1	C_i (2)
3	+1	-1	-1	+1	C_i (3)
4	+1	+1	-1	-1	C_i (4)

Table 1. The matrix of planned experiment

Here C_i (N) is the atomic concentration of i-component in experiment number N. When i equals 1,2,3,4 (i = 1,2,3,4) the concentration corresponds to (Bi,Pb), Sr, Ca, Cu elements. Arguments' values are both on the upper level (+1) and lower level (-1). Quantitative values of these levels are determined by expert estimation. The regression equation for this experiment looks like this:

$$C_i=b_0X_0 + b_1P_{sum} + b_2P(O_2/Ar) + b_{12}P_{sum} \cdot P(O_2/Ar) \quad (1)$$

Coefficients of variables b_0, b_1, b_2, b_{12} are determined by standard procedures method of planned experiment.

To lay out a plan of the experiment it's necessary to determine a varied range of arguments. The criteria for range determination include the following: stability of magnetron discharge, absence of electrical gaps between the electrodes, reasonable timeframe of the process, and the formation of the depositional layer in a shape of metal oxides. Last criterion is determined by stage characteristics of a crystal structure formation. If the depositional film has oxygen deficit, then the intensive evaporation of Bismuth and lead takes place during the annealing stage. It leads to stoichiometry disturbance.

For the first approximation the following values of arguments were chosen: lower level - $P(O_2/Ar) = 0.2$; $P_{sum} = 5$ Pa and the upper level $P(O_2/Ar) = 0.8$; $P_{sum} = 8$ Pa.

Four experimental tests were performed according to a plan shown in a table 1. Before each experiment the target was sputtered onto a shutter for 50 hours to level out the surface concentration of components. In all four experiments, atomic concentrations of metals did not match the 2223 stoichiometric composition. Thus, it's necessary to search for possible manufacturing regimes (modes) to make the essential mixture. To make it happen we used the regression equation analysis. Standard programs of data analysis let us receive the following equations for atomic percentages of the components:

For Bismuth and lead:

$$C_{Bi,Pb} = 2.26 + 2.28 \cdot P_{sum} - 0.068 \cdot P(O_2/Ar) - 0.443 \cdot P_{sum} \cdot P(O_2/Ar); \quad (2)$$

For strontium :

$$C_{Sr} = 5.32 - 2.38 \cdot P_{sum} - 0.56 \cdot P(O_2/Ar) + 0.239 \cdot P_{sum} \cdot P(O_2/Ar); \quad (3)$$

For calcium:

$$C_{Ca} = 0.73 - 2.6 \cdot P_{sum} + 0.31 \cdot P(O_2/Ar) + 0.62 \cdot P_{sum} \cdot P(O_2/Ar); \quad (4)$$

For copper:

$$C_{Cu} = 1.72 + 2.37 \cdot P_{sum} + 0.31 \cdot P(O_2/Ar) - 0.34 \cdot P_{sum} \cdot P(O_2/Ar); \quad (5)$$

Regression equations can be presented in graphs that let us find visually a compromised version in a choice of pressures. Color "green" represents values of stoichiometric composition components (Fig.6).

To determine the pressure area (P_{sum} and $P(O_2/Ar)$) simultaneously with all film variables corresponding to a required mixture - $(Bi,Pb)_2Sr_2Ca_2Cu_3O_{10}$, the data array analysis was completed, taken from the regression equation. Area configuration where the mixture conforms to required accuracy is shown at Fig. 7.

Our study of the results has shown that in order to maintain film stoichiometry - $(BiPb)_2Sr_2Ca_2Cu_3O_{10}$, the optimal pressure variables are $P_{sum} = 4.7$ Pa, $P(O_2/Ar) = 0.35$. We must mention that this point is situated on the outside of the square, rather than its inside. The square shows conditions of the initial experiment.

To experimentally test our conclusions, the target was adjusted and the sputtering was completed at the following pressures - P_{sum} - 4.7 Pa, $P(O_2/Ar)$ - 0.35. Component concentrations were measured with the x-ray spectral microanalysis method in ten points. Average values corresponded with this formula - $Bi_{1.71}Pb_{0.42}Sr_{2.02}Ca_{1.99}Cu_{3.05}O_x$.

Thus, the goal was accomplished by obtaining a thin layer of oxide mixture on a substrate, in which metals content corresponded with the formula of a high-temperature 2223 superconductor.

We should point out that during our calculation of regression equations we used a variable summing up of bismuth and lead concentrations, which equals in formula to (2). In experimentally produced films we found the right correlation between bismuth and lead concentrations. In particular, the average value of bismuth is 1.7 atomic fracture and of lead is 0.4.

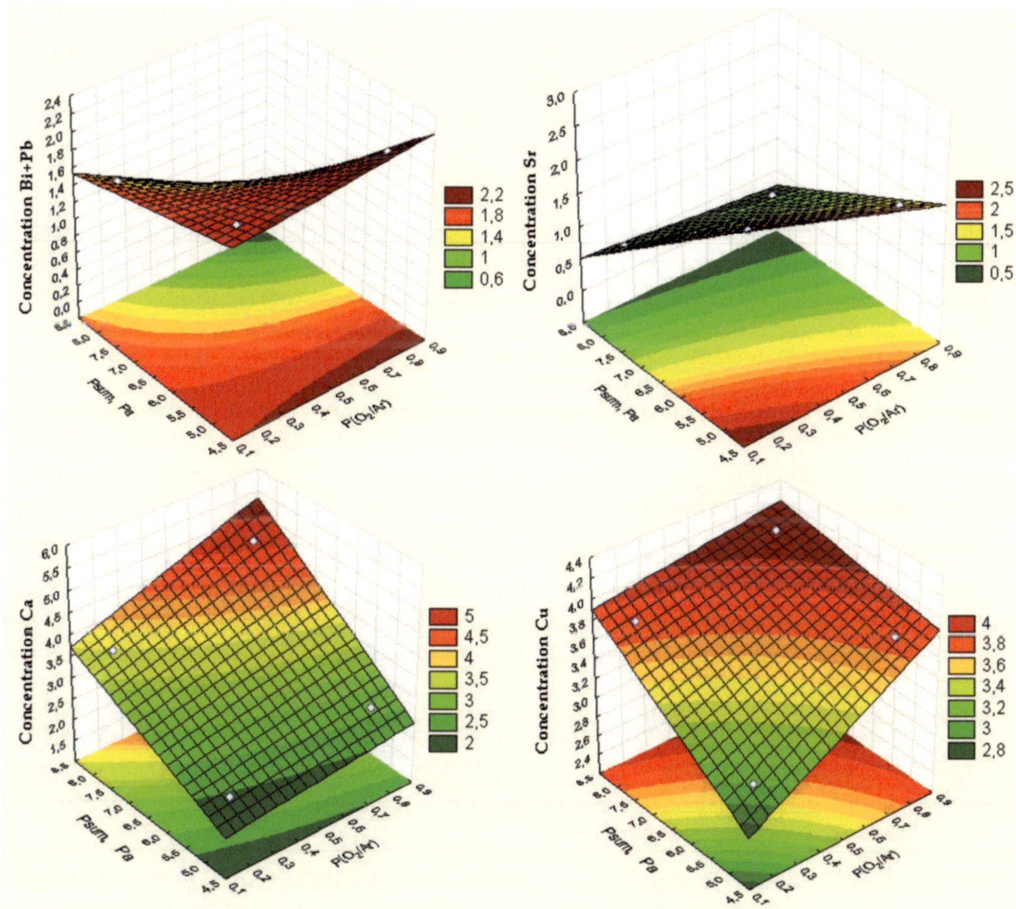

Fig. 6. Dependence of concentration of elements of a film from P_{sum} and $P(O_2/Ar)$.

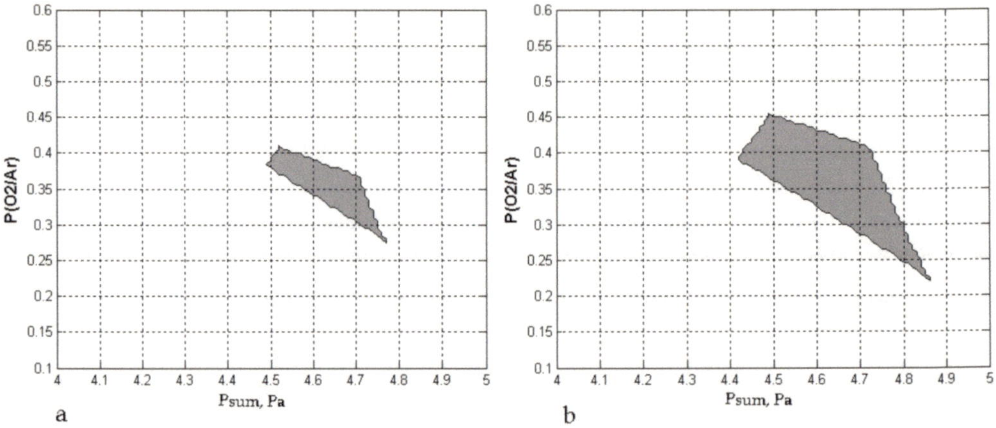

Fig. 7. The area of pressure in which deposited composition corresponds stoichiometric with accuracy: a - 4 % and b - 5 %.

2.3.4 The formation of crystal structure

It was shown in a 2.1 section that in the oxide mixtures of stoichiometric composition and temperatures above 820 C, the formation of superconductive phases takes place. The phase of Bi-2223 composition is formed at higher temperature rate of 860 ^{0}C and a narrow temperature interval. These results are taken to validate the manufacturing technology of thin film superconductive structures. Yet, there are a few considerable differences that must be taken into account.

High-temperature annealing leads to changes in composition not only at its surface, but also at its entire thickness. If the film is annealed at 850.....860 ^{0}C temperatures in an air atmosphere, then it will evaporate from a substrate's surface in a few hours of annealing. Thinly-layered silver sputtering over it doesn't impede this process.

The conducted experiments have shown that the evaporation process depends on both bismuth and lead losses. If the annealing is done in clean oxygen, then it's difficult to form 2223 phase. If the annealing is done in inert atmosphere, then bismuth, copper, and lead evaporate quickly. To stop the evaporation process by regulating the oxygen levels is not possible.

In our research to hold the mixture at substrate's surface, we used oxidized atmosphere with partial pressure of bismuth and lead equal to saturated vapors of these components at annealing temperature. If the process managed to form 2223 phase, then the structure remained stable even at high temperatures.

The Fig 8 shows the surface morphology and film resistivity dependence on temperature after various annealing stages. Low-temperature annealing leads to phase formations that become conductors in 77....300-K temperature range, Fig. 8 a. The annealing at close to optimal conditions forms 2223 phase. Yet, its grains have chaotic orientation. Moreover, there are parts with 2212 phase and other conductive phases, Fig. 8 b.

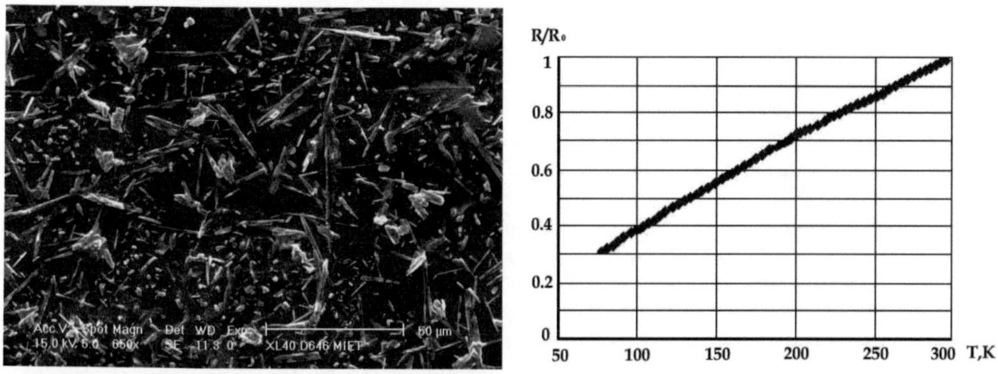

a

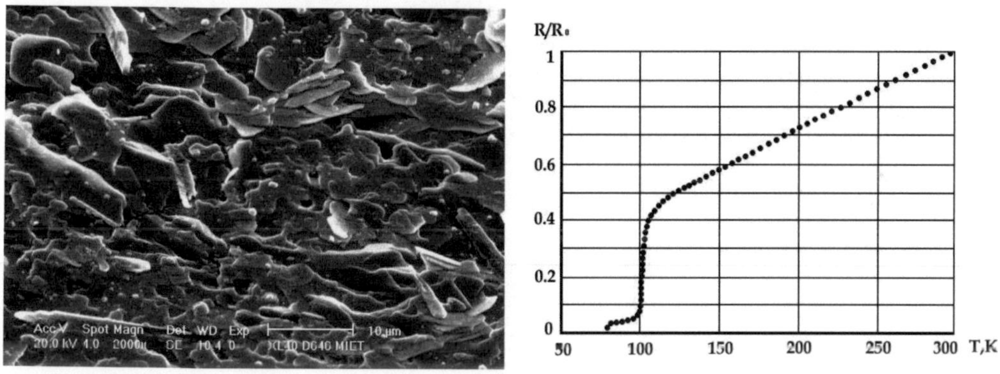

b

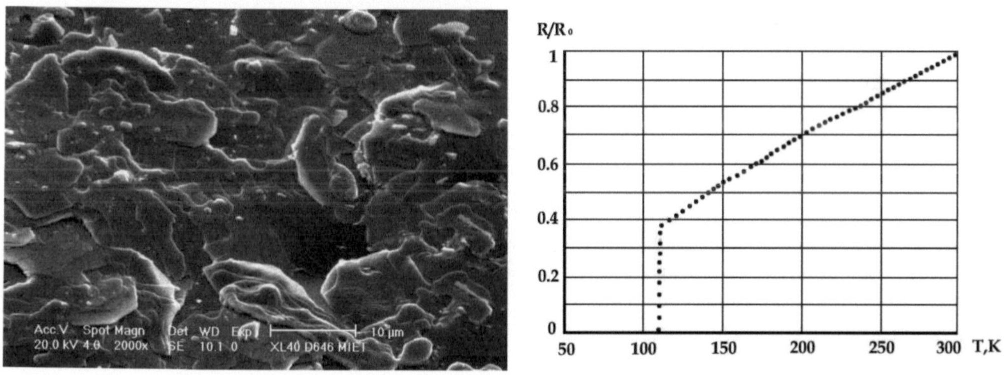

c

Fig. 8. Morphology of a surface and temperature dependence of resistance of films after various modes of annealing

Finally, the optimal annealing modes stimulate 2223 phase growth along the substrate's surface. To obtain such results we used the difference in growth speed of 2223 phase along the axis a, b and c. Grains looked disc-shaped of minor height yet big dimension, Fig. 8 c.

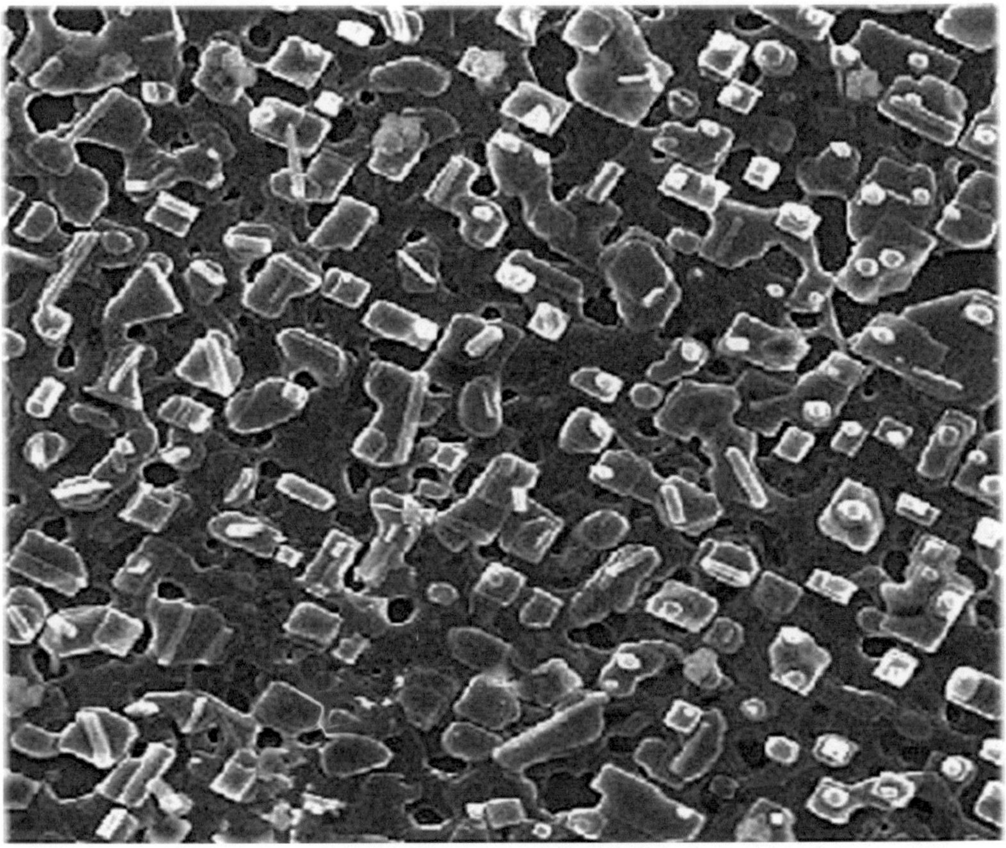

Fig. 9. Morphology of a surface of a film with the oriented grains

Using structures with sputtered films less then 100nm in thickness, the superconductive layer becomes shaped in the polycrystal units. Its orientation repeats the structure of MgO monocrystal. The surface morphology of such samples is very smooth in both electron and tunnel microscopes. During the increase in sputtered layer thickness structures arise having several storey grain formations. The example of such formation is shown on the Fig 9. The sample has continuous superconductive layer with 110 K critical temperature. Small grains located on the surface have orientation of a superconductive crystal.

2.4 Electrical and magnetic characteristics of Bi-2223 thin films

To research both electrical and magnetic characteristics of Bi-2223 films (Grigorashvily, Y.E., Volkov et. al., 2000; Grigorashvili, Y.E., Ichkitidze, L.P., Mingazin, V.T., 2004) using photolithography methods, the bridge-shaped structures were made 0.5....2.0 mm in length and 5....50 µm in width. Wide areas (200x200 μm^2) were formed on the edges of bridges. There were two contact areas made on each field with sputtering and the following silver annealing. Measurements were taken by four probe method at a computer device. In all measurements magnetic field vector B was directed perpendicular to a transport current. Current density varied through sample in a range between 10^1.....10^6 A/cm^2.

All researched films had critical temperature around 110 K. Yet, technical parameter variations could considerably change the size of critical current and magnetic characteristics. The track consists of continuously connected grains of monocrystal superconductor and intergranular boundaries. These two elements behave differently. It was experimentally proven that grains remain to be superconductive at current density 10^5 A/cm^2 and constant magnetic filed 1 T. Intergranular boundaries can be either dialectical, conductive or Josephson elements. If the boundaries are resistive fields, then during the increase in current density, there is a residual current registration at temperatures below critical ones. Every boundary behaves as the Josephson junction in the intermediate condition. Overall, the entire structure can be considered as the Josephson environment. Although such boundaries reduce critical current density, they simultaneously increase the responsiveness to magnetic field. Usually magnetic response value is determined as:

$$S= [R(B)/R(0) - 1]/dB \qquad (6)$$

Here R(0) – is the sample resistance in external magnetic field B_0.

R(B) – is the sample resistance in magnetic field that equals to B_0+dB.

It was experimentally shown that magnetic responsiveness - S depends on sample's temperature, value of a measuring current, and the average value of magnetic field at measurements. All values had extremum in function of third variable at two other fixed variables. Fig. 10 shows magnetic responsiveness dependence on a measuring current in various ranges of measured magnetic fields.

Manufactured films have anisotropic properties in relation to the magnetic field. The Fig. 11 illustrates resistance of the microbridge change depending on the angle between magnetic field direction and substrate's surface. The resistance difference increases with the increase of magnetic field's absolute value.

Value S is managed by technological regimes in broad limits. It's possible to pick up such options when the microbridge in resistance measurement mode can be used as a magnetic field device with high sensitivity. Samples were manufactured with ~ 4 10^3 T^{-1} magnetic sensitivity at 77 K temperature in magnetic fields measuring less than 100 μT. Magnetic flux sensitivity of this device is about 0.1 Φ_0.

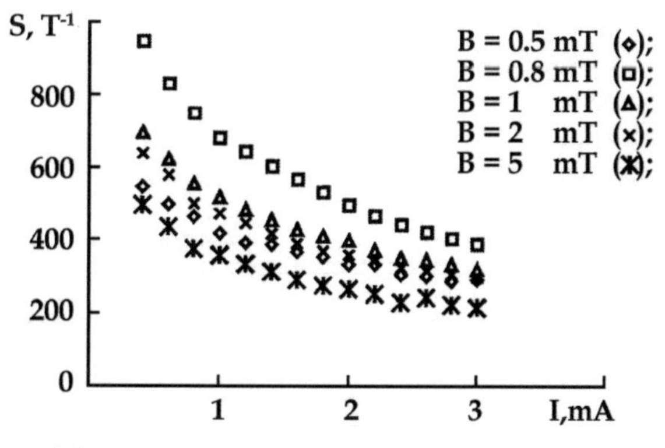

Fig. 10. The dependence of the magnetic sensitivity on the value of the measuring current in static magnetic field. The temperature is 77K.

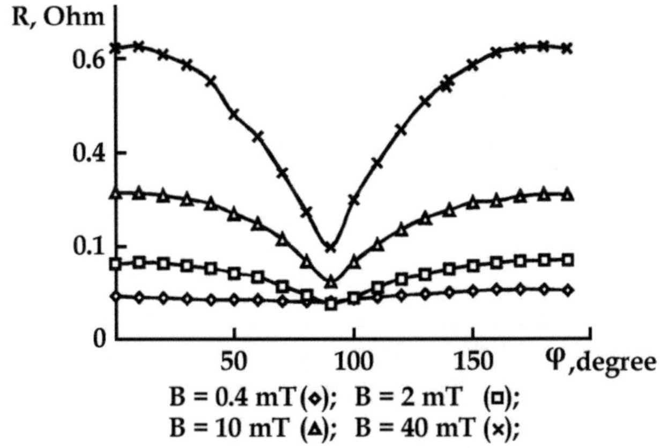

Fig. 11. Structure resistance dependence on the angle between substrate's surface and magnetic field direction. Magnetic field is perpendicular to transport current. The microbridge temperature is 77-K.

3. Conclusion

High-temperature superconductor of $(Bi,PB)_2Sr_2Ca_2Cu_3O_{10}$ (Bi-2223) composition can become a useful material for creation of commercial technologies. This chapter contains research results aiming at creation of methods to form superconductor thin films for electronic structures. Thin films were fabricated on the MgO monocrystal substrate that had the most considerable content of 2223 phase. The result is achieved by precise lead doping, stoichiometry maintenance at all stages of formation, as well as creation of special conditions when growing superconductor film epitaxialy repeats the MgO monocrystal substrate and has the structure of strongly textural polycrystal. Changing technological modes, it is possible to control effectively parameters of boundaries between grains, saving thus critical temperature 110K and changing a critical current with magnetic sensitivity that is used for creation of sensors of physical values.

4. Acknowledgments

This chapter's research results were commissioned by the administration of the Moscow National Research University of Electronic Technology and financed by the Ministry of education and science of the Russian Federation. The author is grateful to the "Superconductive electronics" laboratory employees for their creative input in high-temperature superconductors' problem-solving for many years.

5. References

Adler, Yu.P., Markova, E.V. & Granovskiy, Yu.V. (1976). The design of experiments to find optimal conditions, Publishing house "Science", Moscow

Afanasev, N.V., Chaplygin, Yu. A. (1992). Hysteresis losses in high-temperature superconductors, In: *High-temperature superconductors. Obtaining. Properties. Application,* Grigorashvily Yu. E., pp. 17-27, MIEE, Moscow

Azoulay, J. (1989). Thin-film Bi-Sr-Ca-Cu-O superconductor preparation using a single resistive evaporation source. *J. Phys. Condence. Matter.*, Vol. 1, No. 31, pp. 5305-5307

Basturk, N. (2005). Electrical and Structural Properties of Ex-situ Annealed Superconducting $Bi_2Sr_2CaCu_2O_8$ Thin Films Obtained by Coevaporation of Components. *Turk. J. Phys.*, Vol. 29, pp. 115-118

Cho, A. Y., Arthur, J.R. (1975). Molecular beam epitaxy. *Prog. Solid State Chem.*, Vol. 10, No. 3, pp. 157–191

Grigorashvily, Y.E., Volkov, S.I., Sotnikov, I.L. & Mingazin, V.T.(1999). Formation of thin Bi(Pb)-Ag-Sr-Ca-Cu-O films by ex situ magnetron sputtering. *Supercond. Sci. Technol.*, Vol. 12, No. 5, pp. 270-273

Grigorashvily, Y.E., Volkov, S.I., Sotnikov, I.L., Mingazin, V.T. (2000). On-orbit investigation of the electrical properties of HTSC film structures. *Physica C: Superconductivity*, Vol. 336, No. 1-2, pp. 19-26

Grigorashvily, Y.E., Ichkitidze, L.P. (2001). Transport and magnetic properties of $Bi_{1,7}Pb_{0,4}Sr_2Ca_2Cu_3O_x$ ceramic in magnetic field (≤5 mT), *10-th International Workshop on Critical Currents*, Gottingen, Germany June 4-7, 2001, pp. 308-309

Grigorashvili, Y.E., Ichkitidze, L.P., Mingazin, V.T. (2004). Magnetic resistance of films of HTSC material composed of $Bi_{1,7}Pb_{0,4}Sr_2Ca_2Cu_3O_x$. *High-Temperature Superconductors and Novel Inorganic Materials Engineering. MSU-HTSC YII. 7-th International Workshop*, Moscow, Russia, June 20-25, 2004., p. P9

Grigorashvili, Y.E. (2005). High temperature superconductors in nanoelectronic, In: *Nanotechnology in electronics*, Chaplygin, Yu.A., pp. 361-380, Technosfera, ISBN 5-94836-059-8, Moscow

Grigorashvily, Yu.E., Ichkitidze, L.P. & Volik, N.N. (2006). Magnetomodulation sensor of a weak magnetic field based on HTS $(Bi,Pb)_2Sr_2Ca_2Cu_3O_x$ ceramics. *Phyzica C: Superconductivity*, Vol. 435, No. 1-2, pp. 140-143

Grigorashvili Y.E., Bukhlin, A.V. & Veryuzhskii, I.V. (2010). Fabrication of Nanosized Superconductor Structures with a Critical Temperature Exceeding 100 K. *Semiconductors*, Vol. 44, No. 13, pp. 1714–1717

Ivanov, Z. et al. (1989). Superconducting (Bi,Pb)-Sr-Ca-Cu-O Thin Films Prepared in Situ by Laser Ablation. *Applied Physics Letters*, Vol. 55, No. 20, pp. 2123-2125

Jannah, A.N. (2009). Superconducting Properties of BSCCO Thin Films By Pulsed Laser Deposition. European Journal of Scientific Research, Vol. 29, No. 4, pp. 438-446, ISNN 1450-216X

Kimura, T., Nakao, H. et al. (1991). Bi-Sr-Ca-Cu-O Film on Sapphire Grown by Plasma-Enhanced Halide CVD. IEEE Transaction on Magnetics, Vol.27, No.2, pp.1211-1214

Kuroda, K., Kojima, K. et al. (1991). As-Grown Superconductivity of Bi(Pb)SrCaCuO Thin Films Formed by Consecutively Sputtering Three Pb-Doped Targets of Slightly Different Compositions. Supercond. Sci. Technol., Vol.4, No.1S, pp. 265-267, ISSN 0953-2048

Mongomery, D.C. (1980). *Planning of experiment and data analysis: translate from English*, Publishing house "Shipbuilding", Moscow

Pandey, et al., The Texas A & M University System (College Station, TX). (1994). Method for Forming Single Phase, Single Crystalline 2122 BCSCO Superconductor Thin Films by Liquid Phase Epitaxy, US. Pat. 5314869

Patil, J.M., Bhangale, A.R. et al. (1993). Fabrication of BSCCO Thin Films Using Thermal Evaporation Technique. *Chinese Journal of Physics*, Vol. 31, No. 6-11, pp. 1251-1254

Pilosyan, S.H. (1990). Bismuth containing films with Tc>100 K. Superconductivity: physics, chemistry, technics, Vol. 3, No. 4, pp. 689-692, ISNN 0131-5366

Schultz et al., Symyx Technologies (The Regents of the University of California). (2001). Combinatorial Synthesis of Novel Materials, US.Pat. 6326090

Silver, R.M., Ogawa, E.T., Pan. S. de Lozanne, A.L. (1991). In-situ formation of BSCCO thin films by plasma assisted thermalevaporation. *Magnetics, IEEE Transactions on*, Vol. 27, No 2, pp. 1215-1218

Steinbeck, J. (1989). Superconducting Thin Films of BiSrCaCuO Made by Sequential Electron beam Evaporation. *IEEE Transaction on Magnetics*, Vol. 25, No 2, pp. 2429-2432

Stejskal, J., Leitner, J.J. et al. (2000). Growth of BiSrCaCuO Thin Films by MOCVD. Journal of Crystal Growth, Vol.210, No.4, pp. 587-594, ISSN 0022-0248

Sunshine S., et al. (1988). Structure and physical properties of single crystals of the 84-K superconductor $Bi_{2.2}Sr_2Ca_{0.8}Cu_2O_{8+\delta}$. *Phys. Rev. B*, Vol. 38, No. 1, pp. 893-896

Varilci, A., Altunbas, M. et al. (2002). Production of BiPbSrCaCuO Thin Films and Ag/MgO Substrates by Electron Beam Deposition. *Physica Status Solidi (a)*, Vol. 194, No. 1, pp. 206-215

Volpyas, V.A. & Golman, E.K. (2000). Model of quasirigid spheres at modeling of processes of scattering of particles. Technical Physics, Vol. 70, No. 3, pp. 13-18

Volpyas, V.A. & Kozirev, A.B. (1997). *Physics of weakly ionized plasma*, TOO "Skladen", ISBN 5-89028-005-8, St.-Petersburg

Yoon, D.H., Agung, I., Saito, M., Yoshizawa, M. (1997). Growth of Bi-Sr-Ca-Cu-O Films by Molecular Beam Epitaxy. *Journal of the Korean Physical Society*, Vol. 31, No 3, pp. 48-51

Zakharov, N.D., Hoffschulz, H. et al. (1997). Preparation of BiSrCaCuO thin films by atomic layer-by-layer molecular-beam-epitaxy and high-resolution transmission electron microscopy analysis of the film/substrate interface and of growth defects. *Journal of Alloys and Compounds*, Vol. 251, No. 1-2, pp. 44-47

12

Properties of $YBa_2Cu_3O_{7-\delta}$ Superconducting Films on Sr_2YSbO_6 Buffer Layers

Omar Ortiz-Diaz[1], David A. Landinez Tellez[2] and Jairo Roa-Rojas[2]
[1]*Grupo de Fisica de Materiales, Universidad Pedagogica y Tecnologica de Colombia*
[2]*Grupo de Fisica de Nuevos Materiales, Universidad Nacional de Colombia*
Colombia

1. Introduction

In a previously published work we have shown that the Sr_2YSbO_6 material can be successfully used as a buffer layer for the epitaxial growth of $YBa_2Cu_3O_{7-\delta}$ films with high density current value in self-field at 77 K (Ortiz-Diaz et al., 2010). The layer of Sr_2YSbO_6 material plays the role of a buffer layer because the negative effects of MgO over the superconducting properties of $YBa_2Cu_3O_{7-\delta}$ films were eliminated, and because it evidences an excellent structural matching with $YBa_2Cu_3O_{7-\delta}$ and with MgO. These results show that Sr_2YSbO_6 can be an excellent substrate material for the $YBa_2Cu_3O_{7-\delta}$ layers in coated conductors, using the IBAD–MgO templates. Besides which, with the Sr_2YSbO_6 material the architecture of the coated conductors can be simplified, because only a single Sr_2YSbO_6 buffer layer is used.

The importance of this work is the fact that during the last decade, significant advances in the performance levels of high–temperature superconducting (HTS) wire have made it suitable for commercially viable applications such as electric power cables, fault current limiters, motors, and generators (Maguire&Yuan, 2009; Shiohara et al., 2009). For instance, both the United Sates Department of Energy and private industry have been developing a key superconductor cable and fault current limiter projects (Maguire&Yuan, 2009), and there is a five year Japanese national project for materials and power applications of coated conductors, which was started in 2008 (Shiohara et al., 2009). These power applications share a common requirement: that the superconducting material is formed into a long, strong, and flexible conductor so that it can be used like the copper wire it is intended to replace. And this is where the problems began, because the HTS materials are ceramics that are more like a piece of chalk than the ductile metal copper (Foltyn et al., 2007). The first solution to this problem, the so–called first generation wire, was a tape that was made packing Bi–Sr–Ca–Cu–O (BSCCO) superconducting powder into a silver tube, following a series of rolling and heating steps (Heine et al., 1989). In spite of successful applications, this type of conductor is expensive for most commercial applications due to the use of silver.

Further, BSCCO is not suitable for applications such as motors and magnets at liquid nitrogen temperature; it loses its ability to carry super current in a magnetic field (Foltyn et al., 2007; Heine et al., 1989). High critical current density in more successfully multifilamentary BSCCO wires, with magnetic field, happen at helium liquid temperatures (Goyal et al., 1996; Shen et al., 2010). The alternative approach, known as the second generation wire, uses the epitaxial growth of an $YBa_2Cu_3O_{7-\delta}$ superconducting coating on a thin metal tape. The advantages of this wire are that very little silver is needed, making it inexpensive, and that the compound $YBa_2Cu_3O_{7-\delta}$ retains much higher current carrying ability in a magnetic field.

Despite these advantages of superconductors, the ability to carry current without loss is limited to current densities lower than a critical value, *Jc*. In order to carry a higher current in a wire, the objective of research efforts is to increase *Jc*. In this context, the preparation of bi–axially textured substrates and subsequent epitaxial buffer layers is very important for the realization of long–length $YBa_2Cu_3O_{7-\delta}$ coated conductors. The buffer layer should not only satisfy chemical stability, but structural matching with $YBa_2Cu_3O_{7-\delta}$ as well because the alignment of the superconductor is required for high *Jc* values (Foltyn et al., 2007; Ying et al., 2009).

Different oxide materials have been successfully used as a buffer layer to fulfill these requirements (Foltyn et al., 2007; Huhne et al., 2006; Jia et al., 2002; 2003; Nishikawa et al., 2003; Parans et al., 2003; Sathyamurthy et al., 2003; Wang et al., 2005; Wee et al., 2005). However, most of them are really multilayer architectures, which significantly increase the complexity as well as the cost of production (Ying et al., 2009). Therefore, the development of a single buffer layer is of great interest, as this might simplify the preparation process and lead to a more cost–effective fabrication of coated conductors. To fabricate templates of great length, the most promising approach is, generally, with ion beam assisted deposition, IBAD yttria–stabilized zirconia (YSZ), $Gd_2Zr_2O_7$ or MgO (Arendt et al., 2004; Groves et al., 2002; Hanyu et al., 2007; Koa et al., 2007). Of these, the best is IBAD–MgO because very good biaxial texture can be obtained with films only 10 nm thick, which reduces the production costs (Arendt et al., 2004; Foltyn et al., 2007; Koa et al., 2007). Section 3 is extended a bit the issue of buffer layers and materials used in them.

Sr_2YSbO_6 appears to be a promising material for fulfill these requirements as a buffer layer, as it has a lattice parameter exhibiting a good lattice match with $YBa_2Cu_3O_{7-\delta}$ (mismatch between *a* and *b* $YBa_2Cu_3O_{7-\delta}$ parameters and *a* of Sr_2YSbO_6 is 5%). Previously we have showed that Sr_2YSbO_6 has been applied effectively as a buffer layer for Sr_2YSbO_6 film growth by DC sputtering (Ortiz-Diaz et al., 2010). This superconducting film has a *Jc* value 10^3 times higher than one grown on MgO. The Sr_2YSbO_6 films were deposited over MgO single–crystal substrate, because Sr_2YSbO_6 has a good match with MgO, which is the material of the IBAD–MgO tapes. Other applications for the Sr_2YSbO_6 material are in a Josephson junction because they are an insulating material for the deposition of superconductor films in microwave applications, and for the elaboration of crucibles for the preparation of superconductors due to their chemical non reactivity with $YBa_2Cu_3O_{7-\delta}$.

The Sr_2YSbO_6 material was chosen because we had been working to find new substrates for $YBa_2Cu_3O_{7-\delta}$ within the perovskite family $A_2BB'O_6$ since by means of substitutions they permit adjusting the lattice parameters (Ortiz-Diaz et al., 2004a). Preliminary studies of the material in polycrystalline form showed that it was viable as a substrate for the growth of $YBa_2Cu_3O_{7-\delta}$ films (Ortiz-Diaz et al., 2004b; 2010).

The goal of this chapter is to show a summary of these works, which were taken from a period of 2004–2010, showing the technology of manufacturing, since synthesis and evaluation of a material as a potential substrate for superconducting films, until the effective utilization as a substrate for films with excellent superconducting properties.

Section 2 begin with definition of elementary concepts of critical temperature, critical current density *Jc* and critical magnetic field. One of these topics concerns about the temperature effect on *Jc*, based on unpublished experimental results of curves current-voltage (*I-V*), which permit to calculate the critical current. Also is shown the relevancy of the critical parameters for applications of superconducting films and layers. Section 3 present a short overview of state of the art about the coated conductors (CC), relevance of the buffer layers for the increase of *Jc* in the second generation of superconducting wires. Special emphasis is made on importance the single buffer layers in order to simplify the fabrication process of CC. We are show the relevance of our Sr_2YSbO_6 perovskite for this goal. In section 4 We will describe details of the evaluation of the properties of Sr_2YSbO_6 polycristalline material, such as crystallographic matching and chemical stability between Sr_2YSbO_6 and $YBa_2Cu_3O_{7-\delta}$; also We show details of the fabrication of Sr_2YSbO_6 films on MgO single crystal by magnetron sputtering, besides the structural matching of $Sr_2YSbO_6$with MgO. Section 5 describes the fabrication of $YBa_2Cu_3O_{7-\delta}$ target and the growth proccess of films on MgO and $SrTiO_3$ (STO) single crystals and on the Sr_2YSbO_6 buffer layer. Also is shown the preparation of the conductor path, for electrical measurements, by means of photolitography, and show the results of critical temperature for all the films, and the measurements *I–V* based on which were determined the critical current density.

2. Superconductivity elementary notions and relevant properties for applications

A superconductor is the material that show two important properties: zero DC resistivity and magnetic induction zero inside the material when it is cooled below temperature *Tc*, known as critical temperature. For temperatures $T > Tc$, the superconductor it is in the normal state, and is like normal metal with poor conductivity (Cyrot & Pavuna, 1992; Pool et al., 1995). For a superconductor with a high *Tc* value, in the normal state, the resistance depends linearly on temperature, such as a typical conductor, as shown in figure 1(a). On the other hand, zero magnetic induction in the superconducting state means that the magnetization takes negative values below *Tc* and it is usually zero for temperatures above *Tc*, as shown in figure 1(b).

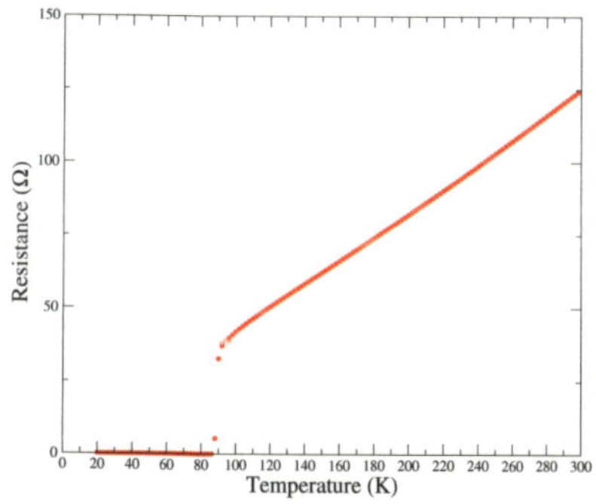

(a) Resistance.

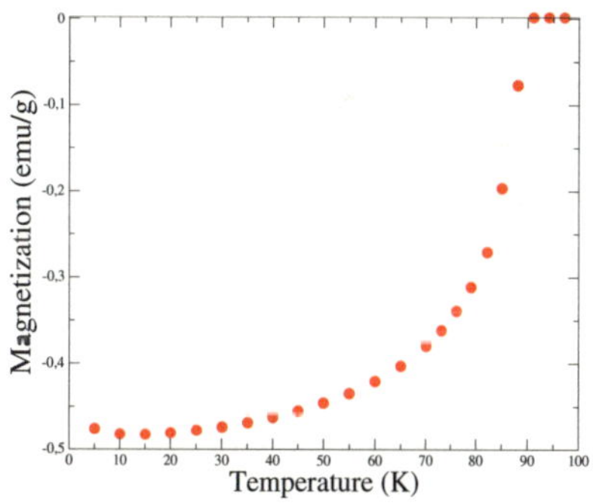

(b) Magnetization.

Fig. 1. Resistance and magnetization curves for a $YBa_2Cu_3O_{7-\delta}$ samples, in function of temperature.

The second property of the superconductors is called diamagnetism. When it is perfect a superconducting material does not permit an externally applied magnetic field to penetrate into its interior. However, superconductivity dissapears and the material returns to the normal state if one applies an external magnetic field of the strength greater than some critical value, B_c, called the critical thermodynamic field. It worth to say that we are calling magnetic field to the magnetic induction field $\boldsymbol{B}$. The superconducting state can also destroyed by passing an excessive current greater than some value I_c, called critical current, which creates a magnetic field at the surface. This limits the maximun current value that the material can sustain and it is a crucial problem for applications of superconducting materials. Those materials that totally exclude an applied magnetic flux, for $T < Tc$, are known as type I superconductors.

There are other superconductors, known as type II superconductors, that totally exclude magnetic field when the applied magnetic field is low, but only partially exclude it when the applied field is higher. For this superconductors there are tow critical fields: the lower B_{c1} and the upper B_{c2}. The flux is totally expelled only up to the lower critical field B_{c1}. Then, in applied fields smaller than B_{c1}, the type II superconductor behaves like a type I superconductor below B_c. Above B_{c1} the flux penetrate into the material until the upper critical field B_{c2} is reached. Above B_{c2} the material returns to the normal state.

Between B_{c1} and B_{c2}, the type II superconductor is said to be in the mixed state. The Meissner effect is only partial. For this range of fields magnetic flux partially penetrates the superconducting sample in the form tiny filaments known as vortices. A vortex consists of a normal core in which the magnetic field is large, surrounded by a superconducting region in which flows supercurrent.

Critical magnetic field and, obviously, critical current depends on temperature. Because of that it is important to report the temperature value at which the measures of critical current are carried out. Figure 2(a) shows a typical curve of voltage V in function on applied current I at temperature of 82 K for a $YBa_2Cu_3O_{7-\delta}$ film which was growth on Sr_2YSbO_6 buffer layer. Note that this temperature is higher than 77 K, usually reported for critical current measurements in high temperature superconductors. In these measurements one applies current I and takes a voltage value between two electrodes. In the so–called criterion of a microvolt/cm (μV/cm), the current value for which the voltage drop in a 1 cm length is 1μ V is defined as a critical current value. Figure 2(b) shows an expanded region of the figure 2(a), in which the red dashed line is a visual guide to determine the current value for which the drop voltage is 1μV. For this example, the critical current value is 4.14 mA, from which by dividing by the cross–sectional area of the sample, yields a relatively high value of $Jc \sim 0.2 \times 10^6$ A/cm^2.

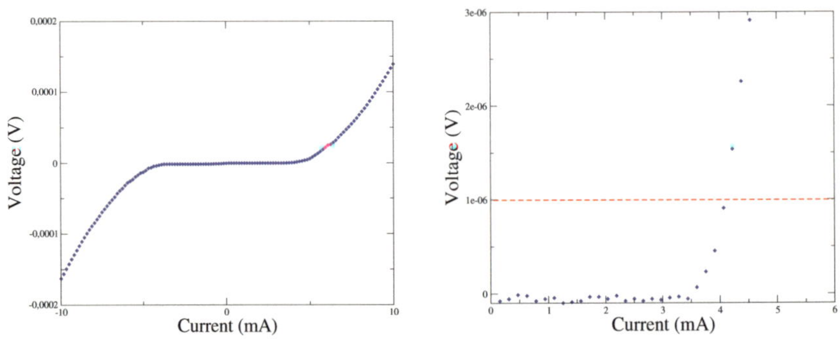

(a) Complete curve.

(b) Detailed section of curve.

Fig. 2. Voltage in function of current, at 82 K for a $YBa_2Cu_3O_{7-\delta}$ film growth on Sr_2YSbO_6/MgO substrate.

Figure 3(a) shows the voltage curves as a function of current for $YBa_2Cu_3O_{7-\delta}$ film on MgO at different temperatures from 5 K to 82 K and, Figure 3(b), the behavior of the density critical

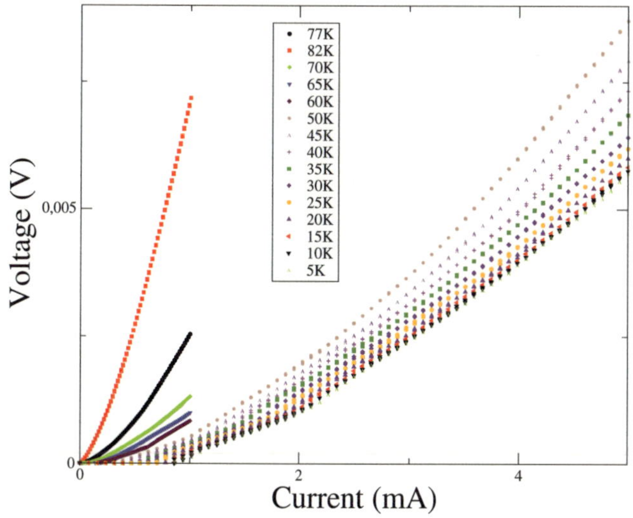

(a) *I–V* curves.

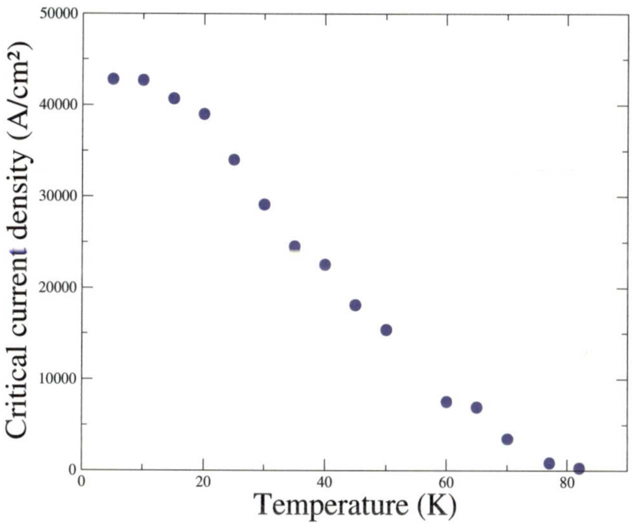

(b) Critical current.

Fig. 3. *I–V* curves for several temperatures and critical current in function of temperature for $YBa_2Cu_3O_{7-\delta}$ on MgO single crystal.

current as a function of temperature, calculated for each temperature in the same way as was done for the data in Figure 2. As the temperature increases, the current value for which the potential drop is $1\mu V$ decreases dramatically, so that at 82 K the curve approaches the linear behavior as a sign that is disappearing superconducting state. This is equivalent to the reduction of critical current density with increasing temperature as shown in Figure 3(b). In contrast, as shown in Figures 2, for this temperature of 82 K the superconductor film grown on Sr_2YSbO_6 buffer layer maintains a relatively high critical current value. From applications viewpoint, is important to determine the critical current value for at determined temperature. For HTS superconductors often the critical current is measured at the reference value of 77 K.

It worth noting that the issue of the temperature is crucial factor for the cost of practical uses of superconductors. For the called conventional superconductors is required the liquid helium coolant which have very higher cost. Nitrogen becomes liquid at 77 K; it is far less costly to liquefy this gas than to liquefy helium. For any application in which liquid nitrogen can replaced liquid helium, the refrigeration cost will be about 1000 times less (Sheahen, 2002). High-*Tc* oxides have critical temperatures above liquid nitrogen. For instance, the $YBa_2Cu_3O_{7-\delta}$ has been found to be superconducting up to $\sim$ 90 K; then liquid nitrogen is sufficient to cool $YBa_2Cu_3O_{7-\delta}$ in the superconductor range.

The maximum current density, *Jc*, that a superconductor can carry without exceeding a voltage drop criterion (1μ V/cm) depends both on the superconductor temperature and the applied magnetic field. From the practical point of view the critical temperature, *Tc*, is the temperature at which a very small *Jc* may flow at zero applied field. At a fixed temperature less than *Tc*, *Jc* will usually drop very rapidly above some threshold value of applied field. Thus, it is common to see a critical *B* plotted as a function of *T*, defined at some value of *Jc*. For an HTS, this critical magnetic field is called the irreversibility field, B_{irr}, which is the field at which flux lines or vortices to flow and causes dissipation. This value is often considerably lower than the B_{c2} of the material (Hull, 2003). Figure 4(a) shows B_{c2} in function of temperature for some common superconductor materials. For $YBa_2Cu_3O_{7-\delta}$ and BSCCO B_{c2} has high values at liquid nitrogen temperature. Figure 4(b) shows B_{irr} in function of temperature for some materials. The B_{irr} curve for BSCCO is noteworthy because it is relatively low at 77 K but starts to rise rapidly as temperatures drop below about 30 K. Thus, at temperatures near 77 K, applications for BSSCO will be limited to low magnetic field applications, such as transmission lines. Higher field applications, such as motors and generators, will require the expensive helium cooling of the BSSCO. This mean that the $YBa_2Cu_3O_{7-\delta}$ is an ideal candidate for these applications at liquid nitrogen temperature, which appears as a dashed line in figure 4(b). This is the main reason why the $YBa_2Cu_3O_{7-\delta}$ is the preferred material for applications involving high magnetic fields, and therefore, the reason why we decided to use $YBa_2Cu_3O_{7-\delta}$ and no other superconductor material for our study.

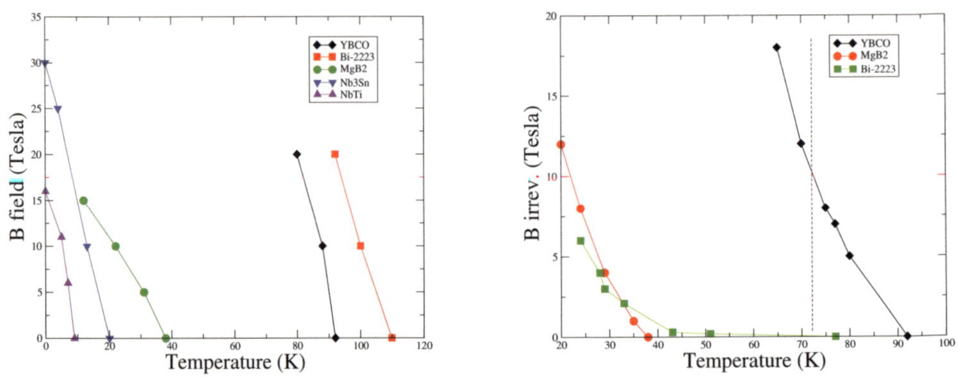

(a) Magnetic field. (b) Irreversivility field.

Fig. 4. Magnetic field in function of the temperature for some superconductors. Data adapted from (Hull, 2003).

Structural and transport properties of the HTS oxide materials is highly anisotropic. For this reason it is helpful, for the understanding of the properties of the $YBa_2Cu_3O_{7-\delta}$, a description of its structure. The unit crystallographic structure of $YBa_2Cu_3O_{7-\delta}$, shown in figure 5, can be seen as a stacking of three perovskites. So, crystallographers classify the structure of $YBa_2Cu_3O_{7-\delta}$ as the perovskite type with vacancies of oxygen; the oxygen contents it is in the range 6 to 7, as indicates the subscript $7-\delta$. The lattice parameter c is around three times larger than a and b parameters.

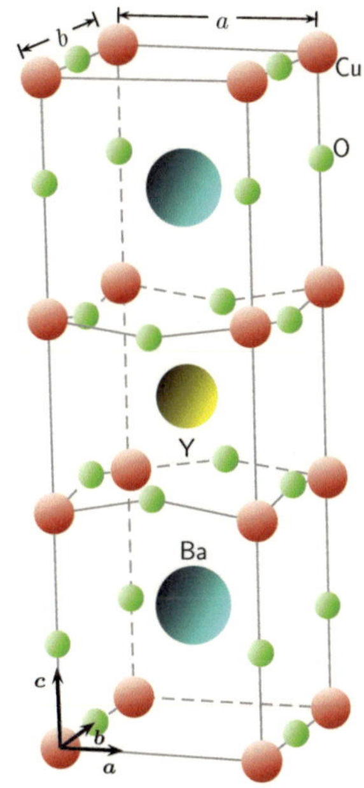

Fig. 5. Schematic representation of $YBa_2Cu_3O_{7-\delta}$ structure.

Thus, its structure consists of a sequence of oxide layers perpendicular to the c-axis separated by Y and Ba atoms, which can be seen in figure 6. Figure 6(a) corresponds to $\delta = 1$, that is, to $YBa_2Cu_3O_6$ and, figure 6(b) is the representation of $YBa_2Cu_3O_7$. The sequence of oxide layers are as follows:

- a Cu–O layer with two vacancies of oxygen for figure 6(b); each Cu is surrounded by 4 O. Thus the typical octahedral coordination has been replaced for a square planar coordination. So along the b-axis are formed the called chain copper oxygen connect by oxygen atoms. In the $YBa_2Cu_3O_6$ there is not chains because each copper ion has surrounded by two oxygen,
- a barium layer,
- a Cu_2O layer where each copper exhibit fivefold pyramidal coordination. The basis of pyramids linked by oxygen atoms, forms the called copper oxygen planes through the ab-plane,
- a layer of Y with for oxygen vacancies sandwiched between two Cu–O planes.

These sequence is duplicated in the upper half of the structure.

The tetragonal $YBa_2Cu_3O_6$ compound is an insulator. By increasing of oxygen concentration the ab-planes are gradually doped with holes and eventually it reaches the $YBa_2Cu_3O_7$ composition in which there are not vacancies of oxygen. The maximun $Tc = 94\,\mathrm{K}$ is obtained for $YBa_2Cu_3O_{6.93}$ and for $YBa_2Cu_3O_7$ the critical temperature is lower (Cyrot & Pavuna, 1992;

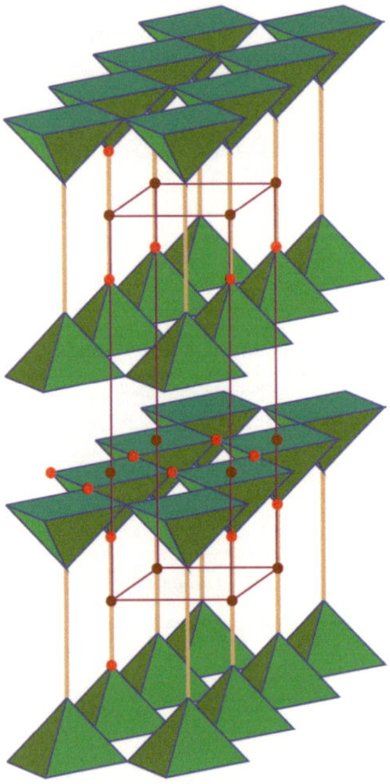

(a) Tetragonal $YBa_2Cu_3O_6$.

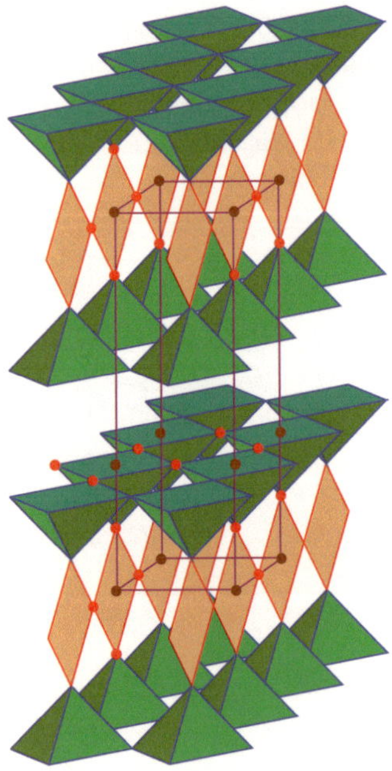

(b) Orthorombic $YBa_2Cu_3O_7$.

Fig. 6. Schematic representation of $YBa_2Cu_3O_{7-\delta}$ structures, shown the planes and chains.

Pool et al., 1995). Those compounds $YBa_2Cu_3O_{7-\delta}$ with optimally oxygen contents have orthorombic structure and are superconductors. They are called $YBa_2Cu_3O_{7-\delta}$ optimally doped.

Conductivity in *ab*-planes are around 100 times than *c*-axis conductivity. For this fact is accepted that superconductivity essentially takes place within quasi dimensional planes, and $YBa_2Cu_3O_{7-\delta}$ is considered as Cu–O planes separated by a charge reservoir (the chains).

As the conductivity, the Jc of the ceramic oxide materials is too highly anisotropic. For this reason a good alignment of the crystalline axes of all the grains is essential to pass high current density from one superconducting grain to another. For polycristalline specimens the Jc is very low, of order hundred A/cm^2. In contrast, higher Jc values, around $\sim 10^7$ A/cm^2, are obtained in single crystals and a good films.

To explain the low value of Jc in polycrystalline samples and understand how to improve these values by using the superconductor in the form of films, it is worth noting, as mentioned in section 4 that for the production of a polycrystalline sample in the form of pellet, one mix a precursor components, which are subjected to heat treatment and appropriate oxygenation for optimal superconducting properties. However, even the best polycrystalline samples consist of many randomly oriented polycrystalline grains, each of them with their anisotropic layered structure. It also contains grain boundaries, structural defects such as twins, voids, which act as non conducting grains. All this conspires against applications that require high critical current densities. In contrast, with single crystals and epitaxially grown films are obtained large values of Jc at 77 K and self field, because superconductivity occurs through the *ab* planes.

To produce films is necessary to use a substrate, which must satisfies basic requirements to promote epitaxial growth and micro structural properties to obtain high current density values: must have good matching of lattice constants with superconductor; must be chemically non reactant with $YBa_2Cu_3O_{7-\delta}$ even under heat treatment at temperature of film deposition and; should not affect the values of critical temperature and critical current density of the superconductor. It worth to say that the recent analyses, about current limiting defects in coated conductors, have shown that the major local current–limiting defects are *a*–axis grains, which exert a big effect on current flow and local dissipation in the coated conductor samples studied (Li et al., 2012). By reducing the amount of grains grown along the *a* axis, the quality of the superconducting layers substantially improves. To achieve this purpose, besides optimizing the deposition conditions, also, the substrate plays a crucial role.

The crystallography and preparation of the substrate is of primary importance in determining the quality of the film deposition. Some of the highest critical current densities at 77 K and highest film uniformities have been achieved in $YBa_2Cu_3O_{7-\delta}$ films grown on commercially available (1 0 0) $SrTiO_3$ substrates (Huhne et al., 2006; Jia et al., 2002; 2003; Ortiz-Diaz et al., 2004a; Sathyamurthy et al., 2003; Varanasi et al., 2008; Wang et al., 2005; Wee et al., 2005; Wu et al., 2005; Ying et al., 2009). However, $SrTiO_3$ is fairly expensive, has high dielectric constant, and undergoes phase transition at $\sim$ 110 K. $LaGaO_3$ with $\sim$ 0.5% mismatch it is neither cheap nor twin free (Cyrot & Pavuna, 1992). Other commercial substrate, $LaAlO_3$ was reported as a problematic substrate for the deposition of thin $YBa_2Cu_3O_{7-\delta}$ films, especially when a thick substrate is required, because detrimental extended defects are developed. The use of $LaAlO_3$ layers of 0.5 mm is advantageous, provided the deposition temperature is kept as low as possible (Koren & Polturak, 2002).

MgO single crystal substrates have been widely used for its economy and acceptable matching of lattice constants. However, it is widely accepted that MgO affects the superconducting properties. In particular, *Tc* and *Jc* values for $YBa_2Cu_3O_{7-\delta}$ and $REBa_2Cu_3O_{7-\delta}$ films are generally lower when are deposited on MgO than the values for films grown on other commercial substrates (Hollmannt et al., 1994; Wee et al., 2005). This has been attributed to chemical reaction between the superconductor and the MgO, which causes the formation of third phases, which act as "non-superconducting dead layers" and; grains (0 0 1) growth rotated 45° with respect to MgO (1 0 0) surface.

Sapphire is another cheap substrate, free twining. However, has relatively high mismatch with $YBa_2Cu_3O_{7-\delta}$ and reacts chemically with superconductor, thus, the interfacial $BaAl_2O_4$ layer is formed, which affects the epitaxial growth of the film and result in poor superconducting films (Hollmannt et al., 1994).

The list of materials tested as substrates for superconducting films is broad and includes a variety of compounds and crystal structures, ranging from oxides and simple perovskites to complex perovskites and pyrochlore structures. For instance, Yttrium Stabilized Zirconium Oxide (YSZ) and cerium oxide, CeO_2 have been widely used, sometimes as a combination of layers. Among simple perovskites that have been proposed, perhaps the most successful are $SrTiO_3$, $LaAlO_3$ and $SrRuO_3$, which are commercially availables as single crystals. Examples of other simple perovskites proposed as substrate materials are $YAlO_3$, $GdAlO_3$, $EuAlO_3$, $SmAlO_3$, $LaGaO_3$, $NdGaO_3$, $PrGaO_3$. Also, were proposed to make substitutions with the idea that variation in the composition allows the lattice parameters vary, and then the matching can be improved, but question arises about the chemical compatibility with $YBa_2Cu_3O_{7-\delta}$ (Hollmannt et al., 1994); thus, are tested as potential substrates, for

$YBa_2Cu_3O_{7-\delta}$ applications in microwaves and electronics, many complex perovskites with formula $A_2BB'O_6$, besides, some pyrochlore structures. Some examples of complex perovskites proposed are Sr_2AlTaO_6 for superconducting devices (Findikoglu et al., 1992; Ying & Hilbert, 1994); Ca_2GaNbO_6; Sr_2GaTaO_6 (Brandle, 1996); Ba_2LaNbO_6 (Pai et al., 1997); $DyBa_2SnO_{5.5}$ (Koshy et al., 1995); $DyBa_2ZrO_{5.5}$ (Yadava et al, 1998); $HoBa_2SbO_6$ (Aguiar et al., 1998) and Ba_2NdTaO_6 (Kurian et al., 2002; Kurian & Morishita, 2003).

As can to see, the search for new substrates for the deposition of superconducting films is continuous, since it is hard to find the substrate to produce the perfect film.

3. Buffer layers for superconducting films

As was say in previous section, there are many potential substrates for $YBa_2Cu_3O_{7-\delta}$ films deposition, but only a few commercially available in single crystal form. Also, each substrate has a problem or disadvantage to the film that has to grow on. To solve this problem without having to wait for the manufacture of new mono–crystalline substrates, some groups began to use these materials in the form of film on any of the commercial MgO and $SrTiO_3$ substrates for growth of superconducting films. The idea was to eliminate the negative effect of single–crystal substrate by using a buffer layer that acts as a chemical barrier that eliminates the reaction between the crystal and the superconductor. In addition, the buffer layer plays the role of template to facilitate the epitaxial growth of superconducting film with improved superconducting and microstructural properties. RF sputtered and laser ablated films of $YBa_2Cu_3O_{7-\delta}$ on a MgO single crystal substrate using a buffer layer of $SrTiO_3$ began to be manufactured in the 1990's.

In 1993 $SrTiO_3$ buffer layers were grown on MgO (1 0 0) substrates to provide a better match to RF sputtered $YBa_2Cu_3O_{7-\delta}$ films. This heterostructure allows a highly textured growth to be achieved over thickness as high as 1 μm. The improvement of lattice matching makes the critical current density increase from 10^3 A/cm^2 for 1 μm films grown to 4×10^5 A/cm^2 for films grown with a $SrTiO_3$ buffer layer (Lucia et al., 1993).

Laser ablated thin films of $YBa_2Cu_3O_{7-\delta}$ on a MgO substrate using a $SrTiO_3$ buffer layer was grown, which is perfectly oriented with respect to the MgO substrate. Superconductivity is improved in a spectacular manner with respect to $YBa_2Cu_3O_{7-\delta}$ directly deposited on MgO. A critical temperature $Tc = 92$ K and a critical current density of 4×10^6 A/cm^2 at 82 K are reached for the first time for films deposited on MgO substrate (Proteau et al., 1995).

In 1996 were grown epitaxial $YBa_2Cu_3O_{7-\delta}$ films by pulsed laser ablation on $SrTiO_3$ buffered (1 0 0) MgO and was found that the $SrTiO_3$ buffer layer provide a better lattice match to the $YBa_2Cu_3O_{7-\delta}$ film and play a crucial role to prevent the interaction between $YBa_2Cu_3O_{7-\delta}$ and MgO. Thus, were obtained $YBa_2Cu_3O_{7-\delta}$ films with $Jc \sim 10^6$ A/cm^2 at 77 K and zero magnetic field (Boffa et al., 1996).

A comparative study of $NdBa_2Cu_3O_{7-\delta}$ films deposited, by laser ablation, on different substrates showed that the transport Jc value of 3.5×10^6 A/cm^2 at 77 K in self field was obtained from the $NdBa_2Cu_3O_{7-\delta}$ on $SrTiO_3$ with Tc value of 91.2 K, while the $NdBa_2Cu_3O_{7-\delta}$ on MgO with Tc of 86.8 K exhibited the low Jc value of 0.25×10^6 A/cm^2. On the other hand, Jc value of 1.55×10^6 A/cm^2 was obtained from $NdBa_2Cu_3O_{7-\delta}$ film on $SrTiO_3$ buffered MgO(1 0 0) substrate with Tc of 91.5 K. The low Jc value of the $NdBa_2Cu_3O_{7-\delta}$ on MgO can be attributed to both depressed Tc value and the existence of 45° rotated grains with (0 0 1)$NdBa_2Cu_3O_{7-\delta}$(1 1 0)MgO (Wee et al., 2005). $NdBa_2Cu_3O_{7-\delta}$ film properties were improved too using a buffer layer of Ba_2NdTaO_6 on MgO (Kurian & Morishita, 2003).

The above example is a sample of the use of compounds different to the commercial substrates as a material for buffer layers. Also, with the emergence of the second generation of superconducting wires took great importance to study and search for new buffer layers. For instance, Biaxially textured $La_2Zr_2O_7$ with CeO_2 and $Ce_{0.7}Gd_{0.3}O_3$ buffer layers were grown on Ni–RABiTS (rolling-assisted biaxially textured substrates) to obtain suitable buffer layer architectures for $YBa_2Cu_3O_{7-\delta}$ coated conductors (Knoth et al., 2005). More recently, it has been used as textured substrate for coated conductors a NiW alloy with a $Gd_2Zr_2O_7$ buffered $Ce_{0.9}La_{0.1}O_{2-y}$ for the growth of $YBa_2Cu_3O_{7-\delta}$ films with $Jc \sim 1 \times 10^6$ A/cm^2 (Zhao et al., 2012). This work is an example of current interest in improving the manufacturing process of superconducting tapes with different combinations of deposition technique and combinations of buffer layers.

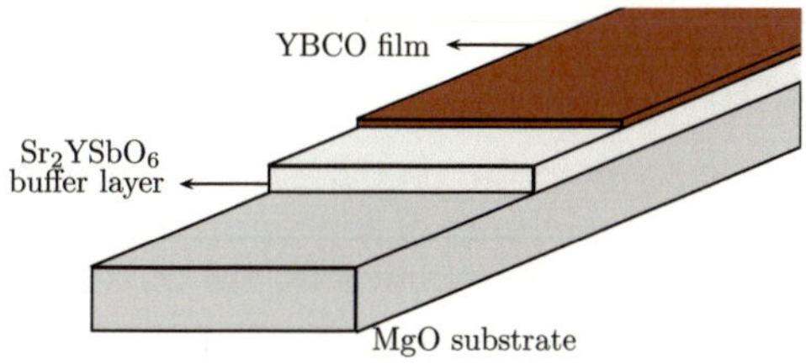

Fig. 7. Schematic representation of the $YBa_2Cu_3O_{7-\delta}$ film on Sr_2YSbO_6 buffer layer.

Film deposition	Buffer layer	Jc ($\times 10^6$ A/cm^2)	Reference
Sputtering RF	$SrTiO_3$	0.4	(Lucia et al., 1993)
PLD	$Sr_2AlTaO_6/LaAlO_3$	1.3	(Findikoglu et al., 1992)
PLD	CeO_2/IBAD–YSZ	2.2	(Li et al., 2012)
PLD	SrO/Sr_2AlTaO_6	~ 0.5	(Takahashi et al., 2003)
PLD	Ba_2LaNbO_6	5×10^6	(Pai et al., 1997)
PLD	$SrTiO_3$	4	(Proteau et al., 1995)
PLD	$Ce_{0.9}La_{0.1}O_{2-y}/Gd_2Zr_2O_7$	~ 1	(Zhao et al., 2012)
Sputtering RF	Sr_2YSbO_6/MgO	0.86	(Ortiz-Diaz et al., 2010)

Table 1. Values of critical current density for $YBa_2Cu_3O_{7-\delta}$ films on some buffer layers at 77 K and self–field. PLD means Pulsed Laser Deposition.

We have mentioned a few examples of materials used for buffer layers; a more extensive list can be consulted by the interested reader in an excellent review of coated conductors (Foltyn et al., 2007). With all this it can see two things: first, the problem of the epitaxial growth appears overcomed and; second, the optimization and reduction of cost of deposition can be obtained with the uses of buffer layer sandwiched between the single substrate and superconducting film, as is shown in figure 7 (Foltyn et al., 2007; Hanyu et al., 2007; Nishikawa et al., 2003; Ortiz-Diaz et al., 2010; Parans et al., 2003).

Most promising applications of type II superconductors are in the power area, where the most advantageous superconductor material is $YBa_2Cu_3O_{7-\delta}$, because high critical current densities needed for this applications can be achieved with it, with acceptably low energy dissipation, even with magnetic fields, as was shown in section 2. These applications require that the superconductor be formed as a long coated conductor tape so that it can be used like the copper wires. The architecture of the superconducting tapes is like the buffered $YBa_2Cu_3O_{7-\delta}$ films showing in figure 7, but with additional protective layer of gold or silver and with the single substrate replaced for a flexible tape of commercially alloy textured with an oxide layer. The most promising technology for the manufacture of coated conductor tape is based on Ion Beam Assisted Deposition (IBAD) (Arendt et al., 2004; Foltyn et al., 2007), where MgO and $Gd_2Zr_2O_7$ are successfully used for texturing. Intermediate oxide layers are epitaxially grown on this textured substrate. The architecture of the tapes can be simplified and, then, the cost of manufacture can be reduced if these intermediate layers are replaced with a single buffer layer. It is the principal motivation for the evaluation of new materials, such as Sr_2YSbO_6. Also, the evaluation is made using the MgO single substrate, because the MgO is one of the oxides successfully used in IBAD technology. Generally, the route followed to evaluate a material as a possible constituent of the buffer layer in a superconducting tape begins with tests on a small sample of single crystalline substrate. That is why in our work the films are grown on single crystal, instead of on a textured tape directly.

4. Evaluation of Sr_2YSbO_6, step by step

The evaluation of Sr_2YSbO_6 stars with the preparation, which was made by the solid state reaction method. Stoichiometric mixtures of high purity (99.99%) commercial precursor oxides Y_2O_3, SrO and Sb_2O_3 in adequate amounts are mixed thoroughly, pelletized and calcined at 1100°C for 18 h. The calcined material was reground, pressed as circular discs and sintered at 1090°C for 135 h. All the above processing was carried out in ambient atmosphere.

For the compatibility studies, single phase $YBa_2Cu_3O_{7-\delta}$ superconducting material was prepared by the solid state reaction method. High purity (99.99%) constituent commercial chemicals Y_2O_3, $BaCO_3$, and CuO were mixed in stoichiometric ratio. The mixed powder was finely ground and calcined at 900°C for 24 h at ambient atmosphere. The calcined material was reground, pressed as circular discs at a pressure of 1.6 ton/cm^2. The pellets were sintered at 930°C for 24 h, followed by slow cooling up to 500°C for 13 h and annealing at this temperature for 24 h at O_2 atmosphere. Samples were finally furnace cooled to room temperature over a span of 12 h.

For the study of the structural characteristics of the materials, X Ray Difraction (XRD) patterns of the samples were recorded by a Siemens D5000 X Ray diffractometer, using Cu-Kα radiation (λ = 1.5406 Å) and studied by Rietveld method with the programs EXPGUI and GSAS (Larson & Dreele, 2000; Toby, 2001).

Powder XRD pattern of Sr_2YSbO_6 is shown in figure 8. It consists of strong peaks which are characteristics of a primitive cubic perovskite plus a few weak line reflections arising from the superlattice. Thus, the whole XRD pattern of Sr_2YSbO_6 can be indexed in a $A_2BB'O_6$ cubic cell with the cell parameter a = 8.2561 Å.

Taking into account the doubling of the basic perovskite lattice parameter, the lattice constant a = 8.2561 Å ($a/2$ = 4.128 Å) of Sr_2YSbO_6 is comparable to lattice constants a and b of $YBa_2Cu_3O_{7-\delta}$. Thus, Sr_2YSbO_6 has a lattice parameter a which presents a good match with the lattices parameters a and b of $YBa_2Cu_3O_{7-\delta}$ superconductors. This first result was crucial for work purposes. First, the Sr_2YSbO_6 show acceptable fit with $YBa_2Cu_3O_{7-\delta}$, one of the basic requirements of a substrate. On the other hand, the matching is also suitable with MgO, which would ensure the epitaxial growth of Sr_2YSbO_6 films.

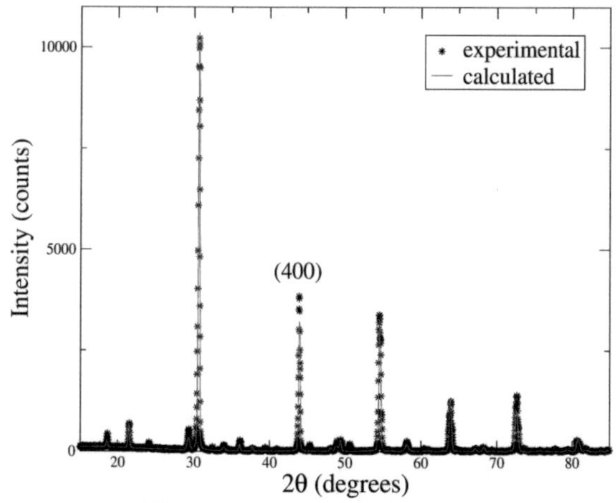

Fig. 8. XRD pattern of Sr_2YSbO_6 sample after the Rietveld refinement.

For the study of chemical and physical compatibility, we have synthesized several composites of Sr_2YSbO_6-$YBa_2Cu_3O_{7-\delta}$ with 15 to 90 vol% of Sr_2YSbO_6 component in the composite. For synthesis of each composite, the component materials were mixed in desired vol% ratios and mixture was pelletized as circular discs at a pressure of 1.6 ton/cm^2. These pellets were heat treated at 900°C for 10 hour in oxygen and cooled down slowly at a rate of 0.5°C/min for proper oxygenation. Chemical stability of Sr_2YSbO_6 with $YBa_2Cu_3O_{7-\delta}$ was examined by X–ray diffractometry of Sr_2YSbO_6-$YBa_2Cu_3O_{7-\delta}$ composites. XRD patterns of Sr_2YSbO_6-$YBa_2Cu_3O_{7-\delta}$ composites are shown in figure 9. As seen from these XRD patterns, all the XRD peaks could be indexed for either Sr_2YSbO_6 or $YBa_2Cu_3O_{7-\delta}$ and there is no extra peak corresponding to impurity phase. Within the accuracy of the XRD technique, these results show that there is no chemical interaction between these materials and Sr_2YSbO_6 is chemically compatible with $YBa_2Cu_3O_{7-\delta}$ superconductors.

The effect of Sr_2YSbO_6 addition on the superconductivity of $YBa_2Cu_3O_{7-\delta}$ superconductors was investigated by measuring dc magnetization of Sr_2YSbO_6-$YBa_2Cu_3O_{7-\delta}$ composites in the temperature range 5–100 K using a Quantum Design SQUID magnetometer. Figure 10 shows the magnetization for Sr_2YSbO_6-$YBa_2Cu_3O_{7-\delta}$ composites for 0, 45 and 90 vol% of Sr_2YSbO_6. As shown from these figures all the Sr_2YSbO_6-$YBa_2Cu_3O_{7-\delta}$ composites gave a superconducting transition temperature $Tc \sim 90$ K as that of the pure $YBa_2Cu_3O_{7-\delta}$ superconductor. A saturated diamagnetic transition is clearly observed in every sample at temperatures well bellow Tc. However, with decreasing $YBa_2Cu_3O_{7-\delta}$ superconductor volume fraction the magnitude of magnetization decreases in all Sr_2YSbO_6-$YBa_2Cu_3O_{7-\delta}$ composite samples.

These remarkable results guarantee that indeed the material does not react chemically with $YBa_2Cu_3O_{7-\delta}$, in spite of the severe heat treatment made to the composites at temperatures above the deposition of $YBa_2Cu_3O_{7-\delta}$ films. For this reason the Sr_2YSbO_6 material can be proposed as a potential substrate for the growth of superconducting films.

It worth noting that the Sr_2YSbO_6 material has good structural matching with MgO. Thus, next step was the epitaxial growth of Sr_2YSbO_6 films on (1 0 0) MgO commercial single crystal substrate, which were performed by magnetron sputtering (13.56 MHz, 70 watt) using a polycrystalline target, which was fabricated, like the firsts samples, by the solid state reaction

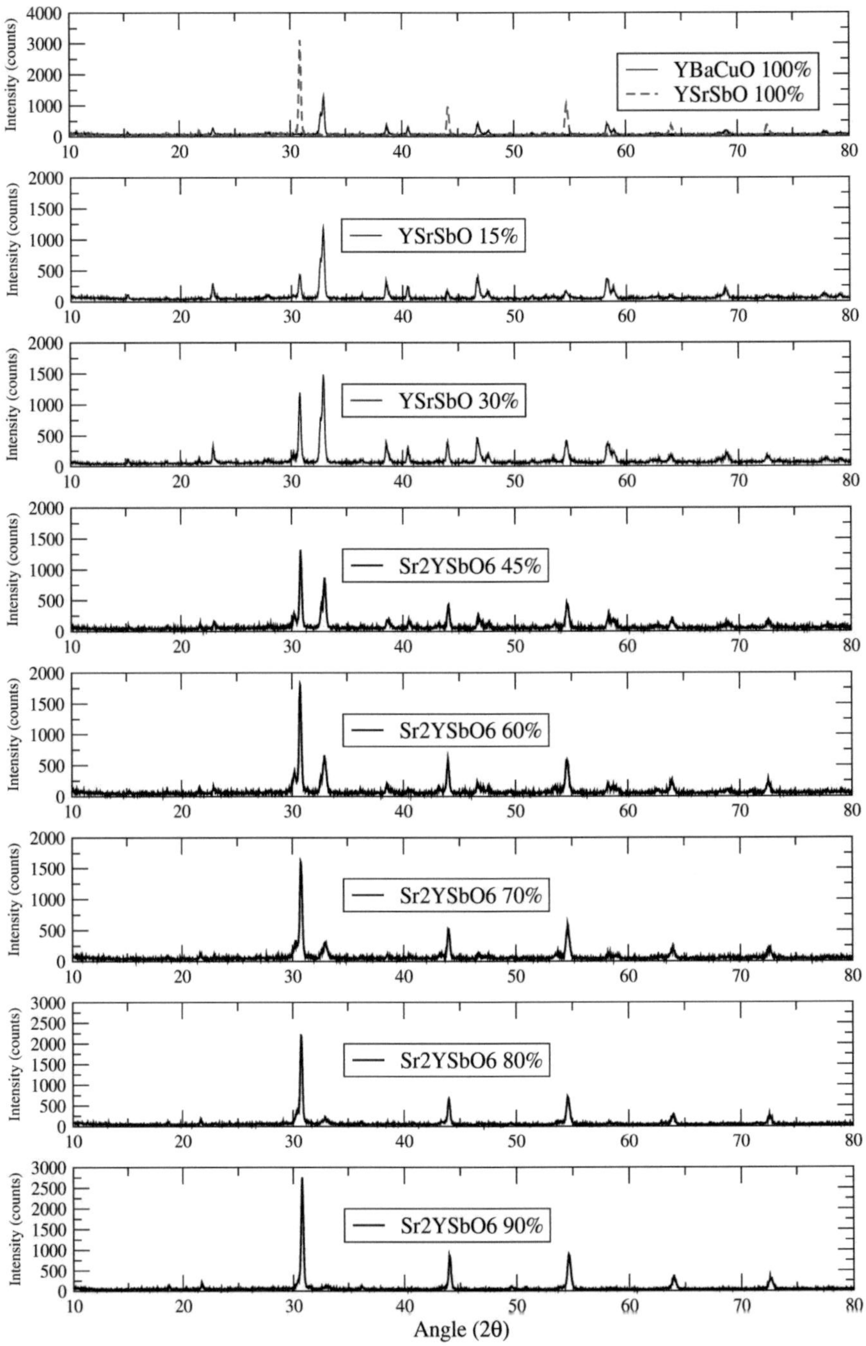

Fig. 9. XRD patterns for several composites Sr_2YSbO_6–$YBa_2Cu_3O_{7-\delta}$. First plot show, as a reference, XRD patterns of single phases of $YBa_2Cu_3O_{7-\delta}$ and of Sr_2YSbO_6.

method, based on SrO, Sb_2O_3 and Y_2O_3, powder oxides. The substrate temperature and oxygen pressure for the Sr_2YSbO_6 growth were kept at 800°C and 7×10^{-3} mbar, respectively.

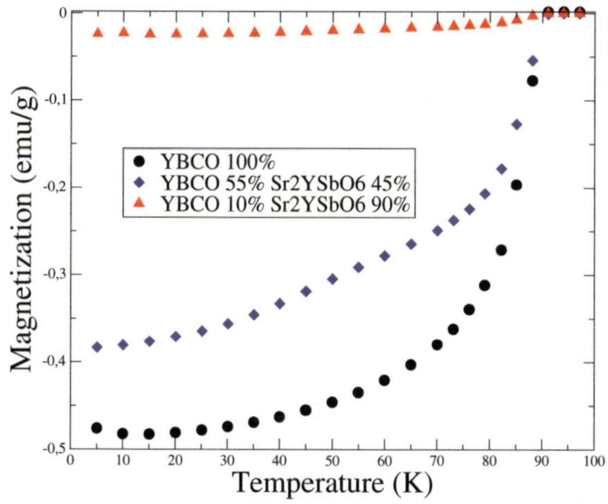

Fig. 10. Magnetization in function of temperature for composites Sr_2YSbO_6–$YBa_2Cu_3O_{7-\delta}$.

The X–ray diffraction pattern for films, were recorded by a PHILLIPS PW1710 diffractometer using Cu–Kα radiation (λ = 1.5406 Å). Figures 11 show XRD pattern for 2θ between 10 and 80 degrees, and a short detailed scan for 2θ between 41 and 46 degrees. Pattern of figure 11(a) consists of strong peaks (2 0 0) of MgO and (4 0 0) of Sr_2YSbO_6. Figure 11(b) shows the MgO peak in $2\theta = 43°$ and the Sr_2YSbO_6 peak in $2\theta = 43.1°$. This result reveals the successfully epitaxial growth of Sr_2YSbO_6 films on MgO (1 0 0) substrate.

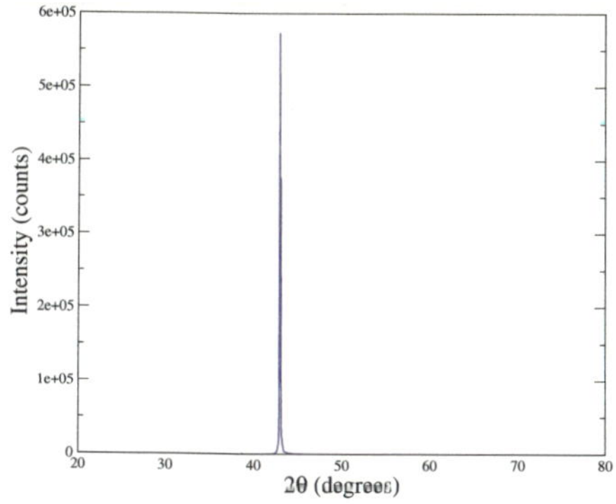

(a) XRD for $10° < 2\theta < 90°$.

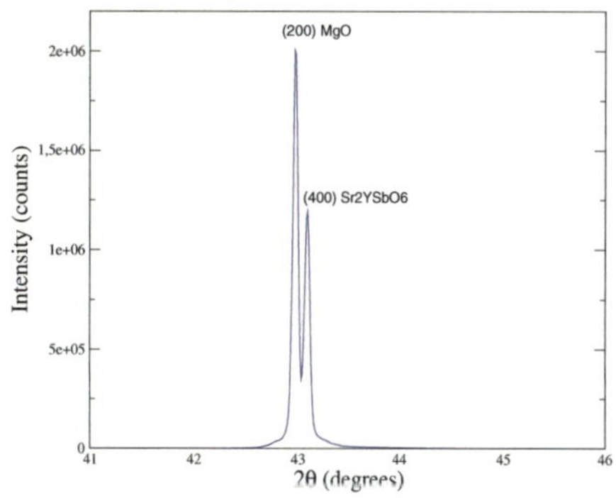

(b) Short XRD.

Fig. 11. XRD patterns for Sr_2YSbO_6 film growth by magnetron sputtering on MgO.

In conclusion, in this section we have showed the preliminary study of structural characteristics of a complex ordered perovskite Sr_2YSbO_6 for its use as substrate material for the fabrication of $YBa_2Cu_3O_{7-\delta}$. Sr_2YSbO_6 has a fairly good lattice match (lattice mismatch $\sim$ 8%) with this superconductor. X-ray diffractometry and magnetic measurements made on Sr_2YSbO_6-$YBa_2Cu_3O_{7-\delta}$ composites show that Sr_2YSbO_6 is chemically and physically compatible with $YBa_2Cu_3O_{7-\delta}$ material, even after a severe heat treatment at 900°C, processing temperature of $YBa_2Cu_3O_{7-\delta}$. These favorable characteristics of Sr_2YSbO_6 show that it can be used as a buffer layer for deposition of superconductor $YBa_2Cu_3O_{7-\delta}$ films using MgO single substrate.

5. Sr_2YSbO_6 as a successfully buffer layer for $YBa_2Cu_3O_{7-\delta}$ films

In the previous section We show that Sr_2YSbO_6 satisfies the requirements for $YBa_2Cu_3O_{7-\delta}$ films deposition. For this reason, was made the growth and characterization of $YBa_2Cu_3O_{7-\delta}$ film on buffered layer of Sr_2YSbO_6. This film, and additional films on commercial substrates of MgO and STO, for comparative studies, were carried by sputtering DC ($\sim$ 30 watt) at an optimized substrate temperature of 850°C an O_2 pressure of 3.5 mbar for 1 hour, followed by cooling up to 550°C in 30 min at O_2 pressure of $\sim$ 850 mbar and therefore were annealed at 550°C for 30 min at the same O_2 pressure.

Figures 12 show the XRD pattern of $YBa_2Cu_3O_{7-\delta}$ film over Sr_2YSbO_6 buffer layer, for 2θ between 10 and 90 degrees. It consists of peaks $(0\,0\,l)$ of $YBa_2Cu_3O_{7-\delta}$, besides the MgO and Sr_2YSbO_6 peaks, such as is detailed in figure 12(b). These result reveals the epitaxial growth of $YBa_2Cu_3O_{7-\delta}$ over Sr_2YSbO_6/MgO buffered substrate.

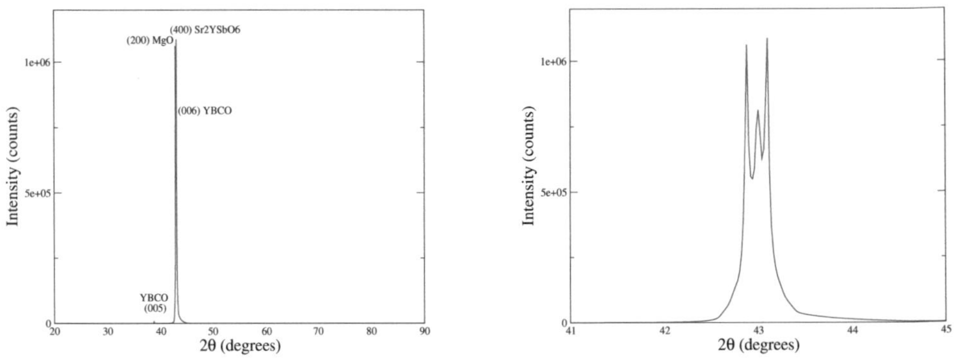

(a) XRD for $10° < 2\theta < 90°$. (b) Short XRD.

Fig. 12. XRD patterns for $YBa_2Cu_3O_{7-\delta}$ film growth on Sr_2YSbO_6 buffer layer.

The superconducting properties of $YBa_2Cu_3O_{7-\delta}$ films were characterized by measurements of the transition temperature (Tc) and critical current density (Jc) at 77 K in self field, by means of ACT measurements (bias AC current of 30 Hz) with four probe method, using the PPMS system of Quantum Design. These measurements were performed on $YBa_2Cu_3O_{7-\delta}$ micro bridges, with 20 μm of width and 100 nm of thickness, which were prepared by UV photo lithography.

For the photolithography, the films were coated with a layer of photolack. Then, the coated surface was put over a mask and was irradiated for 12 min with UV radiation. Irradiated film was submerged and moved into mix of H_2O and NaOH, until we can saw a marks of mask. Then, the film was pasted to another recipient with 60 drops of H_2O and five drops of phosphoric acid, and move until the $YBa_2Cu_3O_{7-\delta}$ superconductor paths that were used for electrical measurements were clears. Finally, film was retired and cleaned. Figures 13 show a $YBa_2Cu_3O_{7-\delta}$ film after the photolithography and detailed microscopic image of the micro brigde. Films of $YBa_2Cu_3O_{7-\delta}$ were contacted by means of indium leads such as is show in figure 14, by measurements both resistance in function of temperature and voltage V in function of current I.

Figures 15 to 17 show the behavior of $YBa_2Cu_3O_{7-\delta}$ films resistance as a function of temperature. For films growth over Sr_2YSbO_6 the curve exhibits linear behavior up to a transition temperature Tc. In figure 16 the measurements corresponding to a $YBa_2Cu_3O_{7-\delta}$

(a) $YBa_2Cu_3O_{7-\delta}$ film after photolithography.

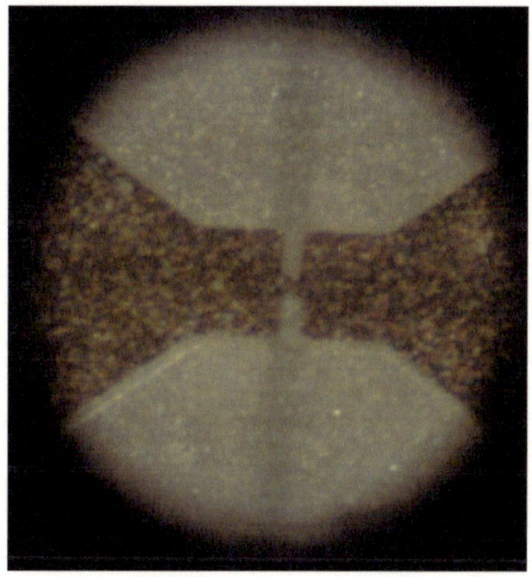

(b) $YBa_2Cu_3O_{7-\delta}$ micro bridge.

Fig. 13. $YBa_2Cu_3O_{7-\delta}$ film after the photholitographic process for preparation of micro bridge.

Fig. 14. $YBa_2Cu_3O_{7-\delta}$ with contacts for superconducting measurements.

films growth over MgO and on $SrTiO_3$, with the same conditions, are shown. Figure 17 is shown for comparison of the resistance behavior.

It worth to say about the ways of defining the sharpness and superconducting transition temperature. There is two criteria widely used. Some authors talk in terms on the onset, 5%, 10%, midpoint, 90%, 95%, and zero resistance points (Pool et al., 1995). The onset, or 0% point is where the experimental curve begins to drop below the extrapolated linear behavior. With this criterion, the Tc value is midpoint at which the resistivity has decreases by 50% below onset. The point at which the first derivative of the resistive transition curve reaches its maximum value could be selected as defining Tc, since is the point of most rapid change from the normal to superconducting phase. Also, the width ΔT between the half-amplitude points of the first derivative curve is good quantitative measure of the width of the transition.

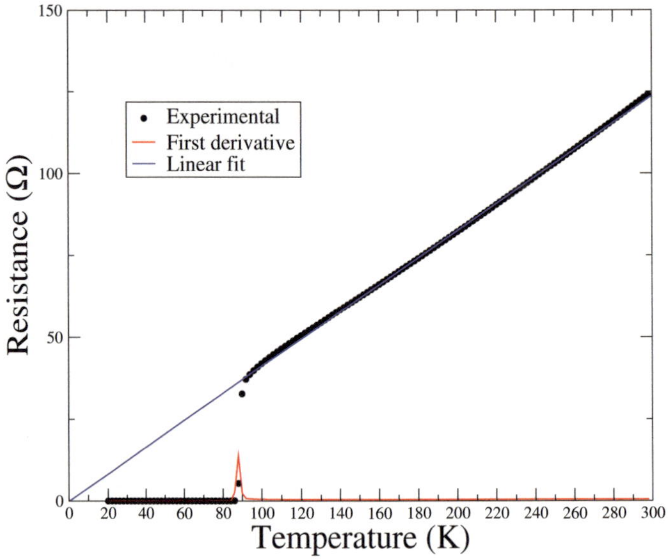

Fig. 15. Resistance in function of temperature for $YBa_2Cu_3O_{7-\delta}$ film on Sr_2YSbO_6 buffer layer.

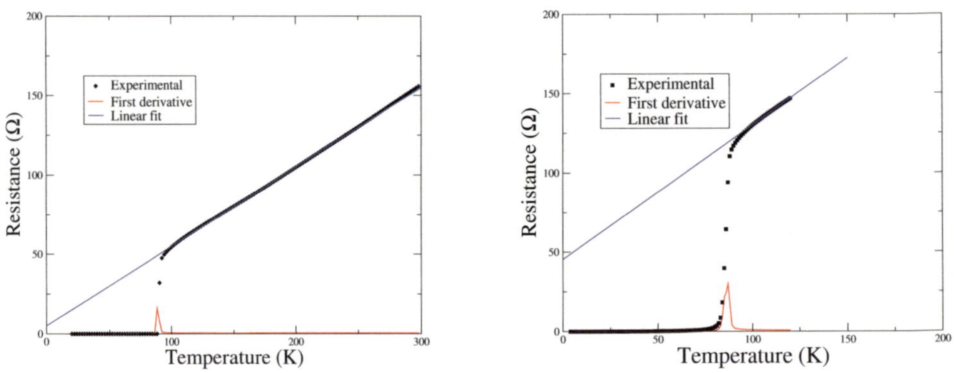

(a) $YBa_2Cu_3O_{7-\delta}$ on MgO.

(b) $YBa_2Cu_3O_{7-\delta}$ on STO.

Fig. 16. Resistance in function of temperature for $YBa_2Cu_3O_{7-\delta}$ films on STO and MgO single crystals.

Based on the first derivative criterion, we determined the Tc values as $Tc = 86.6 \pm 6.6\,\text{K}$ for $YBa_2Cu_3O_{7-\delta}$ on MgO; $Tc = 88 \pm 3\,\text{K}$ for $YBa_2Cu_3O_{7-\delta}$ on STO and; $Tc = 88 \pm 2\,\text{K}$ for $YBa_2Cu_3O_{7-\delta}$ on Sr_2YSbO_6 buffer layer.

Linear fit for 140 K< T <280 K; 99 K< T <147 K and 150 K< T <250 K were made for R–T data for $YBa_2Cu_3O_{7-\delta}$ on Sr_2YSbO_6, MgO and STO, respectively. The linear fit results are

$$R = -0.11 + 0.41T, \tag{1}$$

for $YBa_2Cu_3O_{7-\delta}/Sr_2YSbO_6$ film,

$$R = 45.4 + 0.85T, \tag{2}$$

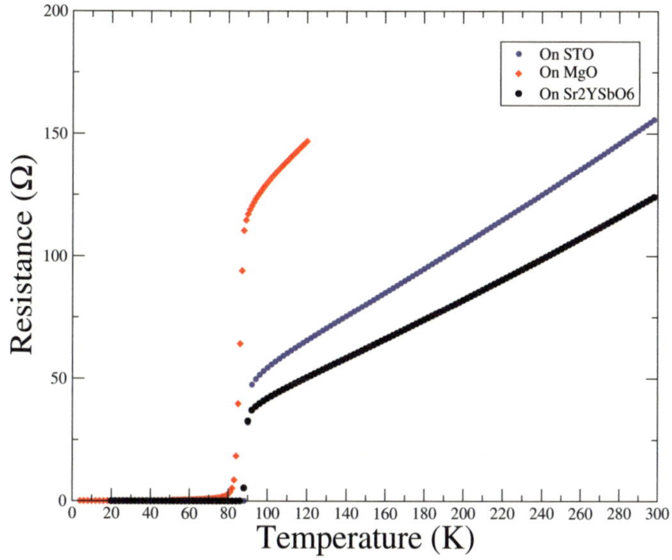

Fig. 17. Resistance in function of temperature for $YBa_2Cu_3O_{7-\delta}$ films.

for $YBa_2Cu_3O_{7-\delta}/MgO$ film, and

$$R = 5.1 + 0.50T, \tag{3}$$

for $YBa_2Cu_3O_{7-\delta}/STO$ film.

Although the *Tc* values are similar for films over STO and over Sr_2YSbO_6 buffer layer, the resistance in the normal zone is less for the $YBa_2Cu_3O_{7-\delta}$ growth on buffer layer; also, the extrapolated residual resistance for this film is less than the residual resistance of $YBa_2Cu_3O_{7-\delta}$ film growth on $SrTiO_3$.

Results of measurements for voltage *V* in function of current *I* (*I*–*V* curves) are shown in figure 18 for the films of $YBa_2Cu_3O_{7-\delta}$ over buffered substrate Sr_2YSbO_6/MgO. Based on this *I*–*V* data and on the *I*–*V* curves for $YBa_2Cu_3O_{7-\delta}/MgO$, of the figure 3, with the $1\,\mu V/cm$ criterion, the critical current values, at 77 K, were determined in ~ 0.013 mA for $YBa_2Cu_3O_{7-\delta}$ on MgO and ~ 17 mA for $YBa_2Cu_3O_{7-\delta}$ on Sr_2YSbO_6 buffer layer. So, the critical current density value for $YBa_2Cu_3O_{7-\delta}$ films growth over Sr_2YSbO_6 buffer layer is $Jc \sim 0.86 \times 10^6\,A/cm^2$ which is three order of magnitud times the *Jc* of $YBa_2Cu_3O_{7-\delta}/MgO$ films. The *Jc* value for $YBa_2Cu_3O_{7-\delta}/Sr_2YSbO_6/MgO$ film growth over buffer layer appears to be less than those reported in the literature ($Jc \sim 10^7\,A/cm^2$). However, it is worth saying that the value for $YBa_2Cu_3O_{7-\delta}/MgO$ is less too in comparison with references ($Jc \sim 10^6\,A/cm^2$). Thus, the sputtering deposition conditions perhaps are not yet optimized, and we believe that with other methods of deposition, such as laser ablation, we could improve the *Jc* results. The results reported in the literature are for films deposited in wealthy laboratories that have optimized deposition conditions.

The figure 18 also shows the voltage curves as a function of the applied current for temperatures of 82, 85 and 90 K. We observe a decrease in the value of critical current with increasing temperature and how at 90 K the linear behavior is similar to of conductors at this temperature, which is a sign that the $YBa_2Cu_3O_{7-\delta}$ is in the normal state.

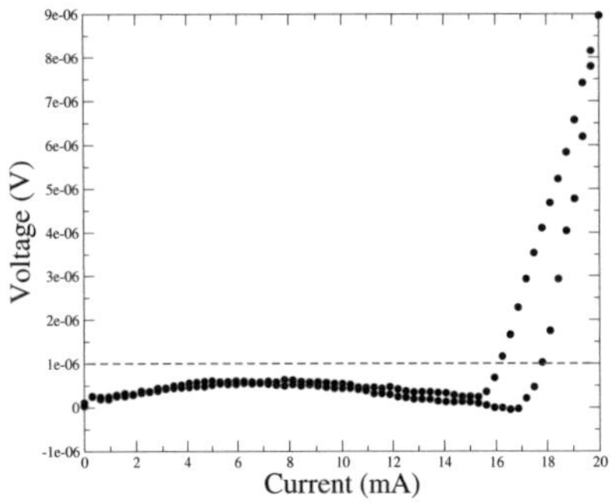

(a) I–V curve at 77 K.

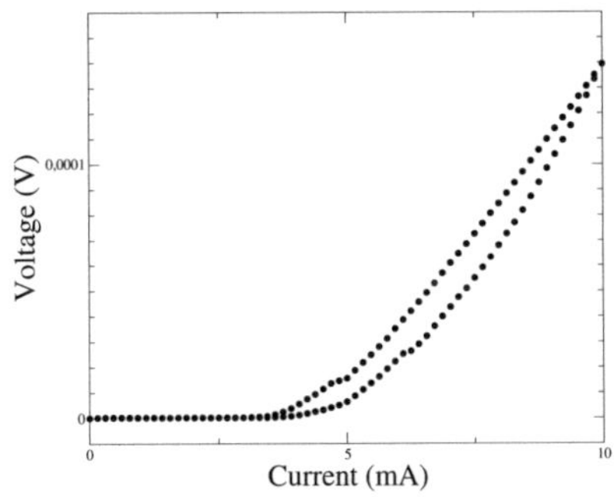

(b) I–V curve at 82 K.

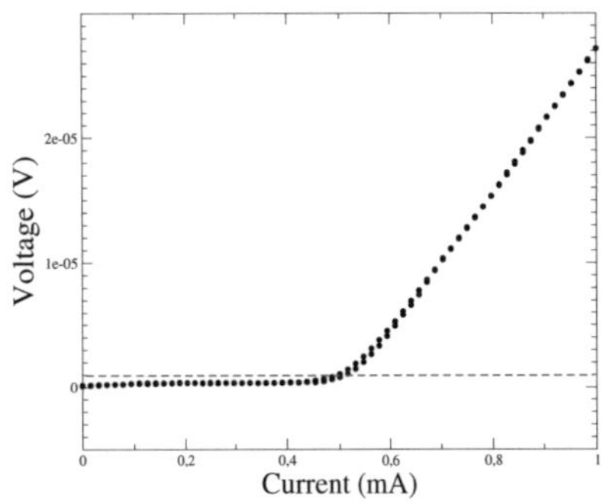

(c) I–V curve at 85 K..

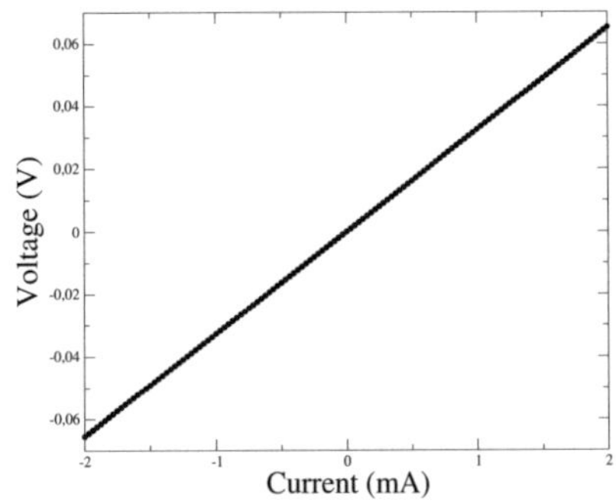

(d) I–V curve at 90 K..

Fig. 18. I–V curves for $YBa_2Cu_3O_{7-\delta}$ films on Sr_2YSbO_6 buffer layer at different temperatures.

6. Conclusion

In this chapter We have showed some relevant properties of type II superconductors for applications that requires high critical current densities even with applied magnetic fields. Special attention was dedicated to structural properties of the substrates for $YBa_2Cu_3O_{7-\delta}$ films in order to improve the Jc values. In this context, in this chapter was showed a case of study: the evaluation of the Sr_2YSbO_6 as a potential material for buffer layer in growth of superconducting films, since preliminary studies of polycrystalline samples until the effective application of this material for deposition of high quality superconducting films. It worth noting that this chapter concerns only on structural issues that limits the Jc values, which is a problem practically solved. Nothing were said about the flux pinning, another property that can be improved in order to obtain Jc values higher that reported at the present.

There has been show a review of the role of substrate in the successful deposition of superconducting films for applications. Furthermore, it has illustrated the convenience of using buffer layers for the growth of superconducting layers. For these purposes also We showed the different steps in the manufacture of superconducting film on the Sr_2YSbO_6 buffer layer used, with a focus on techniques to evaluate the material as a potential substrate for the successful growth of $YBa_2Cu_3O_{7-\delta}$ films.

Previous studies on a polycrystalline Sr_2YSbO_6 material showed an acceptable structural matching with the $YBa_2Cu_3O_{7-\delta}$ and the MgO, allowing the epitaxial growth of Sr_2YSbO_6 film on MgO and subsequently, the superconducting film. Studies on chemical non-reactivity with $YBa_2Cu_3O_{7-\delta}$ despite the severe heat treatment applied to the two materials in contact with each other, at temperature above the deposition of the films, were crucial to ensure that the Sr_2YSbO_6 buffer layer to fulfill its role of chemical barrier to eliminate the negative effect of MgO on the superconducting properties of YBCO film. In fact, the critical current density at 77 K of the film on the Sr_2YSbO_6 buffer layer was three orders of magnitude larger than Jc of the films deposited on MgO directly under the same conditions, as has been widely mentioned in the literature.

The value of critical current density of $YBa_2Cu_3O_{7-\delta}$ film on the Sr_2YSbO_6 buffer layer is smaller than other values reported in the literature, as is shown in table 1, which can be explained by the deposition conditions not yet optimized and by use of magnetron as a technique of deposition. To support this assumption is worth mentioning, for example, that the laser ablated $YBa_2Cu_3O_{7-\delta}$ films on $LaAlO_3$ have 2×10^6 A/cm_2 and the quality of this films is better than of the films grown by sputtering DC (Koren & Polturak, 2002). In this reference Koren mentions that substrate thickness influences the formation of structural defects in the superconducting film, besides the size of superconducting grains is one order of magnitude lower for films deposited by laser ablation than for deposited with DC magnetron, so that their quality improvement. The optimum temperature of deposition is another factor to review, as there is evidence that it may affect the Jc value. For example, buffered substrate $SrRuO_3$/MgO was used for growth of $YBa_2Cu_3O_{7-\delta}$ films with $Jc \sim 2.5 \times 10^6$ A/cm^2 at temperature of 770°C, while for deposition at 790°C Jc decreased to 6×10^5 A/cm^2 (Uprety et al., 2004). Also, in a study by Takahashi is reported the effect of buffer layer thickness on the superconducting properties of $YBa_2Cu_3O_{7-\delta}$. There is a minimum thickness that ensures chemical isolation between $YBa_2Cu_3O_{7-\delta}$ and other layers, but an increase of buffer layer thickness results in decreased critical current density. There is an optimum thickness of this layer that produces the films with the highest value of Jc (Takahashi et al., 2003). In our study, Sr_2YSbO_6 film was deposited with any thickness, which is not necessarily optimal.

As the emphasis was made on the application in coated conductors, it is worth noting that, once found the optimal conditions of manufacture, the Sr_2YSbO_6 could be used as a single buffer layer, thereby simplifying the manufacturing process of the superconductor tapes. Although it appears that the workhorse to discuss possible applications of our material is the field of superconducting tapes, it is interesting to note that the material could be used in other application fields such as electronic devices and Josephson junctions.

7. References

Aguiar, J. A.; de Souza Silva, C. C.; Yadava, Y. P.; Landinez Tellez. D. A.; Ferreira, J. M.; Guzman, J. & Chavira, E. (1998). Structure, microstructure, magnetic properties and chemical stability of $Ho_2Ba_2SbO_6$ with $YBa_2Cu_3O_{7-\delta}$ superconductor. *Physica C*, Vol(307) 189–196

Arendt, P. N.; Foltyn, S. R. Civale, L.; DePaula, R. F.; Dowden, P. C.; Groves, J. R.; Holesinger, T. G.; Jia, Q. X.; Kreiskott, S.; Stan, L.; Usov, I.; Wang H. & Coulter, J. Y. (2004). High critical current YBCO coated conductors based on IBAD MgO. *Physica C*, Vol(412–414), 795–800

Boffa, V.; Petrisor, T.; Ciontea, L.; Gambardilla, U. & Barbanera, S. (1996). Properties of in–situ laser–pulsed deposited YBCO thin films on MgO with $SrTiO_3$ buffer layer. *Physica C*, Vol(260) 111–116

Brandle, C. D. (1996). Preparation of perovskite oxides for high *Tc* temperature. *J. Mater. Res.*, Vol(5)No 5, 2160

Cyrot, M. & Pavuna D. (1992). *Introduction to superconductivity and high-T_c materials*, World Scientific

Desgardin, G.; Monot, I. & Raveau, B. (1999). Texturing of High–*Tc* superconductors. *Supercond. Sci. Technil.*, Vol(12), R115

Findikoglu, A. T.; Doughty, C.; Bhattacharya, S.; Li. Q.; Xi, X. X. & Venkatesan, T. (1992). Sr_2AlTaO_6 films for multilayer high–temperature superconducting devices, *Appl. Phys. Lett.*, Vol(61) 1718–1720

Foltyn, S. R.; Civale, L.; MacManus-Driscoll, J. L.; Jia, Q. X.; Maiorovl, B.; Wang, H. & Maley, M. (2007). Materials science challenges for high–temperature superconducting wire. *Nat Mater*, Vol(6), 631–642

Goyal, A.; Norton, D. P.; Budai, J. D.; Paranthaman, M.; Specht, E. D.; Kroeger, D. M.; Christen, D. K.; He, Q.; Saffian, B.; List, F. A.; Lee, D. F.; Martin, P. M.; Klabunde, C. E.; Hartfield E.; & Sikka, V. K. (1996). High critical current density superconducting tapes by epitaxial deposition of $YBa_2Cu_3O_{7-\delta}$ thick films on biaxially textured metals *Appl. Phys. Lett.*, Vol(69) No 12, 1795–1797

Groves, J. R.; Arendt, P. N.; Foltyn, S. R.; Jia, Q. X.; Holesinger, T. G.; Kung, H.; DePaula, R. F.; Dowden, P. C.; Peterson, E. J.; Stan L.; & Emmert, L. A. (2002). Recent progress in continuously processed IBAD MgO template meters for HTS applications. *Physica C*, Vol(382), 43–47

Hanyu, S.; Iijima, Y.; Fuji, H.; Kakimoto K. & Saitoh, T.. (2007). Development of 500 m–length IBAD–$Gd_2Zr_2O_7$ film for Y–123 coated conductors. *Physica C*, Vol(463–465), 568–570

Heine, K.; Tenbrink, J. & Thoner, M. (1989). High–field critical current densities in $Bi_2Sr_2Ca_1Cu_2O_{8+\delta}$/Ag wires. *Appl. Phys. Lett.*, Vol(55) No 23, 2441–2443

Huhne, R.; Selbmann, D.; Eickemeyer, J.; Hanish J. & Holzapfel, B. (2006). Preparation of buffer layer architectures for $YBa_2Cu_3O_{7-\delta}$ coated conductors based on surface oxidized Ni tapes. *Supercond. Sci. Technol.*, Vol(19), 169–174

Hollmannt, E. K.; Vendik, O. G.; Zaitsev, A. G. & Melekh, B. T. (1994). Substrates for High-*Tc* superconductor microwave integrated circuits. *Supercond. Sci. Technol.*, Vol(7) 609–699

Hull, J. R. (2003). Applications of high–temperature superconductors in power technology. *Rep. Prog. Phys.*, Vol 66, 1865–1886

Jia, Q. X.; Foltyn, S. R.; Arendt, P. N.; Groves, J. R.; Holesinger T, G.; & Hawley, M. E. (2002). Role of $SrRuO_3$ buffer layers on the superconducting properties of $YBa_2Cu_3O_{7-\delta}$ films grown on polycrystalline metal alloy using a biaxially oriented MgO template. *Appl. Phys. Lett.*, Vol(81) No 24, 4571–4573

Jia, Q. X.; Foltyn, S. R.; Arendt, P. N.; Holesinger, T.; Groves J. R. & Hawley, M. (2003). Growth and Characterization of $SrRuO_3$ Buffer Layer on MgO Template for Coated Conductors, *IEEE Trans. Appl. Supercond.*, Vol(13) No 2, 2655–2657

Knoth, K.; Schlobach, B.; Huhne, R.; Schultz, L. & Holzapfel, B. (2005). $La_2Zr_2O_7$ and Ce–Gd–O buffer layers for YBCO coated conductors using chemical solution deposition. *Physica C*, Vol(426–431) 979–984

Koa, K. P.; Hab, H. S.; Kimc, H. K.; Yuc, K. K.; Kob, R. K.; Moonc, S. H.; Ohb, S. S.; Parka C. & Yoo, S. I. (2007). Fabrication of highly textured IBAD–MgO template by continuous reel–to–reel process and its characterization. *Physica C*, Vol(463–465), 564–567

Koren, G. & Polturak, E. (2002) Is $LaAlO_3$ a viable substrate for the deposition of high quality thin films of $YBa_2Cu_3O_{7-\delta}$? *Supercond. Sci. Technol.*, Vol(15), 1335–1339

Koshy, J.; Kumar, K. S.; Kurian, J.; Yadava, Y. P. & Damodaran, A. D. (1995). Transpor properties of the superconducting $Bi_2Sr_2Ca_2Cu_3O_{10}$–$DyBa_2SnO_{5.5}$. *Phys. Rev. B*, Vol(51) 9096–9099

Kurian, J.; Takahashi, Y.; Amemura, T.; Susuki, T. & Morishita, T. (2002). The growth and characterization of epitaxial Ba_2NdTaO_6 dielectric ceramic thin films on (1 0 0) $SrTiO_3$. *J. Phys. D: Appl. Phys.* Vol(35) 3002–3006

Kurian, J. & MOrishita, T. (2003). Epitaxial growth of $NdBa_2Cu_3O_{7-\delta}$ films on (1 0 0) MgO buffered with Ba_2NdTaO_6, a potential dielectric buffer layer for $NdBa_2Cu_3O_{7-\delta}$ films. *Supercond. Sci. Technol.* Vol(16) 422–425

Larson, A. C. & Von Dreele, R. B. (2000) General Structure Analysis System (GSAS). *Los Alamos National Laboratory Report LAUR* 86–748

Li, P.; Abraimov, D.; Xu, A. & Larbalestier D. (2012). Observation of important current–limiting defects in a recent high pinning force MOCVD IBAD–MgO coated conductor. *Supercond. Sci. Technol.*, Vol(25) 025002 doi:10.1088/0953-2048/25/2/025002

Liu, X. Z; He, S. M.; Wu, D. J.; Duan, K. Y. & Li, Y. R. (2006). Heteroepitaxial growth of YBCO/STO thin films for voltage tunable varistors. *Physica C*, Vol(433) 147–153

Lucia, M. L.; Santamaria, J. Iborra, E.; Hernandez–Rojas, J. L. & Sanchez–Quesada, F. (1993). Texture improvement of sputtered $YBa_2Cu_3O_{7-\delta}$ films on MgO (1 0 0) with a $SrTiO_3$ buffer layer. *Physcia C*, Vol(218) 59

Maguire, J. F.& Yuan, J. (2009). Status of high temperature superconductor cable and fault current limiter projects at American Superconductor. *Physica C*, Vol(469), Issue 15–20, 874–880

Nishikawa, H.; Hontsu, S.; Nakamori, M.; Tabata H. & Kawai, T. (2003). Preparation of Perovskite Type Manganite on Al_2O_3 Substrate as an Excellent Buffer Layer for $YBa_2Cu_3O_{7-\delta}$ Growth. *IEEE Trans. Appl. Supercond.*, Vol(13) No 2, 2725–2728

Ortiz–Diaz, O.; Roa–Rojas, J.; Landinez Tellez D. A. & Aguiar, J. Albino. (2004). Evaluation of Sr_2YSbO_6 as a new substrate for $YBa_2Cu_3O_{7-\delta}$ superconductor thin films. *Mod. Phys. Lett. B*, Vol(18), 1035–1042

Ortiz–Diaz, O.; Lopez Carreno, L. D.; Albino Aguiar, J.; Roa–Rojas J. & Landinez Tellez, D. A. (2004). Structural ordering, chemical stability and percolative effect analysis in $YSr_2SbO_6/YBa_2Cu_3O_{7-\delta}$ complex perovskite composites, *Physica C*, Vol(408–410), 886–888

Ortiz-Diaz, O.; Saldarriaga, W; Lopera, W.; Reyes, D.; Cortes, A.; Caicedo, J. M.; Landinez Téllez, D. A. & Roa-Rojas, J. (2011). Sr_2YSbO_6 as a buffer layer for $YBa_2Cu_3O_{7-\delta}$ superconducting films. *J. Mater. Sci.*, Vol(46), 688-692

Pai, S. P.; Jasudasan, J.; Apte, P. R.; Pinto, R.; Kurian, J.; Sajith, P. K.; James, J. & Koshy, J. (1997). $YBa_2Cu_3O_{7-\delta}$ films with high critical current density on epitaxial films of Ba_2LaNbO_6, a new perovskite substrate for $YBa_2Cu_3O_{7-\delta}$ superconductor. *Europhys. Lett.*, Vol(39) No 6, 669–673

Parans Paranthaman, M.; Aytug, T.; Kang, S.; Feenstra, R.; Budai, J. D.; Christien, D. K.; Arendt, P. N.; Stan, L.; Groves, J. R.; DePaula, R. F.; Foltyn S. R. & Holesinger, T. G. (2003). Fabrication of High *Jc* $YBa_2Cu_3O_{7-\delta}$ Tapes Using the Newly Developed Lanthanum Manganate Single Buffer Layers. *IEEE Trans. Appl. Supercond.*, Vol(13) No 2, 2481–2483

Poole, Ch, P.; Farach, H. A. & Creswick, R. J. (1995). *Superconductivity*, Academic Press

Prouteau, C.; Hamet, J. F.; Mercey, B.; Hervieu, M.; Raveau, B.; Robbes, D.; Coudrier, L. & Ben G. (1995). Significant improvement of superconductivity of laser ablated "$YBa_2Cu_3O_7$/MgO" thin films: introduction of a $SrTiO_3$ buffer layer. *Physica C*, Vol(248) 108–118

Sathyamurthy, S.; Parans Paranthaman, M.; Zhai, H. Y.; Kang, S.; Christien, H. M.; Cantor, C.; Goyal A. & Martin, V. (2003). Solution Processing of Lanthanum Zieconate Films as a Single Buffer Layers for High *Ic* YBCO Coated Conductors. *IEEE Trans. Appl. Supercond.*, Vol(13) No 2, 2658–2660

Sheahen, T. P. (2002). *Selected topics in superconductivity, Introduction to High–temperature superconductivity*, Kluwer academis publishers

Shen, T.; Jiang, J.; Kametani, F.; Trociewitz, U. P.; Larbaliester, D. C.; Schwartz, J.& Hellstrom, E. E. (2010). Filament to filament bridging and its influence on developing high critical current density in multifilamentary $Bi_2Sr_2CaCu_2O_x$ round wires. *Supercond. Sci. Technol.*, Vol(23), 025009–025018

Shiohara, Y.; Fujiwara, N.; Hayashi, H.; Nagaya, S.; Izumi, T. & Yoshizumi (2009). Japanese efforts on coated conductor processing and its power applications: New 5 year project for materials and power applications of coated conductors (M-PACC). *Physica C*, Vol(469), 863–867

Takahashi, Y.; Wakana, H.; Ogawa, A.; Morishita, T. & Tanabe, K. (2003). Effects of thin SrO buffer layer on dielectric and superconducting properties of $YBa_2Cu_3O_{7-\delta}/Sr_2AlTaO_6/YBa_2Cu_3O_{7-\delta}$ multilayers. *Jpn. J. Appl. Phys.*, Vol(42) L283–L286

Toby, B. H. (2001). EXPGUI, a graphical user interface for GSAS. *J. Appl. Cryst.*, Vol(34), 210–213

Uprety, K. K.; Ma, B.; Koritala, R. E.; Baurceanu, R. M.; Weber, T. P.; Fisher, B. L.; Dorris, S. E.; Erck, R. A.; Maroni, V. A. & Balachandran, U. (2004). Growth of YBCO film on $SrRuO_3$–buffered MgO substrate. *Supercond. Sci. Tecnol.*, Vol(17) 671–675

Varanasi, C. V.; Burke, J.; Lu, R.; Wu, J.; Brunke, L.; Chuck, L.; Smith, H. E.; Maartense, I. & Barnes, P. N. (2008). Biaxially textured $YBa_2Cu_3O_{7-\delta}$ films deposited on polycrystalline flexible yttria–stabilized zirconia ceramic substrates. *Physica C*, Vol(468), 1070–1077

Wang, H.; Foltyn, S. R.; Arendt, P. N.; Jia, Q. X.; Li, Y. & Zhang, X. (2005) Thickness effects of $SrTiO_3$ buffer layers on superconducting properties of $YBa_2Cu_3O_{7-\delta}$ coated conductors. *Physica C*, Vol(433), 43–49

Wee, S. H.; Moon, S. H.; Park, C. & Yoo, S. I. (2005) Effects of substrates on the $NdBa_2Cu_3O_{7-\delta}$ films fabricated by the PLD process. *Physica C*, Vol(426-431), 996–1000

Wu, C. H.; Chen, M. J.; Hsu, M. H.; Chen, J. C.; Chen, K. L.; Chen, J. H.; Jeng, J. T.; Lai, T. S.; Horng, H. E. & Yang, H. C. (2005). Optimization of step–edge substrates for high–TC superconducting devices. *Physica C*, Vol(433) 108–114

Yadava, Y. P; Landinez Tellez, D. A.; Melo, M. T.; Ferreira, J. M. & Aguiar, J. A. (1998). Structural ordering and chemical stability of a complex perovskite oxide $DyBa_2ZrO_{5.5}$ with $YBa_2Cu_3O_{7-\delta}$ superconductors. *Appl. Phys. A*, Vol(66) 455–458

Ying, Q. Y. & Hilbert, C. (1994). All–$YBa_2Cu_3O_{7-\delta}$ trilayer tunnel junctions with Sr_2AlTaO_6 barrier. *Appl. Phys. Lett.*, Vol(65) 3005–3007

Ying, L. L.; Lu, Y. M.; Liu, Z. Y.; Fan, F.; Gao, B.; Cai, C. B.; Thersleff, T.; Reich, E.; Huhne R. & Holzapfel, B. (2009). Thickness effect of $La_2Zr_2O_7$ single buffers on metallic substrates using pulsed laser deposition for $YBa_2Cu_3O_{7-\delta}$–coated conductors. *Supercond. Sci. Technol.*, Vol(22), 095005–095009

Zhao, Y,; Li, X-F.; Khoryushin, A.; He, D.; Andersen, N. H.; Hansen, J. B. & Grivel. J-C. (2012). Development of all chemical solution derived $Ce_{0.9}La_{0.1}O_{2y}/Gd_2Zr_2O_7$ buffer layer stack for coated conductors: influence of the post annealing process on surface crystallinity. *Supercond. Sci. Technol.*, Vol(25) 015008 doi:10.1088/0953-2048/25/1/015008

13

Structural Characteristic and Superconducting Performance of MgB_2 Fabricated by Mg Diffusion Process

Minoru Maeda[1], Jung Ho Kim[2] and Shi Xue Dou[2]
[1]*Department of Physics, College of Science and Technology, Nihon University*
[2]*Institute for Superconducting and Electronic Materials,*
Australian Institute of Innovative Materials, University of Wollongong
[1]*Japan*
[2]*Australia*

1. Introduction

The phenomenon of superconductivity, which arises from an unusual quantum mechanism below a superconducting critical transition temperature (T_c), shows unique physical properties, namely, exactly zero direct current (dc) electrical resistance, the Meissner effect associated with nearly perfect diamagnetism, and the Josephson effect in two weakly coupled superconductors which may have applicability in high-speed, high-sensitivity, low-voltage sensors and switches. Thanks to these valuable characteristics, today, superconducting materials are having a considerable influence on the development of the next generation of medical services, electrical power systems, transportation, communication technology, and integrated circuits. Thus, our energy-hungry world increasingly requires those technologies and superconductors are becoming indispensable materials.

Superconductivity was discovered by Heike Kamerlingh Onnes in 1911. It is one of the greatest discoveries in the history of science. After that, he tried to fabricate a superconducting magnet using lead (Pb), but it failed. From the failure of his experiment, he learned of the existence of the critical field of superconductors, which completely suppresses superconductivity. His dream was that a high magnetic field could be excited by magnets using superconductors, but it was left unfulfilled. In 1947, a compact helium liquefier was invented by Collins and cowokers, and it made possible some advance in theoretical and experimental studies related to superconducting magnets. From 1960 to 1990, niobium titanium (NbTi) and niobium tin (Nb_3Sn) superconductors were employed in many kinds of approaches towards technical development for real applications. These efforts formed the basis of technical application for magnets using metallic superconducting materials, and even now, many researchers are still making steady progress on it. As a result, superconducting materials have become widely applied in not only research magnets for laboratory use, but also various other instruments, which is more than what was expected. Specifically, magnetic resonance imaging (MRI) equipment provides image contrast between different kinds of soft issues in the human body and it is essential to confirm the diagnosis of spinal disc herniation, anterior cruciate ligament injury, tumour, and stroke. Nuclear magnetic resonance (NMR) equipment is utilized on the cutting edge of genome-related research. In addition, superconducting magnetic energy storage (SMES), magnetic levitation systems for transportation, and large-scale magnets for confinement of plasma in fusion reactors are expected as the next generation of devices for energy production, its storage, and its distribution, even though they are still at the experimental stage of development. On the other hand, cooling technology, which is crucial for superconducting applications, has also maintained steady progress for the past 15 years. The most recent helium liquefier has smaller scale, longer operation life, and higher energy-saving compared with those a decade ago. Small, convenient cryocoolers without liquid helium have been put to practical use and developed into lower power consumption versions. A unique small pulse tube refrigerator without moving pistons in the low-temperature part of the device is also coming into actual utilization and can easily achieve a low temperature of 10 ~ 20 K. Thus, the development of cooling technology has given an impulse to superconducting applications.

Despite the recent progress of cooling technology, today, the majority of current practical superconducting devices are still made of NbTi and Nb_3Sn superconductors, which have the low T_c values below 20 K. MRI is one of the conventional superconducting devices and has been critical to our high-quality healthcare. However, only 10 % of human beings utilize the superconducting technology in the world. The most serious cause of disturbing its largely widespread use is to need liquid Helium for activating MRI magnets at 4.2 K. The cryogen is expensive and unmanageable, and its resource is exhaustible. In order to promote its widespread use and reduce operating cost, the next generation of superconducting technology must eliminate the use of liquid cryogen bath cooling and its real application has been strongly required. Currently, the most promising materials are thought to be bismuth strontium calcium copper oxide (BSCCO), yttrium barium cooper oxide (YBCO), and magnesium diboride (MgB_2). MgB_2, a metallic compound, shows T_c of ~ 39 K, the highest value that has been reported for conventional metallic superconductors, and its superconductivity was discovered by Jun Akimitsu's group in 2001 (Nagamatsu et al., 2001). Owing to its large thermal margin than Nb based superconductors, MgB_2 superconducting materials have been judged potentially capable of meeting the needs of the next-generation superconducting applications operated at 15~30 K. MgB_2 materials also show superior characteristics for industrial applications including that (1) the manufacture method is much easier to that of NbTi wires, (2) the raw materials are much cheaper than that of YBCO thin films and BSCCO wires, and the manufacturing cost would potentially be lower than that of NbTi wires, (3) the density is one third of NbTi and it can yield triple length wire using the same mass of raw materials, (4) the charging rate of the magnet is faster than that of NbTi magnet, (5) the lower anisotropy and the lager coherence length do not make the weak link problem severe compared with YBCO thin films and BSCCO wires (Larbalestier et al., 2001). Although the T_c values of YBCO and BSCCO show 93 K and 110 K, respectively, and those are much higher than that of MgB_2, the prices of those conductors seem not to be reduced in spite of enormous efforts for more than 20 years. In these regards, therefore, MgB_2 materials have been energetically studied for industrial applications in the world since the discovery of the superconductivity.

2. MgB_2 material

2.1 Crystal and electrical structure

In 1954, the crystal structure of MgB_2 was first identified by an X-ray diffraction investigation using nickel-filtered Cu Kα radiation (Jones & Marsh, 1954). MgB_2 possesses the simple AlB_2 type structure, and its corresponding space group, crystal system, and Laue-symmetry class are P6/mmm, hexagonal, and 6/mmm, respectively. The structure consists of alternating layers of close-packed ionized Mg^{2+} atoms, which are separated by honeycomb-type boron sheets. In the case of pure MgB_2 samples without external forcing such as hydrostatic pressure, the *a*-lattice parameter is estimated to range from ~ 3.083 to ~ 3.086 Å (Jones & Marsh, 1954; Jorgensen et al., 2001), with the variation most likely to be due to disorder in the honeycomb lattice. By contrast, the *c*-lattice parameter shows nearly the same value of ~ 3.521 Å. This may reflect the fact that in the case of close-packed hexagonal structure, the slip directions are essentially $[\bar{1}\bar{1}0]$, $[2\bar{1}0]$, and $[\bar{1}20]$ in the *ab* plane, while expansion and contraction along the *c*-axis direction hardly occurs. The fractional coordinates (x, y, z) for Mg and B atoms are allocated to the 1*a* site of (0, 0, 0) and the 2*d* site of (1/3, 2/3, 1/2). The site occupancies of Mg and B are observed to be ~ 0.95-1.00 and 1.00, respectively (Mori et al., 2002; Tsirelson et al., 2003; Kazakov et al., 2005). The atomic displacement parameters U of Mg and B atoms are also observed to be ~ 0.0049-0.0060 $Å^2$ and ~ 0.0042-0.0080 $Å^2$, respectively (Mori et al., 2002; Tsirelson et al., 2003; Kazakov et al., 2005).

Band structure calculations on MgB_2 (An & Pickett 2001; Kortus et al., 2001) suggest that the bands at the Fermi level (E_F) mainly originate from the orbitals of boron due to substantially ionized Mg, and there are four conduction bands, namely, two σ bands delivered from the σ bonding $2s/2p_{x,y}$ orbitals and two π bands delivered from the π-bonding and the anti-bonding $2p_z$ orbitals. The former bands are hole types and occur in two-dimensional (2D) cylindrical forms. The latter bands are both hole type and electron type, and form three-dimensional (3D) tubular networks. The σ states are localized and confined in the boron layers, while the π states are delocalized in all directions. The σ bands within boron sheets are strongly covalent, whereas the π bands over the whole crystal show metallic behaviour, reflecting the lack of covalent bonding (Mazin & Antropov, 2003). The 2D strongly covalent states and the 3D metallic-type states mainly contribute to the total density of states (DOS) at the Fermi energy (E_F).

Calculations of both the band structure and the lattice dynamics for MgB_2 indicate that there are four distinct phonon modes at the Brillouin zone center (Γ), namely the A_{2u}, B_{1g}, E_{1u}, and E_{2g} modes (Bohnen et al., 2001; Yildirim et al., 2001). The A_{2u} singly degenerate mode involves the vibration of Mg and B layers in opposite directions along the c axis. The B_{1g} singly degenerate mode is responsible for the vibrations of B atoms in opposite directions along the c axis, with the Mg atoms stationary. On the other hand, the others are doubly degenerate modes and involve only in-plane motions along the x or y axis. For the E_{1u} mode, the Mg and B layers move in opposite directions. The E_{2g} mode involves the vibration of the B ions in opposite directions, with the Mg ions stationary. The E_{2g} in-plane boron phonons show giant anharmonicity, are strongly coupled to the σ bands, and thus change the $2s/2px,y$ orbital overlap. The strong coupling between the holes of the σ bands and the E_{2g} mode is responsible for the high T_c in MgB_2. The π band charge carriers also become superconducting and contribute to unusual superconducting behaviour in MgB_2 which is not observed in conventional superconductors. Since the two properties of the σ bands are very similar and those of the π bands are very similar, even through one π band is hole type and the other is electron type, the electronic structure is usually simplified to one σ band and one π band. Therefore, MgB_2 is known and described as a two-band superconductor (Liu et al., 2001).

2.2 Unique characteristic caused by two-band and two-gap structure

MgB_2 is known to be theoretically and experimentally identified as the first superconductor where two superconducting gaps clearly exist (Xi, 2008; Tajima, 2005). The unique two-band and two-gap structure has an influence on the resistivity behaviour in the normal state. In the year of discovery, it became well known that the residual resistivity ratio (RRR) was independent of T_c in polycrystalline MgB_2 samples. Specifically, many MgB_2 samples showed widely different RRR values despite having similar T_c values around ~ 39 K (Buzea & Yamashita, 2001). In the case of the high T_c, the disorder effects on it (Eisterer, 2007) due to poor crystallinity and lattice strain can be negligible, and thus, the RRR value calculated from the residual resistivity was thought to depend on impurity contamination in a sample. This indicates that lower RRR is caused by a larger amount of impurity phase, even though it should be noted that the residual resistivity is partially attributable to extrinsic effects such as lack of connectivity associated with cracks, voids, and low mass density (Rowell, J. M. 2003). In the case of the two-gap superconductivity model (Golubov & Mazin, 1997), non-magnetic impurities may contribute to pair breaking, which is similar to the effects of magnetic impurities in a conventional *s*-wave superconductor. Thus, it was difficult to theoretically understand the observed unclear correlation between the RRR and the T_c. Regarding this issue, Mazin et al. clearly explained the mechanism by pointing out the very weak impurity scattering between the σ and the π bands compared with the intraband impurity scattering between the $\sigma\sigma$ and $\pi\pi$ sub-bands, due to the particular electronic structure in MgB_2 (Mazin et al., 2002). It was also found that the $\pi\pi$ scattering rate is larger than the $\sigma\sigma$ scattering rate. In addition, they indicated that the T_c is given by the maximum eigenvalue of the superconducting coupling constant λ for the $\sigma\pi$ and the $\pi\sigma$ interband coupling, and the $\sigma\sigma$ and $\pi\pi$ intraband coupling, while the conductivity is the sum of all the conducting channels responsible for the interband and intraband couplings. The estimated λ for the $\sigma\sigma$ intraband coupling showed the highest value, suggesting that the T_c is affected by the σ band. These results provide the conclusion that because impurities have a stronger effect on the scattering rate of the π band in preference to the σ band, and thus the residual resistivity, they may not contribute to the interband scattering and T_c.

The upper critical field behaviour of MgB_2 was also found to be very different from that of a conventional phonon-mediated Bardeen, Cooper, and Schrieffer (BCS) superconductor with a single-band structure. A unique upward curvature of B_{c2}(T) parallel to the *ab* plane near T_c was observed in single crystal MgB_2 samples (Zehetmayer et al., 2002; Lyard et al., 2002). This upward curvature can be well explained by using the Ginzburg-Landau theory, suggesting that the B_{c2} is determined by the Ginzburg-Landau parameter and the condensation energy (Eisterer, 2007). In the case of MgB_2, which has the two-band structure, the total condensation energy depends on the sum of all the energies arising from the σ band, the π band, and the interaction between the two bands (Eisterer et al., 2005). These condensation energies in the temperature range from absolute zero to T_c show different behaviour from each other. In particular, the contribution of the π band to the total condensation energy at 0 T in temperatures above 30 K rapidly increases, while that below 30 K is negligible. This is the reason why the B_{c2}(T) of MgB_2 shows a positive curvature near T_c. The unique slope can make the zero-temperature $B_{c2}(0)$ significantly higher than that estimated from a single-gap dirty superconductor model (Gurevich, 2003). In addition, an interesting scenario relating to B_{c2}(T) for dirty two-gap superconductors was pointed out by Gurevich (Gurevich, 2007). It is assumed that MgB_2 can be identified as a simple bilayer sample where two thin films corresponding to the σ and π bands are separated by a Josephson contact, which reflects the weak interband coupling in MgB_2. From the two-gap dirty-limit theory for the BCS matrix constants (Golubov et al., 2002), the temperature dependence of B_{c2} is mostly determined by whether the σ film or the π film is dirtier. Provided that the π film is much dirtier than the σ film, the global B_{c2}(T) of the bilayer at higher temperature is mainly controlled by the σ film with the higher B_{c2}, while the π film takes over at lower temperatures. This leads to a considerable upturn in the global B_{c2} curve at low temperatures, which can not be estimated from a single-gap dirty-limit superconductor. This interesting scenario for B_{c2} at low temperatures delivered from the simple bilayer model is indirectly demonstrated by experimentally observed B_{c2} curves for alloyed MgB_2 thin films (Braccini et al., 2005). Specifically, an anomalous upward curvature of B_{c2}(T) at low temperatures observed in a carbon-alloyed thin film provides a record high value of B_{c2}(4.2K) = 51 T parallel to the *ab* plane. The extrapolation of the B_{c2}(T) curve to T = 0 K may reach over ~ 60 T, approaching the paramagnetic limit of 65 T which is calculated from the conventional BCS theory (Sarma, 1963). This value is significantly higher than those for NbTi and Nb_3Sn superconductors, which are widely used in superconducting applications now. Moreover, Gurevich pointed out that given the case of electron-phonon coupling, the paramagnetic limit of MgB_2 is estimated to be ~ 130 T (Gurevich, 2007). This means that there is still large room for a further increase in B_{c2} from optimum dirty conditions of the intraband and the interband behaviour. Therefore, we can extract a conclusion that the two-gap superconductor MgB_2 is suitable for superconducting applications in higher operating temperatures and fields, which can not be reached via the use of Nb-based superconducting materials.

2.3 Current status in development for superconducting application

Two major critical issues facing MgB_2 polycrystalline materials such as wires and bulks for the next generation of superconducting applications are weak grain connectivity and lack of pinning centers (Rowell, J. M. 2003; Eisterer, 2007). In order to improve rapid J_c degradation caused by its weak pinning behaviour in pure MgB_2 wires and bulks, so far, many doping materials have been tested since the discovery of its superconductivity. The first successful dopant for this was silicon carbide (SiC) and its additive improved the irreversibility field (B_{irr}) and J_c (Dou et al., 2002). Numerous researchers focused on the impressive results and then started to do many studies related to carbon (C) containing dopants for real superconducting applications. Even now, SiC is still known to be one of the most effective dopants for MgB_2 materials. Specifically, B_{c2} values near 0 K for the SiC-doped MgB_2 wires fabricated by the *in situ* method (Sumption et al., 2005) and coated-conductors on SiC fibers prepared from hybrid physical-chemical vapor deposition (HPCVD) (Ferrando et al., 2005) were 33 T and 55 T, respectively. These values are much higher than those of NbTi and Nb_3Sn. Similar doping materials, nano-carbon (Soltanian et al., 2003), carbon nanotubes

(Yeoh et al., 2004), and boron carbide (Yamamoto et al., 2005a) were also found to effectively improve the in-field J_c in MgB_2 materials. Thus, it has been found that the carbon (C) element is indispensable for superconducting applications of MgB_2.

From structural analysis of C-doped MgB_2 single crystals (Kazakov et al., 2005), the *a*-lattice parameter was observed to decrease drastically due to the C doping effect. The lattice shrinkage is most likely due to C substitution at B sites and to the difference in atomic radius between these two atoms, which may yield substitutional defects in the crystal lattice. This is because the C replacement effect accumulates in the B layer, and dislocations and its associated stacking faults may form when the accumulated strain exceeds the critical value. The existence of lattice defects in C-doped MgB_2 samples has been proven by experimental results (Kazakov et al., 2005; Avdeev et al., 2003; Zhu et al., 2007; Kim et al., 2012). Specifically, the considerable increase in the atomic displacement parameter at the Mg and B sites in C-doped single crystals reflects the enhancement of lattice disorder (Kazakov et al., 2005). The reason why reliability factors for Rietveld refinement in C-doped samples show high values compared with those for pure samples may be explained by the presence of local disorder in the lattice, which can not be described using conventional refined parameters (Kazakov et al., 2005; Avdeev et al., 2003). The C-doped thin films deposited by HPCVD, which shows a high B_{c2} at 0 K value of 70 T, are found to have intensive structural disorder caused by dislocations (Zhu et al., 2007a). In the case of SiC-doped tapes showing relatively low mass density compared with MgB_2 thin films, the MgB_2 grains are also found to show a large amount of interior contrast generated by intragrain defects that are likely due to lattice disorder (Zhu et al., 2007b). In addition, the relationship between lattice change and not only substitutional, but also interstitial defects in C-doped MgB_2 materials can be theoretically explained by Vienna *ab* initio calculations (Bengtson et al., 2010). Recently, boron vacancies were clearly observed in C-doped MgB_2 wires and it has been found that those vacancies generate intrinsic stacking faults within the MgB_2 grains, together with associated lattice distortion (Kim et al., 2012). Thus, in C-doped MgB_2 materials, the reduced unit cell volume caused by the lattice shrinkage leads to introduction of lattice disorder effects, which result in increasing scattering of charge carriers, reducing the mean free path, and thereby enhancing B_{c2} and J_c (Kazakov et al., 2005; Eisterer, 2007).

However, in the case of inorganic carbon compounds as C doping materials, due to its high melting points or its high decomposition temperature, high-temperature reaction is usually required to effectively cause the lattice shrinkage of the *a*-lattice parameter. These sintering conditions generally offer the disadvantage of grain growth associated with weak grain boundary pinning, because this pinning mechanism is also known to be effective in MgB_2 materials (Larbalestier et al., 2001; Eisterer, 2007). To address the problem, organic compounds as C doping materials with relatively low decomposition temperatures have been intensively studied from around 2006 (Kim et al., 2006; Hossain et al., 2009; Kim et al., 2010; Maeda et al., 2011a). One of the best known organic dopants that has been suggested to date is malic acid ($C_4H_6O_5$) (Kim et al., 2006). The advantage is that the carbohydrate melts and decomposes at a low temperature below the temperature of MgB_2 formation. The decomposition produces highly reactive and fresh C on the atomic scale for C substitution on B site in the lattice. Due to these excellent characteristics, the *a*-lattice parameter decreases even under low-temperature sintering conditions, and grain growth is suppressed, resulting in considerable enhancement of J_c and B_{c2}. Specifically, the J_c value at

4.2 K and 6 T for *in situ* processed $C_4H_6O_5$-doped MgB_2 wires exceeds 1.0 × 10^5 A/cm^2, which is comparable to that for commercial NbTi wires (Kim et al., 2010). The application of cold high pressure densification to those MgB_2 wires yielded a high J_c value of 4.0 × 10^4 A/cm^2 at 4.2 K and 10 T, which is the record value among conventional *in situ* powder-in-tube (PIT) processed MgB_2 wires (Hossain et al., 2009).

According to recent advanced microscopic analysis by using scanning transmission electron microscopy (STEM) and electron energy loss spectroscopy (EELS), un-reacted carbon agglomerates are found to be formed at grain boundaries in C-doped MgB_2 (Kim et al., 2012). The residual impurity arises from inhomogeneous mixing of starting materials including carbon compound as dopant and impedes supercurrent flow. In order to address the issue, recently, a carbon containing gas diffusion method by using oxygen-free pyrene ($C_{16}H_{10}$) was proposed (Maeda et al., 2011b). Pyrene is an aromatic hydrocarbon and its boiling temperature is 404 °C lower than the temperature of MgB_2 formation. Owing to the characteristic, molecular carbon delinked from pyrene gas homogenously distributes into dense MgB_2 materials and it causes intensive structural disorder without any severe deterioration of grain connectivity. In addition, more recently, an alternative carbon doping method by using carbon encapsulated boron nanopowder made from a radio frequency plasma process was also proposed and the resultant C-doped MgB_2 wires showed high J_c performance due to very thin carbon layers on boron surface for its carbon homogenous distribution (Kim et al., 2011). Thus, the weak flux pinning problem has been well addressed by these recent advanced carbon doping methods, resulting in considerable improvement of high-field J_c of MgB_2.

On the other hand, the other major issue, the weak grain connectivity, has also known to be critical for the superconducting performance of MgB_2 since the discovery of its superconductivity (Rowell, J. M. 2003; Eisterer, 2007). This is more serious compared with the former major issue and it occurs in MgB_2 polycrystalline materials prepared from not only *in situ* reaction of Mg and B mixtures but also *ex situ* fabrication through the use of MgB_2 powder. The *in situ* PIT method is the most popular methods for fabricating MgB_2 conductors and the J_c performance is currently higher than that of *ex situ* processed conductors. However, the *in situ* processed conductors show numerous voids caused by MgB_2 formation (Kim et al., 2007). This is because the theoretical volume of the mixture (V_{Mg} + V_{2B}) is larger than the volume of the reacted MgB_2. The resultant mass density is limited to ~ 50 % of the theoretical density, and the porous structure significantly reduces the supercurrent-carrying area and the J_c for MgB_2 materials. In order to counter the severe problem, *in situ* Mg diffusion methods have been proposed (Canfield et al., 2001; Kang et al., 2001) and it seems to be the most effective method even though there is still room for real application.

2.4 Mg diffusion method

Mg diffusion into a B matrix for fabricating dense MgB_2 materials was suggested by Canfield et al. and Kang et al. around the same time (Canfield et al., 2001; Kang et al., 2001), soon after the discovery of superconductivity in MgB_2. Canfield et al. successfully fabricated the first MgB_2 wire using the Mg diffusion method (Canfield et al., 2001). The specific technique was to expose boron filaments to Mg vapour or liquid at 950 °C for 2 hours. The obtained conductor showed a very low residual resistivity of ~ 0.38 μΩcm and a high density above 80 % of theoretical. Even now, MgB_2 conductors prepared by conventional *in situ* or *ex situ* PIT methods do not reach either value. On the other hand, Kang et al. first made high-quality MgB_2 thin films via a similar method (Kang et al., 2001). The successful technique was to deposit amorphous B thin films and sinter them in Mg vapor at 950 °C for 10-30 minutes. The self-field J_c at 5 K for the obtained film showed a high value of ~ 6.0 × 10^6 A/cm^2 and a low residual resistivity of ~ 2.0 μΩcm, which was likely due to its highly dense structure. Thus, the Mg diffusion method was found to yield higher mass density of MgB_2 matrix in materials compared with conventional *in situ* PIT method, and it has been modified by many researchers for further enhancement of J_c and real applications of various materials such as wires and bulks since it was first discovered (Giunchi, 2003; Ueda et al., 2005; Togano et al., 2009).

For the synthesis of dense MgB_2 bulks, the reactive liquid Mg infiltration (RLI) technique was proposed by Giunchi (Giunchi, 2003). The method has the advantage that desired forms can be fabricated on a large scale and in an inexpensive way without hot pressing methods. The following is a brief procedure of the fabrication method. First, compacted B powders in contact with an appropriate amount of Mg bulk pieces are inserted into steel containers with desirable shapes. The containers are arc-welded in Ar atmosphere using a high-temperature resistant technique and then sintered at 750-950 °C. After sintering, the MgB_2 bulks are obtained by removal of the containers. The mass density of the obtained samples is ~ 2.40 g/cm^3, and the transport J_c values at 4.2 K and 10 T reach 3.0 × 10^3 A/cm^2. Moreover, the RLI technique in combination with *in situ* PIT methods can be applied to the fabrication of monofilament and multifilament MgB_2 hollow wires (Giunchi et al., 2003). It is found that hollow cores are formed in the sintered wire due to Mg diffusion into the B matrix. The formed MgB_2 core was observed to have highly dense structure from SEM observation.

In order to address the problem of the low J_c at self field in MgB_2 bulks, the powder-in-closed-tube (PICT)-diffusion method was suggested by Ueda et al. (Ueda et al., 2005). The bulks were prepared by the following procedure. The starting Mg and B powders in appropriate amounts were separately packed into a stainless steel tube. The tube was deformed into a tape-shaped sample and both ends were sealed by bending and pressing. The samples were vacuum-sealed in a silicon dioxide (SiO_2) tube and then sintered at 800-900 °C. The obtained sample sintered at 800 °C for 60 hours exhibits well-connected small grains with sizes of 30-100 nm. The mass density is above 95 % of the theoretical density, and the self-field J_c value at 20 K reaches ~ 8.6 × 10^5 A/cm^2 due to its good connectivity.

To achieve much less MgO secondary phase, as well as highly dense structure in MgB_2 conductors, an interface diffusion process was proposed by Togano et al. (Togano et al., 2006). The conductors were fabricated from an interface diffusive reaction between an Fe-Mg alloy substrate and B at 950 °C for 12 hours. The formed MgB_2 layer was observed to be composed of densely packed equi-axial grains of submicron size and without oxygen contamination. This group also suggested an internal Mg diffusion (IMD) process for further enhancement of J_c in MgB_2 wires (Hur et al., 2008). Specifically, 19-filament dense MgB_2 wires are fabricated by the internal Mg diffusion (IMD) process and the transport J_c values at 4.2 K and 10 T reach ~ 1.0 × 10^5 A/cm^2 (Togano et al., 2009). This performance is the best J_c reported so far in MgB_2 wires. However, despite these attractive performances of Mg diffusion processed materials, very little work has been done on the effects of sintering temperature on the phase composition, lattice parameters, mass density, grain size, T_c, B_{c2}, and J_c in MgB_2 prepared from Mg diffusion process. In this study, therefore, we have systematically evaluated structural and superconducting properties in Mg diffusion processed MgB_2 bulks sintered under various heat-treated conditions.

3. Experimental

3.1 Sample preparation via Mg diffusion and evaporation processes

All samples were fabricated via Mg diffusion and evaporation processes, as shown in Fig, 1.

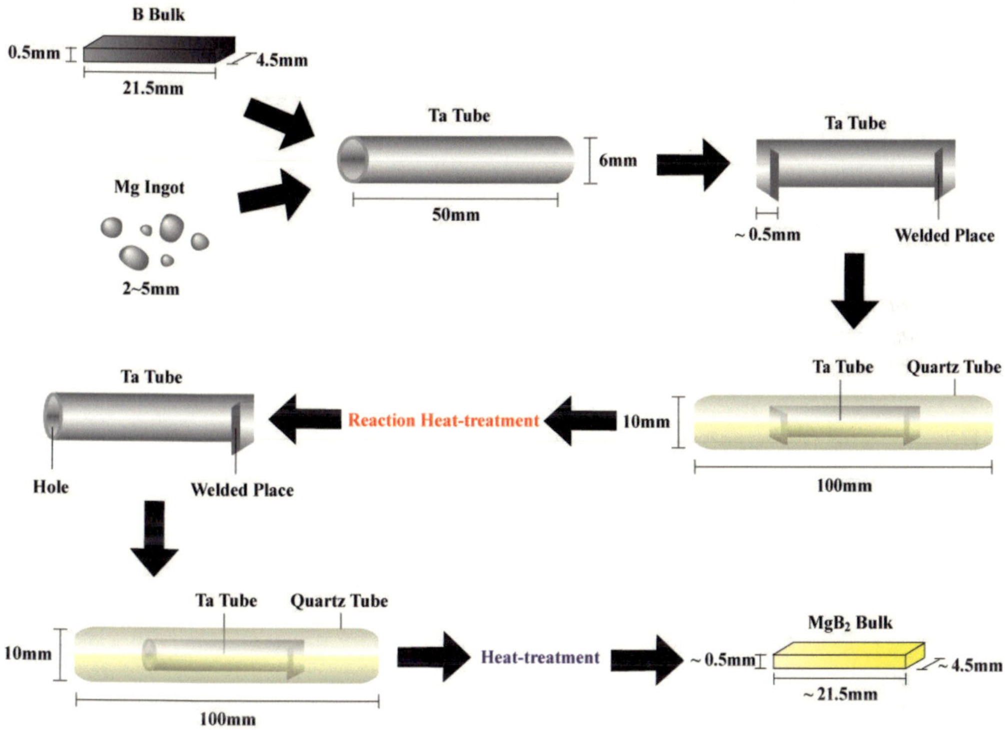

Fig. 1. Schematic diagram showing a fabrication method of MgB_2 via Mg diffusion and evaporation processes.

Mg ingots (purity 99.99 %, size 2-5 mm) and amorphous B powder (purity 99.995 %, size 0.25 μm) were used as starting materials and the molar ratio was chosen to be 1 : 1. The reason for using the large amount of Mg ingots is to eliminate Mg loss caused by MgO formation and to more homogeneously diffuse Mg into B matrix in comparison with what occurs in a Mg diffusion process with the stoichiometric molar ratio. Amorphous B powder was pressed into slabs with dimensions of 4.5 x 21.5 x 0.5 mm under a pressure of 100 MPa. The compacted B powder and Mg ingots were placed in a Ta tube. Both ends of the Ta tube were sealed by a metal inert gas (MIG) welding method in a glove box filled with high purity Ar gas. During the MIG welding process, the Ta tube was heat-sunk with ice-cooled copper block to avoid heating the slab and Mg ingots. The sealed Ta tube was vacuum-sealed in a quartz tube and then heat-treated under four major types of sintering conditions that are (1) an one-step sintering process at 1100 °C which is higher than the boiling point of

Mg, (2) an one-step sintering process at 660 °C which is higher than the melting point of Mg, (3) a two-step process sintered at 1100 °C for a short period of 6 minutes followed by second sintering at 660 °C, and (4) a two-step process sintered at 660 °C for 12 hours followed by second sintering at 1100 °C. After each sintering process for MgB_2 formation, a large amount of unreacted Mg remains in the sealed Ta tubes. In order to remove the residual Mg from MgB_2 sample, the Ta tube was pierced through and again vacuum-sealed in a quartz tube to undergo an Mg evaporation process at 680 °C for 0.5 hour.

3.2 Sample characterization

The phase composition was investigated from X-ray powder diffraction (XRD) patterns which were collected with a Rigaku RINT 2000 diffractmeter using CuKα radiation. Rietveld refinement of the crystal structure was carried out with the RIETAN-2000 program (Izumi & Ikeda, 2000) and the structural parameters were refined by fitting the observed XRD patterns. The morphology and grain size were examined using a HITACHI S-4500 Scanning Electron Microscopy (SEM). The real and the imaginary component curves of ac susceptibility and the dc magnetization loops were evaluated by using a SQUID magnetometer (MPMS-5T, Quantum Design). The T_c and B_{c2} were determined by ac susceptibility measurements at f = 76.97 Hz with $\mu_0 H_{ac}$ = 10 μT and $\mu_0 H_{dc}$ = 0, 0.2, 0.5, 1, 2, 3, 4, and 5 T. The J_c values were derived from the magnetization hysteresis loops using the Bean critical state model (Chen & Goldfarb, 1989).

4. Results and discussion

4.1 Morphology and density

Figure 2 shows the SEM images of the surface of Mg diffusion processed samples via various sintering conditions. For reference, compacted amorphous B powder was also observed, as can be seen in Fig. 2(j). It was found that the one-step sintering process at 1100 °C yields well-developed crystalline grains, as shown in Fig. 2(a), (b), and (c). The crystal grains were also found to grow rapidly with an increase in sintering time. Specifically, the average grain sizes of samples sintered at 1100 °C for 0.5, 1, and 3 hours were observed to be approximately 0.8, 1.5, and 3.0 μm, respectively. The mass densities of the corresponding samples are 2.2, 2.4, and 1.9 g/cm^3, respectively, and close to the theoretical density of MgB_2. However, the drastic decrease in mass density from 2.4 g/cm^3 to 1.9 g/cm^3 is most likely due to the observed rapid growth caused by the increase in sintering time. It is well known that grain boundaries can act as pinning sites in MgB_2 materials (Larbalestier et al., 2001; Eisterer, 2007). The considerably large grains offer the disadvantage of lowering in-field J_c, even though high mass density is easily achieved by the one-step process at 1100 °C for the short sintering time. Contrary to the grain growth, the one-step sintering process at 660 °C for long sintering time ranging from 24 hours to 96 hours was found to leave grain size more or less the same, as can be seen in Fig. 2(d), (e), and (f). It is important to note that the average grain size is much smaller than that of Mg diffusion processed samples at 1100 °C and the grain shape shows poorly crystallized form. It is also well known that crystallinity associated with structural disorder in the lattice as well as grain boundaries can enhance the flux pinning effect (Yamamoto et al., 2005b; Eisterer, 2007). The low-temperature sintering process is advantageous in terms of both poor crystallinity and small grains.

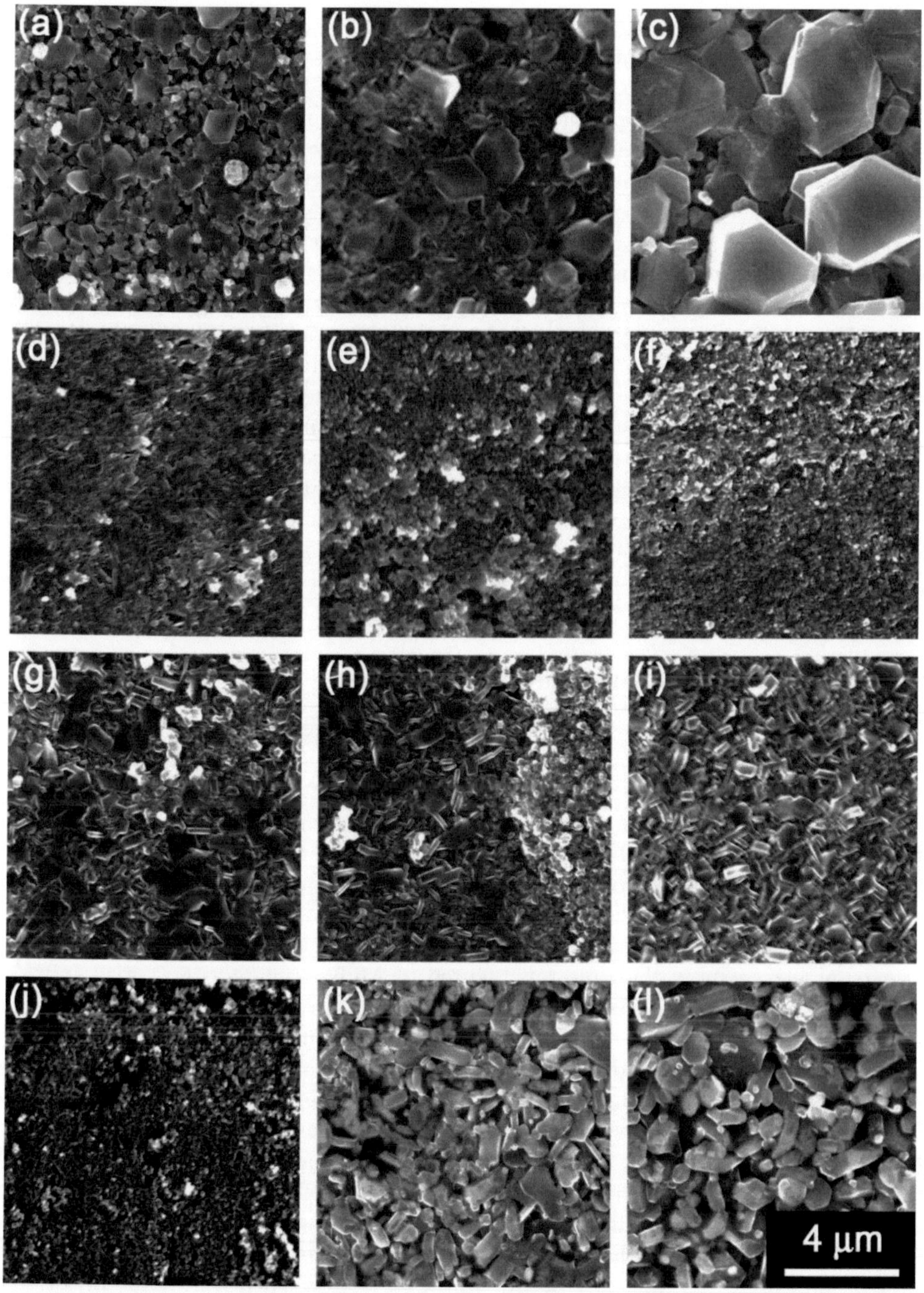

Fig. 2. Microstructure of Mg diffusion processed samples via various sintering conditions including (a) 1100 °C 0.5 h, (b) 1100 °C 1 h, (c) 1100 °C 3 h, (d) 660 °C 24 h, (e) 660 °C 48 h, (f) 660 °C 96 h, (g) 1100 °C 0.1h - 660 °C 3 h, (h) 1100 °C 0.1 h - 660 °C 6 h, (i) 1100 °C 0.1h - 660 °C 12 h, (k) 660 °C 12 h - 1100 °C 0.25 h, and (l) 660 °C 12 h - 1100 °C 0.5 h. (j) Microstructure of compacted B powder before sintering.

However, the Mg diffusion processed samples at 660 °C were brittle and the mass densities are ~ 1.5 g/cm³. The value is almost half of the theoretical density, even though the SEM images of the corresponding samples seem to be dense. The porous structure caused by the low density may intensively limit grain connectivity and hence supercurrent flow. On the other hand, the two-step process first sintered at 1100 °C for 0.1 hour followed by second sintering at 660 °C for 3 -12 hours was found to form well-developed crystalline grains, as can be seen in Fig. 2(g), (h), and (i). The average grain size is below 1 μm and smaller than that of Mg diffusion processed samples via the one-step sintering at 1100 °C. However, unlike Mg diffusion processed samples via the one-step sintering at 660 °C which have small grains, the mass densities are ~ 2.4 -2.5 g/cm³ and those are above 90 % of the theoretical density. The two-step sintering process achieves both high density and small grains, which arise from the natures of high-temperature sintering and low-temperature sintering, respectively. In contrast to the highly dense structure related to the well-connected small grains, the two-step process first sintered at 660 °C for 12 hours followed by second sintering at 1100 °C for 0.25 - 0.5 hour was found to result in loosely connected grains, as can be seen in Fig. 2(k) and (l). The mass densities are ~ 1.7 g/cm³ and this result also confirm its porous structure. The corresponding samples were observed to have well-developed crystalline and relatively large grains. The two-step sintering process yields low-density structure and large crystal grains due to low-temperature sintering and high-temperature sintering, respectively.

4.2 Phase composition and lattice structure

All Mg diffusion processed samples were observed to have the same phase composition, which consists of MgB_2 with a trace amount of MgO from XRD θ-2θ scans. Hereafter the samples will be referred as MgB_2 samples. In order to examine deeper insight of microstructure, structural parameters of four MgB_2 samples sintered via four major conditions were refined and its result is listed in Table 1. It was found that the a-lattice parameter is slightly changed owing to various sintering effects, while leaving the c-lattice parameter the same. The a-lattice parameter of the MgB_2 sample sintered at 660 °C for 48 hours shows 3. 0850 Å, and it is the lowest value of the four samples. The decrease in the a-lattice parameter reduces the unit cell volume of MgB_2. The reduction may introduce structural disorder in the lattice and decrease the DOS at the E_F responsible for a considerable change in T_c and thus B_{c2}. The XRD full width at half maximum (FWHM) is well known to be closely related to crystallite size and structural disorder and it is also listed in Table 1. The one-step sintering process at 1100 °C and the two-step process first sintered at 660 °C followed by second sintering at 1100 °C were found to obtain relatively narrow FWHM reflecting well-crystallized and large grain. This is consistent with the SEM result, as can be seen in Fig. 2. On the other hand, the FWHM values of the rest of MgB_2 samples were observed to be much broader than those of the other MgB_2 samples. The broad FWHM of the MgB_2 sample sintered at the one-step process at 660 °C is caused by poorly crystallized and small grains, as also indicated by the SEM observation. However, it should be noted that the FWHM of the MgB_2 sample first sintered at 1100 °C followed by second sintering at 660 °C is comparable to that of the sample sintered at 660 °C for 48 hours, although the corresponding sample seems to have well crystallized grains which are distinctly lager than that of MgB_2 samples sintered at the one-step process at 660 °C. This indicates that thanks to the second sintering at the low temperature, the two-step process could achieve both a high level of disorder and small grains in the high-density MgB_2 samples, which are essential for enhancing the flux pinning effect. The MgO nanoparticles are also well known to act as pinning sites in MgB_2 thin films (Eom et al., 2001; Singh et al., 2008), whereas it is difficult to control the size in polycrystalline MgB_2 samples and large MgO grains could impede supercurrent flow (Maeda et al., 2011a). In this respect, it is necessary to eliminate the MgO impurity from polycrystalline MgB_2 materials. However, the MgO secondary phase of the four MgB_2 samples is not sensitive to sintering conditions. This indicates that even Mg diffusion process could not avoid MgO formation.

Sintering condition	1100°C1h	660°C48h	1100°C0.1h - 660°C6h	660°C12h - 1100°C0.5h
	X-ray powder diffraction data			
Radiation source	CuKα			
λ (Å)	1.54060 (CuKα_1), 1.54443 (CuKα_2)			
Δ2θ (°)	0.01			
Temperature (K)	298			
	Goodness of fit			
s	1.32	1.27	1.23	1.23
	Lattice parameters			
a (Å)	3.0864 (2)	3.0850 (2)	3.0859 (2)	3.0868 (2)
c (Å)	3.5210 (1)	3.5210 (1)	3.5210 (1)	3.5211 (2)
	Unit cell volume			
V (Å^3)	29.047 (2)	29.020 (3)	29.038 (3)	29.054 (3)
	Reflection of MgB_2 (101)			
FWHM(°)	0.15	0.31	0.29	0.10
	Weight fraction			
MgB_2 (wt%)	95.9	95.8	96.3	96.1
MgO (wt%)	4.1	4.2	3.7	3.9

Table 1. Results of Rietveld refinement on XRD data for Mg diffusion processed samples via four major sintering conditions that are 1100 °C for 1 hour, 660 °C for 48 hours, 1100 °C for 0.1 hour followed by second sintering at 660 °C for 6 hours, and 660 °C for 12 hours followed by second sintering at 1100 °C for 0.5 hour. The fitting quality between the observed pattern and the calculated one was evaluated by goodness of fit *s*.

4.3 Influence on T_c

Figure 3 shows the real and imaginary components of the ac susceptibility for four MgB_2 samples sintered via four major conditions. It was found that the onset of diamagnetism at the dc field of 0 T for the MgB_2 sample sintered at 660 °C for 48 hours is ~ 38 K and 1 K lower than those for the other samples. The decrease in T_c is most likely due to lattice disorder caused by the reduction of unit cell volume, as indicated by the refinement result listed in Table 1. It is noteworthy that although the other three samples show ~ 39 K of the onset diamagnetism at the dc field of 0 T, the superconducting transition curves are different. Specifically, the transition width of the MgB_2 sample first sintered at 1100 °C followed by second sintering at 660 °C is ~ 1.5 K and approximately two times broader than that of the MgB_2 sample sintered by the one-step process at either 1100 °C or 660 °C. It was reported that the broad transition width arises from the nature of two kinds of superconducting MgB_2 phases, which are good ordering phase and a relatively lower level of ordering phase owing to the two-step sintering process (Maeda et al., 2008). The lower level of ordering phase could be a cause for the broad FWHM responsible for structural disorder, as observed in the XRD pattern (Table 1). Contrary to this, the transition width of the MgB_2 sample first sintered at 660 °C followed by second sintering at 1100 °C is ~ 0.3 K, suggesting that the sample consists of nearly perfect crystallite MgB_2 grains without any lattice disorder.

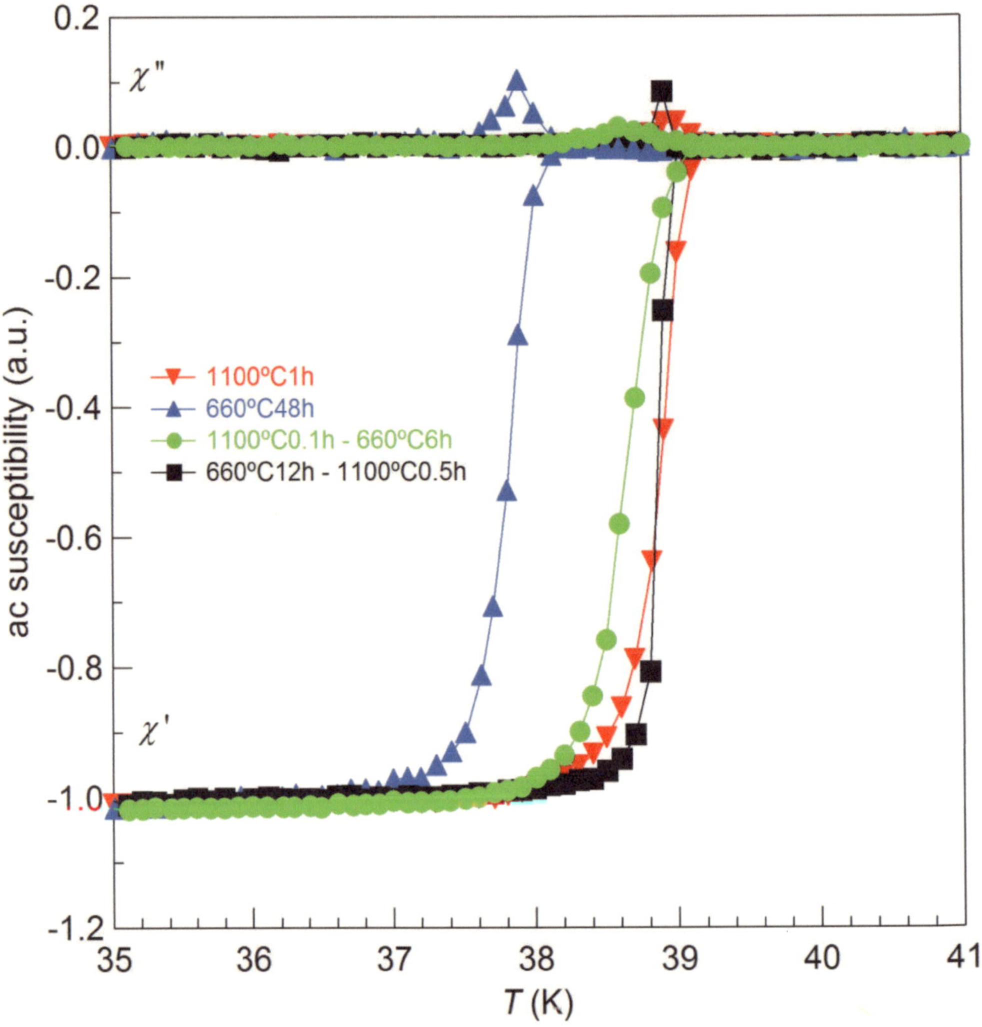

Fig. 3. The real and imaginary components of ac susceptibility with f = 76.97 Hz, $\mu_0 H_{ac}$ = 10 μT, and $\mu_0 H_{dc}$ = 0 T for Mg diffusion processed samples via four major sintering conditions that are 1100 °C for 1 hour, 660 °C for 48 hours, 1100 °C for 0.1 hour followed by second sintering at 660 °C for 6 hours, and 660 °C for 12 hours followed by second sintering at 1100 °C for 0.5 hour.

4.4 Influence on B_{c2}

Figure 4 shows the temperature dependence of B_{c2} for MgB_2 samples sintered via various conditions.

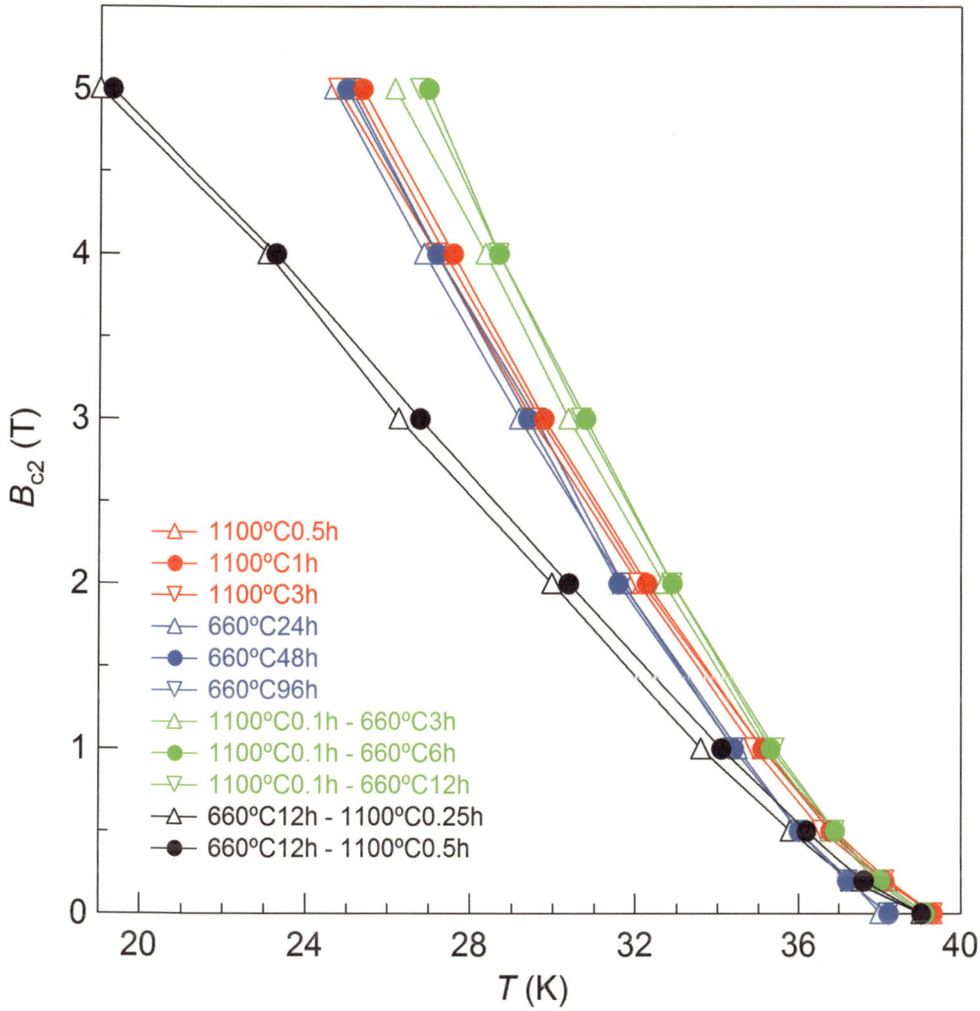

Fig. 4. The temperature dependence of B_{c2} derived from ac susceptibility with f = 76.97 Hz, $\mu_0 H_{ac}$ = 10 μT, and $\mu_0 H_{dc}$ = 0, 0.2, 0.5, 1, 2, 3, 4, and 5 T for MgB_2 samples sintered via various conditions.

It was found that although all MgB_2 samples has positive curvatures of B_{c2} at $T \sim T_c$ reflecting suppression of the diffusivity in the σ bands compared to that in the π bands (Gurevich, 2003), the behaviours of curves at lower temperature are considerably different. The B_{c2} slopes of the MgB_2 samples first sintered at 1100 °C followed by second sintering at 660 °C or sintered only at 660 °C are steeper in comparison with those of the MgB_2 samples sintered at 1100 °C. The structural disorder of both samples may be responsible for this. By contrast, the B_{c2} slopes of the MgB_2 samples sintered at 660 °C followed by second sintering at 1100 °C are intensively shelving compared to those of the other MgB_2 samples. This indicates that the MgB_2 samples do not have any lattice defects, as suggested by the result of the narrow superconducting transition (Figure 3).

4.5 Influence on J_c

Figure 5 shows the magnetic field dependence of J_c at 20 K for MgB_2 samples sintered via various conditions. It was found that the J_c curves are highly sensitive to sintering conditions which lead to considerably differences in MgB_2 structure, as shown in the results of SEM observation, XRD analysis, and ac susceptibility. The most noticeable feature shown in this figure is that the in-field J_c of the MgB_2 samples sintered at 1100 °C followed by

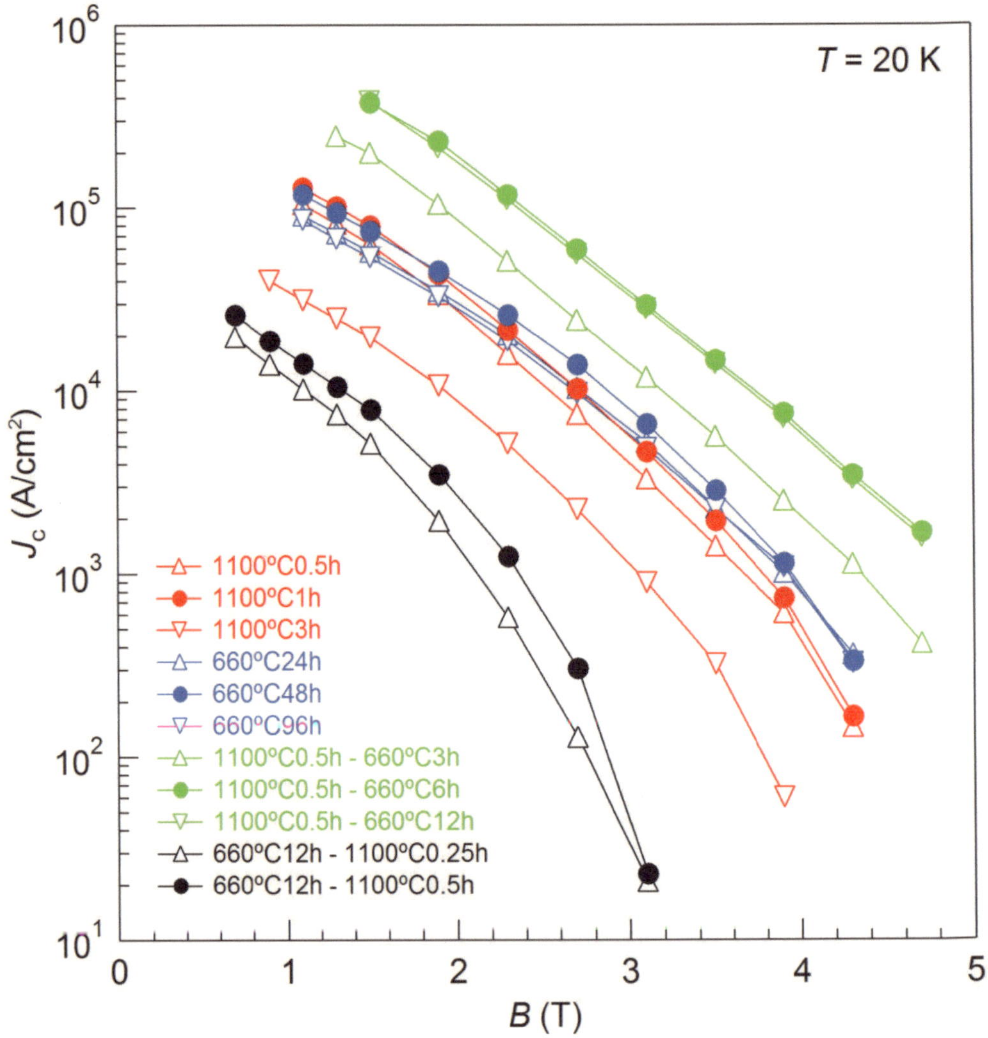

Fig. 5. Magnetic field dependence of J_c at 20 K for MgB_2 samples sintered via various conditions.

second sintering at 660 °C is significantly higher than those of the other MgB_2 samples. The enhancement of J_c at not only low field but also high field is clearly due to the two-step sintering process which yields well-connected small grains with a high level of disorder in highly dense MgB_2 samples. It is also interesting to note that even though the J_c of the MgB_2 sample sintered at 1100 °C for 1 hour is nearly the same as that of the MgB_2 sample sintered at 660 °C for 48 hours, the behaviours of the curves are somewhat different. Specifically, the low-field J_c of the MgB_2 sample sintered at 1100 °C for 1 hour is slightly higher than that of the MgB_2 sample sintered at 660 °C for 48 hours, whereas the corresponding behaviours of high-field J_c are distinctly opposite. This results from high density and lack of pinning sites caused by good crystallinity, which are responsible for the increase in self-field J_c and the decrease in high-field J_c, respectively. Owing to the low density and nearly perfect crystallization, the MgB_2 samples sintered at 660 °C followed by second sintering at 1100 °C show the lowest values of in-field J_c in all MgB_2 samples.

5. Conclusion

In summary, this chapter has introduced structural characteristic and superconducting performance of pure MgB_2 polycrystalline samples fabricated by Mg diffusion processes. The Mg diffusion processes easily yield MgB_2 phase and a trace amount of MgO phase in polycrystalline samples. The microstructure and mass density of the MgB_2 samples can be controlled by Mg diffusion processes via various sintering conditions. The two-step sintering process, namely, the sintering for a short period of 0.1 hour at 1100 °C first, followed by the second sintering at 660 °C, produces highly dense MgB_2 samples characterized by well-connected small grains with a high level of disorder. The unique feature results in considerable improvement of B_{c2} and in-field J_c. Therefore, Mg diffusion process is critical for controlling the structure of MgB_2 polycrystalline samples, and there is great potential for further enhancement of J_c by careful optimisation of sintering conditions.

6. References

An, J. M. & Pickett, W. E. (2001). Superconductivity of MgB2: Covalent bonds driven metallic. *Physical Review Letters*, 86(19): 4366-4369, ISSN 0031-9007.

Avdeev, M.; Jorgensen, J. D.; Ribeiro, R. A.; Bud'ko, S. L. & Canfield, P. C. (2003). Crystal chemistry of carbon-substituted MgB2. *Physica C - Superconductivity and Its Applications*, 387(3-4): 301-306, ISSN 0921-4534.

Bengtson, A. K.; Bark, C. W.; Giencke, J.; Dai, W. Q.; Xi, X. X.; Eom, C. B. & Morgan, D. (2010). Impact of substitutional and interstitial carbon defects on lattice parameters in MgB2. *Journal of Applied Physics*, 107(2): 023902, ISSN 0021-8979.

Braccini, V.; Gurevich, A.; Giencke, J. E.; Jewell, M. C.; Eom, C. B.; Larbalestier, D. C.; Pogrebnyakov, A.; Cui, Y.; Liu, B. T.; Hu, Y. F.; Redwing, J. M.; Li, Q.; Xi, X. X.; Singh, R. K.; Gandikota, R.; Kim, J.; Wilkens, B.; Newman, N.; Rowell, J.; Moeckly, B.; Ferrando, V.; Tarantini, C.; Marre, D.; Putti, M.; Ferdeghini, C.; Vaglio, R. & Haanappel, E. (2005). High-field superconductivity in alloyed MgB2 thin films. *Physical Review B*, 71(1): 012504, ISSN 1098-0121.

Canfield, P. C.; Finnemore, D. K.; Bud'ko, S. L.; Ostenson, J. E.; Lapertot, G.; Cunningham, C. E. & Petrovic, C. (2001). Superconductivity in dense MgB2 wires. *Physical Review Letters*, 86(11): 2423-2426, ISSN 0031-9007.

Chen, D. X. & Goldfarb, R. B. (1989). Kim Model for Magnetization of Type-II Superconductors. *Journal of Applied Physics*, 66(6): 2489-2500, ISSN 0021-8979.

Dou, S. X.; Soltanian, S.; Horvat, J.; Wang, X. L.; Zhou, S. H.; Ionescu, M.; Liu, H. K.; Munroe, P. & Tomsic, M. (2002). Enhancement of the critical current density and flux pinning of MgB2 superconductor by nanoparticle SiC doping. *Applied Physics Letters*, 81(18): 3419-3421, ISSN 0003-6951.

Eisterer, M. (2007). Magnetic properties and critical currents of MgB2. *Superconductor Science & Technology*, 20(12): R47-R73, ISSN 0953-2048.

Eisterer, M.; Zehetmayer, M.; Weber, H. W. & Karpinski, J. (2005). Reversible magnetization of the two-band MgB2 superconductor: A phenomenological approach. *Physical Review B*, 72(13): 134525, ISSN 1098-0121.

Eom, C. B.; Lee, M. K.; Choi, J. H.; Belenky, L. J.; Song, X.; Cooley, L. D.; Naus, M. T.; Patnaik, S.; Jiang, J.; Rikel, M.; Polyanskii, A.; Gurevich, A.; Cai, X. Y.; Bu, S. D.; Babcock, S. E.; Hellstrom, E. E.; Larbalestier, D. C.; Rogado, N.; Regan, K. A.; Hayward, M. A.; He, T.; Slusky, J. S.; Inumaru, K.; Haas, M. K. & Cava, R. J. (2001). High critical current density and enhanced irreversibility field in superconducting MgB2 thin films. *Nature*, 411(6837): 558-560, ISSN 0028-0836.

Ferrando, V.; Orgiani, P.; Pogrebnyakov, A. V.; Chen, J.; Li, Q.; Redwing, J. M.; Xi, X. X.; Giencke, J. E.; Eom, C. B.; Feng, Q. R.; Betts, J. B. & Mielke, C. H. (2005). High upper critical field and irreversibility field in MgB2 coated-conductor fibers. *Applied Physics Letters*, 87(25): 252509, ISSN 0003-6951.

Giunchi, G. (2003). High density MgB2 obtained by reactive liquid Mg infiltration. *International Journal of Modern Physics B*, 17(4-6): 453-460, ISSN 0217-9792.

Giunchi, G.; Ceresara, S.; Ripamonti, G.; Di Zenobio, A.; Rossi, S.; Chiarelli, S.; Spadoni, M.; Wesche, R. & Bruzzone, P. L. (2003). High performance new MgB2 superconducting hollow wires. *Superconductor Science & Technology*, 16(2): 285-291, ISSN 0953-2048.

Golubov, A. A. & Mazin, I. I. (1997). Effect of magnetic and nonmagnetic impurities on highly anisotropic superconductivity. *Physical Review B*, 55(22): 15146-15152, ISSN 1098-0121.

Golubov, A. A.; Kortus, J.; Dolgov, O. V.; Jepsen, O.; Kong, Y.; Andersen, O. K.; Gibson, B. J.; Ahn, K. & Kremer, R. K. (2002). Specific heat of MgB2 in a one- and a two-band model from first-principles calculations. *Journal of Physics-Condensed Matter*, 14(6): 1353-1360, ISSN 0953-8984.

Gurevich A. (2003). Enhancement of the upper critical field by nonmagnetic impurities in dirty two-gap superconductors. *Physical Review B*, 67(18): 184515, ISSN 1098-0121.

Gurevich A. (2007). Limits of the upper critical field in dirty two-gap superconductors. *Physica C - Superconductivity and Its Applications*, 456(1-2): 160-169, ISSN 0921-4534.

Hossain, M. S. A.; Senatore, C.; Fluekiger, R.; Rindfleisch, M. A.; Tomsic, M. J.; Kim, J. H. & Dou, S. X. (2009). The enhanced Jc and Birr of in situ MgB2 wires and tapes alloyed with C4H6O5 (malic acid) after cold high pressure densification. *Superconductor Science & Technology*, 22(9): 095004, ISSN 0953-2048.

Hur, J. M.; Togano, K.; Matsumoto, A.; Kumakura, H.; Wada, H. & Kimura, K. (2008). Fabrication of high-performance MgB2 wires by an internal Mg diffusion process. *Superconductor Science & Technology*, 21(3): 032001, ISSN 0953-2048.

Izumi, F. & Ikeda, T. (2000). A Rietveld-analysis program RIETAN-98 and its applications to zeolites. *Materials Science Forum*, 321-324: 198-205, ISSN 0255-5476.

Jones, M. E. & Marsh, R. E. (1954). The Preparation and Structure of Magnesium Boride, MgB2. *Journal of the American Chemical Society*, 76(5): 1434-1436.

Jorgensen, J. D.; Hinks, D. G. & Short, S. (2001). Lattice properties of MgB2 versus temperature and pressure. *Physical Review B*, 63(22): 224522, ISSN 0163-1829.

Kang, W. N.; Kim, H. J.; Choi, E. M.; Jung, C. U. & Lee, S. L. (2001). MgB2 superconducting thin films with a transition temperature of 39 Kelvin. *Science*, 292(5521): 1521-1523, ISSN 0036-8075.

Kazakov, S. M.; Puzniak, R.; Rogacki, K.; Mironov, A. V.; Zhigadlo, N. D.; Jun, J.; Soltmann, C.; Batlogg, B. & Karpinski, J. (2005). Carbon substitution in MgB2 single crystals: Structural and superconducting properties. *Physical Review B*, 71(2): 024533, ISSN 1098-0121.

Kim, J. H.; Dou, S. X.; Shi, D. Q.; Rindfleisch, M. & Tomsic, M. (2007). Study of MgO formation and structural defects in in situ processed MgB2/Fe wires. *Superconductor Science & Technology*, 20(10): 1026-1031, ISSN 0953-2048.

Kim, J. H.; Heo, Y. U.; Matsumoto, A.; Kumakura, H.; Rindfleisch, M.; Tomsic, M. & Dou, S. X. (2010). Comparative study of mono- and multi-filament MgB2 wires with different boron powders and malic acid addition. *Superconductor Science & Technology*, 23(7): 075014, ISSN 0953-2048.

Kim, J. H.; Oh, S.; Heo, Y. U.; Hata, S.; Kumakura, H.; Matsumoto, A.; Mitsuhara, M.; Choi, S.; Shimada, Y.; Maeda, M.; MacManus-Driscoll, J. L. & Dou, S. X. (2012). Microscopic role of carbon on MgB2 wire for critical current density comparable to NbTi. *NPG Asia Materials*, 4: ISSN(online) 1884-4057.

Kim, J. H.; Oh, S.; Kumakura, H.; Matsumoto, A.; Heo, Y. U.; Song, K. S.; Kang, Y. M.; Maeda, M.; Rindfleisch, M.; Tomsic, M.; Choi, S & Dou, S. X. (2011) Tailored materials for high performance MgB2 wire. *Advanced Materials*, 23(42): 4942-4946, ISSN 0935-9648.

Kim, J. H.; Zhou, S.; Hossain, M. S. A.; Pan, A. V. & Dou, S. X. (2006). Carbohydrate doping to enhance electromagnetic properties of MgB2 superconductors. *Applied Physics Letters*, 89(14): 142505, ISSN 0003-6951.

Kortus, J.; Mazin, I. I.; Belashchenko, K. D.; Antropov, V. P. & Boyer, L. L. (2001). Superconductivity of metallic boron in MgB2. *Physical Review Letters*, 86(20): 4656-4659, ISSN 0031-9007.

Larbalestier, D.; Gurevich, A.; Feldmann, D. M. & Polyanskii, A. (2001). High-Tc superconducting materials for electric power applications. *Nature*, 414(6861): 368-377, ISSN 0028-0836.

Liu, A. Y.; Mazin, I. I. & Kortus, J. (2001). Beyond Eliashberg superconductivity in MgB2: Anharmonicity, two-phonon scattering, and multiple gaps. *Physical Review Letters*, 87(8): 087005, ISSN 0031-9007.

Lyard, L.; Samuely, P.; Szabo, P.; Klein, T.; Marcenat, C.; Paulius, L.; Kim, K. H. P.; Jung, C. U.; Lee, H. S.; Kang, B.; Choi, S.; Lee, S. I.; Marcus, J.; Blanchard, S.; Jansen, A. G. M.; Welp, U.; Karapetrov, G. & Kwok, W. K. (2002). Anisotropy of the upper critical field and critical current in single crystal MgB2. *Physical Review B*, 66(18): 180502, ISSN 1098-0121.

Maeda, M.; Kim, J. H.; Kumakura, H.; Heo, Y. -U.; Zhao, Y.; Nakayama, Y.; Rindfleisch, M. & Dou, S. X. (2011a). Influence of hydrogen-containing argon gas on the structural parameters and superconducting properties of malic acid-doped MgB2 wires. *Scripta Materialia*, 64(11): 1059-1062, ISSN 1359-6462.

Maeda, M.; Kim, J. H.; Zhao, Y.; Heo, Y. U.; Takase, K.; Kubota, Y.; Moriyoshi, C.; Yoshida, F.; Kuroiwa, Y. & Dou, S. X. (2011b). In-field Jc improvement by oxygen-free pyrene gas diffusion into highly dense MgB2 superconductor. *Journal of Applied Physics*, 109(2): 023904, ISSN 0021-8979.

Maeda, M.; Zhao, Y.; Dou, S. X.; Nakayama, Y.; Kawakami, T.; Kobayashi, H. & Kubota, Y. (2008). Fabrication of highly dense MgB2 bulk at ambient pressure. *Superconductor Science & Technology*. 21(3): 032004, ISSN 0953-2048.

Mazin, I. I. & Antropov, V. P. (2003). Electronic structure, electron-phonon coupling, and multiband effects in MgB2. *Physica C - Superconductivity and Its Applications*, 385(1-2): 49-65, ISSN 0921-4534.

Mazin, I. I.; Andersen, O. K.; Jepsen, O.; Dolgov, O. V.; Kortus, J.; Golubov, A. A.; Kuz'menko, A. B. & van der Marel, D. (2002). Superconductivity in MgB2: Clean or dirty?. *Physical Review Letters*, 89(10): 107002, ISSN 0031-9007.

Mori, H.; Lee, S.; Yamamoto, A.; Tajima, S. & Sato, S. (2002). Electron density distribution in a single crystal of Mg1-xB2 [x=0.045(5)]. *Physical Review B*, 65(9): 092507, ISSN 1098-0121.

Nagamatsu, J.; Nakagawa, N.; Muranaka, T.; Zenitani, Y. & Akimitsu, J. (2001). Superconductivity at 39 K in magnesium diboride. *Nature*, 410(6824): 63-64, ISSN 0028-0836.

Rowell, J. M. (2003). The widely variable resistivity of MgB2 samples. *Superconductor Science & Technology*, 16(6): R17-R27, ISSN 0953-2048.

Sarma, G. (1963). On the influence of a uniform exchange field acting on the spins of the conduction electrons in a superconductor. *Journal of Physics and Chemistry of Solids*, 24(8): 1029-1032.

Singh, R. K.; Shen, Y.; Gandikota, R.; Carvalho, C.; Rowell, J. M. & Newman, N. (2008). Effect of oxygen incorporation on normal and superconducting properties of MgB2 films. *Applied Physics Letters*, 93(24): 242504, ISSN 0003-6951.

Soltanian, S.; Horvat, J.; Wang, X. L.; Munroe, P. & Dou, S. X. (2003). Effect of nano-carbon particle doping on the flux pinning properties of MgB2 superconductor. *Physica C - Superconductivity and Its Applications*, 390(3): 185-190, ISSN 0921-4534.

Sumption, M. D.; Bhatia, M.; Rindfleisch, M.; Tomsic, M.; Soltanian, S.; Dou, S. X. & Collings, E. W. (2005). Large upper critical field and irreversibility field in

MgB2 wires with SiC additions. *Applied Physics Letters*, 86(9): 092507, ISSN 0003-6951.

Tajima, S. (2005). Two-gap superconductivity in MgB2. *Agne Gijutsu Center Solid State Physics*, 40(1): 1-12, ISSN 0454-4544.

Togano, K.; Hur, J. M.; Matsumoto, A. & Kumakura, H. (2009). Fabrication of seven-core multi-filamentary MgB2 wires with high critical current density by an internal Mg diffusion process. *Superconductor Science & Technology*, 22(1): 015003, ISSN 0953-2048.

Togano, K.; Nakane, T.; Fujii, H.; Takeya, H. & Kumakura, H. (2006). An interface diffusion process approach for the fabrication of MgB2 wire. *Superconductor Science & Technology*, 19(6): L17-L20, ISSN 0953-2048.

Tsirelson, V.; Stash, A.; Kohout, M.; Rosner, H.; Mori, H.; Sato, S.; Lee, S.; Yamamoto, A.; Tajima, S. & Grin, Y. (2003). Features of the electron density in magnesium diboride: reconstruction from X-ray diffraction data and comparison with TB-LMTO and FPLO calculations. *Acta Crystallographica Section B -Structural Science*, 59(5): 575-583, ISSN 0108-7681.

Ueda, S.; Shimoyama, J.; Iwayama, I.; Yamamoto, A.; Katsura, Y.; Horii, S. & Kishio, K. (2005). High critical current properties of MgB2 bulks prepared by a diffusion method. *Applied Physics Letters*, 86(22): 222502, ISSN 0003-6951

Xi, X. X. (2008). Two-band superconductor magnesium diboride. *Reports on Progress in Physics*, 71(11): 116501, ISSN 0034-4885.

Yamamoto, A.; Shimoyama, J.; Ueda, S.; Iwayama, I.; Horii, S. & Kishio, K. (2005a). Effects of B4C doping on critical current properties of MgB2 superconductor. *Superconductor Science & Technology*, 18(10): 1323-1328, ISSN 0953-2048.

Yamamoto, A; Shimoyama, J; Ueda, S; Katsura, Y; Horii, S. & Kishio, K. (2005b). Improved critical current properties observed in MgB2 bulks synthesized by low-temperature solid-state reaction. *Superconductor Science & Technology*, 18(1): 116-121, ISSN 0953-2048.

Yeoh, W. K.; Horvat, J.; Dou, S. X. & Keast, V. (2004). Strong pinning and high critical current density in carbon nanotube doped MgB2. *Superconductor Science & Technology*, vol. 17(9): S572-S577, ISSN 0953-2048.

Yildirim, T.; Gulseren, O.; Lynn, J. W.; Brown, C. M.; Udovic, T. J.; Huang, Q.; Rogado, N.; Regan, K. A.; Hayward, M. A.; Slusky, J. S.; He, T.; Haas, M. K.; Khalifah, P.; Inumaru, K. & Cava, R. J. (2001). Giant Anharmonicity and Nonlinear Electron-Phonon Coupling in MgB2: A Combined First-Principles Calculation and Neutron Scattering Study. *Physical Review Letters*, 87(3): 037001, ISSN 0031-9007.

Zehetmayer, M.; Eisterer, M.; Jun, J.; Kazakov, S. M.; Karpinski, J.; Wisniewski, A. & Weber, H. W. (2002). Mixed-state properties of superconducting MgB2 single crystals. *Physical Review B*, 66(5): 052505, ISSN 1098-0121.

Zhu, Y.; Larbalestier, D. C.; Voyles, P. M.; Pogrebnyakov, A. V.; Xi, X. X. & Redwing, J. M. (2007a). Nanoscale disorder in high critical field, carbon-doped MgB2 hybrid physical-chemical vapor deposition thin films. *Applied Physics Letters*, 91(8): 082513, ISSN 0003-6951.

Zhu, Y.; Matsumoto, A.; Senkowicz, B. J.; Kumakura, H.; Kitaguchi, H.; Jewell, M. C.; Hellstrom, E. E.; Larbalestier, D. C. & Voyles, P. M. (2007b). Microstructures of SiC nanoparticle-doped MgB2/Fe tapes. *Journal of Applied Physics*, 102(1): 013913, ISSN 0021-8979.

14

Measurement of Levitation Forces in High Temperature Superconductors

İbrahim Karaca
Department of Physics, Niğde University, Niğde
Turkey

1. Introduction

Magnetic Levitation (MagLev) has become a popular of new technological applications since the discovery of the new superconductor materials, in 1987. MagLev transportation systems have been ignited the human's imagination about levitated motion. MagLev systems have become very important in the transportation and other applications in the future life in the world is generally based on the superconductor technology. These applications are ranging from in an energy efficient prototype of a cryogen transfer line to in space energy storage systems.

The superconducting behaviors of materials originated from two basic physical properties. One of them is disappearance of the electrical resistance below the transition temperature (T_C) in the superconducting materials. The another of them is special magnetic properties in an applied magnetic field (H) including perfect diamagnetism at low fields (low temperature superconductors (LTS) and high temperature superconductors (HTSs)) and the penetration of quantized magnetic flux (vortices or flux lines) at higher fields regions (HTSs). The electrons in the superconducting state can be described by a macroscopic wave function which is undisturbed by scattering and, in the absence of applied fields and currents, uniform in space over macroscopic distances. This coherence of the superconducting state is preserved even in the presence of weak currents and magnetic fields (below the upper critical field (H_{C2}), which is a necessary condition for the practical use of the high temperatures superconductors (HTSs) in strong magnetic fields. The additional condition for new applications in these superconductors materials have been used to avoid any motion of flux lines. This can be achieved through the introduction of pinning centers in the superconductor, interacting, in most cases, with the normal conducting core of the flux lines. A flux line is composed of a normal conducting core regions and a surrounding circulating supercurrents area where the magnetic field and supercurrents fall off within penetration depth. The critical current density (J_C) increases with the applied magnetic field (H) increases and penetration depth (λ) decreases. The force-distance (permanent magnets and superconductor) hysteresis loops during the descending and ascending process expanded with critical current density increases.

The mechanism of superconductivity in high temperature superconductor (HTS) is still controversially discussed. The Bardeen-Cooper-Schrieffer (BCS) concept of a macroscopic quantum state of Cooper pairs remains the common basis in almost all theories proposed for HTS. However, there is no consensus about the origin of the pairing so far. The lack of an isotope effect in HTS of highest T_C would favor a non-phonon pairing mechanism as, for instance, d-wave pairing, whereas its presence in several HTS with lower T_C suggests a phonon-mediated s-wave pairing. The most experimental features favor d-wave pairing. A direct test of the phonon mechanism for pairing in HTSs is to check the sign change in the energy gap $(\Delta E_g(k))$ as function of momentum $(\hbar k)$. A d-state energy gap is composed of two pairs of positive and negative leaves forming a four-leaf clover. In contrast, the magnitude of an s-state may be anisotropic but always remains positive.

2. Levitation with superconductors

Superconducting levitation by using the permanent magnet (PM) is a fascinating property of the magnetic behavior of these materials. It can be mainly defined in two regimes for HTSs. In the first regime, the magnetic induction applied an external magnetic field to samples are expelled from the superconductor and do not penetrate inside of the superconductor materials, which is defined as Meissner effect. This effect is possible only if the applied magnetic field does not exceed the first critical field (so called H_{C1}) of the HTS. To obtain experimentally levitation in the Meissner state, it is convenient to close the magnet very carefully, after cooling the superconductor. In the second regime of the levitation, i.e., the mixed state, a vortex penetration through for HTSs, occurs only for fields between the first (H_{C1}) and the second critical magnetic field (H_{C2}) of the HTSs. This is so because magnetic lines could be pinned if they penetrate in the sample and the superconductor would be switched to the mixed state. It is clearly that the levitation measurement is finding in the mixed state acceptable. Magnetic levitation effects are seen Meissner states as the same in HTSs and LTSs. But, magnetic levitation effects originated from flux lines for only HTSs.

When a superconductor is in the presence of a changing applied magnetic field, supercurrents (I_C) are induced inside the sample producing a diamagnetic response of the material. The interaction of the supercurrents with an inhomogeneous magnetic field can produce stable levitation. The flux lines in the small currents densities ($J < J_C$ for a certain critical current J_C) for the real HTSs can be pinned by inhomogenities. Under this condition, when an external applied field increases or decreases, the magnetic flux moves within the superconductor towards its interior or exterior until a critical slope is reached in the flux profile. The current density in this critical state attains its maximum value J_C. The critical slope behavior has been modeled by Bean's critical-state model and its extensions, phenomenological. Brandt showed qualitatively how from the pinning of flux lines some properties of levitation of HTSs can be explained in detail (Brandt, 1998).

Moon *et al.* measured the vertical forces on a levitating superconductor showing some of the main properties (Moon, 1988). This measurement has been extended by several groups in order to study the influence of orientation, material, or shape of superconducting samples, comparing, sometimes, thin films with bulk samples (Navau & Sanchez, 1998). On the other hand, a few models have been developed, most of them studying the interaction between a superconductor and a permanent magnet. Hellman *et al.* and Yang presented models based on total flux exclusion of the samples that thus describe the behavior of LTSs or HTSs materials with higher critical current densities (Navau & Sanchez, 1998). Schönhuber and Moon, Chan *et al.*, and Torng and Chen have taken into account the penetration of supercurrents, the latter with an applied field provided by a pair of oppositely wound coils (Schonhuber & Moon, 1994; Chan, *et al.*, 1994; Torng & Chen, 1992). Some papers have considered granular structure for the superconductors when grains are completely penetrated (Johansen, *et al.*, 1994; Riise, *et al.*, 1994).

All over the models mentioned above assume that a superconducting sample is small enough to consider the magnetic field gradient constant along it. Demagnetization field of the sample due to the finite dimensions of the HTSs has been neglected by the same time. Demagnetization effects have been take into only account in theoretical calculation so far Tsuchimoto *et al.* have calculated the dependence of levitation force upon different geometrical parameters of a PM-HTSs system by using the an axisymmetric boundary element analysis (Tsuchimoto, *et al.*,1994). Portabella *et al.* have been used finite element calculations for the levitation force-distance $(F-x)$ hysteresis loop in order to estimate the value of critical current of a HTSs. Navau and Sanchez calculated the dependence of maximum vertical force upon the length of superconductor, considering both the demagnetization effect and the nonuniformity of the field gradient. In spite of these efforts, a systematic treatment of vertical levitation force for which the applied magnetic field can not be assumed to have a constant gradient has not been developed yet. In the scope of this chapter, we have been observed vertical levitation force as a function of distance between PM and HTSs. Moreover, most of the developments consider the superconductor to be in the critical state and describe it by means of Bean's critical-state model considering the critical current (J_C) as constant.

The most HTSs present a dependence of their critical current density (J_C) upon the internal magnetic field. Some analytical expressions have been proposed and successfully applied to describe the magnetic properties of superconductors (Navau & Sanchez, 1998). Therefore it is necessary to include such dependence in the study of the levitation behavior of the HTSs.

Kim's model dependence on the critical current density was developed by Tsuchimoto *et al.* with numerical (Navau & Sanchez, 1998). But, physical properties dependence on critical current density (J_C) in levitation behavior has not been studied in general so far.

To obtain critical current density (J_C), the consideration includes:

i. Character of response to applied field, equations of dependence on properties of materials;
ii. History of applied field;
iii. The initial state of the system;
iv. The boundary conditions.

Models on magnetization of the HTSs as can be seen in below.

a. Bean model,

Bean proposed a so-called critical-state model in which the hysteresis loop is plugged in a macroscopic parameter critical current density J_C (Bean, 1964). According to the Bean model, critical current density J_C presumes a constant and given as

$$J_C(H_i,T) = J_C(T) \quad (1)$$

where H_i and T are the local magnetic field and the temperature with respectively. In the model also pointed out that J_C was directly determined by the microstructure of the superconductors. The relationship of J_C versus temperature can be expressed as $J_C(T) = J_{C0}(T_C - T)/(T_C - T_0)$ (Berger, *et al.* 2007), where T_C is the critical temperature and J_{C0} the critical current density according to the reference temperature (T_0).

b. Kim model,

Anderson and Kim modified the Bean model (Anderson & Kim, 1964; Kim, et al. 1962) and J_C was suggested that it should vary with the local magnetic field and should have the form

$$J_C(H_i,T) = \frac{J_C(T)}{1 + H_i / H_0} \quad (2)$$

where H_0 is a macroscopic materials parameter with the dimension of field. The model was found to fit the experimental results for some conventional superconductors for a solid cylinder shaped superconductors.

According to Kim model the J_C depend on the term H_i / H_0, which may vary considerably among different materials. Watson showed that, for some systems, the condition $H_i >> H_0$ is fulfilled. Eq. (2) shows a simple linear field dependence of J_C, $J_C(H_i) = A - CH_i$, where A and C are constants (Hindley & Watson, 1969).

c. Power law model,

Irie and Yamafuji are focus on the specific pinning mechanisms, a power-law field dependence of J_C was suggested (Irie & Yamafuji, 1967),

$$J_C(H_i,T) = \frac{K(T)}{H_i^n} \quad (3)$$

where K is a materials parameter and n directly reflects pinning strength. An empirical formula based on experimental results developed by Fietz et al. (Fietz, *et al.*, 1964),

$$J_C(H_i,T) = J_C(T)\exp(-H_i / H_0). \quad (4)$$

A general critical-state model was developed for the critical current density. It is obtained from Eq. (1-4) as the next form,

$$J_C(H_i,T)=\frac{J_C(T)}{\left[1+H_i\,/\,H_0(T)\right]^{\beta}} \tag{5}$$

where β is a dimensionless constant. Eq. (5) can be converted to Bean and Kim models by using the Eqs. (1) and (2), respectively if the $\beta=0$ and 1. Eq. (3) is obtained when the condition $-H_i\,/\,H_0 >> 1$ is satisfied in eq. (5) and the K and n are set to $J_C(T)=\mathrm{H}_0^{\beta}$ and β, respectively (Xu, et al., 1990). In the conditions of $-H_i\,/\,H_0 << 1$ and $\beta >> 1$, with $H_0\,/\,\beta << \xi$, where ξ is a finite value with dimension of field, and taking the limit of $J_C(H_i)$ with the preceding conditions and letting $-H_i\,/\,H_0 = x$, it has

$$\lim J_C(H_i,T)=\frac{J_C(T)}{\lim\limits_{x\to 0}\left[1+x\right]^{\beta}}=\frac{J_C(T)}{\lim\limits_{x\to 0}\left[1+x\right]^{(1/x)(H_0/H_0)}}=J_C(T)\exp(-H_i\,/\,H_0) \tag{6}$$

It can be seen that Eq. (6) has the same form as Eq. (4) and that their physical meanings are identical.

d. Yin model,

Material's equation, known as Yin model, has been developed in the form as (Yin *et al.*, 1994),

$$E(J)=2\upsilon_0 H\exp\left[-\left(\frac{U_0+W_V}{kT}\right)\right]\sinh\left[\frac{W_L}{kT}\right] \tag{7}$$

Where υ_0 , U_0 , $W_V=\eta\upsilon P=E(J)HP/\rho_f$, $W_L=JHP$ and P are a prefactor with dimension of velocity, the pinning potential, the viscous dissipation term of flux motion with viscosity coefficient (with $\eta=HH_{C2}/\rho_n=H^2/\rho_f$), the energy due to Lorentz driving force, a product of the volume of the moving flux bundle and the range of the force action, respectively. In the υ term in the W_V corresponds to $\upsilon_0\exp\left[-U(J)/kT\right]$. U represents the activation barrier. Yin model includes pinning barrier term, viscous dissipation term and direction dependent. It provides a unified description which covers all regimes of flux motion: thermally activated flux flow, flux creep, critical state and flux flow. Yin model contain Anderson–Kim model, Bardeen–Steven model and critical state model $(\vec{F}_P=\vec{J}_C\times\vec{H})$ as its specific forms (Yin *et al.*, 1994).

There are different ways of calculation of magnetic field (H_{PM}) of permanent magnets (Moon, 1990; Sanchez & Navau, 2001; Brandt, 1990; De Medeiros, *et al.* 1998),

i. Biot–Savart's law: Calculating B_{PM} with Biot–Savart's law.

ii. Scalar potential method: Calculating scalar potential φ_μ in current-free area, H_{PM} is the gradient of φ_μ.

iii. Vector potential method: Calculating vector potential $A(PM)$, H is the curl of A.

Field expression in cylindrical coordinates (John David Jackson, 1998), is

$$\vec{H}(\vec{r}) = \vec{\nabla} \times \vec{A} = \frac{\mu_0}{4\pi} \iiint_V \frac{\vec{J}(\vec{r}) \times (\vec{r} - \vec{r}\,')}{|\vec{r} - \vec{r}\,'|} dv' \tag{8}$$

Numerical methods have been developed towards to solve the $J_C(H)$ and the levitation force. Analysis and simulation can be carried out based on above models and various parameters.

The levitation forces evaluation of distance between HTS and PM (or air gap) are seen in Fig. 1. As can be seen from this figure, levitation forces in a magnetic media for experimental and theoretical calculation decreases when distance between HTS and PM increases.

There are several methods for the calculation of levitation force in magnetic field. Some of these methods are Lorentz force equation (Chun &Lee, 2002), Maxwell's stress tensor (Amrhein & Krein, 2009) and virtual work methods (Vandevelde& Melkebeek, 2001). The Lorentz force equation method is important to know the exact distribution of the critical current density and the magnetic field. The Maxwell's stress tensor method is calculated by the integration of the Maxwell's stress tensor on an arbitrary surface surrounding the object.

The magnetic force as the virtual work method is given by the derivation of the co-energy with respect to a virtual displacement.

The magnetic levitation forces evolution of the distance of between HTS and PM with experimental and different theoretical models are seen in Fig. 1.

From this figure, levitation force nearly exponentially increase while distance of between HTS and PM decreases. But, the experimental results nearly agree with the Lorentz force equation methods. In other words, from the comparison, the result by the Lorentz force methods is in strict correspondence with the experimental results quantitatively according to the others simulation methods. Agreement between the Lorentz force methods with the experimental results originated perhaps from the penetration of the applied field into HTSs. Because of the penetration of the applied field into the HTS occurs where the vortex state is. This field passes through the HTSs produce a flux tubes. This states in this type studied so called known repulsive magnetic field in general. Finally, compatible Lorentz force method with the experimental result is not surprising.

Since the discovery of high temperature superconductors (HTSs) this interaction has achieved considerable attention, mainly due to its capability of producing stable non-contacting conditions for such (HTS-PM) systems. The magnetic levitation forces can be demonstrated very easily in experimentally. But, their theoretical models are far more difficult. The reason for these difficulties are the below.

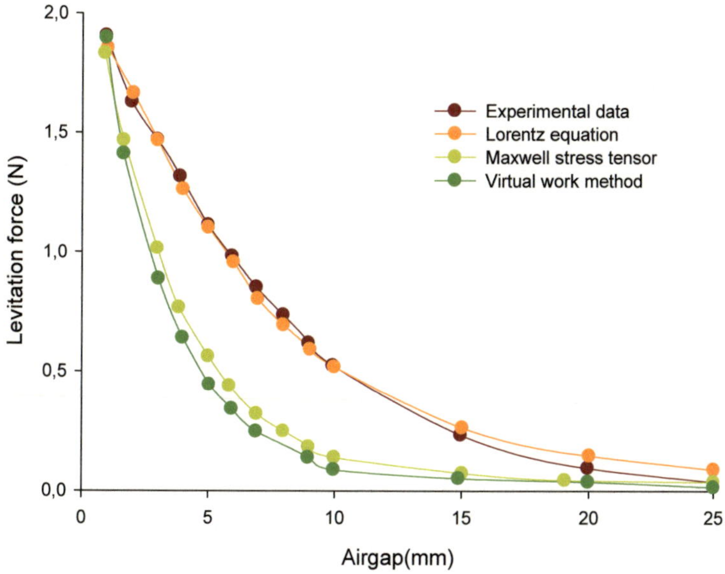

Fig. 1. The levitation forces evolution of the distance of between HTS and PM with experimental and different theoretical models (Chun & Lee, 2002).

1. The HTS's are of HTS with a strongly irreversible magnetic behavior, i.e., their magnetization depends on the prehistory of the external magnetic fields.
2. The levitation phenomenon between HTS and PM materials has been studied for magnetic granularity or polycrystalline bulk materials in general. The magnetic levitation force studies are very poor for single crystalline materials.
3. A finite levitation force is required non uniform magnetic field.

During the past years, many research groups have done extensive studies of levitation forces using HTS bulk materials of various kinds (Takao *et al.* 2011; Lu *et al.*, 2011; Krabbes *et al.*, 2000). The higher levitation force and stable levitation or suspension between a PM and a bulk superconductor, makes superconducting bulks possible for various applications such as superconducting magnetic bearings, flywheel, magnetic levitation transportation systems, generators or motors and permanent magnets.

3. Levitation experimental system and data analaysis

The levitation forces measure between permanent magnet and superconducting high temperature, bulk or film in general. The critical current has the crucial role in the levitation force measuring system. The levitation force between HTS and PM originated from by the interaction between the induced electric current in the HTS and the magnetic field from the PM. The critical current density $J_C(H)$ is closely related with the flux pinning center density and flux pinning force of the sample, and also closely related with the flux pinning, depinning process and the magnetic flux creep rate during the levitation force measurement. The $J_C(H)$ is also a fixed parameter (if we consider the process based on

Bean's model) because the sample is the same, so the levitation force is only related with the total induced electric current in the sample which is depending on magnet moving speed, this is the only variable during the measurement. Superconductors is different from perfect conductor, that if a superconductor is cooled below certain temperature (is called critical temperature, T_C) and an external magnetic field is applied then the superconductor repel the magnetic field because of the supercurrent. This super-current is made of the inside the superconductor via of the induced of the external field as the same Lenz law. This situation, the superconductor has the supercurrent as defined superconductor. Firstly grains effected and a current circulate interior of the grain, which is called internor current, I_C. Secondly all of the interior current sums to surface of the superconductor and they called screening current, I_S; which is repulsive and screening to external magnetic field.

The screening currents produce a mutually repulsing behavior, so the external magnetic field do not penetrate interior of the superconductor, which is called Meissner effect in the superconductor. The German physicists Meissner and Ochsenfeld discovered the phenomenon in 1933 by measuring the magnetic field distribution outside superconducting thin and lead samples. These samples, in the presence of an applied magnetic field, were cooled below what is called their superconducting transition temperature. Below the transition temperature the samples canceled nearly all magnetic fields inside. They detected this effect only indirectly; because the magnetic flux is conserved by a superconductor, when the interior field decreased the exterior field increased. The experiment demonstrated for the first time that superconductors were more than just perfect conductors and provided a uniquely defining property of the superconducting state.

Macroscopic transport properties mechanism of the grain boundary effect on HTS is not entirely understood yet, though on this mechanism several theoretical models have been suggested (Luine & Kresin, 1998; Coble, 2009). In particular, it is assumed that the grain boundary effects are describing as that to following factors: (i) a decrease of electron free path in vicinities of grain boundaries; (ii) stress-field-induced structural disorder and pinnig of magnetron vortices at grain boundaries; (iii) formation of the antiferromagnetic phase within grain cores; (iv) faceting of grain boundaries and d-symmetry of the superconducting order parameter, (v) deviations from bulk stoichiometry in vicinities of grain boundaries.

Figures 2-3 shows a schematic representation of the preparing of the HTS sample with a wet-technique (or ammonium nitrate method) and annealing process of the HTS disk, respectively. The starting material was the powder of nominal $Bi_{1.84}Pb_{0.34}Sr_{1.91}Ca_{2.03}Cu_{3.06}O_{10}$ compound, which was synthesized by a wet-technique using high-purity (>99.99%) Bi_2O_3, PbO, $SrCO_3$, $CaCO_3$ and CuO, We produce our powder by using this method at room temperature. Then, the powder was calcined at 700 °C for 10h, twice as much, and pressed into pellets with 225 MPa. The pellets were placed in a preheated furnace at 850 °C for 100 h then directly cooled to room temperature for quench. After this sintering process pellets regrinding and the powders at various pressures and time were pressed in to cylindrical superconductor disk (CSD: $Bi_{1.84}Pb_{0.34}Sr_{1.91}Ca_{2.03}Cu_{3.06}O_{10}$) with diameter ($2r$) of 13 mm and height ($h$) of 13.52 mm. The CSD was placed in a preheated furnace at 815 °C at a rate of 0.1 °C/min for 100 h then directly cooled to room temperature. The details of producing of HTS material was given in previously our work (Karaca, 2009).

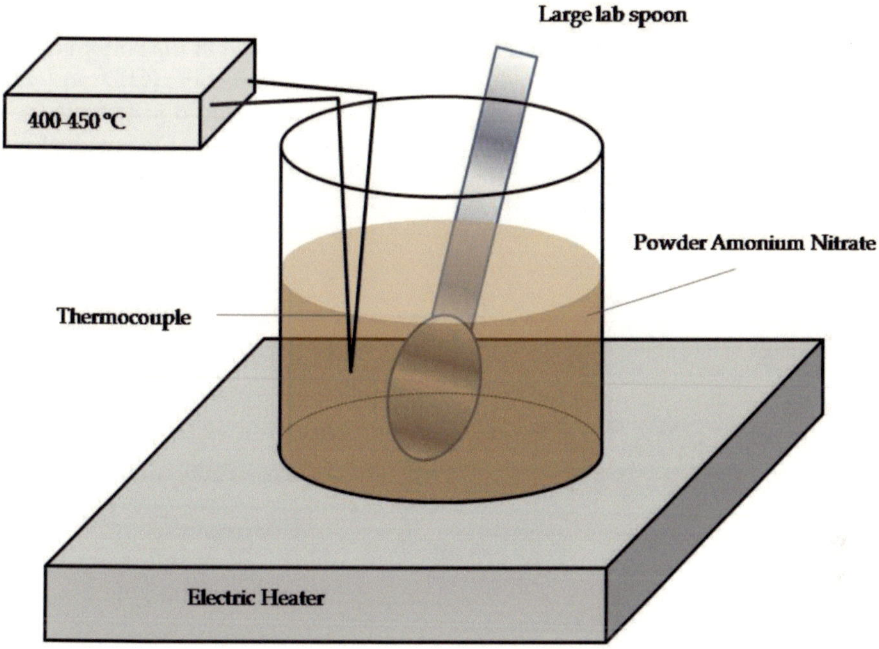

Fig. 2. Schematic representation of the preparing of HTS sample.

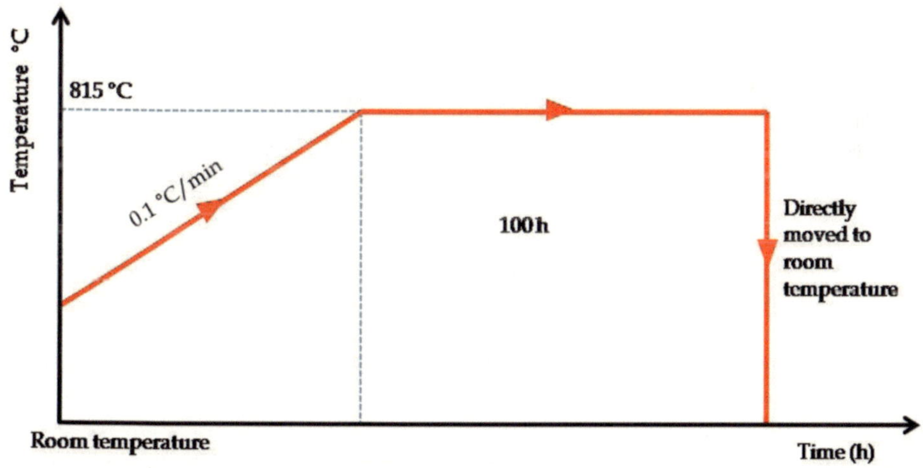

Fig. 3. Schematic representation of the annealing process of the HTS disk.

Magnetic properties were measured by using the magnetic levitation force-distance (between HTS and PM) system at the temperature 77 K in Fig. 4. Measurements were performed in the zero field cooled (ZFC) regimes in applied magnetic field $H_r = 0.15T$.

Magnetic repulsive and attractive forces between the HTS disk and PM were measured using a homemade levitation measuring device (Fig.4). It is figure out in Fig.4 an Nd–B–Fe permanent magnet (1.6x1.4x0.34 cm^3 and $H_r = 0.15T$) was put on the plate of a sensitive electronic scale. CSD sample (13 mm in diameter) was immobilized at the bottom of a liquid nitrogen-filled receptacle which is movable in the vertical direction (CSD and PM are symmetrically fixed on the same line). This receptacle was then placed just above the plate of the balance without touching it. Thus, the observed weight change of the magnet directly reflected the magnetic force induced in this handmade system (Benlhachemi, *et al.*, 1993, Karaca, 2009).

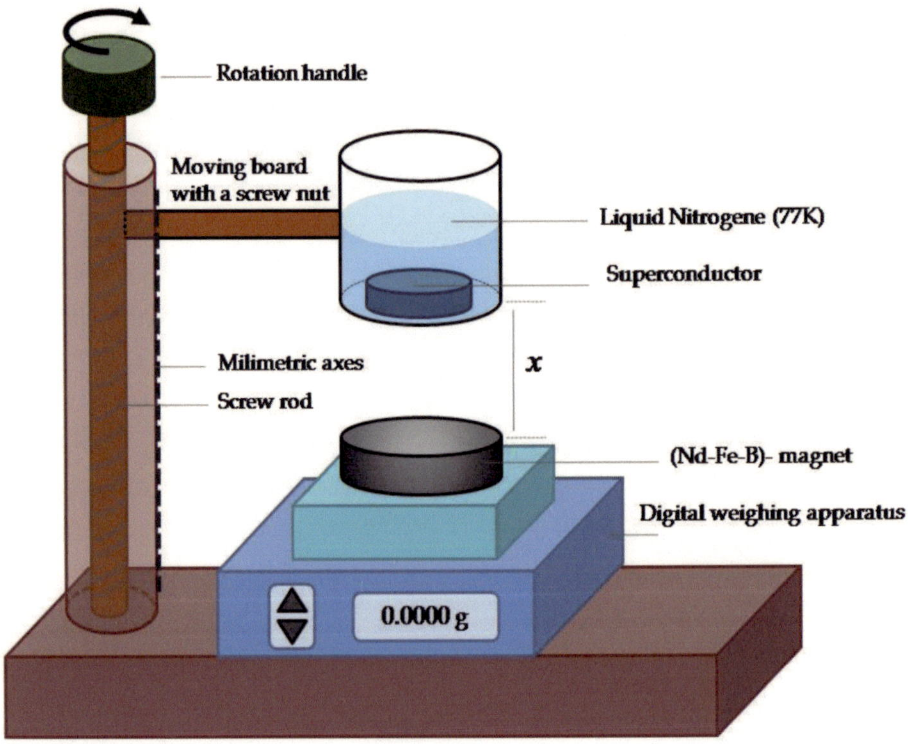

Fig. 4. Schematic representation of the magnetic levitation force-distance (between HTS and PM) hysteresis loop measurement experimental system.

The magnetic levitation force-distance (between HTS and PM) $(F-x)$ hysteresis loop ZFC process during the descending and ascending process is seen in Fig.5. Similarly, the full light green and full dark green circles correspond to the levitation force curves ascending and descending with respectively. The hysteresis curves of magnetic levitation are not the same. In other words, magnetic levitation force when distance ascending faster than magnetic levitation force while distance descending. However, the hysteresis loops strongly depend on the size of grains and their connectivity in the bulk HTS. The magnetic levitation forces nearly the same when distance increases from 20 mm to 36 mm. this behavior corresponds to Meissner effects. The levitation force linearly decreases when distance increases from 0 mm to 6 mm. The external magnetic field fully penetrated HTS disk in this space (0-6 mm), because, there are no hysteretic behavior in this area. The maximum hysteretic phenomena are seen in the distance range 6-20 mm. This phenomena attributed that the vortex state. The results of the levitation force have generally been measured in a ZFC process, a few works were done in the field cooled (FC) process (Yang *et al.*,2003).

The force between HTS and PM system can be written as $F = \mu(dH/dx)$. Here, μ is the magnetic moment related with the bulk magnetization $\vec{M}$ $\left(\vec{M} = \sum_{i=1}^{n} \vec{\mu}_i\right)$. dH/dx is the field gradient produced by the magnet. The trapped field in HTS originated from the PM, an attractive force occurs when the sample is moved away from the PM. This result can be attributed to the number of pinning centers in the HTS sample, which results in an increase of trapped magnetic field inside the samples. In addition, the levitation force is related to the grain size and crystallographic orientation.

The force-distance $(F-x)$ between CSD ($Bi_{1.84}Pb_{0.34}Sr_{1.91}Ca_{2.03}Cu_{3.06}O_{10}$) and PM always hysteresis loop as an example for this chapter during the descending and ascending process are seen in Fig. 5.

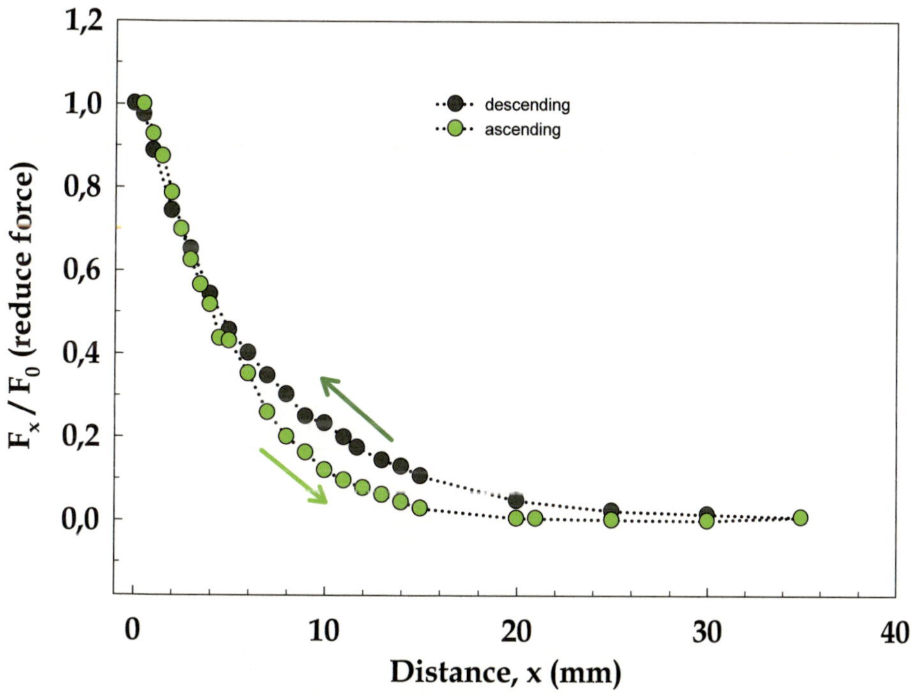

Fig. 5. The interaction force between a CSD and a PM always shows a force-distance ($F-x$) hysteresis loop during the descending (dark green full circle) and ascending (light green full circle) process.

As can be seen from Fig.5, the levitation force while distance increasing smaller than levitation force while distance decreasing. The maximum values between of two levitation forces are seen near the 10 mm. These values are smaller than expectation values for HTS.

The levitation force can be related from the interaction between CSD and PM materials in general. This interaction between CSD and PM are mainly dependent on the properties of the CSD and the magnetic field distribution of the magnet. For a bulk superconductor, the levitation force is dependent on many parameters, such as the critical current density and grain radius, grain-orientation, thickness of the sample, and the cooling temperature. For a magnet, the levitation force is closely related with the magnetic flux density, magnetic field distribution, etc. But for a given pair of bulk superconductor and PM, it is known that the same levitation force can be obtained at a gap distance between the superconductor and the magnet, because of the magnetization history of the superconductor, by a mechanical descending and ascending process of the magnet during the axial levitation force measurement state (Yang *et al.*, 2003).

The low levitation force of sintered superconductor materials can be attributed to two intrinsic material problems of a superconductor in this chapter (Murakami, 1993). The first is the grain boundary weak-link problem and the second is the weak flux pinning problem. In order to resolve these two problems, different material processing techniques have been developed from other workers (Chen *et al.*, 1998).

4. Technological applications of levitation forces in HTS

Temperatures in the range near the 4-30 K and near the 30-180 K are called the low temperature superconductors (LTS) and high temperature superconductors (HTS) regime with respectively. In the scope of this chapter, we have studied levitation force-distance loops for HTS in 77 K. In the recent 30 years, LTS and especially HTS are in a class of superconductor. HTS structures have potential applications in a broad range topical areas, superconducting magnetic energy storage (SMES), power devices including transmission cable, HTS motors and HTS generators, superconducting fault current limiters (SFCL), and high field magnets. See the Fig.6 in below for an example of some application of these type HTS samples.

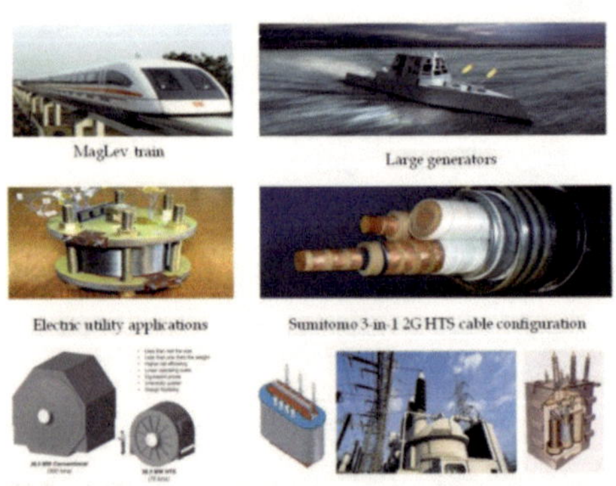

Fig. 6. Some application of the HTS
(www.superpower-inc.com; http://yahoo.brand.edgar-online.com).

The levitation force hysteretic loops applied to new technologies as medical and space technologies in general.

HTS generators are substantially smaller and lighter than copper based machines. These advantages are very attractive in a ship environment. HTS power cables will enable utilities to upgrade power density by 2 to 8 times power corridors because HTS cables transmit power with essential no electrical resistance. Superpower's SFCL technology is enable scaling to high voltage and current with customization to the user's requirements, losses or voltage drop across the device during normal operation, and etc. The HTS technology can be used to make all superconducting magnets with field that will soon exceed 30 Tesla (see for detail, www.superpower-inc.com).

5. Summary

In the scope of this chapter, we have focus on the magnetic levitation force-distance loops for HTS samples. The inter CSD and PM exchange type interaction are determined magnetic levitation forces. The magnetic levitation forces are heavily dependence on grain size, grain connectivity, critical current density, history of applied magnetic field and etc. Combination with hysteresis $(F-x)$ results, these indicate that the Meissner effects, vortex state and metallic behavior are seen the same. The CSD with ZFC process showed that the sample has low quality with weak-link characteristics. A typical hysteresis loop of the axial levitation force between the permanent magnet and the CSCD showed the lower repulsive force in this chapter. This phenomenon is attributed to the fact that the wet technique causes the production of lower dense samples, and is caused by a lower critical current density (Altunbaş et al., 1994; Karaca et al., 1998) and by a lower levitation force. It is considered that the changes in the levitation force are related to the grain orientation, the homogeneity, and the number of the pinning centers in the samples (Yanmaz et al., 2002). Levitation force measurements need to investigate several factors end different materials. In the future life will be need to this investigation results.

6. Acknowledgements

I would like to thanks Mehmet Ali Güzel for help.

7. References

Abrikosov, A. A. (1957). *Zh. Eksp. Teor. Fiz.*, Vol.32, pp. 1442.

Alamgir, A. K., Yamada, M. H., Harada, N., Osaki, K. & Tada, N. (1999). *IEEE Transaction on Appl. Supercond.*, Vol.9, pp. 1864.

Altunbaş, M., Yanmaz, E., Nezir, S., Karal H. & Vidadi Yu A. (1994). *J. Alloys and Compounds*, Vol.215, pp. 263.

Altunbaş, M., Yanmaz, E., Nezir, S., Karal, H. & Vidadi Yu A. (1994). *J. Alloys and Compounds*, Vol.215, pp. 263

Amrhein, M., Krein, P.T. (2009). Force Calculation in 3-D Magnetic Equivalent Circuit Networks With a Maxwell Stress Tensor", *Energy Conversion, IEEE Transactions on*, Vol.24 Issue: 3, pp. 587 - 593.

Anderson, P. W. & Kim, Y. B. (1964). Hard superconductivity: theory of the motion of abrikosov flux lines, *Rev. Mod. Phys.*, Vol. 36, No. 1, pp. 3943.

Bean, C. P. (1962). *Phys. Rev. Lett.*, Vol.8, pp. 250.

Bean, C. P. (1964). Magnetization of high-field superconductors, *Rev. Mod. Phys.*, Vol.36, No. 1, pp. 31-39.

Benlhachemi, A., Golec S. & Gavarri, J.R. (1993). *Physica C: Superconductivity*, Vol.209, pp. 353.

Berger, K., Lévêque, J., Netter, D., Douine, B., & Rezzoug, A. (2007). Influence of temperature and/or field dependences of the E-J power law on trapped magnetic field in bulk YBaCuO, *IEEE Trans. Appl. Super.*, Vol. 17, No. 2, pp. 3028-3031.

Brandt, E. H. (1990). *Appl. Phys. Lett.*, Vol.53, pp. 1554.

Brandt, E.H. (1998). Superconductor Disks and Cylinders in Axial Magnetic Field: I. Flux Penetration and Magnetization Curves, *Phys. Rev. B*, Vol.58, pp. 6506-6522.

Chan, W. C., Jwo, D. S., Lin, Y. F. & Huang, Y. (1994). Magnetic levitation of $YBa_2Cu_3O_y$ single crystals, *Physica C: Superconductivity* Vol.230, pp. 349-353.

Chen, I.G., Hsu, J.C., Janm, G., Kuo, C.C., Liu, H.J., & Wu, M. K. (1998). *Chin. J. Phys.*, Vol.36, pp. 420.

Chun,Y.D. & Lee,J. (2002). Comparison of magnetic levitation force between a permanent magnet and a high temperature superconductorusing different force calculation methods, *Physica C*, Vol.372–376, pp. 1491–1494.

Coble, R. L. (2009). Sintering Crystalline Solids. I. Intermediate and Final State Diffusion Models, *Journal of Applied Physics*, Vol.32 Issue:5, pp. 787 - 792.

Davis, L. C. (1990). *J. Appl. Phys.*, Vol.67, pp. 2631.

De Medeiros, L.H., Reyne G., Meunier, G. (1999). *IEEE Transactions on Magnetics*, Vol.34 (5), pp. 3560–3563.

De Medeiros, L.H., Reyne, G., Meunier, G. (1998). Comparison of Global Force Calculations on Permanent Magnets, *IEEE Transactions On Magnetics*, Vol. 34, No. 5.

De La Cruz, A., Badía, A. (2002). *Physica B: Physics of Condensed Matter*, Vol.321, Issue 1-4, pp. 356-359. DOI: 10.1016/S0921-4526(02)01074-8.

Essmann, U. & Truble, H. (1967). *Phys. Lett. A*, Vol.24, pp. 526.

Fietz, W. A., Beasley, M. R., Silcox, J., & Webb, W. W. (1964). Magnetization of superconducting Nb-25%Zr wire, *Phys. Rev.*, Vol.136, No. 2A, pp. A355-A345.

Gao, W. & Sande, J. B. V. (1992). *Supercond. Sci. Technol.*, Vol.5, pp. 318.

Hellman, F., Gyorgy, E. M., Johnson, D. W., Jr., O'Bryan, H. M., & Sherwood, R. C. (1988). *J. Appl. Phys.* Vol.63, pp. 477.

Hindley, N. K. & Watson, J. H., (1969). Superconducting metals in porous glass as granular superconductors, *Phys. Rev.*, Vol.183, No. 2, pp. 525528.

http://sciyo.com/articles/show/title/wind-power-integrating-wind-turbine-generators-wtg-s-with-energy-storage

http://www.superpower-inc.com (2012)

http://yahoo.brand.edgar-online.com (2012)

Irie, F. & Yamafuji, K. (1967). Theory of flux motion in non-ideal type-II superconductors, *J.Phys, Soc. Jpn.*, Vol. 23, pp. 255-268.

J. Mizia, K. Adamiak, A.R. Eastham, G.E. (1988). *IEEE Transactions on Magnetics* Vol.24 (1), pp. 447–450.

Jin, J. X. (2006). High Tc superconductor theoretical models and electromagnetic flux characteristics, *Journal of Electronic Science and Technology of China*, Vol. 4, No. 3, pp. 202-208.

Jin, J. X., Zheng, Lu H. (2007). *Physica C: Superconductivity*, Volumes 460-462, Part 2, pp. 1457-1458 DOI:10.1016/j.physc.2007.04.151.

Johansen, T. H., & Bratsberg, H. (1993). *J. Appl. Phys.*, Vol.74, pp. 4060.

Johansen, T. H., Bratsberg, H., Riise, A. B., Mestl, H. & Skjeltorp, A. T. (1994). *Appl. Suppercond.* Vol.2, pp. 535.

Johansen, T. H., Yang, Z. J., Bratsberg, H., Helgesen, G., & Skjeltorp, A. T. (1991). *Appl. Phys.Lett.* Vol.58, pp. 179.

Jones, T. E., Mc Ginnis, W. C., Jacobs, E. W., Bossp, R. D., Thibbado, M., Briggs, J. S., & Glad, W. E. (1994). *Physica C*, Vol.201, pp. 279.

Karaca, I. (2009). Characterization of a Cylindrical Superconductor Disk Prepared by the Wet Technique with Microstructure Analysis and Levitation Force Measurements Using a PermanentMagnet, *Chinese journal of Physics*, Vol. 47, No.5, pp. 690-696.

Karaca, I., Celebi, S., Varilci, A. & Malik, A. I. (2003). *Supercond. Sci. Technol.*, Vol.16, pp. 100.

Karaca, I., Ozturk, A., & Nezir, S. (1998). *J. Alloys and Compounds* Vol.268, pp. 256.

Kim, Y.B., Hempstead, C. F. & Strnad, A. R. (1962). Critical persistent currents in hard superconductors, *Phys. Rev. Lett.*, Vol. 9, No. 7, pp. 306-309.

Krabbes, G., Fuchs, G., Schaatzle, P., Gruß, S., Park, J.W., Hardinghaus, F., Stover, G., Hayn, R., Drechsler, S.-L., Fahr, T. (2000). Zn doping of YBa Cu O in melt textured materials: peak effect 2 3 7 and high trapped fields, *Physica C*, Vol.330, pp. 181–190.

Liu, W., Wang, J. S., Liao, X. L., Zheng, S. J., Ma, G. T., Zheng, J., Wang, S. Y. (2010). *Physica C*, Vol.471, Issue 5-6, pp. 156-162. DOI: 10.1016/j.physc.2010.12.016

Lu, Y., Bai X., Ge, Y., Wang, J. (2011). Influence of Thickness on the Levitation Force of High-*Tc* Bulk over a Permanent Magnetic Guideway with Numerical Method, *J Supercond Nov Magn*, Vol.24, pp. 1967–1970.

Luine, J. A. & Kresin, V. Z. (1998). Critical current in high T_c grain boundary junctions, *J. Appl. Phys.* Vol. 84, pp. 3972, DOI:10.1063/1.368576.

Meissner, W., Ochsenfeld, R. (1933). *Naturwissenschaften*, Vol.21 (44), pp. 787–788.

Moon, F. C. (1990). *Int. J. Appl. Electromagn. Mater.* Vol.1, pp. 29.

Moon, F. C., Yanoviak, M. M. & Raj, R. (1988). *Appl. Phys. Lett.*, Vol.52, pp. 1534.

Murakami, M. (1993). *Appl. Supercond.*, Vol.1, pp. 1157.

Navau, C. & Sanchez, A. (1998). *Phys. Rev. B.* Vol.58, pp. 963–970, DOI:10.1103/PhysRevB.58.963.

Qin, M.J., Li, G., Liu, H.K., Dou, S.X. & Brandt, E.H. (2002). *Phys. Rev. B*, Vol.66 , pp. 024516.

Ren, H.T., Xiao, L., Jiao, Y.L., Zheng, M.H. & Chen, Y.X. (2003). *Chin. J. Low Temp. Phys.*, Vol.25 (1), pp. 11.

Riise, A. B., Johansen, T. H. & Bratsberg, H. (1994). The vertical magnetic force and stiffness between a cylindrical magnet and a high-T_c superconductor, *Physica C.*, Vol.234, pp. 108-114.

Riise, A. B., Johansen, T. H., Bratsberg H., Skjeltorp, A. T. (1993). *Physica C*, Vol.218, Issue 3-4, pp. 489-494, DOI: 10.1016/0921-4534(93)90055-U.

Riise, A. B., Johansen, T. H., Bratsberg, H., Koblischka, M. R., Shen, Y. Q. (1999). *Physical Review B (Condensed Matter and Materials Physics*, Vol.60, Issue 13, pp.9855-9861, DOI:10.1103/PhysRevB.60.9855 .

Riise,A.B., Johansen, T. H., Bratsberg, H., Koblischka, M. R. & Shen, Y. Q. (1999). *Phys. Rev. Lett. B*, Vol.60, pp. 9855.

Sanchez A. & Navau C. (2001). *Physica C*, Vol.360, pp. 364–365.

Sanchez, A. & Navau, C. (1997). *Physica C,* Vol.275, pp. 322.

Sanchez, A., Del Valle, N., Pardo, E., Chen, Du-Xing, & Navau C. (2006). Magnetic levitation of superconducting bars, *J. Appl. Phys.* Vol.99, pp. 113904 DOI:10.1063/1.2201767.

Schonhuber, P. & Moon, F.C. (1994). Levitation forces, stiffness and force-creep in YBCO high-tc superconducting thin films, *Applied superconductivity,* Vol.2(7/8), pp. 523-534.

Takao, T., Kameyama, S., Doi, T., Tanoue, N. & Kamijo, H. (2011). Increase of Levitation Properties on Magnetic Levitation System Using Magnetic Shielding Effect of GdBCO Bulk Superconductor, *IEEE Transactions On Applied Superconductivity,* Vol. 21, No.3, pp. 1543-1546.

Torng, T. & Chen, Q. Y. (1992). J. Appl. Phys., Vol.73, pp. 1198.

Tsuchimoto, M., Takeuchi, H. & Honma, T. (1994). *Trans. IEE of Japan* Vol.114-D, pp. 741.

Vandevelde, L., Melkebeek, J.A.A. (2001). A survey of magnetic force distributions based on different magnetization models and on the virtual work principle, *Magnetics, IEEE Transactions on,* Vol.37 Issue: 5, pp. 3405 – 3409.

Weber, H. W. (2001). Pinning F., *in Handbook on the Chemistry and Physics of Rare Earths,* Vol. 31: High Temperature Superconductors – II, ed. by K. A. Gschneider, Jr., L. Eyring, M. B. Maple (Elsevier, Amsterdam, 2001) pp. 187.

Wei, J.C., & Yang, T.J. (1996). *Chin. J. Phys.,* Vol.34, pp. 1344.

Xu, M., Shi, D., & Fox, R. F. (1990). Generalized critical-state model for hard superconductors, *Phys. Rev. B,* Vol. 42, pp. 10773-10776.

Yang, W.M. (2001). *Physica C,* Vol.354, pp. 5.

Yang, W.M., Zhou, L., Feng, Y., Zhang, P.X., Nicolsky, R. & de Andrade Jr, R. (2003). *Physica C,* Vol.398, pp. 141.

Yanmaz, E., Balci, S. & Küçükömeroğlu, T. (2002). *Materials Letters* Vol.54, pp. 191.

Yin, D., Schauer, W., Windte, V., Kuepfer, H., Zhang S. & Chen. J. (1994). *Z. Phys.,* B94, pp. 249.

Yin, D., Schauer, W., Windte, V., Kuepfer, H., Zhang, S., & Chen, J. (1994). A new understanding of the resistive state and the V-I characteristic of High-Tc superconductors, *Z. Phys. B,* Vol.94, pp. 249-254.

Yoshida, K., Sano, Y., & Tomii, Y. (1995). *Supercond. Sci. Technol.,* Vol.8, pp. 329.

Zhao X.F. & Liu Y. (2007). *J. Low Temp Phys.,* Vol.149, pp. 200.

15

Applications of Superconductor/Photosemiconductor Contact Structures in Electronics

Viktor V. Bunda and Svitlana O. Bunda
Transcarpathian State University
Ukraine

1. Introduction

The integration of high temperature superconductors(HTSC's) with conventional semiconductor(SeC) - based technology would have important consequences for micro- and cryophotoelectronics, with the promise of high performance hybrid circuits incorporating the best of what superconductors and semiconductors have to offer as well as the possibility for novel devices (Bunda, 1991). The rapid progress in the fabrication of high quality HTSC thin films on oxygen containing substratum such as $SrTiO_3$, MgO, and $LaAlO_3$ has been observed. However, the results on semiconductors like Si, Ge and GaAs has been also reported.

The high temperature superconductors (HTSC's) are considered to be low carrier density materials. Therefore, the light can penetrate the superconductor and can effectively excite the quasiparticles in it. The study of light detection by a "HTSC – photosemiconductor" hybrid contact structures (HCS's) is very perspective for fabrication of multifunctional photonic circuits - high speed detectors with reasonable sensitivity covering a broad electromagnetic spectrum (Bunda et al., 1996). Therefore, for deeper investigation of such type compounds we need fabrication of the HCS's based on oxygen containing photosemiconductor.

Superconductors (SC) have a wide field of applications due to their charge carrier transport properties without energy loss. The charge carriers in the superconducting state are correlated charge carriers (Cooper pairs) as well as uncorrelated ones (quasiparticles). The special magnetic and transport properties of Cooper pairs provide new applications in the field of tunnel contacts, electronic devices and the generation of very high magnetic fields (Akani et al., 1995). Compared to classical ultrasonic sources, superconducting tunnel contacts emit phonons with higher frequency. Therefore, the properties of different HTSC/photosemiconductor heterojunctions are of great interest for the development of new electronic devices such as diodes with very high rectifying parameters (Bunda et al., 2010a) and transistors based on the Josephson field effect (Akani et al., 1995).

In recent years, two-dimensional (2D) nanostructured materials, such as nanoplates and nanosheets, have attracted much attention because of not only their unique electronic, magnetic, optical, and catalytic properties, which mainly arise from their large surface areas, nearly perfect crystallinity, structural anisotropy, and quantum confinement effects in the thickness (Deng et al., 2008). The potential of 2D nanostructured materials uses for building blocks for advanced materials and devices with designed functions in areas as diverse as lasers, transistors, catalysis, solar cells, light emission diodes, and chemical and biological sensors (Bunda et al., 2010b).

2. Grown and crystal structure of bismuth oxyhalides single crystals

Oxyhalides of bismuth BiOX (X = Cl, Br, I) are very interesting materials which find various applications as X-ray luminescent screens, as anti-Stokes converters, photocatalist, usual luminophors and as photoconductive analyzer of linear polarized radiation in the 0.24 - 1.2μm spectral region. Bismuth oxyhalides provides a very convenient matrix for activation by various rare-earth and 3d-elements: by doping this matrix with different, is possible to obtain a wide variety of optical, luminescent and photoconductive properties. Moreover, these crystals are of essential interest in connection with the research and development work on laser materials. (Bunda et al., 2010b). The great interest for these materials is strongly related to the influence of dimensionality on the behaviour of physical properties (they are 2D structured materials). Bismuth oxyhalides are one of the V-VI-VII group compound semiconductors belonging to the tetragonal system. The structure of BiOX is known to have a layered structure, which is constructed by the combination of the halide ion layer and the bismuth oxygen layer (Bletskan et al., 1973).

The BiOX single crystals are layered 2D structured materials. The crystal structure of BiOX is very similar to the symmetry of space groups P4/*mmm* and/or I4/mmm, where are usual for the $YBa_2Cu_3O_{7-\delta}$ and $Bi(Tl,Hg)_2Sr(Ba)_2Ca_{n-1}Cu_nO_{2n+4}$ HTSC's. Besides, the BiOX crystals, as well as a HTSC's contain oxygen. The thermal expansion and the lattice constants mish-match in the (001) base plane are in good agreement with the same parameters of the HTSC's. Also, the BiOX crystals are good "sparring partners" for the fabrication of "HTSC – photosensitive semiconductor" hybrid contact structure.

Bismuth oxychloride crystals exhibit many intriguing and interesting properties such as photoluminescence, photoconductivity, and thermally stimulated conductivity (Bunda, 1989, 1991, 2005). The luminescent band at 1.6 - 2.2 eV is a result of capture of c-band free electrons by r-centres of recombination. The shape of the absorption edge has been explained by the effect of the laminarity of these crystals. Previous investigations on the crystal structure and morphology of BiOX crystallites showed that they grow usually in the form of platelets with the c-axis normal to the platelets. Investigations on the influence of uniaxial compression on the crystals revealed that they are isotropic along the normal to the layers (Ganesha et al., 1993). Heat conductivity measurements as a function of temperature showed that the electronic component of heat conductivity is negligibly small, since the crystals are nearly insulators in the temperature range 80 -300 K (Bunda, 1991).

The BiOHal samples were prepared in several ways. BiOX pure(starting material for a BiOX single crystals growth) was prepared by dissolving 99,99 pure bismuth oxides in hydrohalides acid, evaporating to form the hydrated halides, dehydrating the latter under vacuum, heating at ~380-700°C in a X_2 atmosphere, cooling to room temperature, and washing the product to remove BiX_3. The BiOX crystals were grown by the vapour gas transport reaction method in closed volume. The compounds BiOX of 99,99% purity was loaded into a high quality polished quartz ampoule (150-165 mm length and 15 mm diameter; combination of cylinder and cone). The ampoule was evacuated under vacuum of 10^{-4} Torr. After pumping the ampoule was filled with a (H_2O and HX) transport agent (TA). The system of chemical gas-transport reactions are as follows (Bunda, 1991):

$$3BiOX_{(s)} \rightarrow Bi_2O_{3(s)} + BiX_{3(g)}; \quad (1)$$

$$BiX_{3(g)} + H_2O_{(g)} \rightarrow BiOX_{(crystal)} + 2HX_{(g)} \quad (2)$$

$$Bi_2O_{3(s)} + 2HX_{(g)} \rightarrow 2BiOX_{(s)} + H_2O_{(g)}. \quad (3)$$

During transport involved three gaseous compounds: H_2O, HX and BiX_3. In the solid state exists a bismuth oxyhalides in polycrystalline and single crystal form. Reaction (1) describes the dissociation process of bismuth oxyhalides. The reaction (2) is a direct gas transport reaction with the water vapour as a transport agent. Reaction (3) describes the halogenation of bismuth oxide to bismuth oxyhalides. The (1-3) reactions cycle is closed and carried to the final transport in a BiOX single crystal of polycrystalline bismuth oxyhalide. The decomposition (1) and hydrolysis (2) reactions of BiOX are endothermic. Therefore, transport is possible only in the $T_2 \rightarrow T_1$ direction ($T_2 > T_1$). A two-zone resistance heated furnace was used for crystal growth. The axis of the ampoule was kept slightly inclined to the horizontal to enhance the transport of the vapour from the source zone to the growth zone during growth. We selected the pressure of transport agent in the ampoule on the basis of optimal mass transfer calculation by the convection method according to the equation (2). The best bismuth oxyhalide crystals were obtained at the 8-10 Torr pressure of transport agent. The some growth parameters of the BiOX crystals obtained are given in table 1.

Substance	Zone temperature (°C)		TA pressure (Torr)	Morphology and size	Growth period (days)
	Source	Growth			
BiOCl	720	600	8-10	Platelets 7x7x3 mm^3	5
BiOBr	700	600	8-10	Platelets 6x7x2 mm^3	6
BiOI	580	450	8-10	Platelets 5x5x1 mm^3	7

Table 1. Temperature of source and growth zones, transport agent pressure and size of BiOX crystals obtained

The crystallinity and morphology of the BiOX crystals was examined by X-ray diffraction (XRD), and field effect-scanning electron microscopy (FE-SEM). The lattice parameters were refined by X-ray powder diffraction data with a Si internal standard (high purity) and by using a least square method. All of the sharp diffraction peaks in the XRD patterns were perfectly indexed as pure phase of BiOCl, in good agreement with the standard JCPDS file (No. 06-0249) of BiOCl. A Laue photograph taken parallel to the large face of the platelets showed clearly a fourfold symmetry axis characterizing (001) planes of a matlockite tetragonal PbFCl--like structure $D_{4h}^7 - P\frac{4}{n}mm$ (No. 129) (Beck & Beyer, 1997) with two formula units in the unit cell and the following atom position (Bercha at al., 1973):

$$2Bi\ in\ position\ (2c)\ (0,1/2,z)(1/2,0,\bar{z})\ of\ 4mm\ (C_{4v}) symmetry$$
$$2O\ in\ position\ (2a)\ (0,0,0)\ (1/2,1/2,0)\ of\ \bar{4}2m\ (D_{2d})\ symmetry \quad (4)$$
$$2X\ in\ position\ (2c)\ (0,1/2,z')\ (1/2,0,\bar{z}')\ of\ 4mm\ (C_{4v}) symmetry..$$

The position (2a) is always occupied by the smaller anion which is coordinated tetrahedrally by the bismuth cations. The Bi^{3+} ion itself is coordinated ninefold capped quadratic antiprism) and site symmetry has C_{4v}. Fig.1 gives the coordination polyhedron of the cation (d) and a projection of the structure along the [010] direction (b).

An examination of the platelets using a polarizing microscope shows no extinction, thus indicating that the tetragonal *c*-axis is perpendicular to the plane of the platelets. In optical micrograph of a BiOCl crystal multiple growth spiral (two on the same sign) can be seen originating from a growth defect. The growth defect is responsible for the formation of the growth spirals. The observed spirals illustrate once again working of the Burton-Cabrera-Frank (BCF) theory of the crystal growth mechanism (Burton et al., 1951). On the other hand, the layers observed under different magnifications in the SEM of BiOBr and BiOI crystals indicate the layer growth mechanism of these crystals (Ganesha et al., 1993).

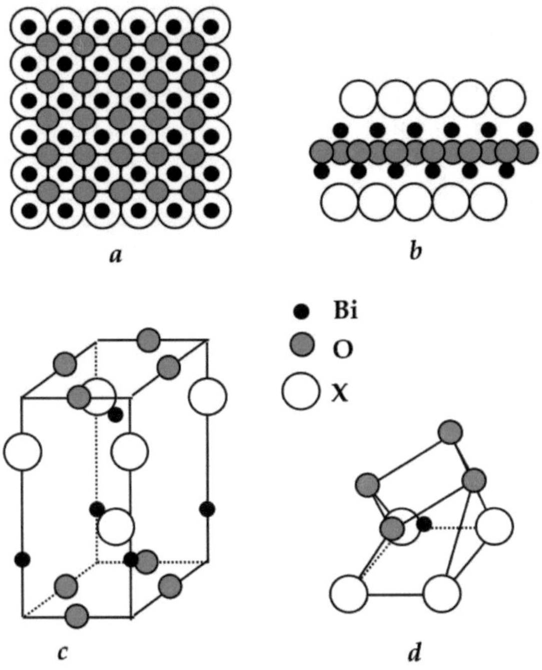

Fig. 1. Crystal structure of bismuth oxyhalides: (a) projection to (001) plane; (b) projection to (100) or (101) plane; (c) clinographic projection of the unit cell; (d) coordination polyhedron of the bismuth

Fig.1a shows the clinographic projection of the unit cell of BiOX crystals. Each Bi atoms is eight-coordinated by four O atoms and four X (halogen) atoms in the form of an asymmetric decahedron (Keramidas et al., 1993). Each atom (Bi, O, X) occupies a special position with multiplicity 2. The faces of the decahedron are 2 rectangles (O-O-O-O, X-X-X-X, which are parallel to the (110) plane and 8 isosceles triangles (four X-O-X and four O-X-O). The decahedra are linked to each other by a common O-X edge along the a and b axes forming infinite layers. The coordination of O and X are follows. Each O atom is linked to four Bi atoms and forming a tetragonal pyramid with the O atom at its apex. Also each X atoms forms with the neighbouring Bi atoms tetragonal pyramid with the X atom at its apex. Neighbouring decahedra form layers along (001) which are connected by common O-X edges. Neighbouring layers of decahedra are connected by common O-O or X-X edges.

The X atom is bonded with four Bi atoms in a planar square to form a pyramid and with its nonbonding (lone pair) electrons pointing to the other side of the square. These nonbonding electrons convert the three-dimensional fluorite-like structure into a two-dimensional layered structure. In the three-dimensional Bi_2O_3, the Bi atoms are coordinated to six O atoms. This structural difference between them is the major reason for BiOX to have a wider optical band-gap. As shown in the fig. 1, the [BiOX] layers are stacked together by the nonbonding (van der Waals) interaction through the X atoms along the c-axis. Therefore, the structure is not closely packed in this direction. When one photon excites zone electron from X np states (n = 3, 4, 5 for Cl, Br, I respectively) to Bi 6p states in BiOX, one pair of a hole and an excited electron appear. The layered structure BiOX can provide the space large enough to polarize the related atoms and orbitals (Zhang et al., 2006). The induced dipole can separate the hole–electron pair efficiently, enhancing photocatalytic activities. BiOX has an indirect-transition band-gap so that the excited electron has to travel certain k-space distance to be emitted to valence band (Table 2). Therefore, the BiOX single crystals can be described as 2D structured materials.

Substance	Lattice constants, Å		Relaxed atomic position		Θ_D, K	E_g, eV
	a	c	z	z'		
BiOCl	3.883	7.347	0.645	0.170	205	3.455
BiOBr	3.915	8.076	0.653	0.154	168	2.924
BiOI	3.984	9.128	0.668	0.132	146	1.89

Table 2. Lattice parameter (a, c), relaxed atomic position (z, z'), Debye temperature (Θ_D) and band gap (E_g) of bismuth oxyhalide crystals (Bunda et al., 2011)

3. Fabrication and classification of HTSC/photosemiconductor hybrid contact structures

In this capture we report the formation process of "ceramics HTSC($YBa_2Cu_3O_{7-\delta}$)/thin film SeC (BiOX)" hybrid contact structures(HCS). The choice of the structure "HTSC-ceramic - SC-thin film" was due to the nonstability of the HTSC- film parameters. The layers of bismuth oxyhalogenides were prepared in the three ways (Bunda, 1991): a) the thermal discrete evaporation in the oxygen atmosphere (p=7-10 Torr); b) the laser impulse evaporation (the laser type of LTYPTSH-500; λ=0.53 μm; ω=50 Hz); c) the deposition by the method of gas chemical transport reactions in the quasiclosed volume. The films were characterized by qualitative X-ray analysis (powder diffraction measurements, using CuK_α-radiation), and by scanning electron microscopy with energy disperse chemical analysis SEM-EDS (for phase composition information). Preliminary $YBa_2Cu_3O_{7-\delta}$ ceramic were overreached in oxygen ions plasma atmosphere for reconstruction of the superconducting properties of disturbed surface layer. The hybrid contact structures were finally annealed at a temperature T=230-250°C (recrystallization and thermal relaxation procedure). Surface factor of prepared samples was controlled by scanning electron microscopy (SEM).

In the "HTSC - SeC" system can be formed three principal types of HCS's (Bunda et al., 1996): tunneling(1), proximity(2) and combined(3). In the case $d \gg \xi_p$ (ξ_p- the value of coherence length in the HTSC; d- the thickness of insulator(I) layer between super- and semiconductor) the electrical properties of "HTSC - SC" junction are mainly determinate by the parameters of the transition layer d. At d ~ ξ_p take place both the HTSC-I-SC junction (HCS of type 1) with high ($\rho_d > 10^3 - 10^5$ Ω·cm) values of the differential resistance ρ_d (at the bias voltage $V_{tr} \rightarrow 0$) and proximity contact(HCS of type 2) with the $\rho_{d \rightarrow} 0$. In the case d ~ ξ_p and $0 < \rho_d < 10^3 - 10^5$ Ω·cm take place HCS of type 3 - combined "proximity- tunnelling" junction (Bunda et al., 1996). The classification of principal possible types of "HTSC-SC" hybrid contact structures is show in the table 3.

Type of structure	Modification		Method of fabrication	ρ_d (Ω·cm)
	HTSC	SeC		
1. Van der Waals press contakt (type 1)	Ceramics	Crystals	Press optical contact with the polished ceramics	$10^3 \div 10^{12}$
2. Heterojunction (type 1, 3)	Ceramics	Thin films	Thermal discrete evaporation; laser ablation	$10^2 \div 10^5$
3. Heterojunction (type 1, 3)	Ceramics	Crystals	Chemical gas transport reactions	$10^1 \div 10^5$
4. Heterojunction (type 2,3)	Films	Films	Laser ablation; magnetron sputtering	$10^{-1} \div 10^2$
5. Heterojunction (type 2)	Films	Crystals	Laser ablation; magnetron sputtering	$10^{-3} \div 10^1$
6. Homogeneous p-p⁺-junction in HTSC ceramics (type 1-3)	HTSC ceramics with oxygen concentra-tion gradient		Oxygen ions implantation	$10^{-2} \div 10^3$
7. Homogeneous p-p⁺-junction in HTSC single crystals (type 1-3)	HTSC single crystals with oxygen concen-tration gradient		Oxygen ions implantation	$10^{-3} \div 10^2$

Table 3. Classification and methods of fabrications HTSC/SeC hybrid contact structures

The realization of "HTSC-SC" HCS's assumes the following conditions (Bunda et al., 1996):

- semiconducting layers should be hole conductivity type (p-type; only for the HCS's of type 1). For the semiconductors with the compensation type of conductivity the degeneracy($\zeta_p = E_F - E_v < 0$) is necessary;
- the thermal expansion and the lattice constants mish-match in the (001) base plane should not exceed the values of 15% and 2-5%, respectively;
- the presence of oxygen ions in the anion sublattice of semiconductor is desirable;
- the crystal structure of semiconductor should be very similar to the symmetry of space groups P4/*mmm* and/or I4/mmm, where are usual for the $YBa_2Cu_3O_{7-\delta}$ and $Bi(Tl,Hg)_2Sr(Ba)_2Ca_{n-1}Cu_nO_{2n+4}$ HTSCs.

Some progress has been made in the growth HTSC thin films on Si and on GaAs (Shiwagan et al., 2002). All successful growths have used intermediate buffer layer between the HTSC films and semiconductor substrates. The major problem in using the semiconductor substrates is the substrate-film inter-diffusion during the in-situ depositions rn addition to the formation of micro cracks in the films due to a relatively large difference rn thermal expansion coefficients between the semiconductor and HTSC. After the deposition of CdSe the critical transition temperature of HgBaCaCuO films was found be increased from 115 K with J_c = 1.7 x 103 A/cm^2 to 117.2 K with J_c = 1.91 x 103 A/cm^2. When the heterostructure was irradiated with red He-Ne laser (2 mW), the T_c was further enhanced to 120.3 K with J_c = 3.7 x 103 A/cm^2 (Shiwagan et al., 2002, 2004). The resistivity is decreased after the laser irradiation, but the nature of variations in resistivity with temperature remains similar to that of without laser irradiated samples. The decrease rn resistivity is attributed to increase in charge carriers in the heterostructure during laser irradiation. The CdSe has the direct band gap 1.74 eV and hence the laser source with energy 1.95 eV greater than band gap of CdSe was selected. Hence during the irradiations, the electron-hole pair generates in the CdSe sample individually at the junction of the heterostructure. As authors (Shiwagan et al., 2004) of work specify, the built-in-junction potential (n) CdSe and (p) HgBaCaCuO heterojunction the electron-hole pairs generated at interface could not recombine, rather holes are transferred to the superconductor region where they can be further trapped into CuO_2 layer and hence there is increase in conductivity.

4. Mechanism of photoconductivity in BiOX (X=Cl, Br, I) single crystals

The spectral dependences of the steady-state photocurrent (photoconductivity) were recorded using a LOMO MDR3 monochromator. Low temperatures were obtained by mounting the sample in an UTREKS continuous flow cryostat. The decay time measurements were performed using pulse UV- laser LGI-21 (pulse width 8-9 ns; excitation wavelength λ=0.337 nm; filters UFS-2 and UFS-6 were used). The change in the intensity was done by a platinum reducer (Bunda, 1989, 2005).

The original specimens had a dark resistivity of ρ_d = 10^{10} to 10^{11} Ω·cm and significant photosensitivity (Bletskan ey al., 1973). At a temperature 293 K and an illumination of 10^4 1x, the ratio ρ_d/ρ_{ph} (ρ_{ph} is the resistivity in the light) reached 10^3; cooling of the specimen led to an increase of 3-4 orders of magnitude in the steady-state photocurrent such that at 90 K, $\rho_d/\rho_{ph} > 10^7$. Thus, there was an intense temperature (T) suppression of the photocurrent (see fig. 2a).

The spectral dependence of the photoconductivity $\sigma_{ph}(h\nu)$ for a typical specimen of BiOCl recorded at 95 K (curve 1) and 293 K (curve 2) is illustrated in Fig. 2b. In addition to the intrinsic maximum $h\nu_1$= 3.54 eV (350 nm) which corresponds to the band gap of E_g = 3.5 eV at T = 100 K a significant impurity photoconductivity is observed due to the transfer of an electron from the *r* centre to the conduction band (Bunda, 1989). On heating, the main maximum is shifted towards the long-wave side at a rate of dE_g/dT= -6.3·10^{-4} eV/K, while the photoconductivity in both the impurity and intrinsic regions falls sharply. The depth of the level which produces the impurity photocurrent maximum was found to be E_{cr}^0=2.2 eV Bunda, 2005)..

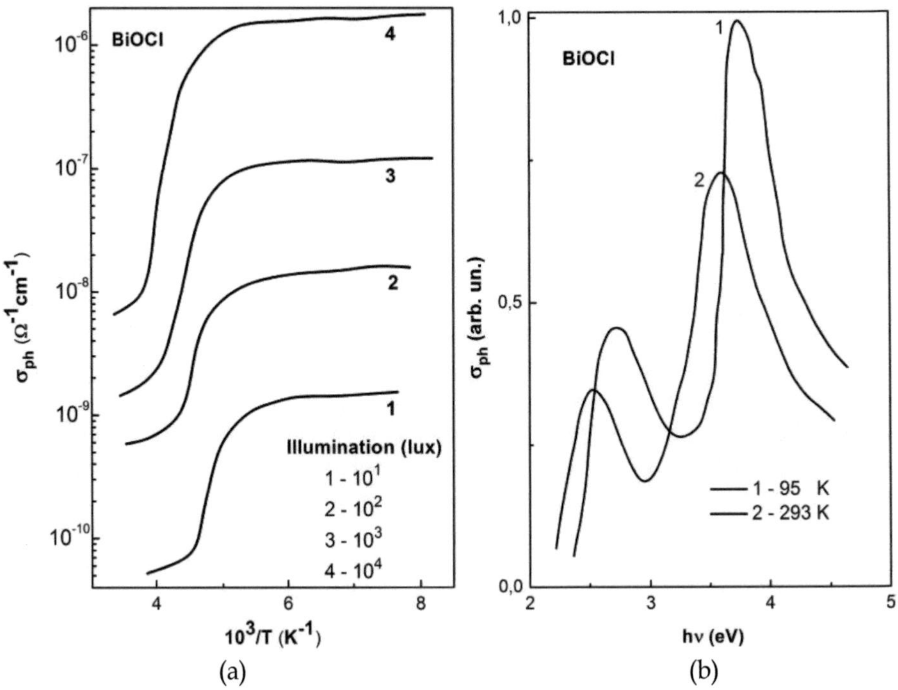

Fig. 2. The temperature dependence (a) and the spectral dependence (b) of photoconductivity BiOCl single crystal

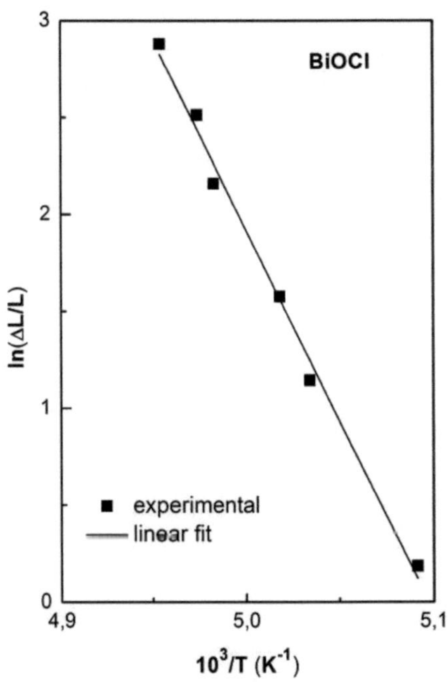

Fig. 3. Temperature dependence of the $\ln(\Delta L/L)$ value

The lux-ampere characteristics of the photocurrent (fig.3) at low temperatures up to the region of T-suppression are linear in the experimental illumination range α=1 [$\alpha = \frac{dL}{dI_{ph}}\frac{L}{I_{ph}}$); $I_{ph} \sim L^{\alpha}$]. In the T-suppression region the lux-ampere characteristics become superlinear and the value of α depends on the concentration of photocurrent carriers n, at a fixed temperature. The maximum value of α observed in the experiment reached α = up 3 to 4. At low temperatures, an optical IR suppression of the photocurrent was observed (Bletskan et al., 1973).

A study of the photoconductivity kinetics was made in the linear regime: $\Delta n \ll n$, where Δn is the change in n caused by a short pulse of light in the intrinsic excitation region. The curve of the decay of the photocurrent during the illumination by a pulse of intrinsic light consisted, of course, of two sections with clear by different times : $\tau_r = (2.5 - 6.6) \cdot 10^{-3}$ and $\tau_s = (0.8 - 4) \cdot 10^{-5}$ s.

The presence of temperature and IR suppression of the photocurrent, the superlinearity of the lux-ampere characteristics, and the two sections observed on the relaxation curves of the photoconductivity, indicate unequivocally that in these particular crystals, the recombination processes are determined by two types of centres: "sensitizing" r centres and fast s centres.

The energy distance of the r centres from the valence band is E_{Vr}^{T} and the thermal value was determined from the dependence on the carrier concentration of the temperature at which the thermal suppression of photocurrent begins as in (Bletskan et al., 1973): E_{Vr}^{T}=1.3 eV.

Another more accurate method of determining E is to measure the dependence of $\Delta L/L$ on 1/T, where ΔL is the change in intensity of the exciting light L which is necessary for maintaining the electron concentration n, with a change in the temperature in the T-suppression region,

$$\frac{\Delta L}{L} = \frac{g_s B}{C_{rn} n} = g_s Q_v \frac{C_{rp}}{n C_{rn}} \exp(-\frac{E_{Vr}^{T}}{kT}), \quad (5)$$

where g_s is the portion of the total carrier-recombination current passing through the s centres; g_r is the portion passing through the r centres; C_{rp} and C_{rn} are the probabilities of hole or electron trapping at an r centre; and Q_v is the static factor of the valence band. From the curve of $\ln(\Delta L/L)$ (Fig. 3) was determined E_{Vr}^{T} = 1.3 eV (Bunda, 1989, 2005).

Next, we present results of the study of photoconducting spectra anisotropy of the BiOX single crystals, obtained with the polarization of electrical vector $\vec{E}$ of the electromagnetic wave parallel and perpendicular to each of the two axes ([100] or [010]) of the unit cell. The BiOHal compounds are broad-band (1.80 - 3.50 eV) photo-semiconductors. The maxima of their photoconductivity are in the interval of 0.24-1.2 μm and theirs photosensitivity's are $S = \sigma_{ph}/\sigma_d = 10^4 - 10^8$ arbitrary units at the temperature 80K (Fig. 4). The anisotropy of photoconducting spectra is believed to be the results of the layer matlockite-type (PbFCl) crystal structure of all bismuth oxyhalide compounds.

The coefficient of photopleochroism was calculated as

$$P = \frac{I_{\parallel} - I_{\perp}}{I_{\parallel} + I_{\perp}} \cdot 100\% , \quad (6)$$

where $I_{//}$ and $I_{\perp}$ is the photocurrent values at the $\vec{E}//[100]$ and $\vec{E}\perp[100]$ light polarization, respectively.

The position of electrical vector $\vec{E}$ can be varied either by rotating of the crystal. Anisotropy in 3 - 5 times was observed for the photosensitive value S. The change in light polarization at the 90 degrees (from [100] to [010] crystallographic direct) leads to the change in the photoconductivity of 50-100 units (Fig. 5).

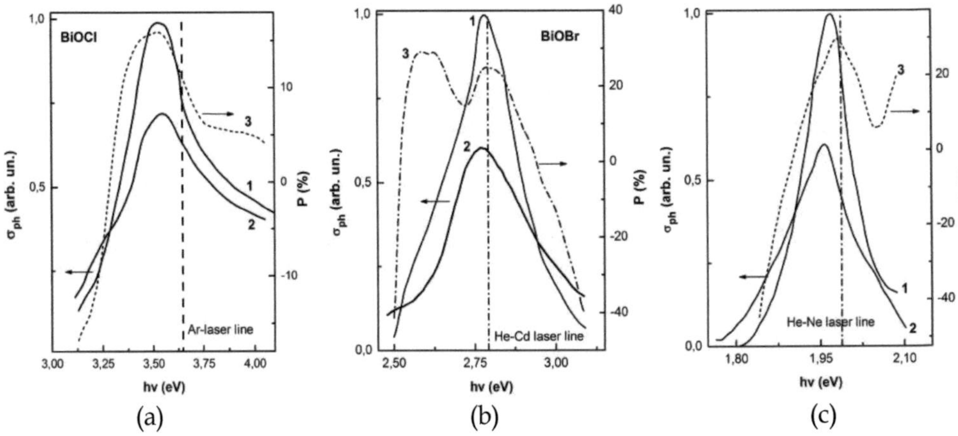

Fig. 4. Photoconductivity and photopleochroism spectra of (a) BiOCl, (b) BiOBr and (c) BiOI single crystals at different light polarization and temperature 80K

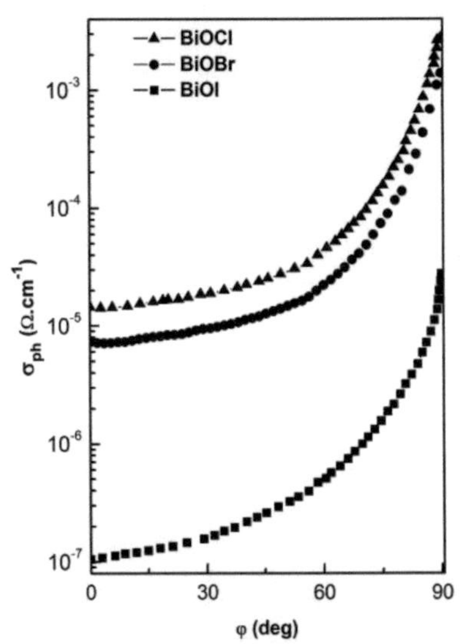

Fig. 5. Angular dependencies of photoconductivity of BiOX single crystals

The temperature region of BiOX based photoresistors is in the range 80-400K. They are stable in air, and their parameters do not depend on the environment, which eliminates the need for sealing. It was established experimentally that the adsorption of oxygen leads to a slight increase even values of photosensitivity (Bunda, 1989, 2005). Other adsorbents do not cause changes in the photoelectric properties. Crystal structure of the sample reduces the intrinsic noise of the radiation detector. The functional photoresistors based on BiOX single crystals, included in electronic circuits as a two-terminal networks, what is their great advantage. They can follow the rotation of polarization plane or work in schemes of the maximum signal finding. Besides, the BiOX crystals can be used as photo-potentiometers. The selectivity of BiOX spectral characteristics make them good radiation detectors of the nitrogen, argon and helium-cadmium lasers, as well as some of the semiconducting light-emitting diodes (LEDs). Therefore, they can be used in optoelectronic circuits in conjunction with these lasers and LEDs.

5. Energy structure and tunnelling effects of HTSC/BiOX hybrid contact structures

The bismuth oxyhalide single crystals are very good "sparring partners" to create a HTSC/BiOX hybrid contact structures. This is for the following reasons:

1. The good thermal expansion and lattice constants mishmash (15% and 3% respectively) for the base planes (001) is observed.
2. The BiOX compounds have a layered structure and can be easily crystallized as an ultra-thin (0.25-0.5 μm) single crystalline layers on the HTSC substratum's using the method of chemical gas transport reactions.
3. The BiOX crystals are broad-band (E_g=1.85-3.65 eV) photosemiconductors. The maximum of their photoconductivity are in the 0. 24 - 1. 0 mkm interval and theirs photosensitivity's are $S = \sigma_{ph}/\sigma_d = 10^4 - 10^8$ at the temperature 80K.
4. The bismuth oxyhalide single crystals can be easily intercalated (5-7 weigh %) and doped (0. 01-3%) by the 3d- and 4f- elements. That leads to the stable p-type conductivity, and the level of degeneration $\zeta = E_v - E_F$ (E_v - valence band; E_F - Fermi level position) can be controlled.
5. The BiOX layered compounds contain not less than two elements - the cation (Bi^{3+}) and anion (O^{2-}), like in superconductors $YBa_2Cu_3O_{7-\delta}$ or $Bi_{2-x}Pb_xSr_2Ca_{n-1}Cu_nO_{2n+4}$. Therefore the contamination process of the heterojunctions during the formation procedure are absent.

The unique temperature behaviour of the HCS's resistance consists in an N-shape anomaly around the superconducting transition temperature (interval 85-105K; T_c~92K; fig.6). The temperature dependence of the BiOCl:Ti semiconductor resistance can be described by the exponential function $\rho(T) = \rho_0 \exp(-E_g/kT)$, where E_g=2.8 eV and ρ_0=10^2-10^7 Ω·cm, depending on the activation regimes.

The detailed consideration of the temperature dependence of the heterojunction resistance in both directions (curves 1 and 2 on fig.6) shows the anomaly of heterojunction resistance in the narrow temperature interval above T_c (ΔT=5K). The resistance first decreases down to ρ_{min} at T=T_c+2K and then rapidly increases to the maximum at T=T_c+5K. Subsequently, the resistance of heterojunction decreases according to the exponential dependence up to the room temperature. So the temperature dependence of $YBa_2Cu_3O_{7-\delta}$/BiOCl:Ti heterojunction resistance were characterized strongly pronounced N-shaped character in the narrow temperature interval above the superconducting transition. Fig. 6 shows the considerable

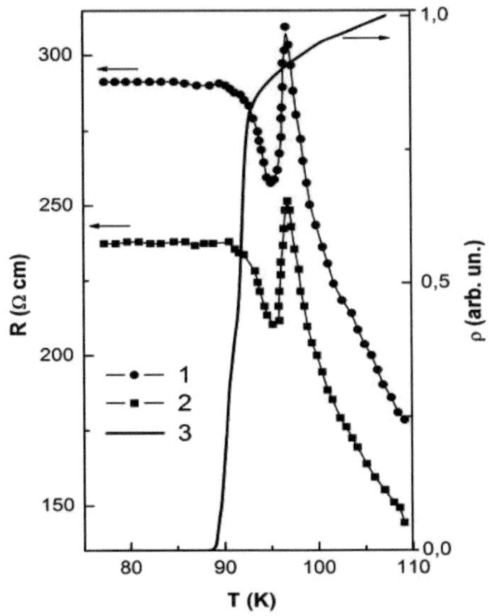

Fig. 6. Dependencies of the resistance R versus temperature for the $YBa_2Cu_3O_{7-\delta}$/BiOCl:Ti HCS along the semiconductor layer in (100) plane (curve 1) and trough the junction (perpendicular to (001) plane; curve 2). The curve 3 characterize the temperature dependence of $YBa_2Cu_3O_{7-\delta}$ ceramics resistance

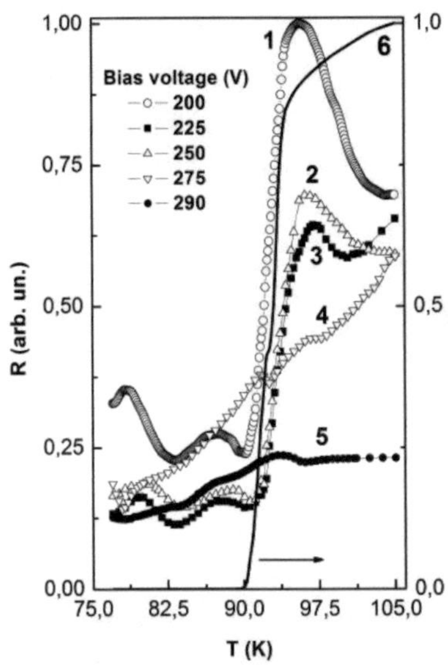

Fig. 7. Temperature dependencies of $YBa_2Cu_3O_{7-\delta}$/BiOI hybrid contact structure resistivity at the different values of bias voltage along the (100) plane. The curve 6 characterize the temperature dependence of $YBa_2Cu_3O_{7-\delta}$ ceramics resistance difference in the values of maximally resistance's (amplitudes of the peaks) and N-anomalous temperature width for the curves 2 and 3 (see fig.6). It is interesting that these anomalies in the $\rho(T)$ dependence appear at the T>T_c, and the appreciable deviation (decreasing) of the differential resistance appeared long before the approach to T_c (at T=85K).

Fig. 7 show the dependencies of resistance versus temperature for the hybrid contact structure $YBa_2Cu_3O_{7-\delta}$ (ceramics)/BiOI (single crystals) along the (001) plane of semiconductor at the different values of the bias voltage V_{tr}. We observed very similar to $YBa_2Cu_3O_{7-\delta}$/BiOCl:Ti heterojunction "resistivity - temperature" behaviour (see Fig.6). Note the characteristic features of these curves. The first, in T_c vicinity the N-like anomaly of electrical resistance is observed (these anomaly is not dependent on the geometry of the measurements). Second, the amplitude jumps of the resistance *R(T)* (and width of derivate negative value $dR(T)/dT$) was strongly dependent on the applied bias voltage. Third, exist the areas of bias voltage where the resistance jumps have maximum amplitude. For planar and through geometry them are 90-100 V and 220-255 V, respectively. Fourth, in the T_c vicinity the anomalies of thermopower are observed.

The red He-Ne laser (with λ = 632.8 nm, E = 1.95 eV and power P = 1.5 mW) was irradiated onto BiOI surface of $YBa_2Cu_3O_{7-\delta}$/BiOI heterostructure, for the 1 hour. The variation in the G_T/G_{300K} normalized conductance during laser irradiation was measured as a function of temperature and is shown in figure 8. It can be excitingly observed that the T_c is further increased from 92K to 94.5 K and critical current value J_c measured at 80 K is $1.85x10^3$ A/cm^2. Here, the increase superconducting parameters can only be attributed to the increase in carrier concentration when the sample was irradiated by laser having the energy greater than the band gap of semiconductor (Shiwagan el al., 2002).

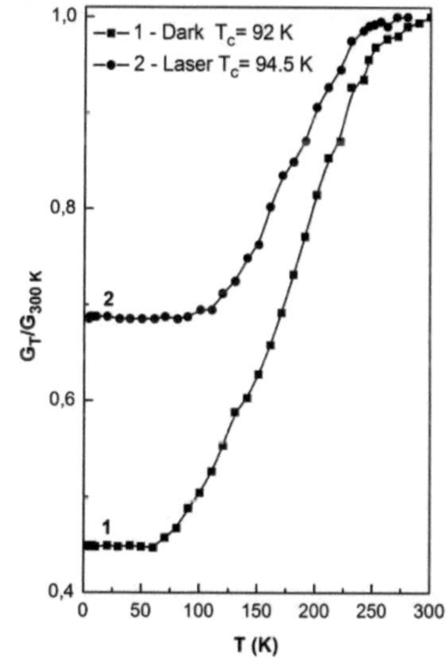

Fig. 8. Variation in normalized conductance of $YBa_2Cu_3O_{7-\delta}$/BiOI heterostructure in (1) dark; and (2) under laser irradiation

The increase in superconducting parameters after laser irradiation is attributed to the microlevel dislocations due to lattice mismatch, attachment to the nonstoichiometric oxygen present on the BiOI surface, development of a diffusion region in semiconductor side of the interface (Shiwagan et al., 2004).

The proposed energy band diagrame for the $YBa_2Cu_3O_{7-\delta}$/BiOCl:Ti hybrid contact structures is shown on the Fig. 9. It is assumed that the semiconductor is characterized by a p-type conductivity (Alfeev et al., 1990,1991). He is in a degeneration state. The degree of degeneration $\zeta_p = E_v - E_F$ is much more than the $2\Delta_0$ twice energy gap value ($\zeta_p \gg 2\Delta_0$).

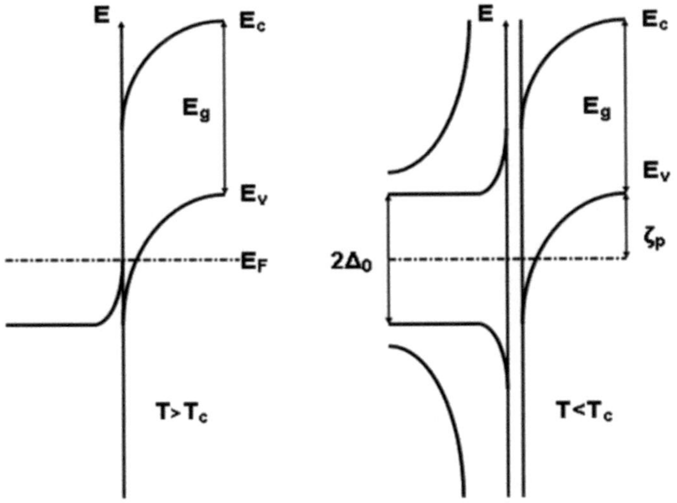

Fig. 9. The energy band diagram of "HTSC/SeC" heterojunctions

At temperatures $T < T_c$ we observed the changes of elementary excitations spectra of HTSC - formation of the $2\Delta_0$ energetic gap near the Fermi level. The unique of the temperature dependence $YBa_2Cu_3O_{7-\delta}$ /BiOCl:Ti junction resistance and negative differential resistance values in the current-voltage characteristics around the superconducting transition can be related to quantum tunnelling of quasiparticles through the junction. The E_F Fermi surface divides basic and excited states of quasiparticles (Cooper's pairs). This provides the possibility for the mutual control of their properties.

Single particle tunnelling can be described by the tunnelling Hamiltonian (Tinkham, 1996)

$$H_T = \sum_{\sigma kq} T_{kq} c_{kq}^* c_{kq} + h.c. \tag{7}$$

In eq. (7) the specific details of the tunnelling barrier are included in the phenomenological tunnelling matrix element T_{kq} with wave vectors k and q of the electrons of one of the electrodes. The tunnelling process itself is described by the annihilator c_{kq} which removes an electron with wave vector k and spin σ, and the creator c_{kq}^* generating an electron with q on the other electrode (Kleine, 2010).

Besides the tunnelling probability given by $|T_{kq}|^2$, the tunneling current is governed by the number of quasiparticles trying to tunnel and on the available states into which they can tunnel. This is accounted for by the density of states of both electrodes $N_{1,2}$ and their occupation given by the Fermi function *f(E)*. Multiplying this with the tunnelling probability leads to the following expression for the tunnelling current

$$I(U) = A|T_{kq}|^2 \int_{-\infty}^{+\infty} N_1(E) N_2(E+eU)[f(E) - f(E+eU)]dE \tag{8}$$

with A as a constant of proportionality (Tinkham, 1996).

In the case of a normal metal/semiconductor (NSeC) tunnel junction applying a voltage shifts the Fermi energy of the electrodes by eU with respect to each other, which results in a net tunnelling current and a normal state conductance $G_{nn} \neq 0$. This current increases with bias since on the one hand more quasiparticles can participate in tunnelling and on the other hand more empty states are available. Therefore, as long as the barrier is not affected by the bias, eq. (8) gives a linear current vs. bias (I-U) characteristic in NSeC systems with $I(U) = U/R_T$. R_T is called tunnelling resistance and is proportional to $1/|T_{kq}|^2$.

Within the BCS theory a temperature dependent energy gap $\Delta(T)$ can be deduced from the excitation spectrum of a superconductor exhibiting its largest value at T = 0. With rising temperature, the energy gap shrinks and vanishes at $T = T_c$. The energy gap $2\Delta(T)$ corresponds to the energy which is necessary to split up a Cooper pair and create two single quasiparticles in the superconductor. These quasiparticles exist above the gap as exited states. The size of the energy gap at T=0 is related to T_c via $\Delta(0) = 1.764kT_c$. Close to T_c the temperature dependence of the energy gap is (Tinkham, 1996)

$$\Delta(T) \approx 1.74\Delta(0)\sqrt{1-\frac{T}{T_c}}. \tag{9}$$

The transition from normal to superconducting state leads to a change of the density of states N as a direct consequence of the emergence of the energy gap. The superconducting density of states N_s for quasiparticles at energy E reads (Buckel & Kleiner, 2004)

$$N_s(E) = \begin{cases} N_n(E_F)\cdot\frac{|E-E_F|}{\sqrt{(E-E_F)^2-\Delta^2}} & for\ |E-E_F| \geq \Delta \\ 0 & for\ |E-E_F| \leq \Delta \end{cases} \tag{10}$$

where N_n is the normal metal density of states at $T > T_c$ and E_F the Fermi energy. It follows from eq. (10) that no single particle states exist in the energy interval $E_F \pm \Delta$, whereas N_s exhibits a singularity at the gap edges and reduces to N_n at larger energies, see Fig. 9.

In superconducting tunnelling spectroscopy one mostly deals with semiconductor/superconductor (SeCSC) tunnel junctions which exhibit nonlinear $I-U$-characteristics. For an applied bias $|U| < \Delta/e$ and $T = 0$ no single particle tunnelling is possible, because no available excitation states do exist in the superconductor ($N_s - 0$), see eq. (10). The characteristic strong increase of the tunnelling current at $|U| = \Delta/e$ is due to the singularity in N_s and the resulting large number of unoccupied states, see Fig. 9. For $|U| \gg \Delta/e$ the $I-U$ - characteristic resembles the one of an SeCN junction since N_s is similar to N_n in this energy range. With eq. (8) the bias dependence of the tunnelling current of an SeCSC junction is described by (Kuznetsov, 2002, 2004)

$$I(U) = A|T_{kq}|^2 N_1(0)\int_{-\infty}^{+\infty} N_{2s}(E)[f(E) - f(E+eU)]dE. \tag{11}$$

When taking the derivative of eq. (11) with respect to the applied voltage, one gains the following expression for the differential conductance (Tinkham, 1996)

$$G_{ns} = \frac{dI_{ns}}{dU} = G_{nn}\int_{-\infty}^{+\infty}\frac{N_{2s}(E)}{N_2(0)}\left[-\frac{\partial f(E+eU)}{\partial(eU)}\right]dE \tag{12}$$

which directly gives N_s for $T = 0$. Temperature dependent STS measurements, as shown in Figures 6 and 7, exhibit smeared features with increasing T, because the energy gap shrinks with T and even with an applied bias $|U| < \Delta/e$ some quasiparticles can tunnel due to thermal energy (Kuznetsov, 2001).

When a superconductor (SC) and a semiconductor (SeC) are in electrical contact and form an SeCSC interface, a variety of different phenomena and transport processes can happen. In a device containing multiple SeCSC interfaces, both local (Andreev reflection) and nonlocal (charge imbalance) processes can occur (Klein, 2010).

6. Conclusion

In the "HTSC - SeC" system can be formed three principal types of HCS's (Bunda et al., 1996): tunneling(1), proximity(2) and combined(3).

The bismuth oxyhalide single crystals are layered 2D structured materials. The crystal structure of BiOX is very similar to the symmetry of space groups $P4/mmm$ and/or I4/mmm, where are usual for the $YBa_2Cu_3O_{7-\delta}$ and $Bi(Tl,Hg)_2Sr(Ba)_2Ca_{n-1}Cu_nO_{2n+4}$ HTSC's. Besides, the BiOX crystals, as well as a HTSC's contain oxygen. The thermal expansion and the lattice constants mish-match in the (001) base plane are in good agreement with the same parameters of the HTSC's. Also, the BiOX crystals are good "sparring partners" for the fabrication of "HTSC – photosensitive semiconductor" hybrid contact structure.

The transition of a high temperature superconductor in the superconducting state leads to a significant increase in the photosensitivity of the semiconductor layer of the heterostructure. This is due to single spin quantum tunneling processes of the Cooper pairs inside the semiconductor layer. Also observed the N - shaped anomalies in the temperature dependences of the conductivity (resistivity) in the vicinity of the superconducting transition.

Thus, we believe that based on "high-temperature superconductor - photosensitive semiconductor" hybrid contact structures can be created entirely new device of cryophotoelectronics.

7. References

Bunda, V.V. (1991). Specification of the temperature behavior of the resistance of heterojunctions from high-temperature superconductors with degenerate semiconductors near T_c. *Pis'ma v Zhurnal Tekhnicheskoi Fiziki (written in Russian)*, Vol. 17, No.22 (December 1990), pp. 15-18, ISSN 0320-0116

Akani, M., Froese, A., Staikov, G., Lorenz, W.J.,Röhner, M., Hopfengärter, R. & Saemann-Ischenko, G. (1995). Low-temperature behaviour of high-T_c superconductor / semiconductor heterojunctions. *Physica C*, Vol.245, No.1-2, (December 1994), pp. 131-138, ISSN 0921-4534

Deng, Zhengtao, Chen,Dong, Peng, Bo & Tang, Fangqiong. (2008). From bulk metal Bi to two-dimensional well-crystallized BiOX (X = Cl, Br) micro- and nanostructures: synthesis and characterization. *Crystal growth & design*, Vol.8, No.8 (May 2008), pp. 2995-3003, ISSN 1528-7483

Bunda, V.V., Bunda, S., Krivsky, V., Butorin, A. & Zihika, V. (1996). Physical properties of "HTSC-photosemiconductor" hybrid contact structures. *Czechoslovak Journal of Physics*, Vol. 46, No. 3 (November 1995), pp. 1361-1362, ISSN 0011-4626

Bunda, V., Bunda, S., Vashchuk, F., Feher, A. & Kovač, J. (2010). Physical properties of "HTSC/ Photosemiconductor" micro- and nanoplates based junctions. *Acta Physica Polonica A*, Vol. 118, No.5 (November 2010), pp. 1040-1041, ISSN 0587-4254

Bunda, V., Bunda, S., Kajnakova, M., Feher, A., Syrkin, E. & Feodosyev, S. (2010). Low temperature specific heat of BiOX (X=Cl, Br and I) single crystals. *Acta Physica Polonica A*, Vol. 118, No.5 (November 2010), pp. 986-987, ISSN 0587-4254

Bunda, V., Bunda, S., Kajnakova, M., Feher, A., Syrkin, E. & Feodosyev, S. (2011). Low-temperature specific heat of single crystal bismuth oxyhalides. *Low temperature physics*, Vol.37, No. 4 (July 2010), pp. 326-328, ISSN 1063-777X

Bletskan, D.I., Kopinets, I.F., Rubish, I.D., Turyanitsa, I.I. & Shtilikha, M.V. (1973). Photoconductivity and photoluminescence in BiOCl crystals. *Soviet Physics Journal*, Vol.16, No.5 (May 1973), pp.646-648 ISSN 1064-8887

Bunda, V.V. (1989). Optical anisotropy of deep recombination centers in the layered crystals of the matlokyte family. *Ukrainskii Fizicheskii Zhurnal (Russian Edition)*, Vol.34, No.5 (December 1988), pp. 663-666, ISSN 0503-1265

Bunda, V.V. (2005). Functional detectors of radiation for the middle ultraviolet area based on bismuth oxyhalide single crystals. *In: Physics of Electronic Materials. 2nd International Conference Proceedings. Kaluga. Russia. May 24-27*, Vol.2, pp. 3940-3942

Ganesha, R., Arivuoli, D. & Ramasamy, P. (1993). Growth of some group V-VI-VII compounds from the vapour. *Journal of crystal growth*, Vol.128, No.1-4 ,pp.1081-1085 ISSN 0022-0248

Burton, V.K., Cabrera, N. & Frank, F.C. (1951). The growth of crystals and the equilibrium structure of their surfaces. Philosophical transaction of the royal society, Vol.243, No.866 (June 1951), pp.299-358, ISSN 1471-2962

Beck, H.P. & Beyer, T. (1997). Lattice potentials as an instrument in crystal chemistry. II. Applications to the crystal bulk. *Zeitschrift für Kristallographie*, Vol. 221, No.8(December 1996), pp. 565-571, ISSN 0044-2968

Keramidas, K. G., Voutsas, G.P. & Rentzeperis, P. I. (1993). The crystal structure of BiOCl. *Zeitschrift für Kristallographie*, Vol. 205, No. Part-1 (January 1992), pp. 35-40, ISSN 0044-2968

Bercha, D.M., Puga, G.D. & Borets, A.N. (1973). Two-phonon processes in crystals of D_{4h}^7 symmtry. Physica Status Solidi (B), Vol.55, No.1 (January 1973), pp.69-78 ISSN 0370-1972

Zhang, Ke-Lei, Liu, Cun-Ming, Huang, Fu-Qiang, Zheng, Chong & Wang, Wen-Deng (2006). Study of the electronic structure and photocatalytic activity of the BiOCl photocatalist. Applied Catalysis B: Environmental, Vol.68, No.3-4 (August 2006), pp.125–129, ISSN 0926-3373

Shivagan, D.D., Shirage, P.M. & Pawar, S.H. (2004). Studies on the fabrication of $Ag/Hg_1Ba_2Ca_1Cu_2O_{6+\delta}/CdSe$ heterostructures using the pulse electrodeposition technique. *Semiconductor Science and Technology*, Vol.19, No.3 (September 2003), pp. 323-332, ISSN 0268-1242

Shivagan, D.D., Shirage, P.M. & Pawar, S.H. (2002). Studies on photoinduced effects in pulse-electrodeposited Ag/Hg-1212/CdSe hetero-nanostructures. *Pramana–Journal of physics*, Vol.58, No.5&6 (May 2002) pp. 1183-1190, ISSN 0304-4289

Alfeev, V.N. & Neustroev, L.N. (1990). The photo-current in the tunnel contact: high-temperature superconductor-semiconductor-semiconductor. *Doclady Akademii Nauk SSSR (Reports of the Soviet Academy of Sciences -in Russian)*, Vol.311, No. 5 (Yanuary 1990), pp. 1106-1110, ISSN 0002-3264

Alfeev, V.N., Men'shov, V.N. & Neustroev L.N. (1991). Photoelectric amplification in the tunneling structure degenerate semiconductor/high-temperature superconductor/insulator/metal. *Fizika Nizkich Temperatur (Low Temperature Physics - in Russian)*, vol.17, No.1 (November 1990), pp. 128-130, ISSN 0132-6414

Kleine, A. (2010). *Experiments on nonlocal processes in NS devices.* Inauguraldissertation zur Erlangung der Würde eines Doktors der Philosophie vorgelegt der Philosophisch-Naturwissenschaftlichen Fakultät der Universität Basel, UNI, Basel, Schwizerland, 108 p.

Kuznetsov, G.V. (2004). Transport of the charge carrier through superconductor - semiconductor contact. *Fizika Nizkich Temperatur (Low Temperature Physics - in Russian)*, vol.30, No.10 (April 2004), pp. 1038-1044, ISSN 0132-6414

Kuznetsov, G.V. (2002) Charge carrier passing through the metal-superconducting semiconductor contact. *Fizika i technika poluprovodnikov (Semiconductors – in Russian)*, Vol.39, No.9 (February 2002), ISSN 0015-3222

Kuznetsov, G.V. (2001) Thermoelectrical current through the metals - superconducting semiconductor contact. Pisma v ZETF (JETP Letters - in Russian), Vol.74, No.10 (October 2001), pp. 556-559, ISSN 0021-3640

Novel Application of YBCO Ring and Closed Loop with DC and AC Magnetic Flux Transformation

Janos Kosa
Kecskemet College, Faculty of Mechanical Engineering and Automation
Hungary

1. Introduction

The chapter presents the theoretical and experimental results of DC and AC magnetic flux transfer with one closed superconducting loop. This method may open a new path for advanced applications by using superconducting ring and perfect closed superconducting loops. I examined a novel arrangement of YBCO ring and continuous loops made of HTS wire. With this solution we can develop a self-limiting transformer. At the same time I also tested the operation of the YBCO ring made from bulk and continuous YBCO loop developed with my own technology. I could carry out these experiments because earlier I had elaborated the machining of the YBCO bulk and wire.

In the first experiment I used YBCO superconducting rings made from a bulk because I managed to drill the YBCO bulk just in some minutes. This technology is low price (Kosa & Vajda, 2007; Kosa et al., 2009). Fig. 1 shows one bulk and two rings drilled from the same bulk. The used YBCO bulk was produced at IPHT Jena, Germany.

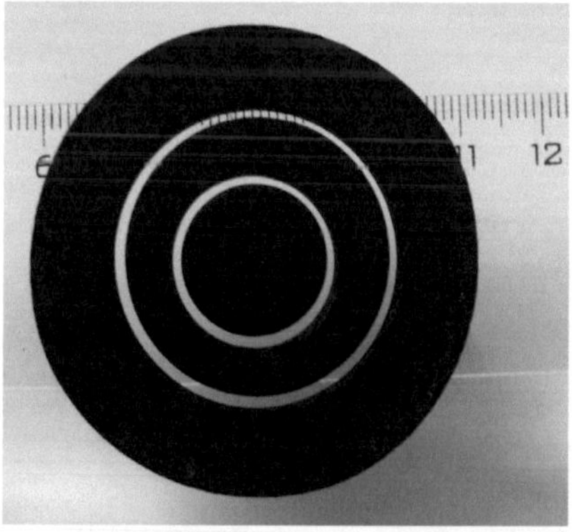

Fig. 1. Rings from YBCO bulk with economical and fast technology

Later I applied superconducting wire because the diameter of the superconducting ring from bulk is small. It is a positive fact there is possibility to soft-solder the YBCO wires to create closed loop, but this method is not optimal solution.

- For the beginning the transfer of the magnetic flux is impossible with DC current between independent iron cores.
- Secondly it is established fact we can use the soldered wire with AC current in different applications but with soldering we cannot ensure appropriate uniform temperature in the superconducting wire when the YBCO loop will change to normal state from superconducting state. In soldered spot the Joule loss is considerably higher than in the wire itself.

This is the reason why I did not apply the closed loops with soldering for my applications. However, we must consider the fact, that even in the case of a really good soldering, the superconducting join will not exist.

A Korean group produced 100 % BSCCO loops for the first time (Hee-Gyoun Lee et al.,2006). In 2009 the other Korean group wrote that the perfectly closed YBCO wire making process needs further development. I modified their technology and managed to develop a different cutting method (Kosa et al., 2010). After development of the technology of the cut I examine the arrangement of the continuous YBCO superconducting ring and loop (100% YBCO join). Fig. 2 shows the photograph of an eleven-turn serial loop. The advantage of this winding is that there is no twist along the longitudinal axis that could cause cracks in the YBCO layer at operating temperature, i.e. at 77 K or a lower temperature. It must be mentioned that with this method we can only produce coils with odd number of turns. The odd number comes from topology of the surface of the cut wire. The applied superconductor wire (2G HTS wire) was produced by SuperPower, Inc., New York, USA. Type: SF 12050.

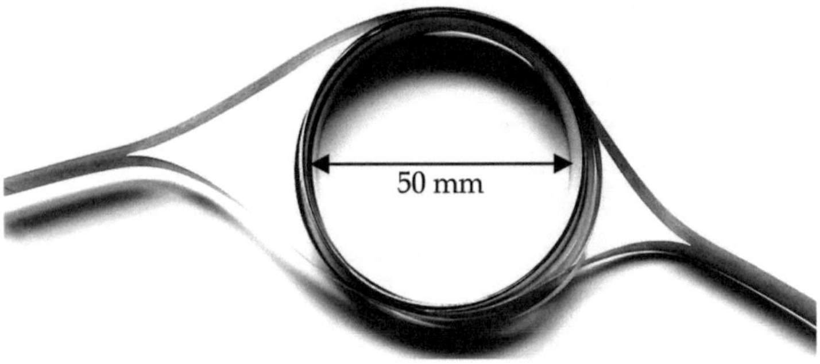

Fig. 2. Eleven-turn serial loop from YBCO wire without soldering and twists along its longitudinal axis (loops with layering)

With used technology we can create parallel and serial loops. This is the task of planning yet. We can apply the cut wires in DC and AC developments.

For example:

- DC flux transformer with one loop (Kosa & Vajda, 2009)
- self-limiting transformer and inductive fault current limiter (AC application)
- different solutions with active magnetic short circuit (DC and AC applications), etc.

With these rings and loops we can transfer the magnetic flux between two independent iron cores inductively from one closed iron core to the other one. This method gives a new possibility to transform the DC and AC magnetic field. This arrangement can result in novel applications in the field of transfer both the energy of DC magnetic field and AC electrical energy. In Fig . 3 we can see the scheme of the connection of the theory and applications in this chapter.

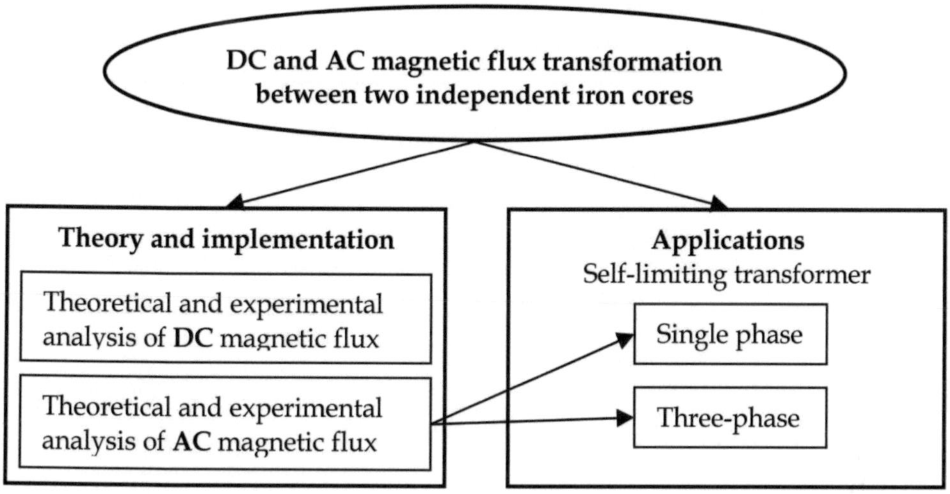

Fig. 3. The scheme of the connection of the theory and applications in this chapter

2. Theoretical and measurement examination of the transfer of the DC magnetic flux with one loop

We can see the experimental arrangement in Fig. 4.

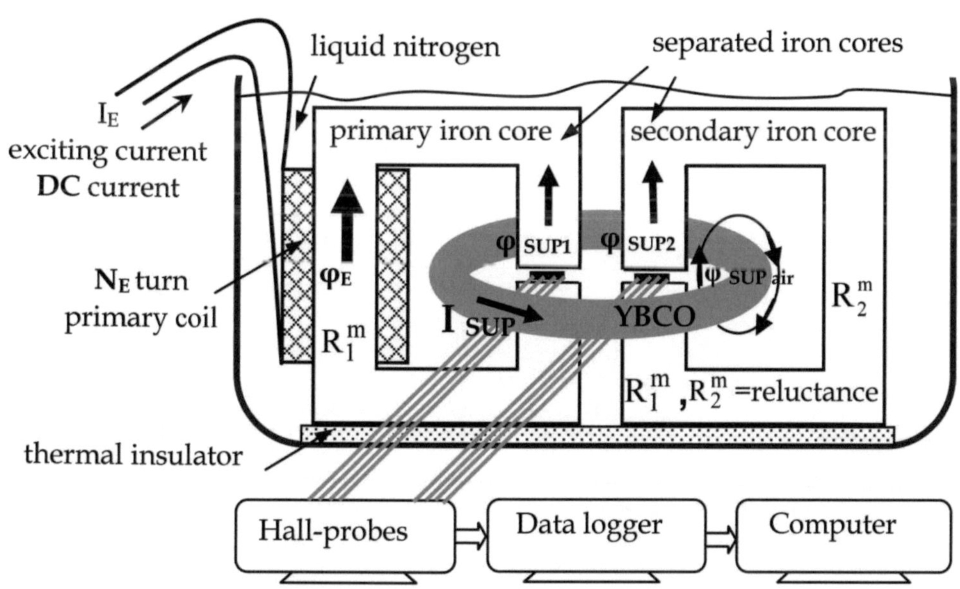

Fig. 4. Scheme of the transformation of the DC magnetic flux with one loop

2.1 Mathematical description of DC flux transfer

We can find a detailed theoretical description of the DC flux transformer with two coils (Orlando & Delin, 1990). In my case I describe a different solution. My DC flux transformer is implemented by using one YBCO ring. First I describe the theoretical examination. The Table 1 shows the nomenclature.

Parameters	Nomenclature
N_E	number of the primary turn (exciter coil)
φ_E	magnetic flux of the exciter coil in the primary iron core without superconductor loop
φ_{SUP1}; φ_{SUP2}	magnetic flux of the superconducting loop in the primary and secondary iron cores
$\varphi_{SUP\ air}$	magnetic flux of the superconducting loop inside the closed superconducting loop in air
R_1^m ; R_2^m	reluctance (magnetic resistance) of the primary and secondary iron cores
R_{air}^m	reluctance (magnetic resistance) of the air around superconducting loop
I_{SUP}	current of the superconducting loop
I_E	exciting current

Table 1. Nomenclature

$$\varphi_E = \varphi_{SUP\,1} + \varphi_{SUP\,2} + \varphi_{SUP\,air} \tag{1}$$

$$\frac{N_E I_E}{R_1^m} = I_{SUP}\left(\frac{1}{R_1^m} + \frac{1}{R_2^m} + \frac{1}{R_{air}^m}\right), \quad \text{where} \quad R = \text{reluctance} \tag{2}$$

$$I_{SUP} = \frac{N_E I_E R_2^m R_{air}^m}{R_2^m R_{air}^m + R_1^m R_{air}^m + R_1^m R_2^m} \tag{3}$$

If $R_2^m \rightarrow \infty$, there is not secondary iron core, then:

$$I_{SUP} = \lim_{R_2^m \rightarrow \infty} \frac{N_E I_E R_2^m R_{air}^m}{R_2^m R_{air}^m + R_1^m R_{air}^m + R_1^m R_2^m} = \frac{N_E I_E R_{air}^m}{R_{air}^m + R_1^m} \tag{4}$$

If $R_2^m \rightarrow \infty$ and $R_{air}^m \rightarrow \infty$, then

$$I_{SUP} = \lim_{R_{air}^m \rightarrow \infty} \frac{N_E I_E R_{air}^m}{R_{air}^m + R_1^m} = \lim_{R_{air}^m \rightarrow \infty} \frac{N_E I_E}{\frac{R_{air}}{R_{air}^m} + \frac{R_1^m}{R_{air}^m}} = N_E I_E \tag{5}$$

If $R_{air}^m \rightarrow \infty$ and és $R_2^m \neq \infty$ then

$$I_{SUP} = \lim_{R_{air}^m \rightarrow \infty} \frac{N_E I_E R_2^m R_{air}^m}{R_2^m R_{air}^m + R_1^m R_{air}^m + R_1^m R_2^m} = \frac{N_E I_E R_2^m}{R_2^m + R_1^m} \tag{6}$$

The operation is based on well known physical law. The possibility of the DC and AC flux transformation with single loop is based on the principle of the magnetic flux constancy in the time in a superconducting closed loop. The mentioned arrangement gives possibilities for the realization of novel industrial applications. The temporary constancy of the resultant flux of the superconductor ring is realized with less ring electrical current if we use secondary iron core.

We can see from Eq. 6 that the current of the superconducting ring decreases if there is a secondary iron core. We can see from equation (6), that the electrical current of the superconductor depends on the geometric data and magnetic permeability.

If the current of superconducting ring is bigger than the critical current of superconductor, there is not magnetic flux in the secondary iron core. In this case the coupling ends between the iron cores. The magnetic flux constancy depends on the current of the superconducting ring or loop because the characteristic (voltage–current) of superconductor. Non-linear effect of the characteristic is bigger in the case of the DC current, than at application of AC current.

2.2 Determination of the equivalent reluctance on the secondary side

The starting equation

$$\varphi_2 = \frac{I_{SUP}}{R_2^m} \tag{7}$$

$$\varphi_2 = \frac{\dfrac{N_E I_E R_2^m R_{air}^m}{R_2^m R_{air}^m + R_1^m R_{air}^m + R_1^m R_2^m}}{R_2^m} \tag{8}$$

$$\varphi_2 = \frac{N_E I_E R_{air}^m}{R_2^m R_{air}^m + R_1^m R_{air}^m + R_1^m R_2^m} \tag{9}$$

$$\varphi_2 = \frac{N_E I_E}{R_1^m + R_2^m + \dfrac{R_1^m R_2^m}{R_{air}^m}} \tag{10}$$

$$\varphi_2 = \frac{N_E I_E}{R_{eq2}^m} \tag{11}$$

The result is as follows

$$R_{eq2}^m = R_1^m + R_2^m + \frac{R_1^m R_2^m}{R_{air}^m} \tag{12}$$

If $R_1^m + R_2^m >> \frac{R_1^m R_2^m}{R_{air}^m}$, then

$$R_{eq2}^m \rightarrow R_1^m + R_2^m \tag{13}$$

In Fig. 5 we can see the equivalent circuit. Consequently so the magnetic flux in the secondary side is with ease definable.

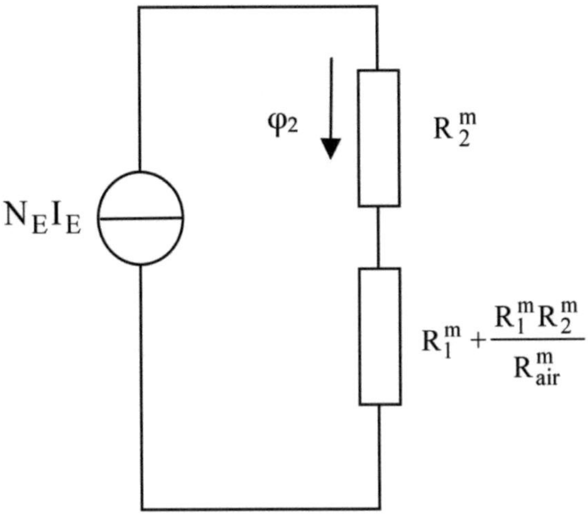

Fig. 5. Equivalent magnetic circuit on the secondary side

The scheme shows importance of proportion of the reluctances for the calculus of the magnetic flux in the secondary side.

2.3 Determination of the equivalent reluctance on the primary side

$$\varphi_1 = \frac{N_E I_E - I_{SUP}}{R_1^m} \tag{14}$$

$$\varphi_1 = \frac{N_E I_E - \frac{N_E I_E R_2^m R_{air}^m}{R_2^m R_{air}^m + R_1^m R_{air}^m + R_1^m R_2^m}}{R_1^m} \tag{15}$$

$$\varphi_1 = \frac{\frac{N_E I_E R_2^m R_{air}^m + N_E I_E R_1^m R_{air}^m + N_E I_E R_1^m R_2^m}{R_2^m R_{air}^m + R_1^m R_{air}^m + R_1^m R_2^m} - \frac{N_E I_E R_2^m R_{air}^m}{R_2^m R_{air}^m + R_1^m R_{air}^m + R_1^m R_2^m}}{R_1^m} \tag{16}$$

$$\varphi_1 = \frac{\frac{N_E I_E R_1^m R_{air}^m + N_E I_E R_1^m R_2^m}{R_2^m R_{air}^m + R_1^m R_{air}^m + R_1^m R_2^m}}{R_1^m} \tag{17}$$

$$\varphi_1 = \frac{N_E I_E (R_{air}^m + R_2^m)}{R_2^m R_{air}^m + R_1^m R_{air}^m + R_1^m R_2^m} \tag{18}$$

$$\varphi_1 = \frac{N_E I_E}{R_1^m + R_2^m \times R_{air}^m} \tag{19}$$

$$\varphi_1 = \frac{N_E I_E}{R_{eq1}^m} \tag{20}$$

$$R_{eq1}^m = R_1^m + R_2^m \times R_{air}^m \tag{21}$$

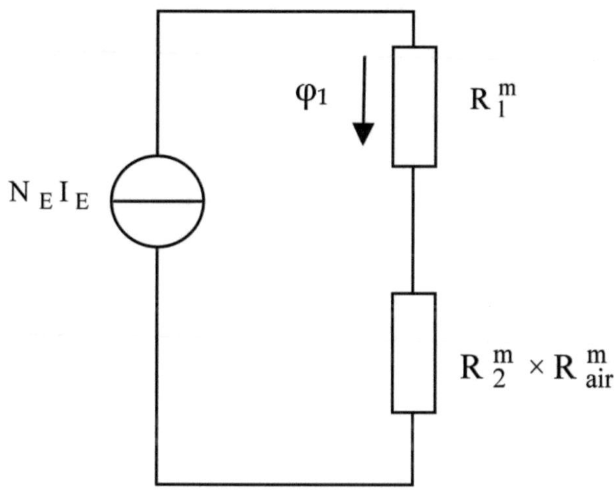

Fig. 6. Equivalent magnetic circuit on the primary side

If $R_{air}^m >> R_2^m$, then

$$R_{eq1}^m \rightarrow R_1^m + R_2^m \tag{22}$$

We can see, if $R_1^m + R_2^m >> \frac{R_1^m R_2^m}{R_{air}^m}$ and $R_{air}^m >> R_2^m$ then

$$R_{eq1}^m = R_{eq2}^m = R_1^m + R_2^m \tag{23}$$

$$R_{eq1}^m = R_{eq2}^m = R_{EQ}^m \tag{24}$$

In this case

$$\varphi_1 = \varphi_2 = \frac{N_E I_E}{R_1^m + R_2^m} = \frac{N_E I_E}{R_{EQ}^m} \tag{25}$$

2.4 General solution in the case of "M" pieces of closed iron core

As long as M pieces of iron core loops the superconductor ring and number of the electrical excitation are K pieces then we can describe the current of the superconducting ring. Fig. 6 shows the arrangement.

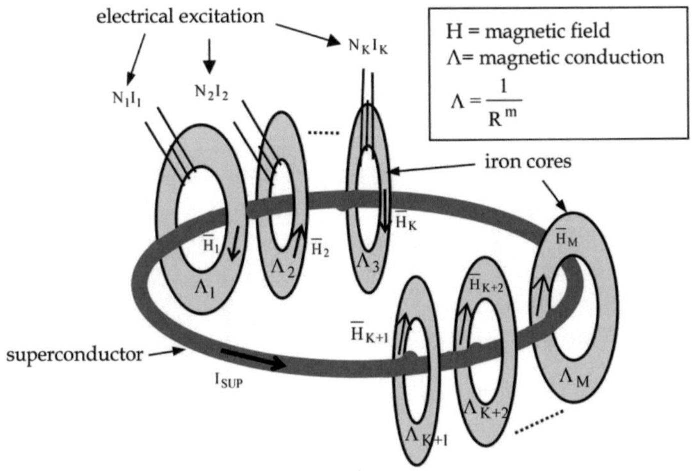

Fig. 7. The arrangement in the case of general solution

We can define numerically the magnetic flux in the iron cores from Ampere's Law.

$$\sum_{i=1}^{K} N_i I_i \Lambda_i = I_{SUP}(\sum_{1}^{M} \Lambda_i) \quad \text{where} \quad \Lambda_i = \frac{1}{R_i^m} \quad \text{and} \quad K \leq M \tag{26}$$

$$I_{SUP} = \frac{\sum_{i=1}^{K} N_i I_i \Lambda_i}{\sum_{i=1}^{M} \Lambda_i} = \frac{\sum_{i=1}^{M} N_i I_i \Lambda_i}{\sum_{i=1}^{M} \Lambda_i}, \text{ as } I_{K+1} = 0 \tag{27}$$

In optional iron core where there is not exciting current:

$$\varphi_X = I_{SUP} \Lambda_X = \frac{\sum_{i=1}^{M} N_i I_i \Lambda_i}{\sum_{i=1}^{M} \Lambda_i} \Lambda_X \tag{28}$$

In optional iron core where there is exciting current:

$$\varphi_y = (N_y I_y - I_{SUP}) \Lambda_y = (N_y I_y - \frac{\sum_{i=1}^{M} N_i I_i \Lambda_i}{\sum_{i=1}^{M} \Lambda_i}) \Lambda_y \tag{29}$$

Equation (29) is the general solution.

2.5 Experimental results of DC magnetic flux transfer with one loop

First I changed the DC current of the coil of the primary side with rapid switches and then slowly. I measured the flux density in the primary and secondary iron core.

2.5.1 Measurement experiments with ZFC method

It can be seen in Fig. 8 that with this solution the change of the magnetic flux of the primary iron core generates opposed flux in the secondary iron core. This is natural because the resultant flux of a closed superconducting loop does not change.

In Fig. 9 and 10 we can see the magnetic flux density of the secondary iron core does not decrease during 2200 seconds.

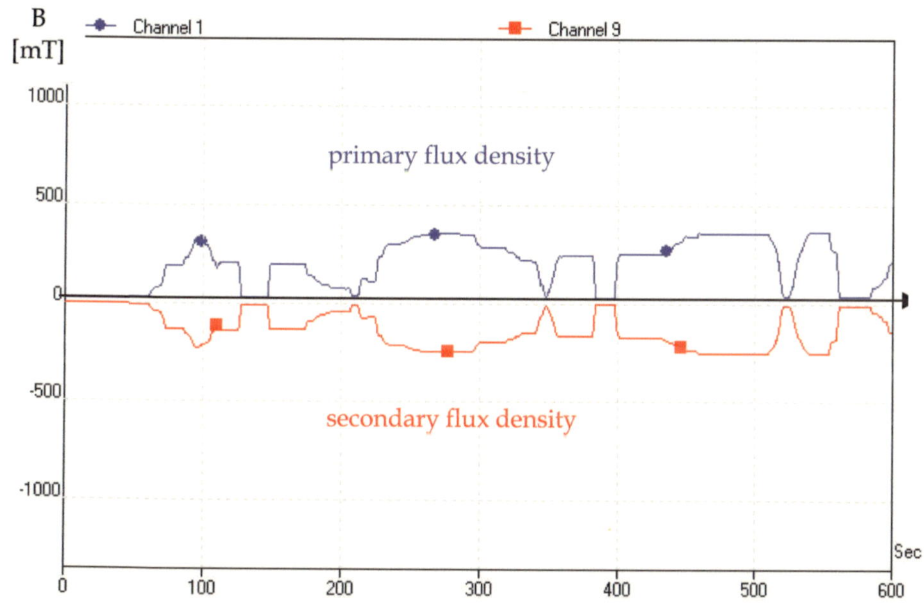

Fig. 8. DC results with one loop

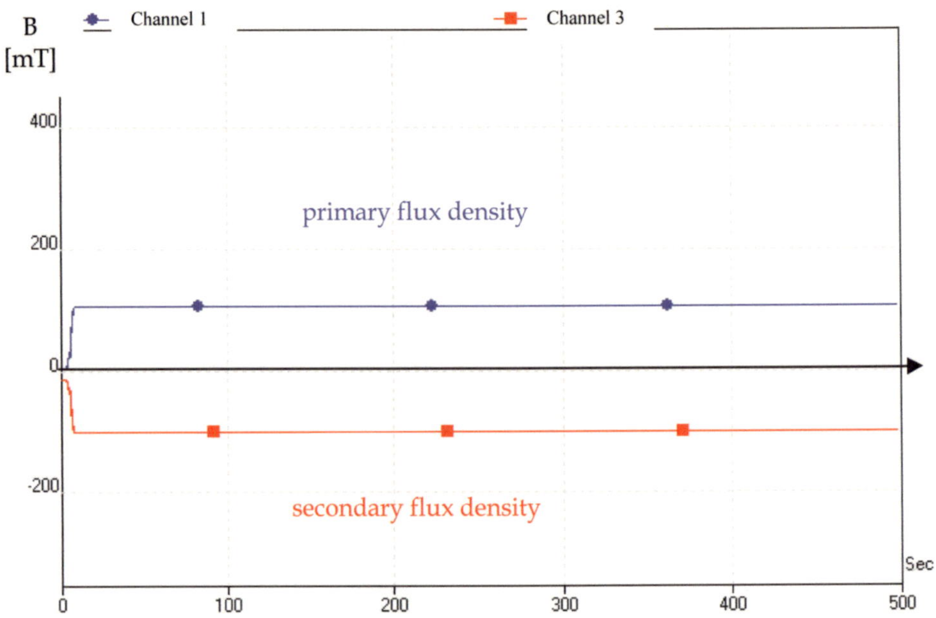

Fig. 9. Measurement result during 2000 seconds I.

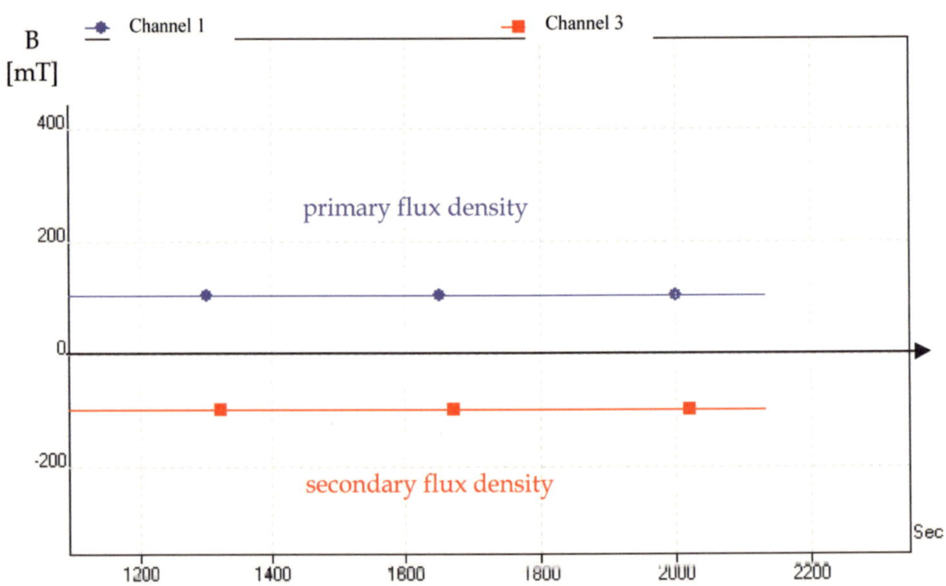

Fig. 10. Measurement result during 2000 seconds II.

The practical application of the transformation of DC magnetic field cannot be carried out with the critical current of superconductor during long time. We have to use a lower superconducting current.

The value of the critical current depends on definitions and even the producers give different definitions. One of the producers gives $1\frac{\mu V}{cm}$; another one gives $0.5\frac{\mu V}{cm}$ as the critical value. This is also influenced by business purposes.

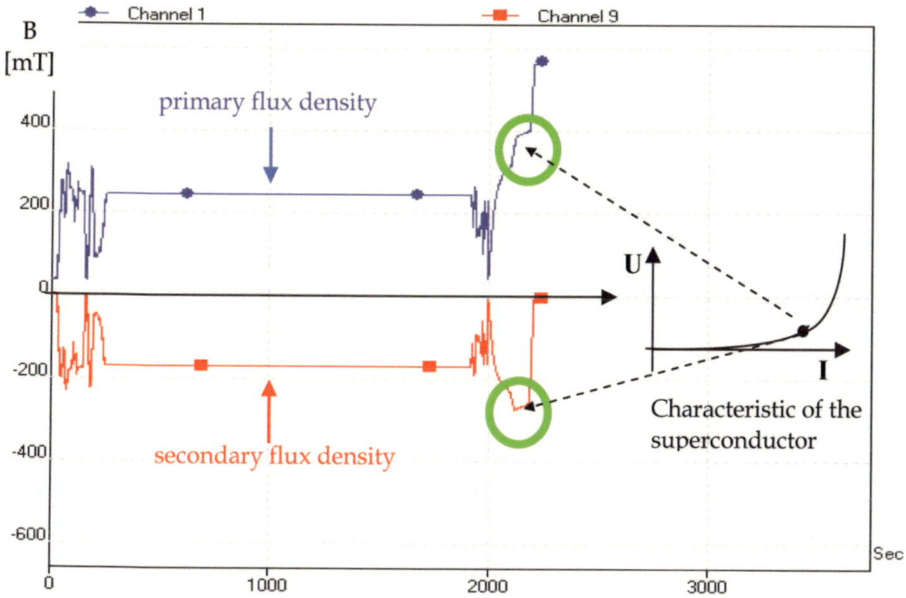

Fig. 11. The decrease of the current of the ring (effect of the non-linear characteristic)

In Fig. 12 we can see a measuring result with a thick copper-ring (instead of superconductor) in liquid nitrogen. Surface of ring was 1.5 cm^2. The flux density of the secondary side decreases very fast.

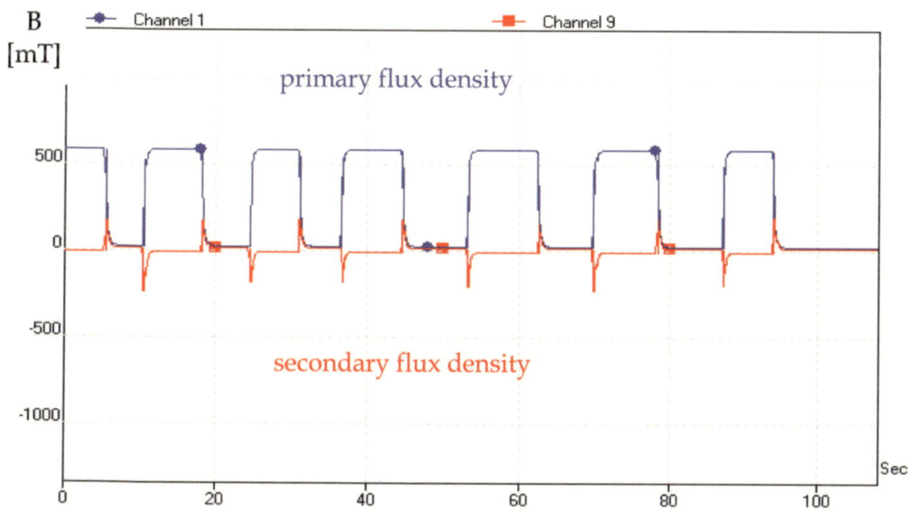

Fig. 12. Measuring result with a thick copper-ring in liquid nitrogen

2.5.2 Measurement experiments with FC method

In the next experiment I have generated magnetic flux only in the primary iron core when the ring was not yet in superconducting state. Then I cooled it down to 77 K and reduced exciting current on the primary iron core to 0 ampere. In parallel in the secondary iron core the flux increased in the opposite direction. After this I changed the polarity of the primary exciting current and the flux increased in the secondary iron core. We can see a measurement result in Fig. 13.

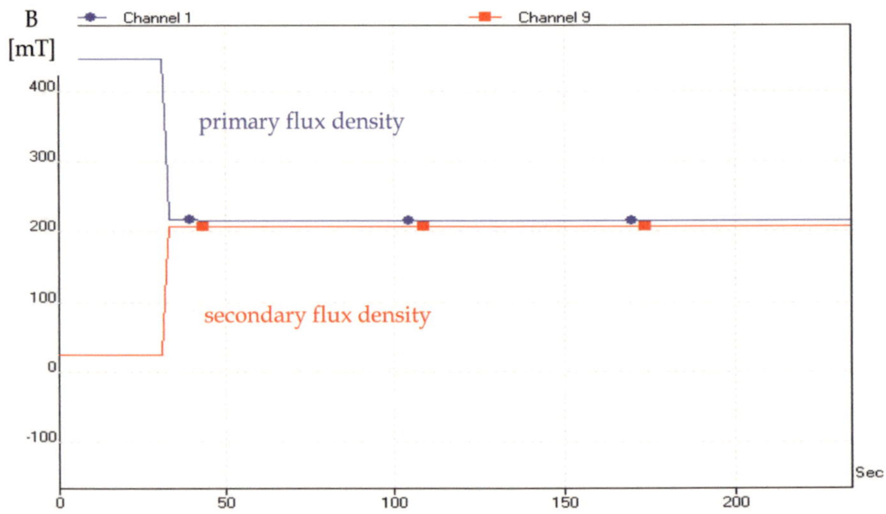

Fig. 13. Measurement result with FC method

Fig. 14 shows a measurement during 5200 seconds.

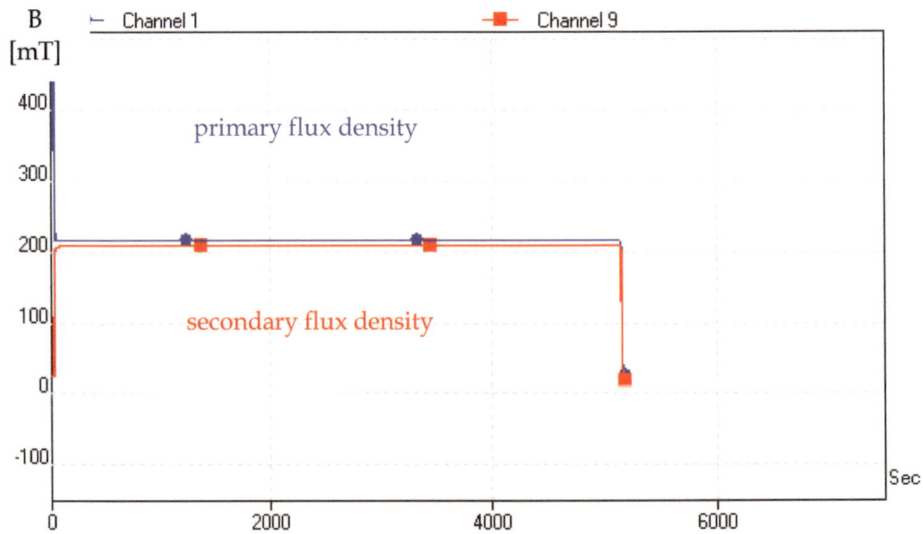

Fig. 14. Measurement result with FC method during 5200 seconds

In the next experiment I have generated magnetic flux in primary and secondary iron core with the same direction when the ring was not yet in superconducting state. The flux density of both iron cores was 0.9 T. After cooling I changed the flux of the primary iron core. We can see the result in Fig. 15. The system is also stable.

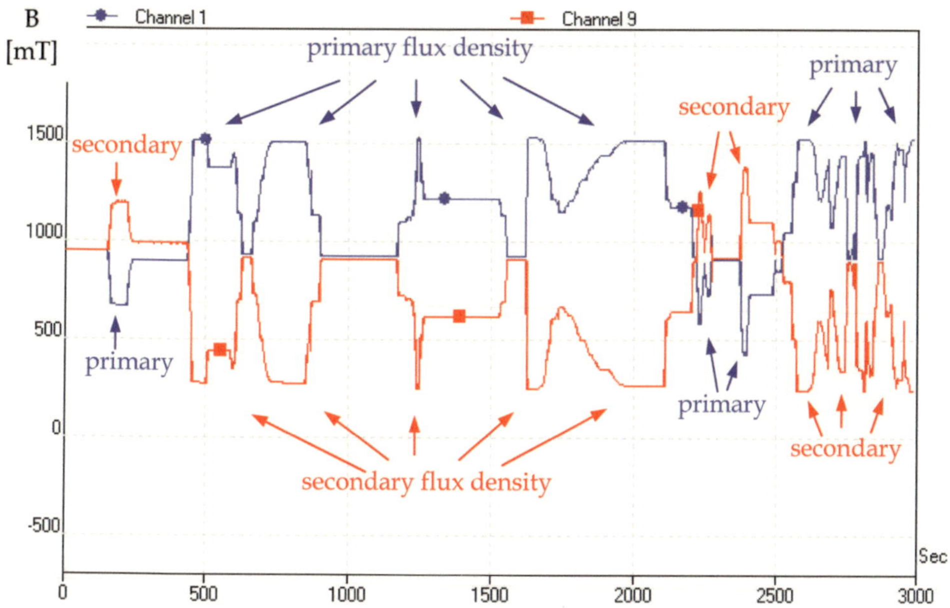

Fig. 15. Exciting current is on both iron cores and after this is cooling (77 K). After this is decrease and increase of the primary current.

3. Theoretical and measurement examination of the transfer of the AC magnetic flux with one loop for single-phase self-limiting transformer

We can see the experimental arrangement in Fig. 16.

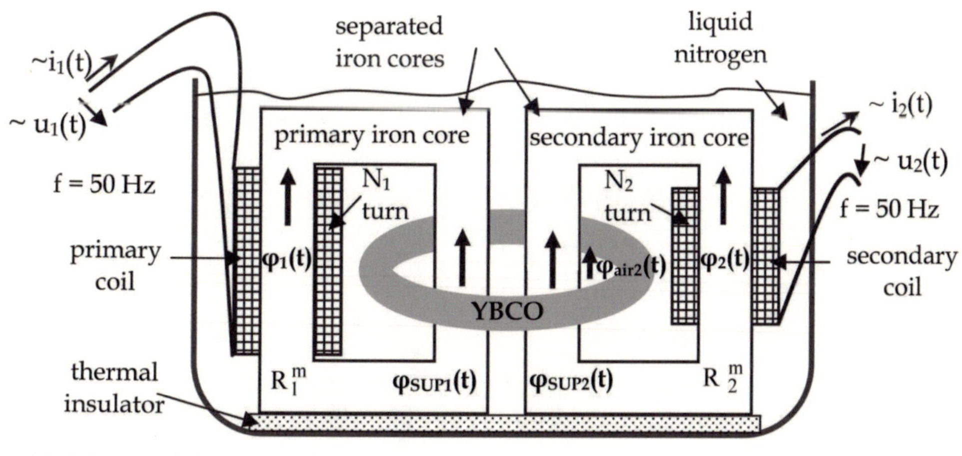

Fig. 16. Scheme of the single-phase self-limiting transformer using by AC flux transformer

The experimental arrangement is a single-phase self-limiting transformer by using AC flux transformer (Kosa et al.) In the case of AC application we can use closed YBCO wire. The superconducting YBCO loop creates the coupling between the primary and secondary coils. This case occurs if the current of superconducting loop is bigger than the critical current of the superconducting loop. The advantage of this solution is that the transformer is able to break the coupling between the primary and secondary coils even if there is not load current in the secondary side, only the primary current increases.

3.1 Mathematical description of single-phase AC flux transfer

$$\varphi_1(t)+\varphi_2(t)=\varphi_{SUP1}(t)+\varphi_{SUP2}(t)+\varphi_{air}(t)$$

$$\frac{N_1 i_1(t)}{R_1^m}+\frac{N_2 i_2(t)}{R_2^m}=\frac{i_{SUP}(t)}{R_1^m}+\frac{i_{SUP}(t)}{R_2^m}+\frac{i_{SUP}(t)}{R_{air}}=(i_{SUP}(t))\left(\frac{1}{R_1^m}+\frac{1}{R_2^m}+\frac{1}{R_{air}}\right) \tag{30}$$

The following functions contain the effect of the inductivities of all transformer elements: $i_{SUP}(t)$, $i_1(t)$, $i_2(t)$.

$$i_{SUP}(t)=\frac{N_1 i_1(t)}{R_1^m\left(\frac{1}{R_1^m}+\frac{1}{R_2^m}+\frac{1}{R_{air}^m}\right)}+\frac{N_2 i_2(t)}{R_2\left(\frac{1}{R_1^m}+\frac{1}{R_2^m}+\frac{1}{R_{air}^m}\right)} \tag{31}$$

$$i_{SUP}(t)=\frac{N_1 i_1(t)}{1+\frac{R_1^m}{R_2^m}+\frac{R_1}{R_{air}^m}}+\frac{N_2 i_2(t)}{1+\frac{R_2^m}{R_1^m}+\frac{R_2^m}{R_{air}^m}} \tag{32}$$

$$i_3(t)=\frac{N_1 i_1(t)}{1+\frac{R_1^m}{R_2^m}+\frac{R_1^m}{R_{air}^m}}+\frac{N_2 I_2 i_2(t)}{1+\frac{R_2^m}{R_1^m}+\frac{R_2^m}{R_{air}^m}}= \tag{33}$$

$$i_{SUP}(t)=\frac{N_1 i_1(t)}{\frac{R_2^m R_{air}+R_1^m R_{air}^m+R_1^m R_2^m}{R_2^m R_{air}^m}}+\frac{N_2 i_2(t)}{\frac{R_2^m R_{air}^m+R_1^m R_{air}^m+R_1^m R_2^m}{R_1^m R_{air}^m}} \tag{34}$$

$$i_{SUP}(t)=\frac{N_1 i_1(t)\ R_2^m R_{air}^m}{R_2^m R_{air}^m+R_1^m R_{air}^m+R_1^m R_2^m}+\frac{N_2 i_2(t)\ R_1^m R_{air}^m}{R_2^m R_{air}^m+R_1^m R_{air}^m+R_1^m R_2^m} \tag{35}$$

If $R_{air}^m \to \infty$, then

$$i_{SUP}(t)=\frac{N_1 i_1(t)\ R_2^m}{R_2^m+R_1^m}+\frac{N_2 i_2(t)\ R_1^m}{R_2^m+R_1^m} \tag{36}$$

If there is not current in the secondary side, then

$$i_{SUP}(t) = \frac{N_1 i_1(t)\ R_2^m}{R_2^m + R_1^m} \qquad (37)$$

3.2 Measurement results of single-phase self-limiting transformer using by flux transformer

Fig. 17 shows a static characteristic of the single-phase self-limiting transformer. This was my first experiment in the case of self-limiting transformer using by flux transformer (Kosa et. al., 2010). I have measured the primary and secondary voltages.

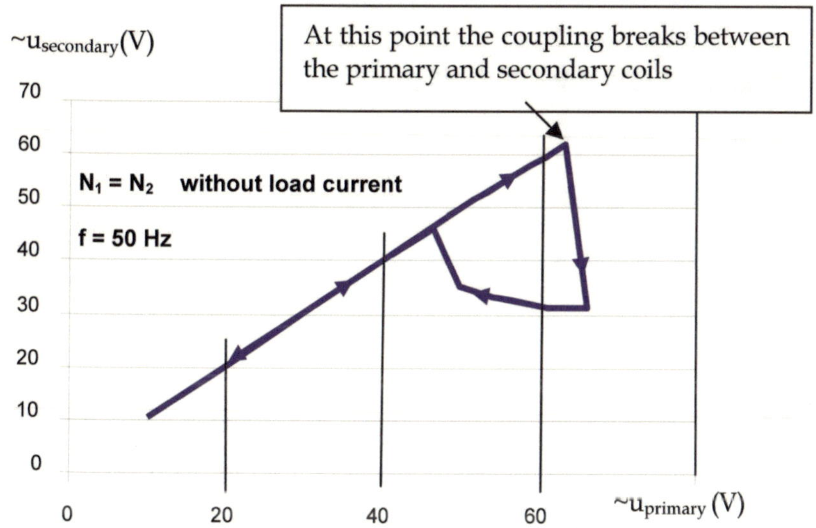

Fig. 17. Static characteristic

We can see a theoretical scheme of the arrangement for dynamic examination in Fig 18.

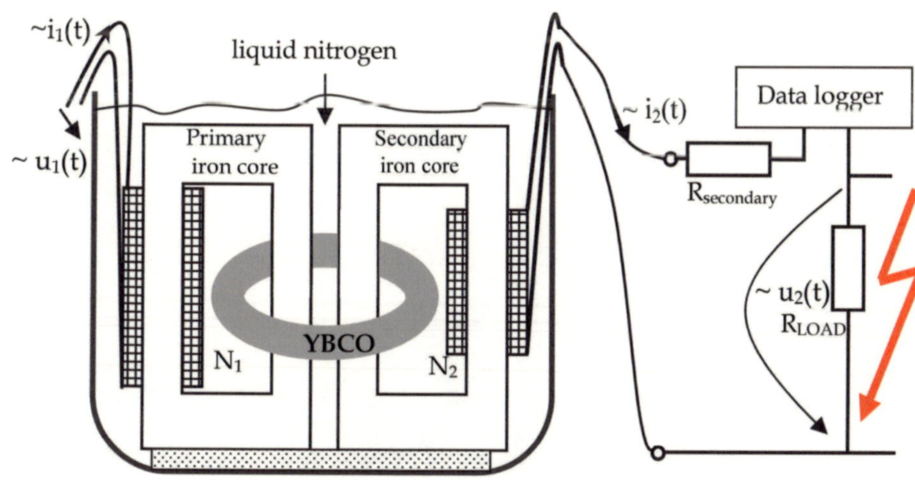

Fig. 18. Measurement arrangement

I generated artificial fault from the operating current of the secondary coil and measured its current. In the experiment, the fault time was 5 periods. The sample rate was 10 kHz.

The current limitation occurs within 3.3 ms (Kosa et al.). This time is less than a quarter of a period (Fig. 19).

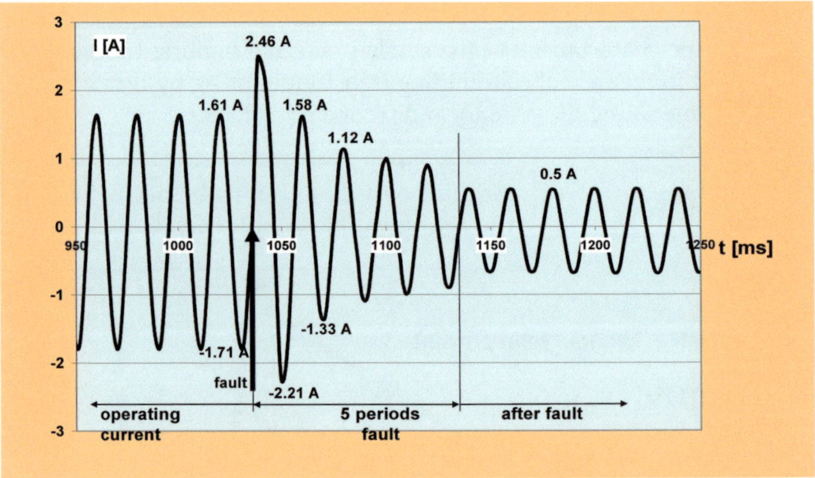

Fig. 19. Dynamic characteristic

3.3 Arrangement of the three-phase self-limiting transformer with flux transfer

The possibility of three-phase application is also based on the principle of magnetic flux constancy in the closed superconducting loop. The solution includes two independent iron cores (Kosa & Vajda, 2011). As I used two iron cores, consequently the loss, the weight, the size, and the price are higher compared to the conventional transformers but this solution has several advantages. For example, the fault power is less, switching is fast. In the case, when there is no load current on the secondary coil and primary voltage increases, the arrangement is able to break coupling between the primary and secondary sides. The developed new type three-phase self-limiting transformer was 400 VA. I realised the primary and secondary windings of the three-phase transformer for this experiment with copper wires. I used SF 12050 wire. This wire (2G HTS wire) was produced 2G HTS wire) was produced by SuperPower, Inc., New York, USA. We can see a photo from device during fitting (Fig. 20).

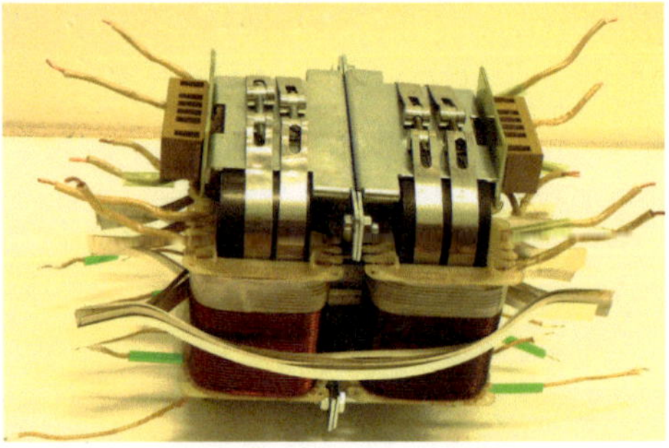

Fig. 20. Photo from the self-limiting transformer during fitting

Table 2 shows the parameters of the experimental arrangement.

Parameters	Data
S_{rated}	400 VA
$U_{primary}$	3 x 400/230 V
$U_{secondary}$	3 x 42/24 V
$N_{primary}$	697
$N_{secondary}$	73
YBCO loops	3 x 7 loops

Table 2. Parameters of transformer

We can see the cutaway view of the device (Fig. 21) and the electrical connection (Fig. 22).

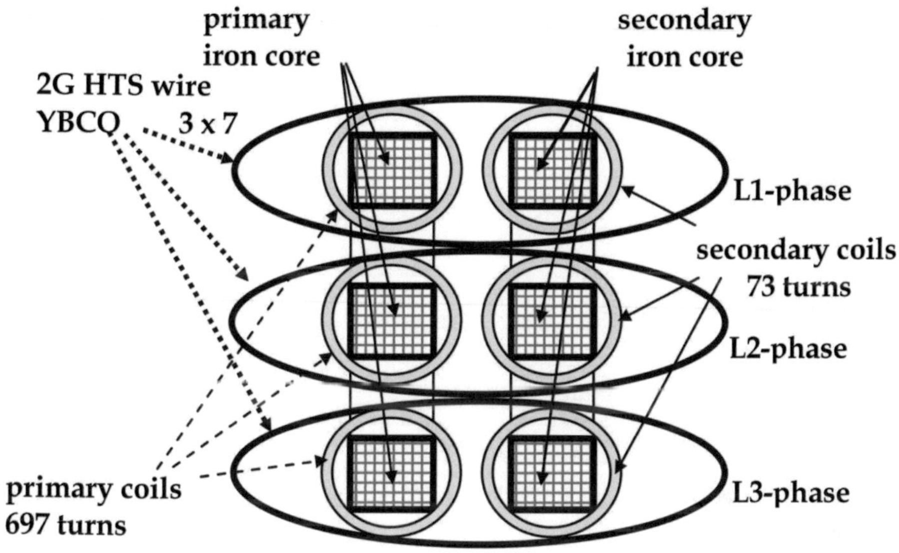

Fig. 21. Cutaway view of arrangement

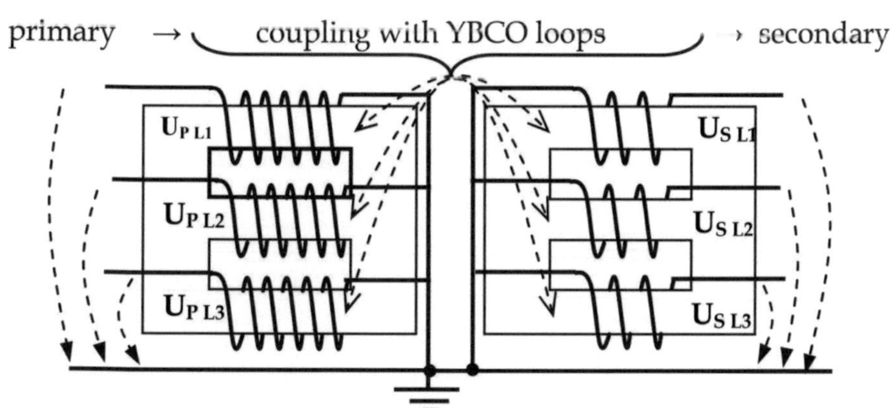

Fig. 22. Electrical connection

3.4 Measurement results

Measurements were carried out using 4-channel data logger. Sample rate = 50 kS/s.

You can see in Fig. 23 and Fig. 24 the secondary voltage of the self-limiting transformer without load-current and with maximum load-current. What is the difference? Hardly anything.

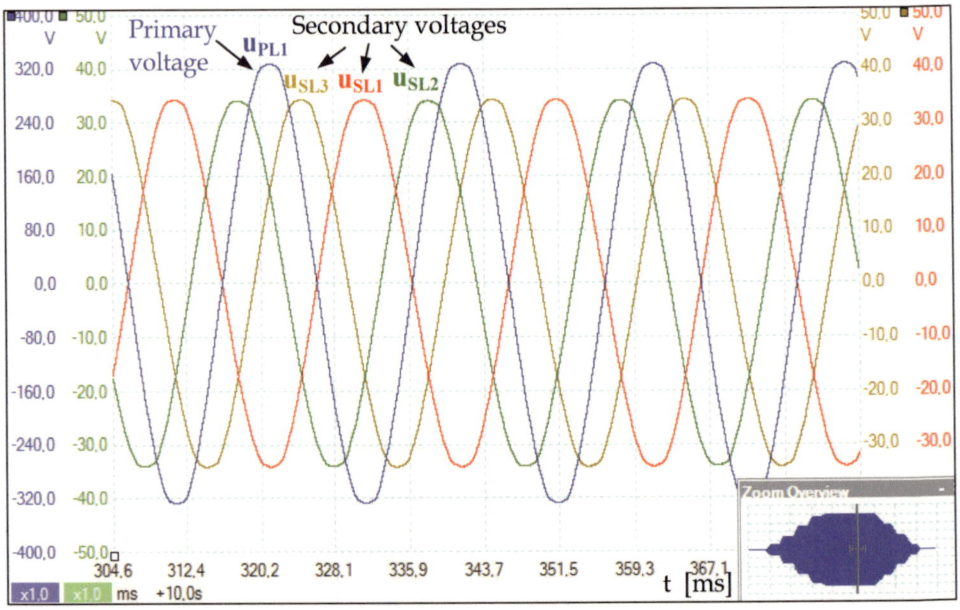

Fig. 23. Measurement result without load-current

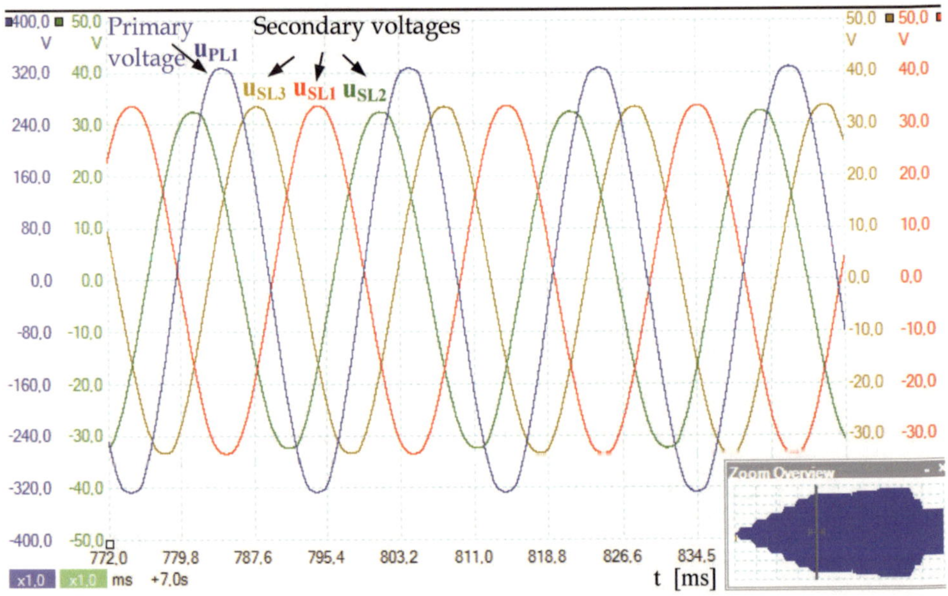

Fig. 24. Measurement result with maximum load-current (3 x 4.5 Ω, P = 400 Watt)

In Fig. 25 we can see a fault. There is load-current on the secondary side (R_{load} = 3 x 4.5 Ω) and single-phase short circuit in L_1 phase.

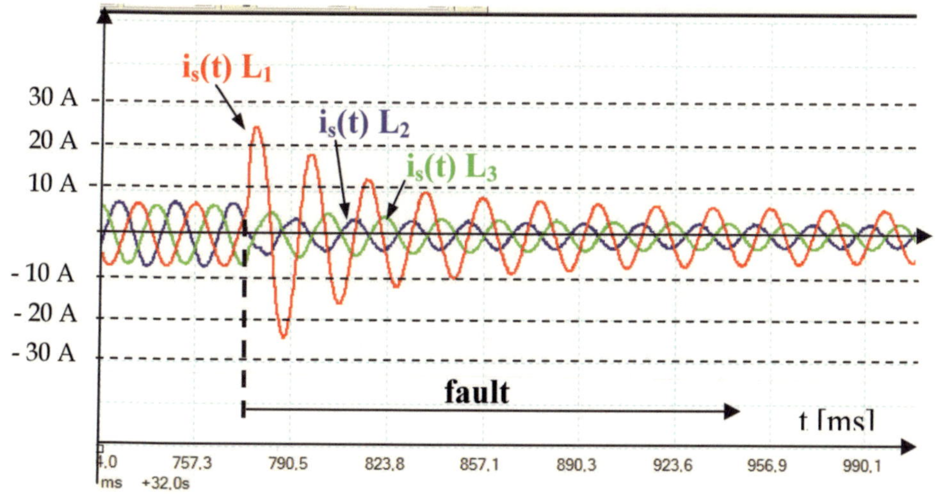

Fig. 25. Single-phase short circuit in L_1 phase

The advantage of the arrangement is that in the case of single-phase short circuit the current will decrease in all the three phases. This can be an appropriate solution for high power machines. In the case of high-power electric motor if there is a single-phase breaking or a single phase short circuit, we can decrease the current in the three phases with this solution.

4. Conclusion

The continuous closed loop made from YBCO tape is suitable for use in such equipment. It is possible to create the novel type 3-phase self-limiting transformer with flux transfer. The short circuit current is less than the operational current within some periods. This is very advantageous for high-power machines (motor). It seems that the arrangement is good also in the case of the single-phase breaking of high-power electric motor. If there is a breaking in one phase, the current increases in both the other phases. And the superconductor will be in the normal state. This means that the current will decrease in an electric motor. The two iron cores' greater size, cost and loss compared to the conventional transformers are facts but the transformer has several advantages e. g. fast switching, the ability of limiting even if the primary current is increased without loading the secondary side.

5. Acknowledgment

The author thanks the SuperPower, Inc. in New York for YBCO wire, facts and figures of the wire and thanks the German colleagues at IPHT, Jena for YBCO bulk.

6. References

Kosa, J. & Vajda, I. (2007). Environmentally friendly machining of ceramic based YBCO bulk superconductor, Journal of Materials Processing Technology, 181, pp. 48-51

Kosa, J.; Vajda, I. & Farkas, L. (2009). Qualification of the Machining and Fitting Precision of YBCO Bulks and Rings Joined Together via the Examination of the Trapped Flux, *IEEE Transactions on Applied Superconductivity*, vol. 19, No. 3, pp. 2182-2185

Hee-Gyoun Lee; Jae-Geun Kim, Sun-Wang Lee, Woo-Seak Kim, Seung-Wook Lee, Kyeong-Dal Chai, Gye-Wan Hong, Tae-Kuk Ko. (2006). Design and fabrication of permanent mode magnet by using coated conductor, *Physica C*, 445-448, pp. 1099-1102

Kosa, J.; Vajda, I. & Gyore, A. (2010) Application Possibilities with Continuous YBCO Loops Made of HTS Wire, *Journal of Physics-Conference Series* 234:(3) Paper 032030, http://www.ewh.ieee.org/tc/csc/europe/newsforum/Contents11.html, ST 153

Kosa, J. & vajda, I. (2009). Transformation of the DC and AC Magnetic Field with Novel Application of the YBCO HTS ring, *IEEE Transactions on Applied Superconductivity*, vol. 19, No. 3, pp. 2186-2189, www.ewh.ieee.org/tc/csc/europe/newsforum/technicalnews.html, ST 74

Orlando, T. & Delin K. (1990). *Foundations of Applied Superconductivity*, Addison-Wesley Publishing Company, pp. 88-89

Kosa, J. & Vajda I. (2011). Novel 3-Phase Self-Limiting Transformer With Magnetic Flux Applied by Perfect Closed YBCO Wire Loops, *IEEE Transactions on Applied Superconductivity*, vol. 21, No. 3, pp. 1388-1392

Superconductor Application to the Magnetic Fusion Devices for the Steady-State Plasma Confinement Achievement

Yeong-Kook Oh*, Keeman Kim, Kap-Rai Park and Young-Min Park
National Fusion Research Institute (NFRI), Daejeon
Republic of Korea

1. Introduction

1.1 History of the superconductor application into fusion devices

The exploitation of the new energy source is necessary in upcoming century due to the rapid increasing of energy consumption and the shortage of the fossil energy sources in the world. There have been a lot of scientific and engineering efforts to make realization of the fusion energy production as a clean and limitless energy source to the mankind. The fusion energy has great advantages in the points of the energy density and the amount of resources, and it is a dominant energy source in the universe like the sun and stars. To achieve a reliable fusion reaction, a good confinement technology is required to overcome the repulsive force between nucleuses of light atoms like hydrogen isotopes as shown in Fig. 1. Among the confinement technologies two kinds of methods are dominant in the fusion researches. The one is the inertial confinement which focusing the high power laser beams in to the small size fuel pellet. The other is the magnetic confinement which confining the fuel gas in plasma state in vacuum by applying the high magnetic fields.

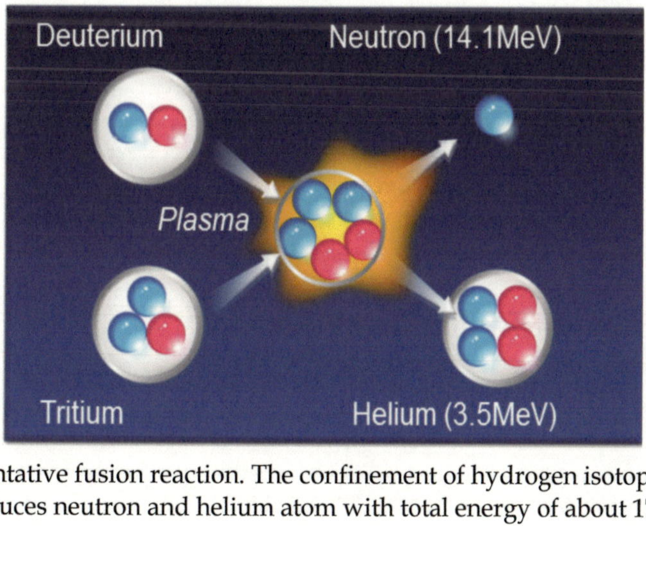

Fig. 1. A representative fusion reaction. The confinement of hydrogen isotopes, deuterium and tritium produces neutron and helium atom with total energy of about 17.6 MeV

The first use of superconducting coils in experimental fusion devices dates back to the mid-seventies. Three main kinds of fusion devices have been studied for plasma confinement using superconducting coils: the mirror machines, the stellerators and the tokamaks. In the initial stage, the application of the superconductor was only for DC or slowly changing in magnetic field like toroidal field (TF) coils in tokamaks or stellerator magnets. The efforts of the last 10 years concentrated on the tokamaks with limited stellerator. The stored magnetic energy and the size of the superconducting fusion devices grew by up to three orders of magnitude in 30 years, from the 20 MJ in the first superconducting tokamak, T-7 to the international thermonuclear experimental reactor (ITER).

There has been a lot of development in the superconducting magnet technologies for the application to the fusion devices, in the superconducting technologies related with the cooling, the superconductor, and the operation mode (Bruzzone, P., 2010). In the cooling method to keep cold a superconducting magnet, the liquid helium bath cooling was the only option in the very first applications. As alternative cooling options, superfluid helium bath cooling at 1.8 K was applied into Tore-Supra in France and forced-flow supercritical helium cooling at about 4.5 K was applied into the recent tokamaks (SST-1, EAST, KSTAR, JT60SA, and ITER) and stellerators (LHD and W7-X). The requirement of mechanical stiffness and reliability of the high voltage electric insulation has ruled out the pool cooling option from the fusion devices of present and future generations. In large devices, with total mass flow rate in the range of several kg/s, the forced flow cooling of supercritical helium is assisted by cold circulation pumps and heat exchangers. Extensive activities on conductor development led to a large variety of NbTi and Nb_3Sn conductor design. With increasing coil size and stored energy, the operating current must be increased to keep low values in the number of turns and the inductance. At operating current over 5 kA, the single multi-filamentary composite is no longer an option because of the ac loss and stability issues. Cable-in-conduit conductor (CICC), both NbTi and Nb_3Sn, became a dominant option in superconductor for large current over 10 kA in fusion device. The advantages of the CICC are direct contact of the conductor with forced-flow helium, structural rigidity is guaranteed by steel jacket surrounding the cable, and the electric insulation by the turn insulation and vacuum pressure impregnation (VPI). In the operation mode of the superconductor, only the DC coils were considered an application target for superconductors. The T7, T15, and Tore Supra tokamaks had copper coil for plasma start-up and shaping. By appearance of the CICC, the superconductor application was expanded into the fast varying central solenoid (CS) or poloidal field (PF) coils in the recent tokamaks, the EAST (Wu, S. et al., 2002), KSTAR (Lee, G.S. et al., 2001), JT-60SA (Yoshida, K. et al., 2008), and ITER (Mitchell, N., 2003). Figure 2 shows the peak field and operation current of the superconducting coils in the fusion devices. NbTi conductor is used in the coils operated at lower field less than 7 T, except Tore-Supra device which is operated at 9 T by cooling with 1.8 K superfluid helium. Nb_3Sn conductor is used in the coils operated higher field. KSTAR TF conductor is operated at 35 kA and is the highest current carrying conductor before ITER operation.

1.2 A recent superconducting tokamak, KSTAR

The operation of the KSTAR device has been started after the integrated commissioning of all superconducting magnets in 2008 (Oh, Y.K. et al., 2009). The mission of the KSTAR project is to develop a steady-state-capable advanced superconducting tokamak, and to establish a scientific and technological basis for an attractive fusion reactor (Lee, G.S. et al., 2001). Figure 3 shows the design of the KSTAR device comparing with ITER device. The

KSTAR device has lots of technical similarities with the ITER device such as using the same material of superconducting conductor, Nb_3Sn CICC, and targeting the steady-state high performance plasma confinement over 300s. KSTAR adopted the most outstanding research results from the present devices and will exploit high performance steady-state operation which will provide the core technology for ITER and for future reactors. The designed parameters of KSTAR, as shown in Table 1, are major radius 1.8m, minor radius 0.5 m, toroidal field 3.5 T, and plasma current 2 MA for 300 s. The specific features in the KSTAR design are full superconducting magnets, passive stabilizers and segmented 3-dimentional in-vessel coils for the fast plasma stabilizations, various kinds of heating and current drive systems for the high-beta and non-inductive current drive, and strong plasma shaping for double-null or single-null configurations.

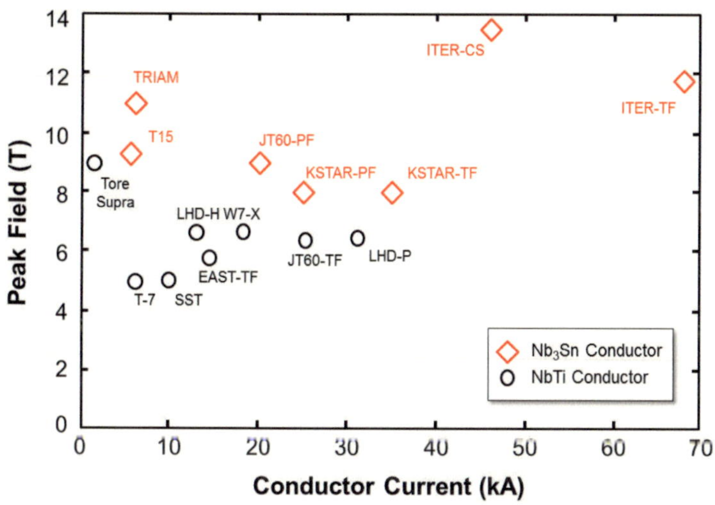

Fig. 2. The progress of the superconductor application in the magnetic fusion devices. The comparison of the operating current and peak field of the conductors in the worldwide superconducting fusion devices.

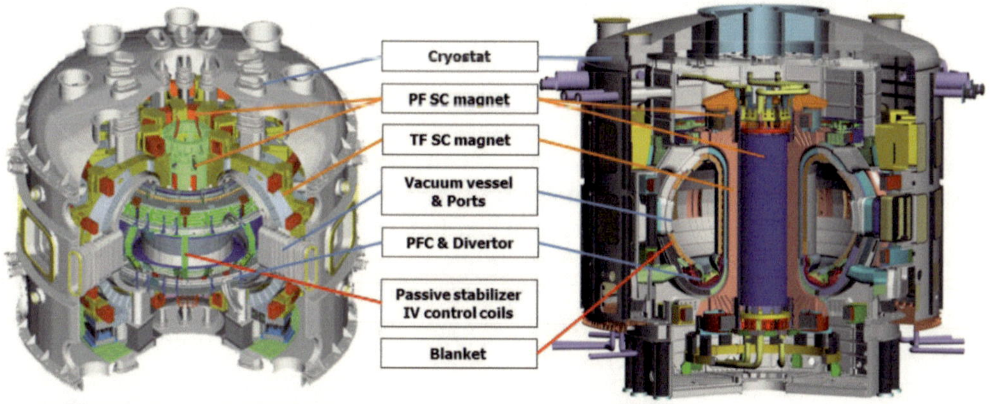

Fig. 3. The design comparison between KSTAR and ITER devices. The dimension is much different but the components and their engineering are similar together.

	KSTAR	ITER
Major radius, R0 [m]	1.8	6.2
Minor radius, a [m]	0.5	2.0
Plasma current, Ip [MA]	2.0	15 (17)
Elongation, κ	2.0	1.7
Triangularity, δ	0.8	0.33
Toroidal field, B_0 [T]	3.5	5.3
Pulse length [s]	300	400
Plasma volume [m^3]	17.8	830
Plasma cross section [m^2]	1.6	22
Plasma shape	DN, SN	SN
Normalized beta	5.0	1.8 (2.5)
Plasma fuel	H, D	H, D, T
Superconductor	Nb_3Sn, NbTi	Nb_3Sn, NbTi
Auxiliary heating /CD [MW]	28	73 (110)

Table 1. Key design parameters of the KSTAR and ITER devices.

In this chapter, the generals about the superconducting technology used in the magnetic fusion devices are described with an example of the KSTAR device, which is the most recently constructed superconducting fusion device and is in operation. The development of a CICC conductor for the fusion application is described in section 2, and Superconducting magnet fabrication and test is in section 3, magnet assembly and interfaces are in section 4, integrated commissioning of the superconducting tokamak is in section 5, and the plasma experiments and consideration of the magnet operation are in section 6.

2. Cable-in-conduit conductor development

Most of the superconducting coils in KSTAR are made of Nb_3Sn superconductor with Incoloy908 jacket, but large PF coils (PF6-7), which operated at lower field, are made of NbTi superconductor with STS316LN jacket.

Usually TF magnet operated in static mode but PF magnet in pulsed mode. So in ITER design, two different type of superconductor was considered. The superconductor for the TF magnet should satisfy the HP-I criteria that is high current density capacity over 750 A/mm^2 at 12 T, 4.2 K and the superconductor for PF magnet should satisfy the HP-II criteria that is low hysteresis loss generation less than 250 mJ/cc at of 3 T cycle. But in KSTAR a new superconductor has been developed to satisfy both criteria simultaneously (HP-III), and constructed TF and PF magnet using same type of superconductor. The cross sectional view of the Nb_3Sn superconductor strand is seen in Fig. 4(a). In the superconductor configuration, the number Nb_3Sn filaments are over 3,000 to get the current density and low hysteresis loss. To reduce the ac loss at the fast varying field environment, each strand is coated with high resistance chrome with a thickness of 1 ± 0.2 μm. The chrome coating is used to survive after the long period of heat treatment of the Nb_3Sn coil.

Large number of strands are cabled together and overwrapped with steel jacket in the form of CICC to have structural rigidity at the large current in the range of several tens of kA and high field over 7 T. KSTAR CICC is shaped into square cross section with round corner as shown in the Fig. 4(b). The cable patterns of the TF and PF conductors are 3x3x3x3x6 of 486 strands and 3x4x5x6 of 360 strands, respectively. Two superconducting strands and one OFHC copper strand are cabled together to become a triplet in the first cabling stage. The cabling pitch of the TF and PF conductors are 40-73-157-227-355 mm and 40-80-145-237 mm, respectively. At the final stage of cable fabrication, the cable is wrapped with a thin stainless-steel strip, 30 mm wide and 0.05 mm thick, with 20 % overlap at each side. The selection of the jacket material is also important factor deciding the Nb_3Sn conductor performance, because the operating current limit is decreased according to the strain applied to the conductor and the strain is mainly come from the difference in coefficient of the thermal expansion (COE) between superconductor and jacket material. In KSTAR, Incoloy 908 is selected as the jacket material for Nb_3Sn conductor.

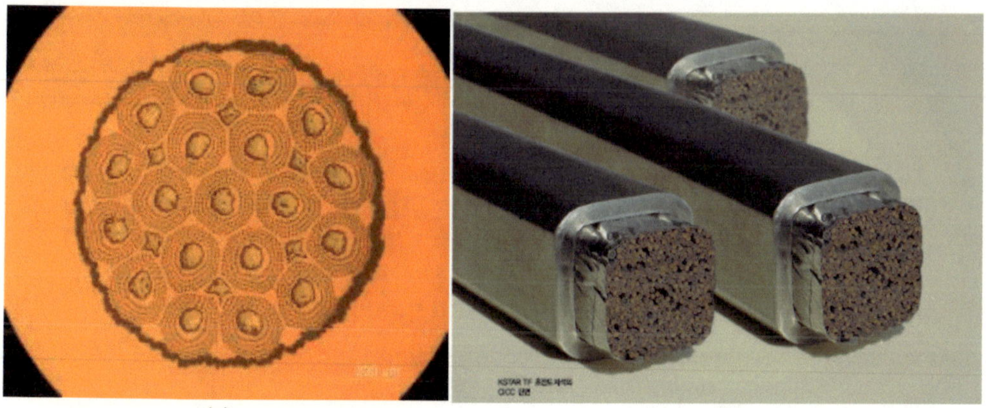

(a) (b)

Fig. 4. The picture of the KSTAR superconductor. Cross sectional view of the Nb_3Sn superconductor (a) and CICC (b).

The length of the conductor is also an important factor in the magnet design. If a longer CICC is available then the number of internal joint could be reduced or excluded. It could guarantee more stable operation of the magnet because lots of troubles come from the joints. In KSTAR, tube-mill process was developed to produce long CICC with various lengths up to 1.8 km to exclude the internal joint. The tube-mill process consists of forming, welding, sizing and squaring procedures. A strip is wrapped around the superconducting cable through a series of progressive roller dies and welded using gas tungsten arc welding (GTAW). The welded sheath is cooled immediately by water and the face-bead of the weld is ground by a bead grinding machine. The shape of the welding back-bead is controlled to be smooth not to damage the superconducting cable. Then, the conductors are formed to the final dimension of the CICC.

The leak tightness is an important qualification of the conductor, because the helium leak from the conductor could hinder the cooling the magnet to the cryogenic temperature due to the increased convective heat transfer. The CICC fabricated by tube-mill process have more possibility of the leak due to the longitudinal welding line. Leak detection of CICC spool was performed in the water chamber by pressurizing the CICC spool with helium at about 50 bars as shown in Fig. 5(a). The locations of the leak were identified by detecting bubbles generated in the water chamber. After machining the leak point with a special grinder, repairing welding was performed carefully with high purity helium supplying inside the CICC as shown in Fig. 5(b). Leak tightness of each CICC spool was guaranteed by repeating the leak detection and repairing process (Lim, B.S. et al., 2005).

(a) (b)

Fig. 5. Picture of qualification of the KSTAR CICC. The leak detection of the CICC spool in the water chamber by pressurizing the conductor with helium gas (a) and grinding the leak point for the repair welding (b).

3. Superconducting magnet fabrication and qualification test

3.1 Fabrication procedure of superconducting magnet

The superconducting magnet system of the KSTAR device consists of 16 TF coils and 14 PF coils as shown in Fig. 6. The design parameters of the TF coils are listed in table 2. The TF magnet system provides a magnetic field of 3.5 T at the plasma centre, with a peak flux density at the TF coils of 7.2 T and the stored energy of 470 MJ. The nominal current of the TF coils is 35.2 kA with all the coils connected in a series. The total cold mass of the TF magnet system is about 150 tons. The coolant used in TF coils is supercritical helium with an inlet temperature of 4.5K and an inlet pressure of 5 bar. There are four cooling channels per TF coil and the design value of the total helium mass flow rate in the 16 TF coils is 300 g/s.

The design parameters of the CS and PF coils are listed in table 3. The PF magnet system, which consists of 8 coils in the central solenoid coil system (PF1-4) and 6 outer PF coils (PF5-7), sustains the plasma current of 2 MA for 20 s, inductively (Kim, K. et al., 2005). The designed peak currents are 25 kA and 20 kA for the Nb_3Sn conductor and the NbTi conductor, respectively. The upper and lower coils of PF1, PF2 and PF7 are connected in a series inside the cryostat and the other coils can be operated separately for single-null and double-null configurations. The CS coils are segmented into four pairs of solenoid coils with different number of turns for the string shaping achievement. The total helium mass flow rate in CS and PF coils is about 300 g/s.

The coil fabrication procedure is as follows: (i) coil winding with numerical dimension control and zirconia bead grit blasting, (ii) attachment of the He feed-throughs and joint terminations, (iii) heat treatment for A15 reaction of Nb_3Sn superconductor, (iv) insulation taping and ground wrapping, (v) vacuum pressure impregnation (VPI), (vi) encasing in magnet structure, and (vii) test and delivery.

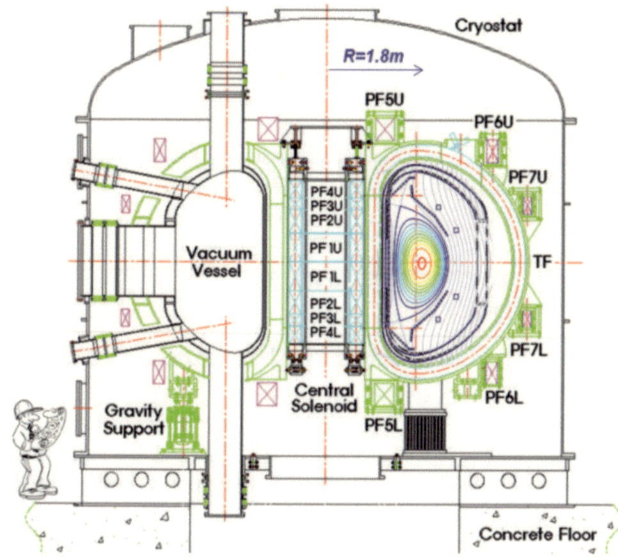

Fig. 6. The elevation view of the KSTAR device and the location of the magnets.

Parameters	**TF coils**
Superconductor	Nb_3Sn
Conduit	Incoloy 908
No. of coils	16
Toroidal field [T]	3.5
Peak field in conductor [T]	7.2
Operating current [kA]	35.2
Stored magnetic energy [Mj]	470
Centering force [MN]	15
No. of turns per coil	56
No. of pancakes	8
No. of turns per pancake	7
Conductor length per coil [m]	610
No. of cooling channels	4
Overall height [m]	4.2
Overall width [m]	3.0

Table 2. Major parameters of KSTAR TF coils.

Parameters	CS coils				PF coils		
	PF1	PF2	PF3	PF4	PF5	PF6	PF7
Superconductor	Nb3Sn				Nb3Sn	NbTi	
Conduit	Incoloy908				Incoloy908	STS316LN	
Number of coils	2	2	2	2	2	2	2
Radial position, R [mm]	570	570	570	570	1073	3090	3730
Axial position, Z [mm]	249	699	1005	1264	229.5	1920	980
Coil width, dR [mm]	230.5	230.5	230.5	230.5	326.3	206.6	158.8
Coil height, dZ [mm]	493.7	398.0	206.6	302.3	398.0	398.0	302.3
No. turns per coil	180	144	72	108	208	128	72
No. pancakes	20	16	8	12	16	16	12
No. turns per pancake	9	9	9	9	13	8	6
CICC length / coil [m]	663	537	283	410	1422	2505	1707
No. cooling channels	10	8	4	6	8	8	6
Weight per coil [ton]	2.34	1.81	0.96	1.39	4.8	8.22	5.65

Table 3. Major parameters of KSTAR PF coils.

The continuous winding scheme is developed to exclude the internal joints. In other words, each coil is wound using a single CICC except PF6U and PF6L coils, which have one internal joint for each coil. Helium feed-throughs are attached per each double pancake. The winding stations for the TF and PF coils are shown in Fig. 7. The TF and PF1-5 coils use Nb3Sn strand and require the reaction heat treatment process. Since PF6 and PF7 coils use NbTi CICC, which does not require the reaction heat treatment process, the helium feed-through attachment and Kapton and S2-glass insulation taping are carried out during the winding process (Park, K.R. et al., 2005).

(a) (b)

Fig. 7. The picture of coil winding stations of the TF coil (a) and PF coil (b).

After the winding, the coils are placed in a structure for the heat treatment and the preparation for heat treatment including the installation of magnet lead holding fixtures, the attachment of feed-throughs and the preparation of joint termination is carried out. In order to minimize the deformation of the winding pack during the heat treatment process, 3-dimensionally machined stainless steel pieces are also installed to the empty space of layer transition area. Heat treatment is conducted in three steps of temperature: first plateau at

460 ^{0}C for 100 hours to remove oxygen and oil contaminants from the cable, second plateau at 570 ^{0}C for 200 hours to enhance the diffusion of Sn to Nb filament and 660 ^{0}C for 240 hours for the A15 reaction of Nb_3Sn. An Argon gas purging system is being operated during the baking process to prevent the Stress Accelerated Grain Boundary Oxidation (SAGBO) of Incoloy 908 and the oxygen content is maintained below 0.1 ppm (Kim, K. et al., 2006).

After the heat treatment, each turn of the coil is individually separated and the CICC is insulated with 50 % overlapped layers of Kapton and S2-glass tapes. S2-glass roving filler is applied at the corner of CICC to minimize the resin rich area. G10 pieces, which are shaped to fill the empty space of layer transition area, are also inserted and the coil bundle is ground insulation wrapped using S2-glass tape. As a quench detection sensor of PF1-5 coils, an externally co-wound voltage tap, which is made of a conducting polymer tape, is wound along the CICC during the turn insulation process. The co-wound voltage tap has been developed to minimize the induced voltage in quench detection line by cancellation of self-inductance. Finally vacuum pressure impregnation (VPI) is performed after placing the coil in a molding frame.

3.2 Qualification test of superconducting magnet

After completion of VPI, acceptance test of each coil is followed including visual and dimensional checks, Jc value, helium flow balance and high voltage insulation test (HIPOT). In the visual check and dimensional inspection, any defect or dimensional deformation is checked as shown in Fig. 8. The results show that all coils are fabricated within the allowable dimensional tolerance of ±2 mm. To qualify the heat treatment of the Nb_3Sn coil, several standard sample strands are located around each coil in vacuum furnace during the heat treatment and the critical current density (J_c) of the strand samples is measured to evaluate the Nb_3Sn magnet performance. This qualification method is very effectively compared to the full current test at cryogenic temperature. Most of the sample showed the current density is over the design criteria of 750 A/mm^2 at 12 T, 4.2 K as shown in Fig. 9. Pressure drop of each channel is measured to assess the flow unbalance among channels during operation. From the measurement, the TF07 coil has the largest flow imbalance among channels, which was about 20%. The averaged values are within approximately 10% (Park, K.R. et al., 2009).

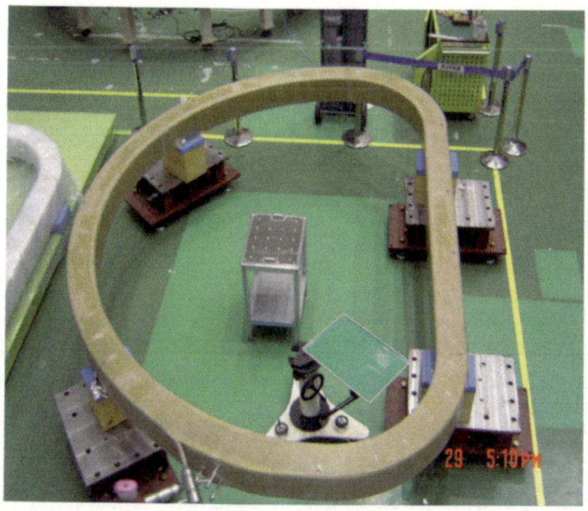

(a)

(b)

Fig. 8. Visual and dimensional inspection of the TF coil (a) and PF coils (b).

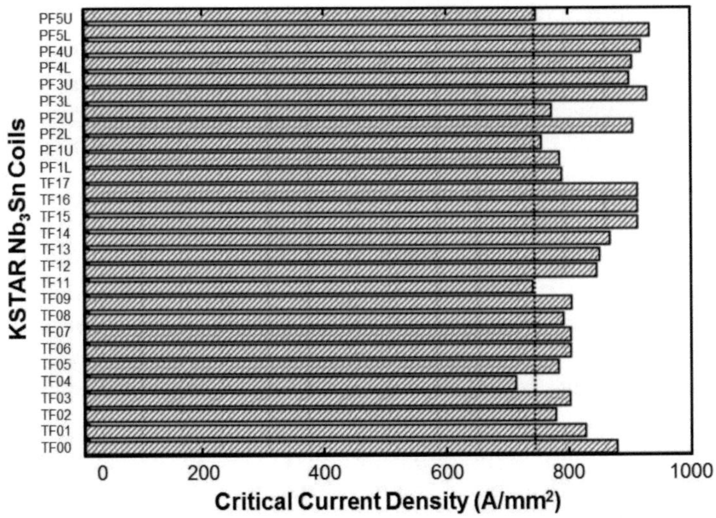

Fig. 9. The measured Jc values of the all Nb_3Sn coil after heat treatment.

The HIPOT tests were carried out up to DC 15 kV and AC 10 kV@60 Hz to ensure the insulation integrity of the coils after the VPI. All coils satisfied insulation requirement of DC insulation resistance above 100 MΩ for the ground insulation. The impulse test was carried out to check the turn-to-turn insulation by applying the pulsed 2.4 kV at a double pancake in a coil. The measured voltage waveforms of all pancakes were compared with each other because the waveform difference is a sign of the turn-to-turn insulation breakage. There was no observed turn-to-turn insulation failure.

Performance test of the full size TF prototype coil, TF00, and a pair of CS model coil were carried at cryogenic temperature in the test facility as shown in Fig. 10(a). The major objective of the test is to confirm the validity of the design and the fabrication process. TF00 was cooled down in 10 days within a maximum temperature variation of 50 K. The residual resistivity ratio (RRR) of the coil was measured to be over 200. The superconducting phase transition of TF00 occurred around 18 K. TF00 was repeatedly charged and discharged with various scenarios. The coil was operated well without any quench up to 30 kA (Oh, Y.K. et al., 2004). The result of repeated current charge tests shows that TF00 is fabricated robustly. But the test was interrupted by the arc occurrence on one of joint due to weakness of the busline supporters and movement of the busline. We could learn from the TF00 experience that the supporting structures of the magnet system should be robust enough to sustain any kinds of force and movement. A pair of CS model coils, which has been designed for the background coil using the same conductor as KSTAR CS coil for the conductor test at 8T with a ramp rate of 3T/s, has been installed in the test facility. The supporting structure of the main background magnetic field coil system was made of glass-fiber reinforced plastic (GFRP) material for the reduction of the eddy current during pulsed operations. CS model coils were cooled down to 5 K in 9 days. However, the temperature of the G10 coil support structure was approximately 30 K due to the low thermal conductivity. The total helium flow rate was ~45 g/s and the pressure drops ~1 bar. The coil was successfully excited to 20 kA, where the peak field is 8.62 T, using the DC power supply for the KSTAR TF coil. In the view point of temperature margin, the operating condition of the coil at 20 kA is a similar operating condition of the KSTAR CS coil at 25 kA. The ac loss measurement was conducted with several cycles of triangular current waveform and with a long pulse sinusoidal waveform with offset to exclude nonlinear control error of the power supply at low current. However, the prototype AC power supply showed a logical error during the crossing of zero current and the AC test was performed with an offset current, which prevents the control failure of the AC power supply near zero current. The coupling time constant, $n\tau$, which was measured by changing the frequency and the amplitude of sinusoidal current as shown in Fig. 10 (b) (Oh, Y.K. et al. 2006, & Lee, S. et al., 2006). It was less than 30 ms, which is approximately half of the design specification of 60 ms. The result of lower coupling loss time constant have a positive expectation of the KSTAR CS coil operation and it could guarantee the fast field variance required for the plasma startup and feedback control of the KSTAR CS magnet during plasma operation. The CS model coil could be applied for the sample conductor test of ITER or for conductor R&D for the fusion reactor (Oh, D.K. et al., 2004).

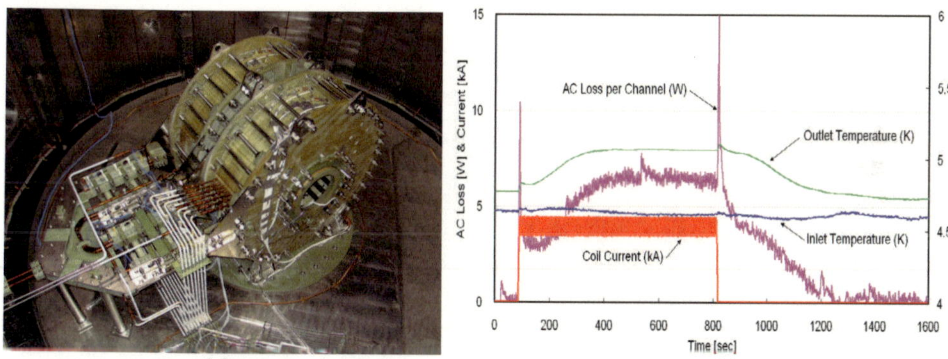

(a) (b)

Fig. 10. CS model coil performance test. Coil installation in the vacuum cryostat (a) and Sinusoidal current waveform for ac loss measurement (b).

4. Assembly of the magnets and interfaces

4.1 TF Magnet structure and assembly

The KSTAR magnet structure consists of 16 TF structures, one CS structure, and 80 PF structures. Most of structures are made of strengthened authentic stainless steel STS316LN. Key design factors are mechanical stiffness under the large amount of Lorentz forces, electric insulation along the toroidal direction to prevent eddy current by the plasma startup and disruption, and effective cooling schemes. TF coils are under large amount of in-plane force up to 15 MN when TF coil is charged at the nominal operating current of 35.2. And there are also out-of-plane forces up to 2.6 MN/m due to by the CS, PF, and the plasma current (Choi, C.H. et al., 2002). To sustain the electromagnetic load on the TF coils, TF structure is designed to have a wedged shape at inboard leg and to locate inter-coil structures which are strongly connected to TF case outboard legs and containing shear keys and bolts. The shape of the inter-coil structure is optimized to reduce the peak stress within allowable according to the analysis (Ahn, H.J. et al., 2002). To prevent eddy current into the toroidal direction at the plasma startup or disruptions, insulation spacer made of G-10 is attached on the entire surface at the gap between TF structure modules. There are 18 cooling tubes embedded inside of the structure around the coil and an additional cooling tube on the TF structure surface to limit the heat influx into the TF coil in case of the hot spot generation in the TF structure due to plasma disruption or PF coil quench. The major fabrication procedure of the TF structures are (i) fabrication of C-shaped coil case welded with inter-coil structure, (ii) fabrication of flat cover plate welded with inter-coil structure, (iii) coil encasing, (iv) final enclosure welding, (v) second vacuum pressure impregnation, (vi) final machining and delivery, and (vii) toroidal insulation attachment (Bak, J.S. et al., 2006). Figure 11 shows the final machining of TF coil case after the coil encasing. A special assembly tools has been developed to locate each magnet at the exact position. Each TF magnet is launched carefully into the 22.5 degree gap of the vacuum vessel using the loading vehicle. It is rotated in the toroidal direction and is aligned at the final position within installation error less than 1 mm as shown in Fig. 12 (Yang, H.L. et al., 2006). To compensate the thermal contraction of the TF magnet structure, whole TF structure was installed with a vertical offset of 5 mm at room temperature.

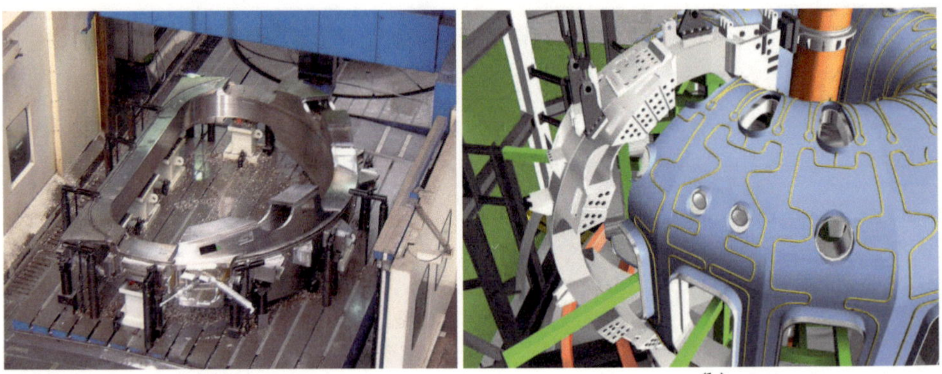

(a) (b)

Fig. 11. TF magnet structure fabrication and assembly. Final machining after coil encasing (a) and the schematic drawing of a TF magnet installation into torus of the vacuum vessel (b).

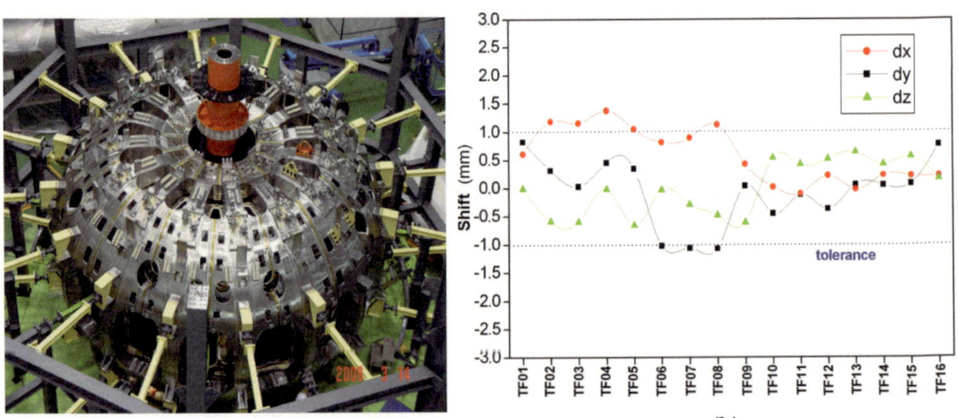

(a) (b)

Fig. 12. TF magnet assembly qualification. Completion of whole TF magnet assembly (a) and installation error of TF magnet within 1 mm in the parallel shift (b).

4.2 CS/PF magnet structure and assembly

The major function of the CS structure is to apply a vertical compression force on the CS coil assembly to prevent a relative motion of CS coils and to sustain the repulsive forces between CS coils during operation (Oh, Y.K. et al., 2002). The CS coils are under compressive force during the reference scenario. However, the peak separation force of 12 MN occurs between PF1 and PF2 coils under several operation conditions, and it is called as maximum repulsive force (MRF) condition (Choi, C.H. et al., 2002; Sa, J.W. et al., 2002). The designed axial compression force is about 15 MN at 5 K. It will be applied partly by assembling the pre-compression structure at room temperature up to 13.4 MN and partly by the thermal contraction difference between CS coils and pre-compression structure during cool down. During the subassembly of the CS magnet structire, the pre-compress structures are heated to apply the pre-compression at room temperature. When heating applied on the structure, the temperature of the epoxy-rich area of the coil also has possibility of damage due to overheat. So, the temperature of the structure was limited at 140 ^{0}C, and the pre-compression on CS coils are estimated to be about 8 MN at room temperature. The sub assembled CS magnet assembly is inserted into the center of KSTAR device and hang on the TF structure shoulder as shown in Fig. 13(a).

The PF coil structures to place the 6 PF coils on the TF coil with a vertically symmetry to the machine mid-plane at 80 locations. All PF coil structures should absorb the difference in the thermal contraction between TF coil structures and PF coils during cool-down to cryogenic temperature, and endure the vertical and radial magnetic forces due to current charging. In order to satisfy these design requirements, the hinges or flexible plates are applied in the design.

(a) (b)

Fig. 13. Pictures of the KSTAR magnet system. The installation of the CS coil assembly into the center of the KSTAR device (a) and the completion of magnet assembly (b).

4.3 Magnet interfaces

Magnet interface components are to supply electric power and cryogen and to monitor the operational status. Figure 13(b) shows the whole magnet system after installation of whole magnet system and its interface components.

SC buslines, which is components to carry the current from the cryostat boundary to the coil terminal joints, are made of NbTi CICC with circular cross section to allow bending in any angle. Each busline was bended before installation according to the 3-dimensional assembly drawing and was insulated by wrapping with pre-pregnant glass-fiber and Kapton film. Two types of joints are installed between magnet and buslines, one is lap type joints which are installed at 136 locations and the other type is strand-to-strand joint which are installed at TF coil only (Park, Y.M. et al., 2009).

Current lead system is an interface component to deliver the large current from the power supply at room temperature to the superconducting magnet at the cryogenic temperature. KSTAR requires 7 pairs of PF leads having the current carrying capacity up to 25 kA for more than 350 s and 2 pairs of TF leads having the current capacity over 17.5 kA for steady state. Usually the heat load on the current lead is major portion of the heat load on the magnet system. So there are lots of researches to reduce the heat load of the current lead. In KSTAR, heat load reduction was achieved by replacing the copper conductor in the current lead material with brass conductor due to the low thermal conduction.

For the cooling of the magnet system and its interface, a cryogenic facility was constructed with a cooling capacity of 9 kW at 4.5 K equivalent. The cryogenic facility consists of gas management and compress station at room temperature, cold box to cool-down the cryogen, and helium distribution box to match the heat load in KSTAR. Figure 14 shows the process flow diagram of the KSTAR cryo-facility. The cryogenic helium is supplied to the magnet system in three phases. Supercritical helium is supplied to the all SC coils and magnet structures at 4.5 K, 5.5 bar with mass flow rate of 600 g/s, liquid helium into current lead at 4.2 K, 1.3 bar, and gaseous helium into thermal shield at about 55 K, 18 bar. (Chang, H.S. et al., 2008; Kim, Y.S. et al., 2009). A thermal damper system to moderate the pulsed heat from PF magnet consists of several heat exchangers immersed in the liquid helium bath and several supercritical helium circulators.

In the magnet operation with large stored energy, normal regions arising in the superconducting coil can lead to permanent damage due to excessive heating. Reliable quench detection and protection is very important for the safe operation of the superconducting magnets. A total of 83 quench voltage detectors are installed and operated independently to detect quenches in the TF and PF coil systems (Chu, Y. et al., 2009). The quench voltage detection system of the KSTAR TF coils uses a conventional balanced bridge method detecting and comparing the voltage signal across each two TF coils with balanced inductance. For the fail-safe operation and to anticipate the malfunction of the detector, two detectors monitor the same section constantly in a simultaneous manner. Most of the self-inductance of each module could be compensated easily. There is some amount of non-linear voltage at low current due to Incoloy908 ferromagnetic material in the Nb_3Sn superconductors. When PF coil is operated together, the induced voltage is detected for about 100ms due to the coupling of the inductance between TF and PF coils. Generally, the holding time was set to prevent the false signal from the voltage spike to trigger the interlock. As the TF QDS holding time is set to be 2 s, the voltage noise due to blip operation was therefore not a problem The PF current abrupt change which is called as the blip operation during the short time span is necessary for making the loop voltages in the toroidal direction and initiating the plasma current. Hence, the noises of the quench detection system should be carefully analyzed and compensated below the proper levels. Two quench detectors were used for detecting the resistive voltage drop across the winding. The primary one uses the co-wound voltage sensing cable, which traces the surface of the CICC closely. If the magnetic coupling The PF current abrupt change which is called as the blip operation during the short time span is necessary for making the loop voltages in the toroidal direction and initiating the plasma current. Hence, the noises of the quench detection system should be carefully analyzed and compensated below the proper levels. Two quench detectors were used for detecting the resistive voltage drop across the winding. The primary one uses the co-wound voltage sensing cable, which traces the surface of the CICC closely. If the magnetic coupling between the coil and the co-wound tape is sufficiently strong, the measured voltage at the co-wound voltage sensor can eliminate the most part of the inductive voltages in a coil. A conductive fabric tape was selected as a co-wound voltage sensing cable. The secondary one is the most commonly used method as shown. However the inductive voltage is detected in both conventional and co-wound sensors due to the field profile in each coil. By adding the differential comparator to subtract unbalanced signals in PF coils, most of the inductive voltage signal could be expelled as shown in the Fig. 15. Beside of the noise handling scheme, the quench detection holding time is introduced. That is, if the measured voltage stays above the threshold voltage over the predefined holding time, then the quench interlock is activated and the magnet energies are dumped into external dump resistors.

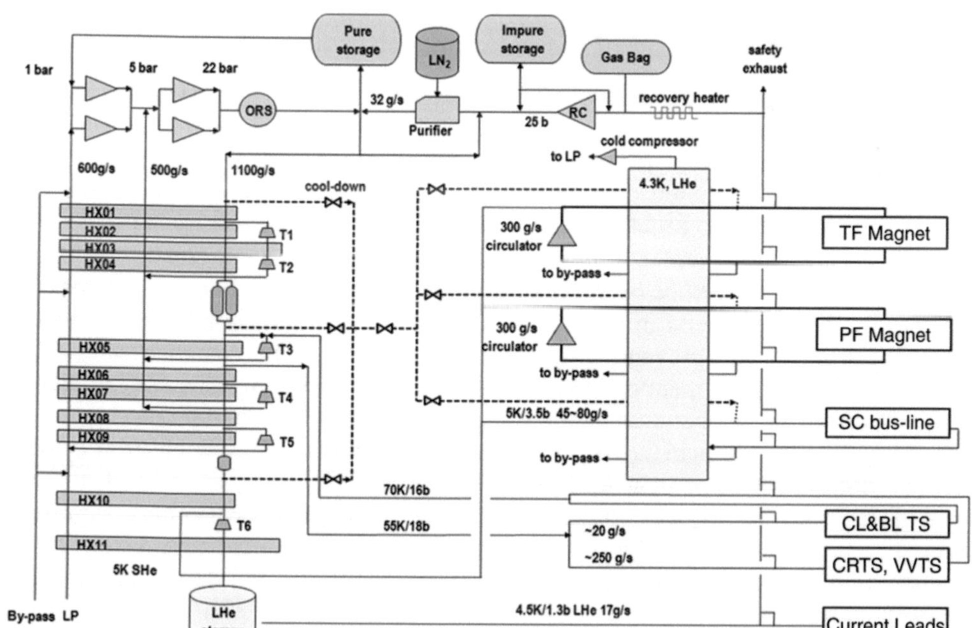

Fig. 14. Process flow diagram of the 9 kW cryo-facility to cool the KSTAR magnet system.

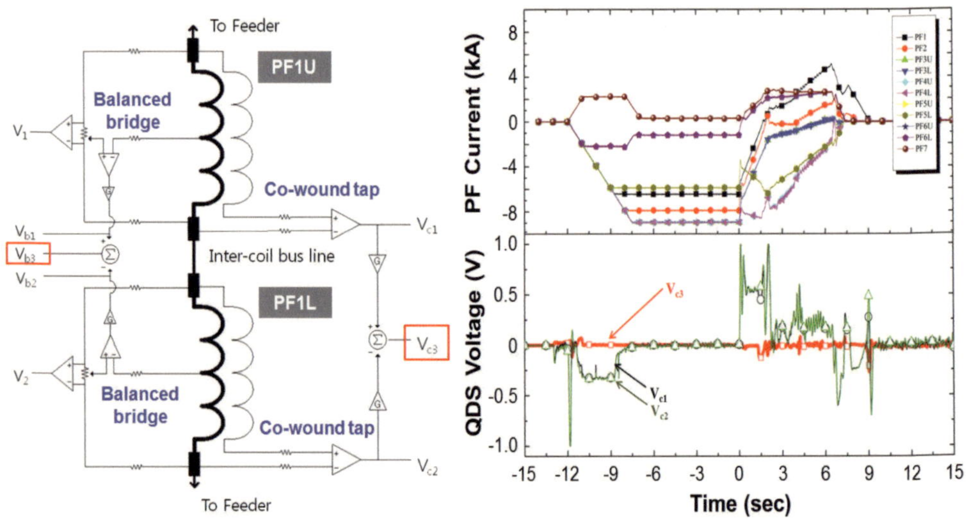

(a) (b)

Fig. 15. Quench detection of KSTAR PF magnets. Quench detection circuit of the PF1 by balanced bridge and co-wound voltage tap (a), and comparison of the inductive voltage signals on the quench detection sensors during the regular PF coil operation (shot #5199). Most of the inductive signal is cancelled at Vc3 signal lines. (b)

5. Integrated commissioning of the magnet system in fusion device

The operation of the superconducting fusion device could be classified into two phases, the integrated commissioning phase to inspect the quality of each component and to reduce unexpected risk occurrences and the normal operation phase to achieve the mission of the fusion device by the integrated operation of all system. There were several cases that the occurrence of damage or problem in the superconducting magnet system during its commissioning or initial operation made delay or prohibit the normal operation in superconducting fusion device or particle accelerator. The commissioning and the initial operation for the first plasma in the KSTAR device have been accomplished successfully without any severe failure preventing the device operation and plasma experiments. All the superconducting magnets operated in stable without quench for long-time dc operation and with synchronized pulse operation by the plasma control system . The first plasma target in the range of 100 kA was achieved by assisted startup using an 84GHz electron cyclotron resonance heating (ECH) system and real-time feedback control (Oh, Y.K. et al., 2009). Figure 16 shows KSTAR device and the overall sequence of the KSTAR commissioning and the first plasma operation and the picture of control room during the commissioning. The commissioning of the KSTAR magnet system is classified into several steps: (i) inspection at room temperature after tokamak assembly, (ii) inspection during and after magnet cool-down, and (iii) stability assessment under operation condition of static or pulsed operation of magnet system.

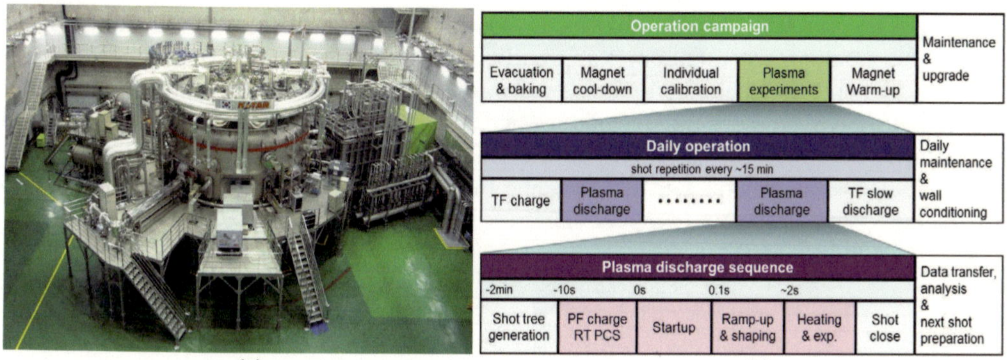

(a) (b)

Fig. 16. Bird-eye view of the KSTAR device in during the 2010 operation (a), and the operation sequence of the plasma experiments (b).

5.1 Inspection at room temperature

After completion of the KSTAR device assembly, a detailed inspection of each component was followed before the cooling down of the magnet system. The major inspection activities at room temperature are leak detection, electric insulation check, and magnetic field survey.

The leak detection is one of the critical activities in the superconducting magnet system. Although most of gas except helium are condensed on the magnet surface after the magnet cool-down, so the leak from the helium circulation line should be identified and repaired for the stable magnet cool-down and long-time operation for several month. The careful inspection of the leak from the magnet cooling lines is conducted by pressurizing the magnet cooling line with helium up to about 20 bar. The leak from the magnet system in KSTAR is not so high and could be repaired well. It is due to the tight quality control of each component before its installation in KSTAR. Usually most of the leaks come from the joint or welding point between different materials, and electrical breaker is the components of high leak possibility due to the connection of stainless steel tube to Ceramic insulator or GFRP insulator. So the leak inspection was conducted of ach electric breakers under high pressure helium after the several tens of thermal cycles from room temperature and liquid nitrogen temperature. To ensure the integrity of insulation of the superconducting system after all sub-systems were completely installed, the dc HIPOT test was carried out up to 10 kV in each coil to diagnose the coil insulation before the cool-down. Conclusively, the insulation resistances of all coils were more than 100 MΩ which is the same criterion for the individual coil (Park, K.R. et al., 2009).

The magnetic field distribution in the vacuum vessel is measured after a low current less than 100 A is applied to each coil at the room temperature environment and compared with the calculated data. When the TF current is 100 A, the B-field at the plasma center is approximately 100 gauss, and the B-field distribution is consistent with the calculated data although there was some discrepancy of the field detection close to the vacuum vessel inboard surface. There was a large discrepancy between measurements and calculations for the PF coils as shown in Fig. 17 (Yonekawa, H. et al., 2009). For instance, PF7 measurements at all the radial positions were approximately 30% lower than the calculated values dismissing Incoloy 908. On the other hand, the PF1 discrepancy gradually increased from 3% to 36% as the measurement point approached the VV inside. The PF coils appeared to be shielded by some magnetic component surrounding the VV as their magnetic field was reduced in the VV whereas the TF coils seemed to be embedded in some magnetic material as their magnetic field rose suddenly at the VV inside close to the TF's windings. Magnet system cool-down was launched after passing the all required room temperature inspection for magnet system and other components.

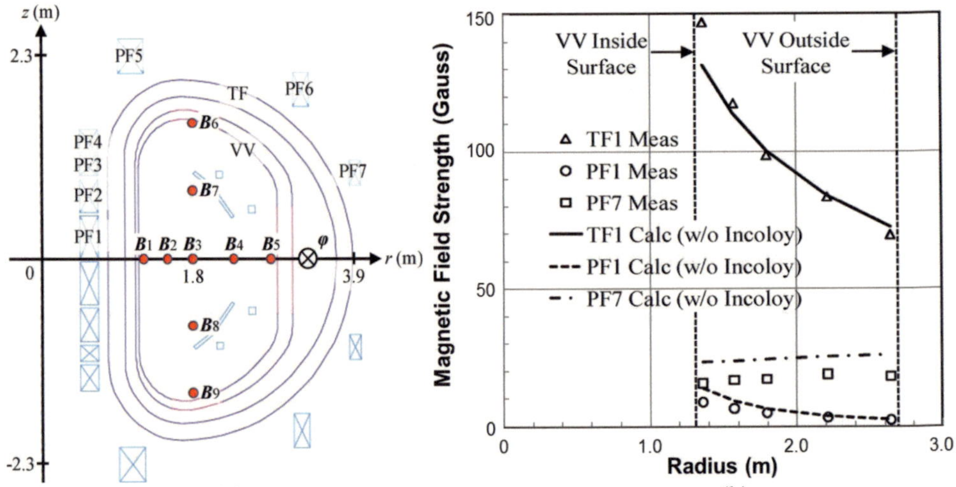

(a) (b)

Fig. 17. The positions of the hall proves to measure the field profile in vacuum vessel (a), and the measured field strength at five radial positions according to the TF, PF1 or PF7 coil current charging up to 100 A (b).

5.2 Inspection during and after cool-down

The total cold mass is about 300 tons. In order to cool-down this massive structure, the 9 kW helium refrigerators, the helium distribution valve box and in-cryostat cooling piping were installed. The cool-down period was about 23 days as shown in Fig. 18 (Kim, Y.S. et al, 2009). The major parameters such as the vacuum, the temperature, and the mass flow rate, were carefully checked during the cool-down. The structural behavior was also monitored because the different thermal shrinkage between the structurally connected bodies may give harmful effects on the structural integrity. The maximum difference of the temperature distribution inside the cryostat was set to be 50 K, which is the value obtained from the experience in many large scale superconducting magnet applications. This criterion is not applied to the thermal shield because the thermal shield is an independent structure without any structural connection with the superconducting magnet system. As the much possibility of the cold leak exists, the residual gas inside the cryostat was continuously analyzed. The superconducting to normal transitions of the KSTAR magnet system were observed during the cool-down. As expected, the TF coil and the PF1-5 coils showed the transition around 18 K. The measured RRR values were about 200 which satisfied the design value of more than 100. The PF6-7 and bus lines which are made of NbTi superconductor showed the transition around 9 K with RRR of about 100.

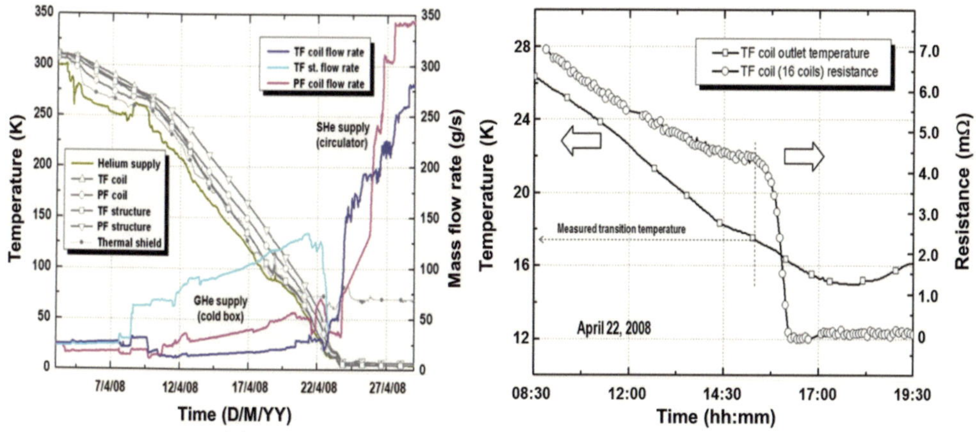

(a) (b)

Fig. 18. The cool down characteristics of the magnet system. The change of overall mass flow rate of helium (a), and phase transition to superconductor at around 18 K (b).

After finishing the cool-down, the HIPOT test was carried out up to 6 kV for the final insulation check. As the operating currents were lower than the designed ones for all the coils in the first operation, the maximum possible voltage generation is less than 2.5 kV between coil terminals (Park, K.R. et al., 2009). Therefore, the insulation test voltage of 6 kV was considered to be sufficient for ensuring the insulation integrity.

Before the high current operation, it is mandatory to check the electrical joints for high currents. The designed resistance value is less than 5 nΩ per one lap joint. As the voltage was measured across the bus line which has the 3–4 lap joints, the averaged joint resistance was used to validate the joint resistances. The joint resistances were measured by V-I characteristic curve by scanning the current in the range of ± 900 A. The resistance was obtained by the slope of the V-I curve. The measured joint resistance is less than 2.5 nΩ per joint and it satisfied the design value of less than 5 nΩ (Park, Y.M. et al., 2009).

5.3 Inspection under the high current operation

The objectives of inspection of the superconducting magnet under the high current operation are to check operational stability of magnets at high current operation conditions and to check the controllability of the coil currents and magnetic field profile for plasma discharges and confinement. The TF system was tested by increasing the current level in steps. The thermo-hydraulic and magnetic parameters were measured for the operating conditions. Figure 19(a) shows that TF magnet operated stable 30 kA for more than 5 hours with the temperature rising less than 0.1 K during the current charge and discharge periods. The quench detection system also operated reliably. During the current charging and discharging time, the detected signal was less than 25 mV and well below the trigger condition of 50 mV and 2 s (Chu, Y. et al., 2009).

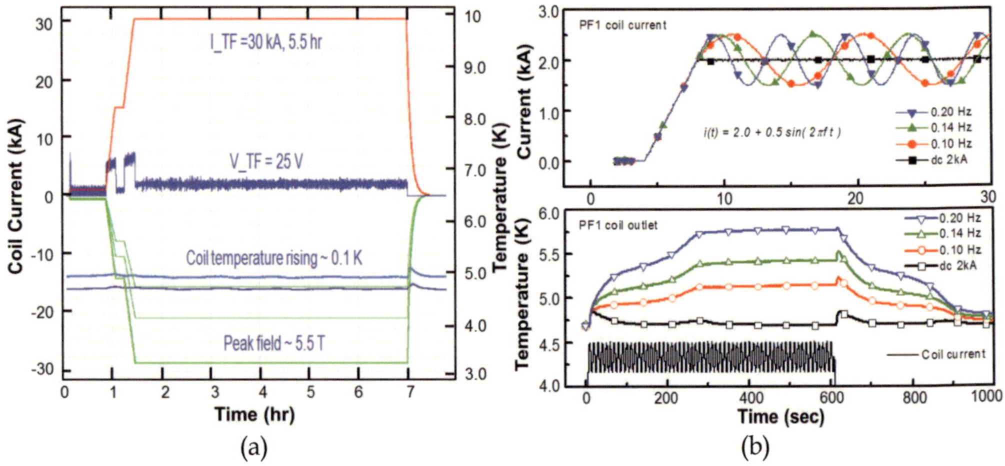

Fig. 19. The operational characteristic of the magnets. Stable operation of TF magnet at 30 kA over 5.5 hours with temperature rising of about 0.1 K (a), and ac loss measurement of PF1 coil by applying sinusoidal current waveform up to 600 s (b).

For KSTAR operation at currents up to 2 MA and long pulses up to 300 s, the ac losses of PF magnets can be a very important factor in deciding the operational capability. Precise measurements of the ac losses of the PF magnet system in KSTAR are difficult due to the limited number of sensors installed in KSTAR. AC loss was measured at the current waveforms of the typical single pulse waveform and of long pulse sinusoidal waveforms which were adopted in the KSTAR CS model coil experiments (Oh, Y.K. et al., 2009). Figure 19(b) shows the current waveform for ac loss measurement of the PF1 coil. The applied current was dc biased 2 kA and sinusoidal ac current with amplitude of 0.5 kA for 600 s according to changing the frequency from 0.1 to 0.2 Hz. The figure shows that thermal parameters reached a steady-state condition after 350 s. The coupling loss of PF1 showed different results according to the waveforms (Lee, H.J. et al., 2010; Oh, D.K. et al., 2010). The coupling loss time constant, $n\tau$ of PF1 measured by the single trapezoidal waveform was about 62.5 ms at the initial operation time and reduced to about 50ms after two months operation. It could be expected that the cyclic electromagnetic force applied to the strands tends to increase the inter-strand contact resistance, and resulted in the reduced coupling loss. When compared with sinusoidal method, the measured time constant was about 35 ms. The difference of the $n\tau$ could come from the non-linear current control at low current less than 500 A. Another interesting result is that the coupling time constant of the PF6 coil is also about 33ms even though it has NbTi conductor. These results could come from the adoption of chrome coating on all three kinds of strands, Nb_3Sn, NbTi, and copper. These results show that the KSTAR construction has advantages for various operational capabilities especially high speed PF6 coil operation as in the first plasma operation. The superconducting magnet commissioning was completed without any severe problems in spite of the fact that individual cool-down tests were not conducted prior to assembly.

6. Plasma experiments and magnet operation

6.1 Plasma confinement and experiments

After the successful integrated commissioning in 2008, KSTAR device has been operated to meet the project mission to establish a scientific and technological basis for the attractive fusion reactor. The operating performance of the device was upgraded every year. In the first campaign, the first plasma was achieved with a plasma current over 100 kA just after the integrated commissioning in 2008 (Oh, Y.K. et al., 2009). In the second and third campaigns, by the reliable control of the plasma current and shape, high performance confinement (H-mode) plasma could be achieved at the plasma current over 500 kA as shown in the Fig. 20. In the 4th campaign, the plasma current reached up to 1 MA and various researches could be conducted such as suppression of the edge local mode (ELM).

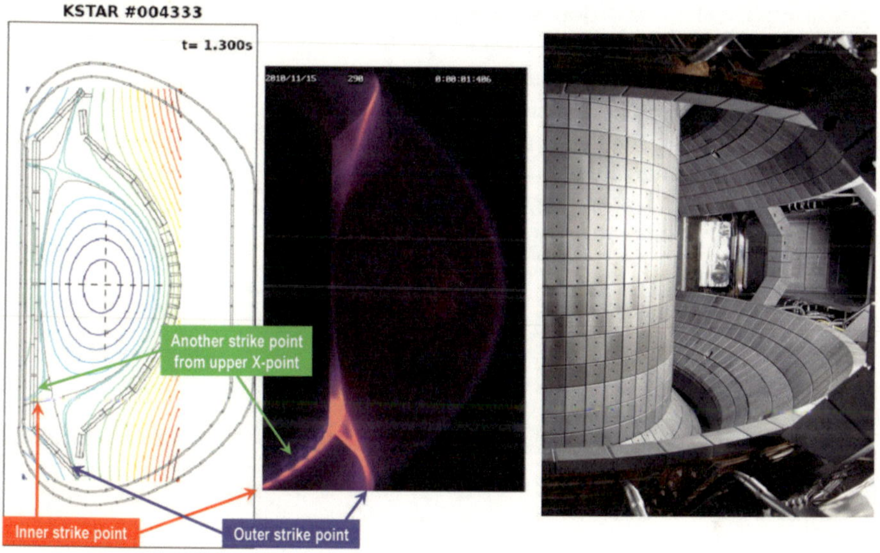

Fig. 20. High performance plasma confinement (H-mode) achievement in KSTAR. Plasma profile reconstruction using EFIT, CCD visual image, and the picture of in-vessel components.

Usually the plasma operation could be classified into several phases; plasma startup to generate a plasma current up to about 100 kA, feedback plasma control for the current ramp up and shaping, plasma heating and experimental research under various operational modes, and plasma current ramp-down or disruption. The plasma current and shape are controlled by the plasma control system which is a feedback control loop with the diagnostics sensors and PF magnet power supplies (Hahn, S.H. et al., 2009).

The basic criteria for plasma startup are formation of the field null and the loop voltage to drive the plasma current rise (Leuer, J.A. et al., 2010). However the startup in the KSTAR was complicated due to the nonlinear magnetic behavior associated with the Incoloy 908 jacket material used in the construction of the Nb_3Sn superconductors. The field distortion was investigated by scanning the field profile using the hall probe array and monitoring the e-beam trajectories according to the TF and PF coil current (England, A.C. et al., 2011). The field deformation due to the ferromagnetic effect was deliberately compensated by including the control of field gradients since the field gradients determined the positional stability of the plasma current channel. As a consequence, plasma start-up became much more robust against the deviation from the expected plasma initial ramp-up. And the field null formation was intentionally delayed until the toroidal electric field reached a sufficient level to make inductive breakdown of the neutral gas. By doing so, we could obtain better breakdown conditions with the same level of loop voltage. Hence purely ohmic discharges were routinely achieved under various toroidal magnetic fields (Kim, J. et al., 2011).

ELMy H-mode discharges have been observed from the third campaign with combined auxiliary heating of neutral beam injection ($P_{NBI} \leq 1.4$ MW) and electron cyclotron resonant heating ($P_{ECRH} \leq 0.2$ MW) in a relatively low density regime. The L–H transitions occurred typically right after the end of the plasma current (Ip) ramp-up to 0.6 MA and after the fast shaping to a highly elongated double-null (DN) divertor configuration (Yoon, S.W. et al., 2011). The ELM is an evidence of the H-mode confinement but it is an issue to be prevented or mitigated to protect the plasma facing components or divertor from the damage by the energetic particles flux. The ELM is one of the critical issues in the ITER device. By applying three-dimensional field on the plasma, ELM suppression or mitigation could be achieved in the 4th campaign. As shown in Fig. 21. KSTAR is targeted to achieve the reliable plasma control at the plasma current of 2 MA, in-inductive steady-state operation up to 300s, and plasma heating using by 28 MW auxiliary heating systems.

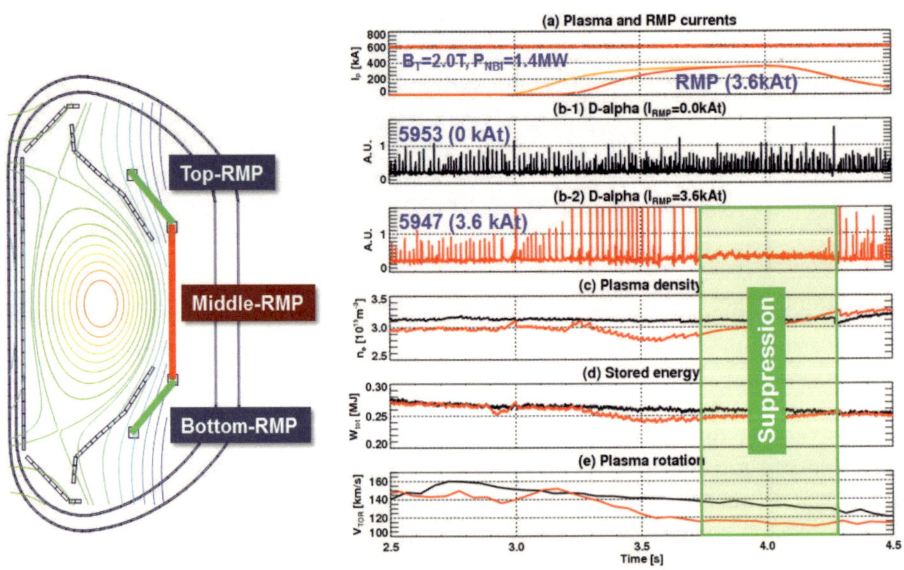

Fig. 21. Edge localized mode (ELM) occurrence in H-mode plasma and its suppression by applying 3D field (n=1 RMP) using in-vessel control coils.

6.2 Cryogenic stability of magnet system

In the KSTAR operation, the TF magnet has been operated up to the designed operating current of 35.2 kA in stable with temperature rising of 0.1 K. The toroidal field was 3.5 T at plasma center and peak field inside conductor was 7.2 T with temperature margin of about 4

K. The superconductor of the KSTAR TF magnet has an operational capability of toroidal field up to 5.0 T with a temperature margin of 1.1 K as shown in Table 4.

Parameters	**Designed operation**		**Extended operation**	
Toroidal field, B_0 [T]	3.5	3.5	**4.5**	**5.0**
Peak field, B_{peak} [T)	7.2	7.2	9.3	10.3
Coil current, I_{TF} [kA]	35.2	35.2	45.3	50.3
Strain on Nb_3Sn [%]	**0.7**	**0.3**	**0.7**	**0.3**
Current sharing Temp, T_{CS} [K]	9.16	10.5	5.86	5.85
Temp. margin, T_{margin} [K]	4.4	5.7	1.1	1.1

Table 4. Estimation of the temperature margin of the TF magnet at designed condition and extended performance conditions.

The PF magnets are designed to be operated for the plasma current startup and ramp up to 2 MA inductively, and plasma of 1 MA current has been achieved in 2011 operation in spite of the limited electric power supply. The designed current of 2 MA is expected to be achieved after the completion of a motor-generator system. Nominal operational scenario of the PF magnets is started from the initial magnetization to store a maximum flux in each magnet and is followed by a synchronized fast flux swing for the plasma startup and finally is continued by the fast feedback control of plasma current, shape, and position. When the PF magnet operated in pulse, large amount of pulsed heat is generated in the magnet due to the ac loss which is deeply related with conductor design and the field swing rate.

The cryogenic system of the PF magnet has been designed to be operated under the pulsed heat load by adopting the thermal damper concept. A cryogenic circulator will be operated in static to supply the supercritical helium in to PF magnets with a total mass flow rate of about 300 g/s with pressure drop of about 2 bar. The transient heat from the PF magnet is extracted by the heat exchangers which are immersed at the liquid helium bath as shown in Fig. 22(a). Most of the PF shot was operated stably within the However the operation range of the circulator. However in some scenario with fast current changing scenario, a large thermal load causes abrupt flow and pressure increase of the helium. The supply and return pressure of the circulator are increased simultaneously with shot. The mass flow rate in the PF 1 magnet was maintained the value of zero for several seconds as shown in Fig. 22(b). And the operation point of the cryogenic circulator moved into the unstable zone and transiently (Lee, H.J. et al., 2011; park, Y.M. et al., 2010). Although the PF magnets are made of with low as loss conductor, additional efforts are required for the stable operation of the cryogenic circulator as well as magnet system itself.

One of the approaches is optimizing the PF magnet operation scenario sharing the burden of ac loss with other magnets. There were similar approaches in the conventional tokamaks to reduce the burden on central solenoid by solenoid-free startup in DIII-D (Leuer, J.A. et al., 2010) or by coaxial helicity injection in NSTX (Raman, R. et al., 2001). Another way is to limit the maximum field change rate within certain range because the ac loss is proportion to the square of the field change rate. The modification the cryogenic circuits could be considered to clamp the cryogenic parameter fluctuation in the cryogenic circulator under the large thermal load in PF magnet. It could maximize the operation capability of PF magnet without increasing overall capacity of the cryo-facility.

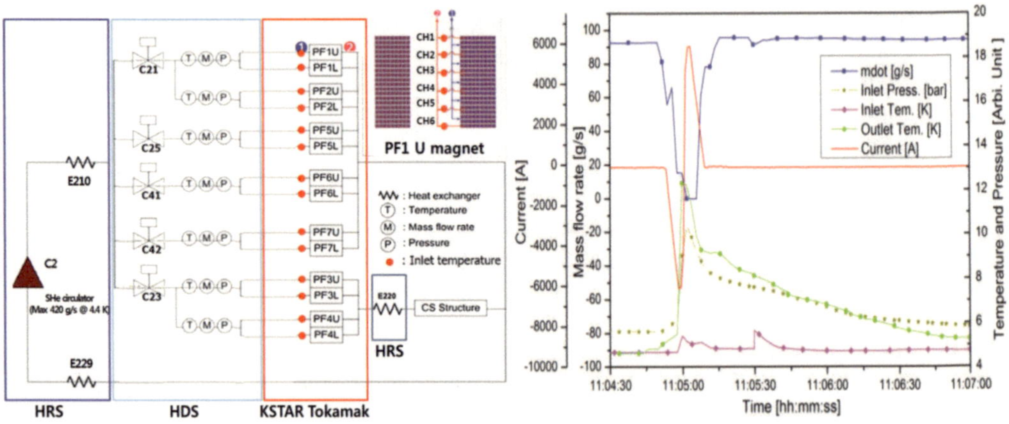

(a) (b)

Fig. 22. Flow diagram of the PF magnet cooling circuits (a), and flow reversal occurrence in some extreme operation of the PF magnet at bipolar operation with 10 kA/s (b)

6.3 Other issues in superconductor application to fusion devices

For superconductor application to the future fusion reactor, more R&D activities are required to develop the conductor designs having stable operation capability under the extreme environments such as higher magnetic field and higher current density operation. However there are some technical issues which hinder the fabrication of the large current Nb_3Sn conductor for ITER.

One of the issues is irreversible degradation in Nb_3Sn CICC (Bruzzone, P. 2010), usually the longitudinal strain applied to the Nb_3Sn makes the decrease the allowable limit of the operation current density. The longitudinal strain mainly comes when the CICC was cooled down from the heat treatment temperature about 600 C to operating temperature of 5 K due to the difference of coefficient of thermal expansion. And it is regarded as reversible phenomena. Another load applied on the conductor is the transverse load by the Lorentz force from the conductor current and self-field. In many sample test of the ITER conductor showed the degradation of the performance of Nb_3Sn CICC by the number of load cycle. The possible interpretation of the degradation is plastic deformation of the superconductor filament by the bending of strand when the cyclic transverse load is applied. The cycled load applied on the Nb_3Sn conductor is helpful for the ac loss reduction but reduction of the dc performance.

ITER TF magnet is designed with a concept of insertion of circular conductor in the pre-machined groove in the radial plate to prevent the accumulation of the transverse force on the conductor. But the conductor length changed after heat treatment mainly due to the residual stress during the jacket compaction and coil winding. So it is not easy to control the gap between TF CICC and the grove in the radial plate.

Those kinds of issues will be explored according to the operation of the present superconducting fusion devices.

7. Conclusion

The application of the superconductor to the magnetic confinement fusion devices is inevitable to achieve a steady-state plasma confinement under the high field environment. There has been a lot of progress in the superconducting magnet technology in fusion application. Initially it was only to the static field magnets by using the Rutherford-type conductor cooled in a liquid helium bath, but now it is applied to the pulsed field magnets also by using the CICC type conductor which is cooled with forced flow supercritical helium and is robust for the large electromagnetic and high voltage conditions. The KSTAR device has been constructed using the fully superconducting magnets which are up-to-date in technologies.

The representative outstanding technologies of the magnet systems in the KSTAR devices are as follows; high performance Nb_3Sn conductor development to be used commonly in TF and PF magnets, reliable chrome coating on strand surface to reduce the ac coupling loss, long-length CICC production up to 1.8 km by tube-mill welding process of jacket, high reliability in the leak detection and repair welding of CICC, continuous magnet winding without internal joint by numerical shape control, successful heat treatment of Nb3Sn magnet with Incoloy908 jacket without failure, turn insulation taping and full penetration VPI technology, new method of TF magnet installation by rotating each one along the surface of the vacuum vessel, applying pre-compression on the central solenoids structure by heating the structure and wedge adjustment, and stable operation and protection of the magnet system. The successful construction and operation of the KSTAR device using the Nb_3Sn superconductor brought a great positive impact in the worldwide fusion reactor preparation including the international experimental fusion device, ITER.

There are several remained issues in the superconducting magnet application in the world-wide superconductor application into the fusion reactors, especially for the large scale fusion reactors. Large fluctuations of the thermo-hydraulic parameters in cryo-facility are mainly due to the ac loss in the PF magnet limit the operational capability of the PF magnets. And the performance degradation of the Nb_3Sn conductor was found in the development for the ITER magnet conductor such as allowable current density decrease under the large amount of the transverse force In the CICC and the current sharing temperature is also decreased after the large number of operation cycle. Those kinds of issues will be explored according to the operation of the present superconducting fusion devices. New optimum operational scenarios of the PF magnets should be developed also to reduce the ac loss and to maximize the static operation range.

8. Acknowledgement

The authors appreciate all the participants and researchers who contributed to the construction, operation and experiments of the KSTAR device, especially to the superconducting magnets system development. This work was supported by the Korean Ministry of Education, Science and Technology under the KSTAR project.

9. References

Ahn, H.J.; et al. (2002). Engineering design status of the KSTAR TF coil structure. *IEEE Trans. on Applied Superconductivity,* Vol. 12, pp. 492-495

Bak, J.S.; et al. (2006). Status of the KSTAR tokamak construction. *Fusion Engineering and Design,* Vol. 81, pp. 2315-2324

Bruzzone, P. (2010). Superconductors for fusion: Achievements, open issues, roadmap to future. *Physica C,* Vol. 470, pp. 1734-1739

Chang, H.S.; et al. (2008). The on-site status of the KSTAR helium refrigeration system. *Advances in Cryogenic Engineering,* Vol. 53, pp. 437-444

Choi, C.H.; et al. (2002). Electromagnetic loads on the KSTAR magnet system. *IEEE Trans. on Applied Superconductivity,* Vol. 12, pp. 534-537

Chu, Y.; et al. (2009). Quench detection based on voltage measurement for the KSTAR superconducting coils. *IEEE Trans. on Applied Superconductivity,* Vol. 19, pp. 1565-1568

England, A.C.; et al. (2011). Tokamak field error measurements with an electron beam in KSTAR. *Fusion Engineering and Design,* Vol. 86, pp. 20-26

Hahn, S.H.; et al. (2009). Plasma control system for "Day-one" operation of KSTAR tokamak. *Fusion Engineering and Design,* Vol. 84, pp. 867-874

Kim, J.; et al. (2011). Stable plasma start-up in the KSTAR device under various discharge conditions. *Nuclear Fusion,* Vol. 51, p. 083034

Kim, K.; et al. (2005). Status of the KSTAR superconducting magnet system development. *Nuclear Fusion,* Vol.45, pp. 783-789

Kim, K.; et al. (2006). Status of the K STAR Project. *Journal of Cryogenic Society of Japan,* Vol. 41, pp. 184-192

Kim, Y.S.; et al. (2009). First commissioning results of the KSTAR cryogenic system. *IEEE Trans. on Applied Superconductivity,* Vol. 19, pp. 1577-1581

Lee, G.S.; et al. (2001). Design and construction of the KSTAR tokamak. *Nuclear Fusion,* Vol. 41, pp. 1515-1523

Lee, H.J.; et al. (2010). The ac loss measurement of the KSTAR PF1 coils during the first commissioning. *IEEE Trans. on Applied Superconductivity,* Vol. 20, pp. 551-555

Lee, H.J.; et al. (2011). Analysis of the reversal flow phenomenon of supercritical helium due to ac losses in KSTAR PF magnets at low current. *IEEE Trans. on Applied Superconductivity,* Vol. 21, pp. 2012-2015

Lee, S.; et al. (2006). Ac loss characteristics of the KSTAR CSMC estimated by pulse test. *IEEE Trans. on Applied Superconductivity,* Vol. 16, pp. 771-774

Leuer, J.A.; et al. (2010). Plasma startup design of fully superconducting tokamak EAST and KSTAR with implications for ITER. *IEEE Trans. on Plasma Science,* Vol. 38, pp. 333-340

Leuer, J.A.; et al. (2011). Solenoid-free startup experiments in DIII-D. *Nuclear Fusion,* Vol. 51, p. 063038

Lim, B.S.; et al. (2005). Jacketing and reapir of the KSTAR CICC. *IEEE Trans. on Applied Superconductivity,* Vol. 15, pp. 1415-1418

Mitchell, N. (2003). Summary, assessment and implications of the ITER model coil test results. *Fusion Engineering and Design,* Vol. 66-68, pp.971-993

Oh, D.K.; et al. (2004). Design and thermo-hydraulic analysis of upgraded PUMA system for the development of a test facility of superconducting CICC. *IEEE Trans. on Applied Superconductivity,* Vol. 19, pp. 1557-1560

Oh, D.K.; et al. (2010). Development of numerical model of KSTAR PF conductor and magnet for the analyses of AC loss on the results of KSTAR PF magnet test run. *IEEE Trans. on Applied Superconductivity,* Vol. 20, pp. 2164-2167

Oh, Y.K.; et al. (2002). Engineering design status of the KSTAR central solenoid structure. *IEEE Trans. on Applied Superconductivity,* Vol. 12, pp. 615-618

Oh, Y.K.; et al. (2004). Cool-down and current test results of the KSTAR prototype TF coil. *Advances in Cryogenic Engineering,* Vol. 49, pp. 241-248

Oh, Y.K.; et al. (2005). Status of the KSTAR Tokamak Construction and Assembly. *Journal of the Korean Physics Society,* Vol. 49, pp. s1-s6

Oh, Y.K.; et al. (2008). Completion of the KSTAR construction and its role as ITER pilot device. *Fusion Engineering and Design*, Vol. 83, pp. 804-808

Oh, Y.K.; et al. (2009). Commissioning and initial operation of KSTAR superconducting tokamak. *Fusion Engineering and Design,* Vol. 84, pp. 344-350

Park, K.R.; et al. (2005). Status of the KSTAR PF6 and PF7 coil development. *IEEE Trans. on Applied Superconductivity,* Vol. 15, pp. 1375-1378

Park, K.R.; et al. (2009). Qualification test results of the KSTAR superconducting coils from the construction to the commissioning steps. *IEEE Trans. on Applied Superconductivity,* Vol. 19, pp. 1561-1564

Park, Y.M.; et al. (2009). Construction and commissioning of the KSTAR current feeder system. *IEEE Trans. on Applied Superconductivity,* Vol. 19, pp. 1587-1591

Park, Y.M.; et al. (2010). Analysis of the helium behavior due to AC losses in the KSTAR superconducting coils. *IEEE Trans. on Applied Superconductivity,* Vol. 20, pp. 530-533

Raman, R.; et al. (2001). Non-inductive current generation in NSTX using coaxial helicity injection. *Nuclear Fusion,* Vol. 41, p. 1081-1086

Sa, J.W.; et al. (2002). Detailed evaluation of insulation stress in the KSTAR central solenoid. *IEEE Trans. on Applied Superconductivity,* Vol. 12, pp. 619-622

Wu, S.; et al. (2007). An overview of the EAST project. *Fusion Engineering and Design,* Vol. 82, pp. 463-471

Yang, H.L.; et al. (2006). KSTAR assembly. *Proceeding of IAEA FEC 2006 conference (October, 2006), http://www-naweb.iaea.org/napc/physics/fec/fec2006/html/node339.htm, FT/2-2*

Yonegawa, H.; et al. (2009). KSTAR magnetic field measurement during the commissioning and remnant field evaluation. *IEEE Trans. on Applied Superconductivity,* Vol. 19, pp. 1573-1576

Yoon, S.W.; et al. (2011). Characteristics of the first H-mode discharge in KSTAR. *Nuclear Fusion,* Vol. 51, p. 113009

Yoshida, K.; et al. (2008). Conceptual design of superconducting magnet system for JT-60SA. *IEEE Trans. on Applied Superconductivity*, Vol. 18, pp. 441-446

18

MgB_2 SQUID for Magnetocardiography

Yoshitomo Harada[1],
Koichiro Kobayashi[2] and Masahito Yoshizawa[2]
[1]*National Institute for Materials Science, Tsukuba,*
[2]*Graduate School of Engineering, Iwate University, Morioka,*
Japan

1. Introduction

The discovery of the superconductivity at a transition temperature (T_C) of 39 K in magnesium diboride (MgB_2) has attracted much attention from many researchers for scientific as well as technical reasons [1]. Compared with Cu-based superconductors (cuprates), MgB_2 has lower anisotropy and larger coherence length, in addition to high T_c [2]. These characteristics of MgB_2 give rise to new applications for superconductor devices that can operate in the temperature range between 20 and 30 K; examples of such devices are Josephson junctions and integrated circuits. This temperature range can be easily achieved by using economical and compact cryocoolers or liquid hydrogen. The use of cryocoolers may transform the superconductor from being specialized and advanced technology into common usage in consumer devices. In the future, hydrogen gas may be widely used for carbon-free power generation such as in fuel cells. Liquid hydrogen would be available for these purposes, and may be utilized for the cooling of low-temperature devices. In addition, MgB_2 is considered to be a clean superconducting material, using neither toxic nor rare earth elements.

Thus far, we have developed SQUID (Superconducting Quantum Interference Device) apparatuses, such as magnetocardiography (MCG) and non-destructive evaluation (NDE) systems, by using Nb and cuprates. Since the discovery of superconductivity in MgB_2 and its promising potential, we have focused our research on developing SQUID as well as a high-frequency filter made of MgB_2. As a first step, we developed a synthesis method for high-quality films, which is the basis of fabrication of such devices. Next, since 2004, we have been working on developing a SQUID and a high-frequency filter device. Finally, we succeeded in measuring the magnetic signal from a human being (MCG signal) by using MgB_2 SQUID with a specially developed control circuit (digital FLL). In this article, we discuss our achievements and progress on the development of superconducting devices using MgB_2 films.

2. Synthesis of high-quality films

Synthesis of high-quality film is a key technology in developing superconductor devices. We would first like to introduce various synthesis methods for MgB_2 films. There have already been many reports on the techniques that have been used far for the preparation of MgB_2

thin films. These preparation methods can be categorized into two types: post-annealing methods and as-grown methods.

We will now introduce the post-annealing method. First, Mg and B together or MgB_X alone are evaporated on the substrates at room temperature in this method. Second, these films are annealed inside an electric furnace in an atmosphere of Mg gas. This process improves the physical properties so that and we can obtain superconducting behavior [3,4]. This method is characterized by achieving a high T_C through high-temperature annealing. However, this method is not preferred for the synthesis of the films used in the devices due to the occurrence of the interfacial reactions and the degradation of the surface flatness by high-temperature annealing.

On the other hand, the as-grown method does not need an annealing process. The as-grown method is suitable for the fabrication of superconducting devices because it provides a flat surface. To date, several groups have reported the synthesis of as-grown superconducting MgB_2 films. In particular, the HPCVD (hybrid physical chemical vapor deposition) [5] method and the MBE (molecular beam epitaxy) method [6-9] have achieved high T_c values that are over 30 K. The HPCVD method, which enables a high-temperature growth process of the film at about 1000 K, succeeded in synthesizing the film with a high T_C value of 39 K by using Mg that has a high degree of volatility.

In contrast to HPCVD, the MBE method adopts a low synthesis temperature of about 573 K and is carried out in a high vacuum. The T_C of MgB_2 films grown by the MBE method is relatively lower than that obtained from other methods and is limited to be around 34 K. The growth conditions for the MBE method adopted by the research groups other than our group are, however, high growth rate and high substrate temperature. The high growth rate compensates for the deficiency of Mg through re-evaporation due to its volatility and prevents the oxidation of Mg. This concept is very similar to that of the HPCVD method. High deposition rate and high growth temperature have been a mainstay in the fabrication of the high-quality MgB_2 films.

We adopted a low substrate temperature, a low deposition rate, and an ultra-high vacuum to obtain high-quality MgB_2 films. The characteristic feature of the MBE method is low-temperature synthesis. According to Liu *et al.*, MgB_2 is synthesized in a wider temperature range, particularly at low temperatures, as the vacuum increases [10]. The growth at low temperature reduces the re-evaporation of Mg. The deposition rate and the supply of Mg and B have to be reduced due to the super-saturation of Mg. Under these conditions, Mg may be oxidized. However, the ultra-high vacuum in an MBE apparatus prevents the oxidation and permits the growth of high-quality films.

In terms of synthesis conditions, our method is conceptually opposite to the conventional fabrication methods for MgB_2 films. Figure 1 shows the typical temperature dependence of the resistivity curves of MgB_2 films fabricated by our method. The growth temperature was set as 473 K, which is about 100 K lower than that in other reports. In our method, the range of the deposition temperature was between 373 K and 523 K. This is the same range as that in the phase-diagram of Lie *et al.* and may be obtained by extrapolating the pressure range to 10^{-7} Pa and 10^{-8} Pa. The Tc of the sample in Fig. 1 was 32 K, and the transition width is smaller than 1.0 K. In our method, T_c can reach a value of up to 37 K by the adoption of various additional process improvements, which will be presented later.

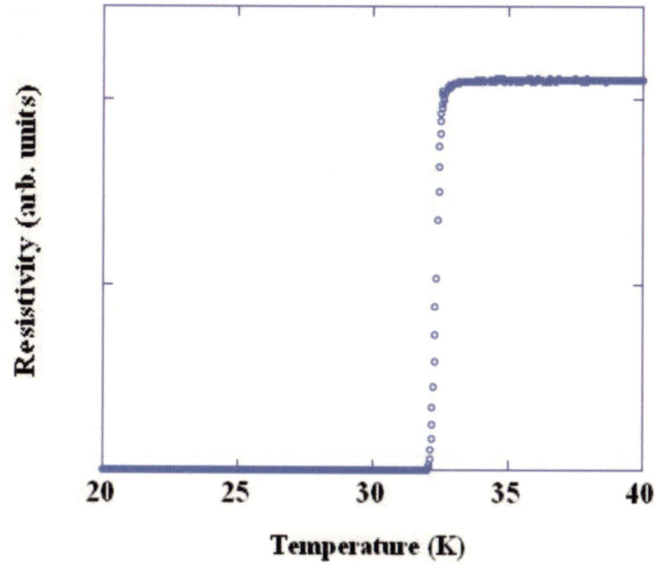

Fig. 1. Temperature dependence of the resistivity of MgB_2 films.

Figure 2 shows the critical current density (J_C) of MgB_2 films as a function of temperature. The measurement was carried out using patterned micro-bridges, which were made by standard photo-lithography and the Ar ion beam milling method. The inset of Fig. 2 is an optical microscope image of the MgB_2 micro-bridge. The dimensions of the micro-bridge were 10 μm (width) × 30 μm (length). The electrical contact was made by gold wire directly bonded to a Cu/MgB_2 contact pad using silver paste. The highest value of J_c is 11.6 MA/cm^2 at 4.2 K on an MgO (100) substrate. The $J_C(0)$ value is 12.5 MA/cm^2. This value is considered to be very high for MgB_2 films.

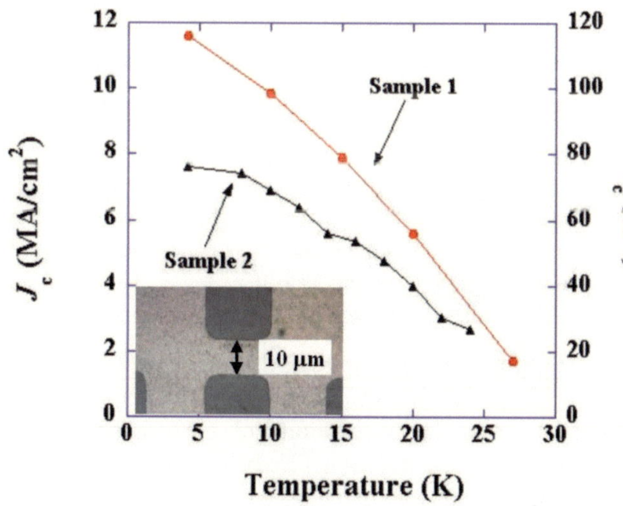

Fig. 2. Temperature dependence of the critical current density (J_C) of MgB_2 micro-bridge. J_C is calculated from the critical current (I_C).

We will discuss the reason why the J_C of our sample synthesized on an MgO substrate is so high. From AFM and SEM measurements, the grain size of MgB_2 films on MgO substrates is 20-50 nm. This value is smaller than that on an Al_2O_3 or Si substrate. The mismatch between the in-plane lattice constant of MgB_2 and MgO is 26.7 %. MgO (100) plane has a square surface structure, whereas MgB_2 has a hexagonal surface structure. These conditions led to a smaller grain size of MgB_2 when it is deposited on MgO. The grain shape of MgB_2 is of the columnar-type, and the grain boundaries work as effective pinning centers. Therefore, the smaller grain size provides more pinning centers and leads to high values of J_C in the perpendicular field.

Next, we tried to further improve the quality of the grown MgB_2 films such that they can be used for the fabrication of superconducting devices. According to the works of Shimane University [11] and NICT [12], the in-plane lattice matching between MgB_2 and the substrate plays an important role in the growth of high-quality MgB_2 films using the MBE method.

The structural quality of MgB_2 films was improved by using AlN ($\Delta d = 1.9\%$) and TiZr ($\Delta d = 3.6\%$) buffer layers having lattice constants close to that of MgB_2. We initially employed ZnO (0001) as the substrate as its substrate closely matched that of MgB_2. The in-plane lattice spacing of the hexagonal ZnO lattice (a = 0.3522 nm) is close to that of the MgB_2 lattice ($\Delta d = 5.4\%$; Δd of Al_2O_3, for example, is 35.2 %) [13]. However, we could not do an in-plane alignment for these films, although their T_c is relatively high at 35 K. The cross-sectional transmission electron microscope (TEM) image showed a large amount of reaction products near the MgB_2/ZnO interface. In addition, intermixing between Zn and MgO was observed near the MgB_2/ZnO boundary, which is ascribed to the free energy difference between the substrate and the film. The free energy of MgO is -547.1 kJ/mol and is lower than the value of -303.3 kJ/mol for ZnO. Therefore, Zn in ZnO is easily replaced by Mg, and MgO is formed.

The best-quality MgB_2 film was obtained by adopting titanium (Ti) film as a buffer layer on a ZnO substrate, which was fabricated by evaporation using an electron-beam gun cell in an MBE chamber [14]. Figure 3 shows the superconducting properties of MgB_2 films when a 15 nm Ti buffer layer is adopted. When the MgB_2 film was deposited on a ZnO (0001) substrate, the T_C increased from 33 K to 36K with the insertion of a Ti buffer layer. By using *in situ* reflection high-energy electron diffraction (RHEED), Ti buffer layers were grown epitaxially in the configurations Ti (0001) | | ZnO (0001) and Ti $\left[1\bar{1}00\right]$ | | ZnO$\left[11\bar{2}0\right]$. The in-plane orientation (ϕ) scans of the $11\bar{2}2$ reflections around the c-axis of the MgB_2 films are shown in Fig. 4, wherein two sets of 6-fold symmetric peaks appear at intervals of every 60°. The peak sets corresponds to two types of domains rotated by 30° from each other about the c-axis. Furthermore, we found that one set of the peaks increased in highest and became predominant with increasing thickness of the Ti buffer layer. This result indicates that the in-plane alignment of the MgB_2 lattice is improved by the presence of the Ti buffer layer.

Nishidate *et al.* investigated the effect of the Ti buffer layer on ZnO substrates prior to the deposition of Mg and B by studying and calculating its molecular dynamics (MD) [15]. Because Ti atoms occupy a particular adsorption point, Mg adsorption was limited on a metal oxide substrate and Mg is used in the formation of MgB_2. In addition, the lattice constant of Ti, a = 0.295 nm, matches that of MgB_2, and the free energy of -849.89 kJ/mol for Ti-O is lower than the corresponding values for both MgO and ZnO. Ti also hinders the formation of MgO.

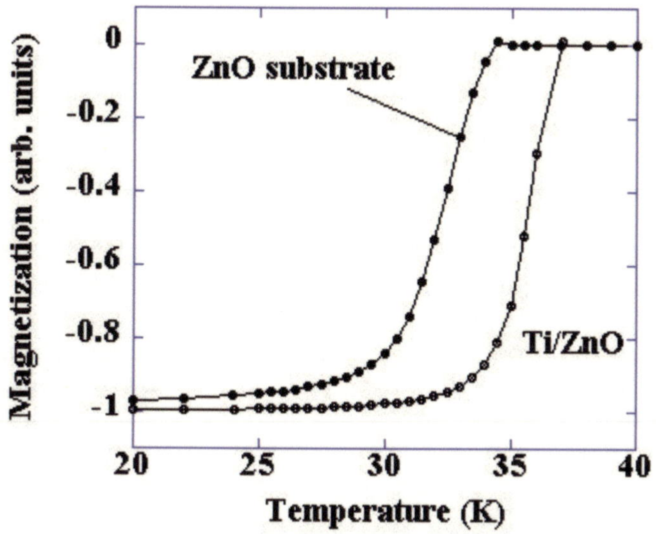

Fig. 3. Temperature dependence of the magnetization of MgB_2 film on the ZnO and Ti/ZnO substrate.

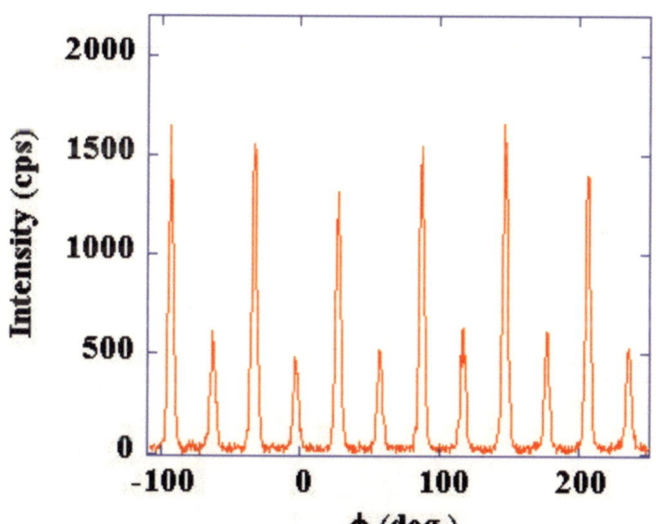

Fig. 4. XRD ϕ scan image of MgB_2 $11\bar{2}2$ peaks from MgB_2/Ti films deposited on ZnO substrate.

We achieved a value of T_C = 37 K by the MBE method, which is the same as that with HPCVD. Fujiyoshi *et al.* report that the J_C of the MgB_2 films on the Ti buffer layer is higher than that of MgB_2 on the substrates [16]. Therefore, the pinning force due to the grain boundaries at the MgB_2 film on the Ti buffer layer is stronger than that of other films.

3. Implementation of Josephson junction and SQUID by lithography technique

There are three types of Josephson junction fabrication methods for SQUID and they are as follows: SIS tunneling junctions, nano-bridge junctions and SNS junctions. The first tunneling MgB_2 junctions were fabricated by the NTT group [17]. We have developed and used an SIS junction [18]. Subsequently, both the Twente University and European groups succeeded in the fabrication of nano-bridge MgB_2 junctions [19]. We have developed nano-bridge MgB_2 SQUIDs devices that are optimized for MCG measurement, which are based on the design of the Twente University group. Nano-bridge junctions were made using focused ion beam (FIB) milling apparatus. Detailed fabrication conditions are reported elsewhere [20]. Figure 5(a) shows the layout of the MgB_2-SQUID device in this experiment. Figure 5(b) shows the scanning ion microscope (SIM) image of the nano-bridge junctions. We designed the size of junction to be 200 nm wide and 200 nm in length. From the SIM image, we observed that the actual junction size matched the design expectations very close.

Figure 6 (a) shows the voltage modulation (ϕ–V curve) of the nano-bridge MgB_2 SQUIDs at 24.4 K. Figure 6 (b) is the I-V curve for this SQUID. The voltage modulation due to the applied magnetic field can be seen to increase with decreasing temperature. An approximate magnitude of 22 μV is recorded at 20 K. The bias current applied was tuned to get the largest voltage modulation; the bias current value was 2280 μA for this sample, giving rise to the maximum modulation of 18.47 μV.

Figure 7 shows the ϕ-V curve of other SQUID with the same nano-bridge size. The measurement temperature is set to 20.81 K. The voltage modulation was measured for the bias current from 2550 to 2585 μV. The maximum modulation voltage was 33.55 μV when the bias current was set to be 2577 μA. The external magnetic field density corresponding to the magnetic flux quantum was 11.4 nT. The measured effective SQUID area (A_{eff}) was estimated to be 0.18 mm^2.

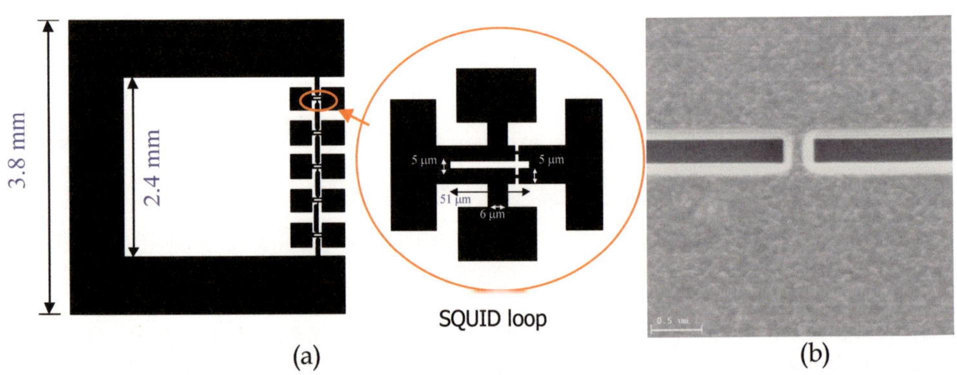

Fig. 5. (a) Layout of an MgB_2-SQUID device. (b) FIB-SIM image of nano-bridge junctions. The size of nano-bridge is 200 nm wide and 200 nm long.

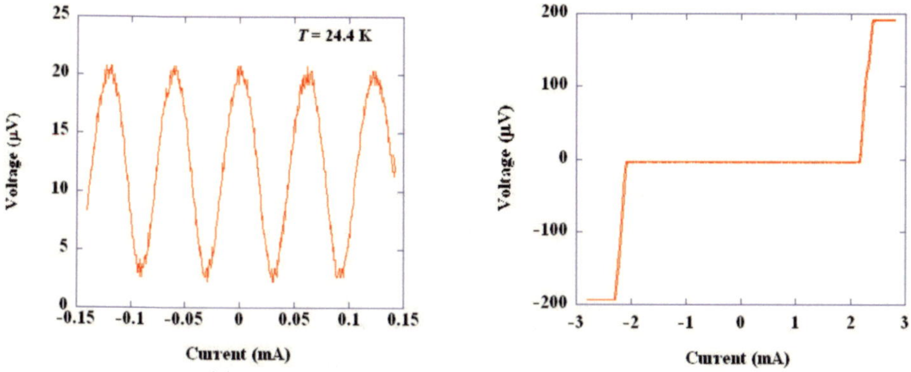

(a) (b)

Fig. 6. SQUID characteristics of 200 nm nano-bridge type MgB_2-SQUID. (a)Voltage modulation, (b) *I-V* property.

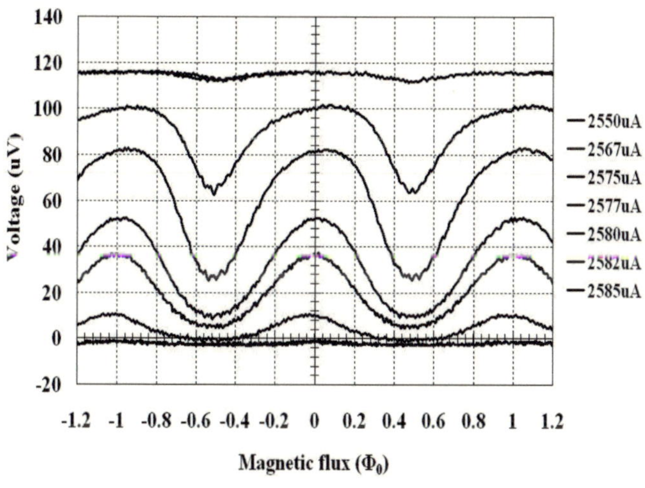

Fig. 7. Magnetic flux dependence of voltage modulation of an MgB_2-SQUID with various bias currents.

We encountered two problems during the development of SQUID. One is the reduction of the yield ratio, due to the small size of the nano-bridge, which is close to the size limitation of the FIB fabrication process. Another is the hysteretic behavior of the *I-V* curve at the transition edge, which appears at low temperatures. This causes instability in the bias current and makes measurements very difficult. The first problem was solved by adopting thinner films. We are able to obtain *I-V* and ϕ-*V* properties and even expand the width of the nano-bridge by decreasing film thickness. At present, we retain a high yield of about 90 % for the fabrication of 65 nm thick film and the 400 nm wide nano-bridges. The second problem was resolved by placing a resistor in parallel to the MgB_2 nano-bridges. We were able to eliminate the effect of hysteresis by inserting the resistor.

4. SQUID electronics

The input-output (Φ- V) characteristic of SQUID is nonlinear as shown in Fig. 8. A sinusoidal output voltage with the appropriate periodicity is generated, when a magnetic flux is applied in SQUID. A control circuit, known as the FLL (flux locked loop) circuit, is necessary to linearize the input-output characteristic so that the SQUID can be utilized as a magnetic sensor. An example of such an FLL circuit is shown in Fig. 9. The FLL circuit provides the feedback necessary to fix the output at the defined value (the lock point) in a sinusoidal SQUID output. Then, by using only the small area in the vicinity of the lock point, we get the desired linear output. The external magnetic field applied to a pick-up coil produces the voltage in the SQUID. The EMF-voltage is amplified, integrated and then fed back to SQUID in order to maintain to the lock point. Then the applied magnetic field is nullified. This integrated value becomes the output signal of the FLL circuit, and it is equal to the input magnetic signal applied to the pick-up coil.

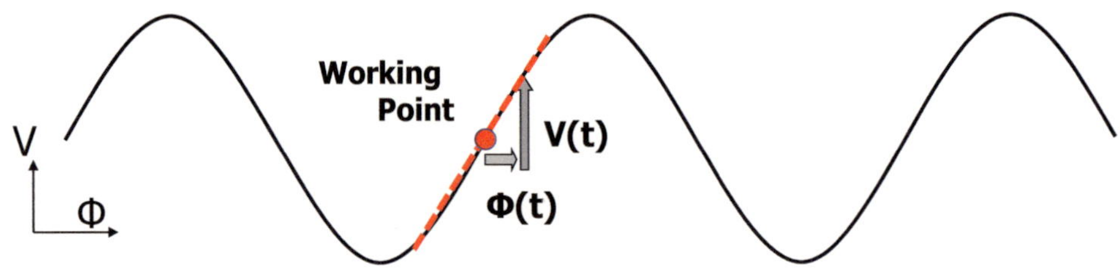

Fig. 8. Input-output (Φ- V) characteristic of SQUID.

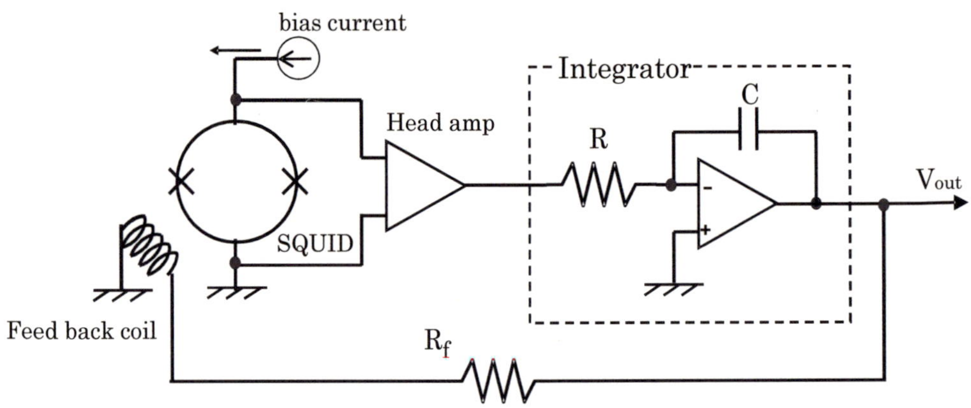

Fig. 9. Example of the FLL circuit

In recent years, it has been shown by several groups that the electronics within the digital SQUID, where the flux-locked loop system is controlled by a digital circuit, can achieve a wide dynamic range [21-23]. These FLL systems can measure biomagnetic signals in a magnetically unshielded environment. We call a digitally controlled flux-locked loop system (D-FLL).

A D-FLL system satisfies unshielded conditions: its dynamic range is expanded by jumping to a lock point on the SQUID's periodic Φ-V characteristic and by keeping a count of the number of such events. This method was named flux-quanta counting (FQC) [24, 25]. The integrator of the D-FLL is reset every time the feedback flux exceeds ±1 Φ_0 and the corresponding Φ_0 steps are counted. Usually, the original input flux is reconstructed on a digital signal processor that is a part of the FLL circuit. The dynamic range depends on the architecture of the digital signal processor, which is typically 32-bit. Furthermore, high-speed signal processing makes a digital feedback loop delay shorter. A high-performance digital signal processor is required to increase the dynamic range and slew rate. So far, it has proven difficult to make a high performance system with the typical single-chip microcontroller that is generally used in such circuits.

We have developed a double-counter equipped D-FLL system in Fig. 10. The double-counter method has two counters on a single-chip microcontroller and a host computer. The digital integrator unit calculates feedback flux data. If feedback flux exceeds ±0.5Φ_0, the integrator unit subtracts 1 Φ_0 from the digital integrator data or adds 1 Φ_0 to it, instead of resetting the FLL. The feedback loop process works more quickly than the data transmission from the one-chip microcontroller into a host computer. The integrator data and Couner1 data is sent at periodic time intervals defined by a timer. Counter1 is reset every time after this transmission is complete and commences the count of Φ_0 steps during the next interval. Counter2 integrates the data transmitted from Counter1. The host computer reconstructs the measured flux data using Counter2 as well as integrator data. Because a host computer reconstructs input flux, the workload of the single-chip microcontroller is decreased, leading to a quicker operation of the feedback loop than that of the system without a double-counter method. The high-performance D-FLL system using a single-chip microcontroller is achieved by a method that employs two counters.

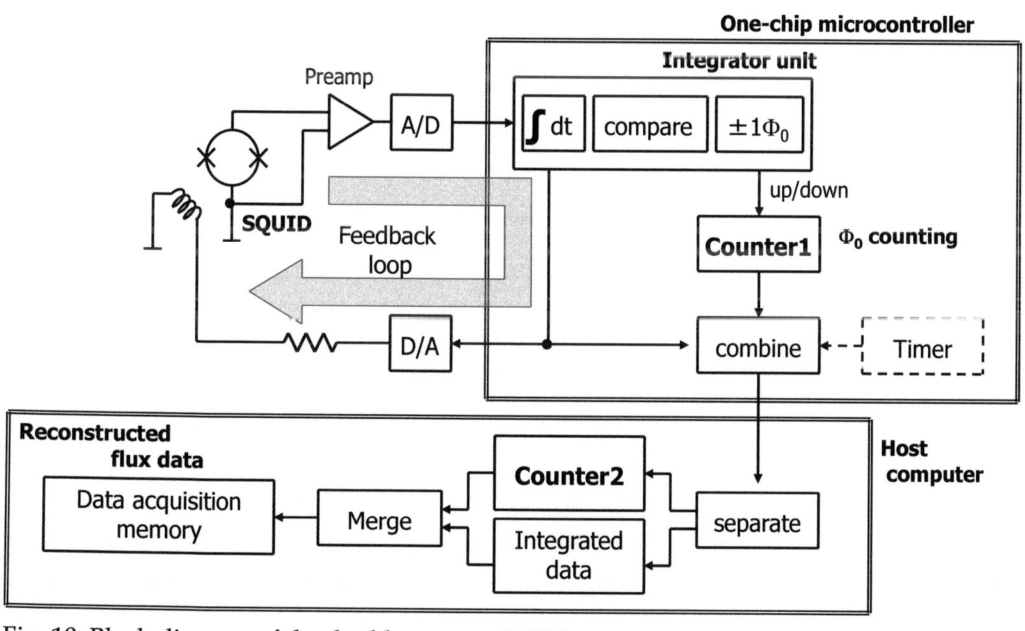

Fig. 10. Block diagram of the double-counter D-FLL system.

Our latest D-FLL system is composed of a commonly used 8-bit single-chip microcontroller, a 12-bit A/D converter, two 16-bit D/A converters and some other inexpensive and easily obtained devices [26]. The measured magnetic flux data is recorded at the rate of 1 kHz via an RS-232 interface. The time delay of the digital feedback loop is 8.4 μs. The first-order gradiometer is made of a low-T_c dc SQUID (niobium:Nb), has a pickup coil with a diameter of 17.8 mm and a baseline of 50 mm, and was used for the characteristic evaluations of the D-FLL system. The sensitivity which is calculated using the relation B/Φ_0 = 0.6 nT/μΦ_0, where B is the flux density in the pickup coil, gives a measure of the flux resolution of 11.35 fT/digit (19 μΦ_0/digit) for this system.

Figure 11 shows an active area of the D-FLL system. The terminology "active area" means that the system can measure magnetic signals just in this area. The right vertical axis is shown as a dynamic range when a 16-bit D/A converter is used for about 1 Φ_0 feedback flux. The measured dynamic range is 218 Φ_0 (141 dB) at 1 Hz, 22 Φ_0 (121 dB) at 10 Hz, and 2.1 Φ_0 (100 dB) at 100 Hz. In addition, the measured slew rate is 1.3 kΦ_0/s. The D-FLL system can operate in a stable manner if power line noise that is usually the main problem encountered with unshielded operations is less than 4.1 Φ_0 in amplitude at 50 Hz.

Figure 12 shows the measured D-FLL output noise spectra both inside and outside a magnetically shielded room (MSR). The lower line is measured in the MSR and the upper line is measured in an unshielded environment. The white noise level used was 36.1 fT/Hz$^{1/2}$ and the $1/f$ corner frequency was 2 Hz in an MSR. The noise spectrum measured outside an MSR contained some unidentified noises beside power line noise that exhibited 465 pT/Hz$^{1/2}$ (0.78 Φ_0/Hz$^{1/2}$) at 50 Hz.

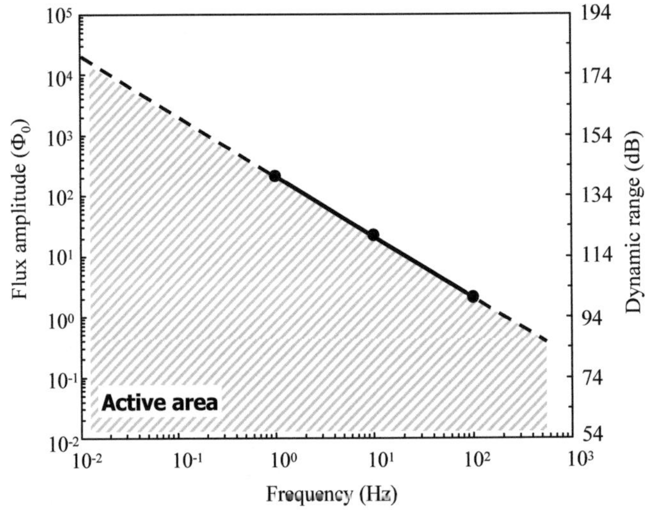

Fig. 11. Active area of the D-FLL system. An "active area" implies that the system can measure magnetic signal just in this area. The right vertical axis is shown as a dynamic range when 16 bit D/A converter is used for 1 Φ_0 feedback flux.

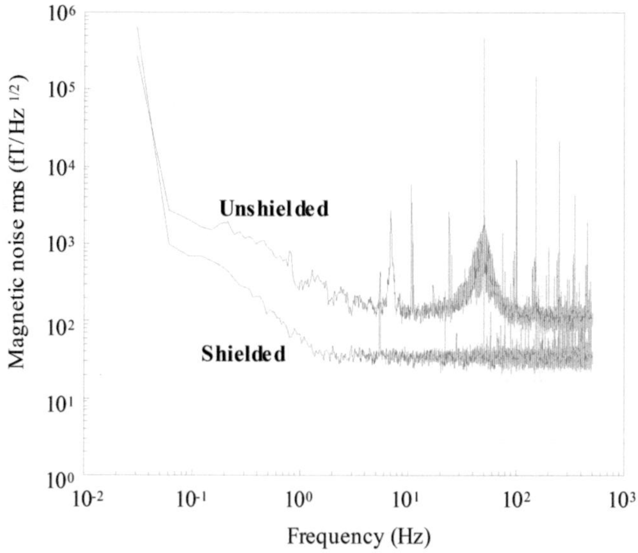

Fig. 12. Magnetic noise spectra measured in an MSR. The lower line is the measurement in an MSR and the upper line is from an unshielded environment.

5. Magnetocardiogram measurements

SQUIDs have been utilized for the detection of weak magnetic fields in our daily life from non-distraction evaluation to bio-magnetic field detection. Recently, the magnetocardiography (MCG) has attracted much attention as a potential application of SQUID, because MCG systems with SQUID provide higher sensitivity than other systems and because cardiac diagnosis plays an important role in the prevention of heart disease. Thus far, we have developed a multi-channel MCG system by using Nb SQUID [27]. Following that effort, we developed a conventional MCG system made of MgB_2.

Figure 13 shows the block diagram of a cryocooler system and the close-up of a cold cylinder. A commercially available pulse-tube cooler (Sumitomo RP-052D) was used. The cold head of the cryocooler is connected to the compressor via a valve unit. Because the valve unit produces magnetic noise, the cold head and the refrigerator component are installed in an MSR, and the valve unit is set outside of it. They are connected by a long copper tube.

The mechanical vibration of the refrigerator is ±3 μm at 1.2 Hz. The cold stage is connected to the refrigerator by a copper column to enable refrigeration. The distance between them is 608 mm.

This refrigerator can achieve a temperature of about 5 K without a load. Our cryocooler has two modes. The first mode is the SQUIDs characterization mode. This is used for characterization of MgB_2 SQUIDs. The cold stage and the cold cylinder are covered by a radiation shield, which is made of copper and has a thickness of 3.4 mm. The MgB_2 SQUID is set underneath the cold stage. The lowest temperature in this mode was 5.6 K.

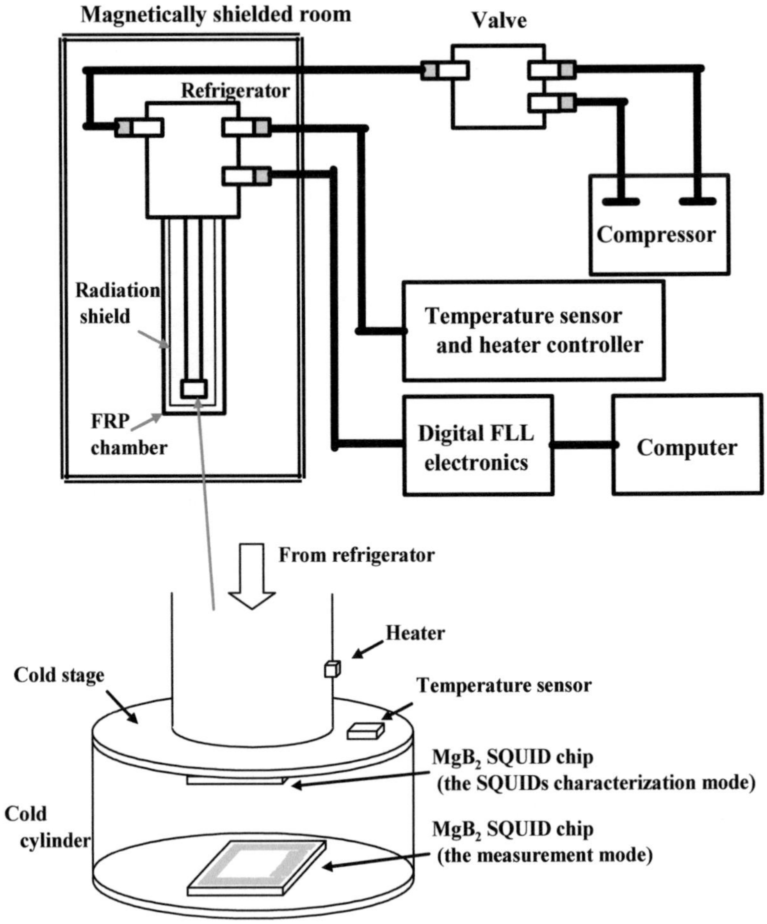

Fig. 13. Block diagram of a cryocooler system and the close-up of a cold cylinder. The MgB_2 SQUID is set under the cold stage or on the bottom plate of the cold cylinder when the system is used in either the SQUIDs characterization mode or the measurement mode.

The second mode is the measurement mode. This mode is used for measurement of an external magnetic field, as in the case of MCG. The cold stage and the cold cylinder are covered by super insulation films instead of a radiation shield. The MgB_2 SQUID is set on the bottom plate of the cold cylinder and has a thickness of 2.0 mm. The lowest temperature achieved in this mode was 11.9 K. A temperature sensor and a heater are placed on the cold stage and the column, respectively. They are connected to a PID controller, with which the temperature of the cold stage is controlled. Both the SQUID characterization mode and the measurement mode are controlled and operated by a computer through the D-FLL electronics.

Figure 14 shows the magnetic flux noise spectrum measured inside a single layer MSR. The white magnetic flux noise was 6.8 $pT/Hz^{1/2}$. The largest peak was caused by power lines of 50 Hz. Other peaks were observed at 1.2 Hz and its harmonics of that frequency. They were caused by the refrigerator and had a magnitude of 13.8 pT at 1.2 Hz.

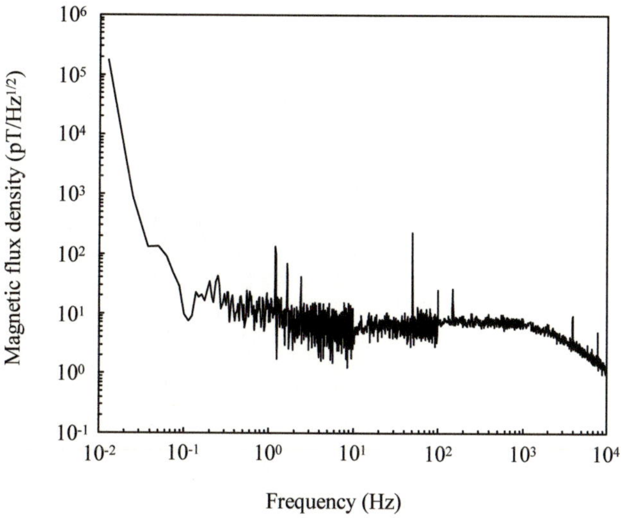

Fig. 14. Magnetic noise spectrum of the system measured inside an MSR.

Figure 15 shows the photograph of the actual measurement and the MgB_2 magnetometer that was used in the experiments. The magnetocardiogram was measured by our MgB_2 SQUID system inside the MSR. This measuring apparatus was cooled by a pulse tube refrigerator instead of liquid helium.

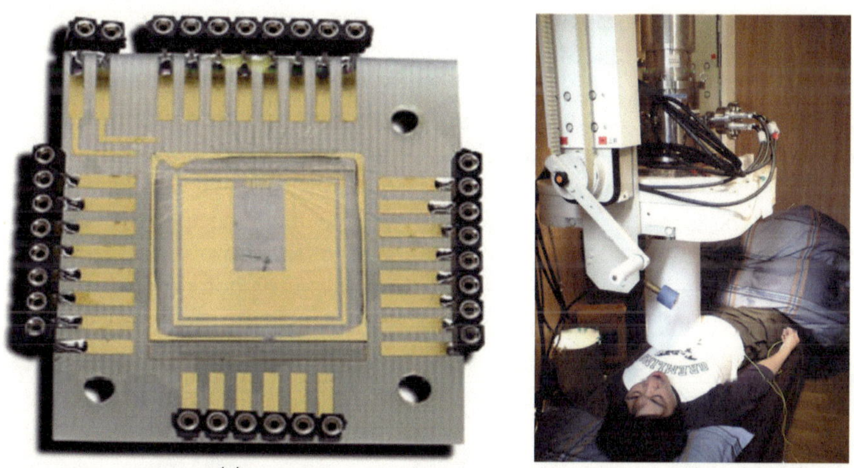

(a) (b)

Fig. 15. Photographs of the MgB_2-SQUID device and magnetocardiogram measurements.

Figure 16 shows a magnetocardiogram waveform of a healthy volunteer. This waveform was obtained by applying a digital low-pass filter whose cutoff frequency was 40 Hz and the waveform was averaged over 536 measurements [28]. We have succeeded in measuring the QRS complex and the T wave which is characterized in the activity of the human heart. This result shows a possible practical use of the MgB_2-SQUID MCG system cooled by refrigeration above 20 K without liquid cryogen, although the system is still under development.

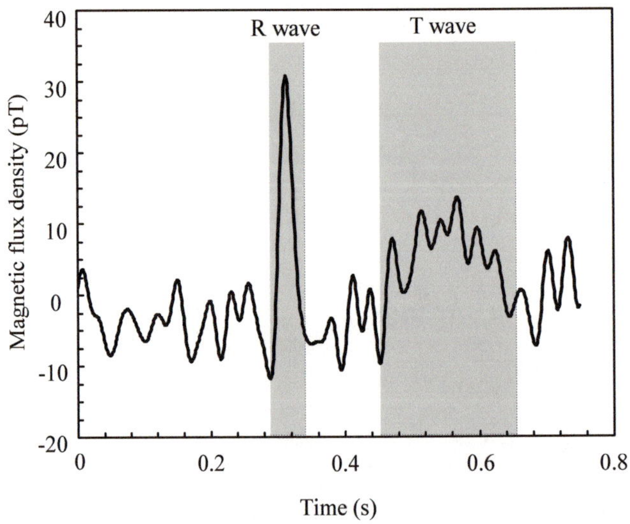

Fig. 16. Magnetocardiogram waveform measured by the MgB_2 SQUID system inside an MSR. The waveform was obtained by applying a software low-pass filter (40Hz) and 536 times averaging.

6. Summary

We built a prototype of a nano-bridge type MgB_2 SQUID device and optimized it for the measurement of magnetocardiogram. The results indicate that our method for the fabrication of SQUID of an MgB_2 magnetometer optimized for MCG measurement is effective. We are currently conducting research to reduce the noise from the refrigerator and to optimize the MgB_2 SQUID MCG system. A multi-channel type MCG system is also under development. The crystalline characteristics of MgB_2 films should be further improved to obtain better device performance. In the near future, the MgB_2 SQUID device is expected to have excellent performance. In addition to the SQUID device, we developed a high-frequency filter device. The detailed results of the filter device are shown in the papers of co-workers at Yamagata University [29]. Figure 17 shows the photographs of the filter device that was developed and studied.

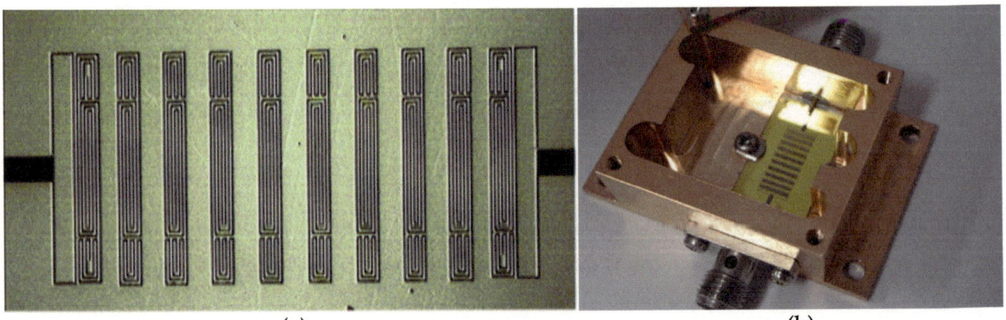

(a) (b)

Fig. 17. Photographs of the high-frequency filter device. The filter device was mounted on a Cu cavity.

7. Acknowledgments

The authors thank H. Yamaguchi, Y. Fujine, D. Oyama, K. Ikeda, S. Goto, T. Nakajima, T. Takahashi, H. Iriuda, T. Oba, A. Okubo, J. Araaki, K. Meguro, T. Abe, H. Endo, Y. Uchikawa and K. Fujisawa for the cooperation of MgB_2 SQUID project, and M. Iitake for his technical assistance of FIB process. This work was supported by the Japan Science and Technology Agency. A part of this work was conducted at the AIST Nano-Processing Facility, supported by "Nanotechnology Network Japan" of the Ministry of Education, Culture, Sports, Science and Technology (MEXT), Japan.

8. References

[1] Nagamatsu, J., Nakagawa, N., Muranaka, T., Zenitani, Y. & Akimitsu, J. (2001). Superconductivity at 39K in magnesium diboride, *Nature* 410: 53.

[2] Buzea, C. & Yamashita, T. (2001). Review of the superconducting properties of MgB_2, *Supercond. Sci. Technol.* 14: R115.

[3] Kang, W. N., Kim, H. J., Choi, E. M., Jung, C. U. & Lee, S. I. (2001). MgB_2 superconducting thin films with a transition temperature of 39 Kelvin, *Science* 292: 1521.

[4] Eom, C. B., Lee, M. K., Choi, J. H., Belenky, J. J., Song, X., Colley, L. D., Naus, M. T., Patnaik, S., Jiang, J. J., Rikel, M., Polyanskii, A., Gurevich, A., Cai, X. Y., Bu, S. D., Babcock, S. E., Hellstrom, E. E., Larbalestier, D. C., Rogado, N., Regan, K. A., Hayward, M. A., He, T., Slusky, J. S., Inumaru, K., Haas, M. K. & Cava, R. J. (2001). Thin film magnesium boride superconductor with very high critical current density and enhanced irreversibility field, *Nature* 411: 558.

[5] Zeng, X., Pogrebnyakov, A. V., Kotcharov, A., Jones, J. E., Xi, X. X., Lysczek, E. M., Redwing, J. M., Xu, S., Li, Q., Lettieri, J., Schlom, D. G., Tian, W., Pan,X. & Liu, Z. K. (2002). In situ epitaxial MgB_2 thin films for superconducting electronics, *Nature Mater.* 1: 35.

[6] Ueda, K. & Naito, M. (2003). In situ growth of superconducting MgB_2 thin films by molecular beam epitaxy, *J. Appl. Phys.* 93: 2113.

[7] Jo, W., Huh, J.-H., Ohnishi, T., Marshall, A. F., Beasley, M. R. & Hammond, R. H. (2002). In situ growth of superconducting MgB_2 thin films with preferential orientation by molecular beam epitaxy, *Appl. Phys. Lett.* 80: 3563.

[8] Erven, A. J., Kim, T. H., Muenzenberg, M. & Moodera, J. S. (2002). Highly crystallized as-grown smooth and superconducting MgB_2 films by molecular beam epitaxy, *Appl. Phys. Lett.* 81: 4982.

[9] Kobayashi, Y., Doi, T., Kitaguchi, H., Okuzono, M., Nagatomo, K., Hamada, S. & Hakuraku, Y. (2005). Preparation of MgB_2 thin films by an electron-beam evaporation technique and post-annealing effect on the as-grown films, *TEION KOGAKU* 40: 7 (in Japanese).

[10] Liu, Z.-K., Schlom, D. G., Li, Q. & Xi, X. X. (2001). As-grown superconducting MgB_2 thin films prepared by molecular beam epitaxy, *Appl. Phys. Lett.* 78: 3678.

[11] Sakata, O., Kimura, S., Tanaka, M., Yata, S., Kato, T., Yamanaka, K., Yamada, Y., Matsushita, A. & Kubo, S. (2004). High-quality as-grown MgB_2 thin-films fabrication at a low substrate temperature using an in-plane-lattice near-matched epitaxial-buffer-layer. *J. Appl. Phys.* 96: 3580.

[12] Tsujimoto, K., Shimakage, H., Wang, Z. & Kaya, N. (2005). Crystallinity and superconductivity of as-grown MgB_2 thin films with AlN buffer layers, *Physica C* 426-431: 1464.

[13] Harada, Y., Takahashi, T., Kuroha, M., Iriuda, H., Nakanishi, Y., Izumida, H., Endo, H. & Yoshizawa, M. (2006). Fabrication of as-grown MgB_2 films on ZnO (0001) substrates by molecular beam epitaxy, *Physica C* 445-448: 884.

[14] Harada, Y., Yamaguchi, H., Oba, Y., Fujine, Y., Goto, S., Iriuda, H. & Yoshizawa, M. (2007). Novel stage in fabrication of as-grown MgB_2 films by adopting Ti seed layer, *Physica C* 463-465: 945.

[15] Nishidate, K., Yoshizawa, M. & Hasegawa, M. (2008). Energetics of Mg and B adsorption on polar zinc oxide surfaces from first principles, *Phys. Rev. B* 77: 035330.

[16] Haruta, M., Fujiyoshi, T., Kihara, S., Sueyoshi, T., Miyahara, K., Harada, Y., Yoshizawa, M., Takahashi, T., Iriuda, H., Oba, T., Awaji, S., Watanabe, K. & Miyagawa, R. (2007). High critical current density under magnetic fields in as-grown MgB_2 thin films deposited by molecular-beam epitaxy," Supercond. Sci. Technol. 20 (2007) L1.

[17] Ueda, K. & Naito, M. (2004). Tunnel junctions on as-grown MgB_2 films, *Physica C* 408–410: 134.

[18] Oba, T., Goto, S., Sasaki, S., Nakanishi, Y., Fujino, Y., Harada, Y., Nakamura, M., Saito, A. & Yoshizawa, M. (2008). Fabrication of all-MgB_2 Josephson junctions using MgO insulator layer, *Physica C: Superconductivity* 468: 1892.

[19] Gonnelli, R. S., Daghere, D., Calzolari, A., Ummarino, G. A., Tortello, M., Stepanov, V. A., Zhigadlo, N. D., Rogacki, K., Portesi, C., Monticone, E., Mijatovic, D., Veldhuis, D. & Brinkman, A. (2006). Recent achievements in MgB_2 physics and applications: A large-area SQUID magnetometer and point-contact spectroscopy measurements, *Physica C* 435: 59.

[20] Harada, Y., Oyama, D., Fujine, Y., Goto, S., Ikeda, K., Nakashima, T., Meguro, K., Kobayashi, K. & Yoshizawa, M. (2008). Fabrication of DC-SQUIDs of As-grown MgB_2 films, *TEION KOGAKU* 43: 354 (in Japanese).

[21] Drung, D. (2003). High- Tc and low-Tc dc SQUID electronics, *Supercond. Sci. Technol.* 16: 1320.

[22] Ludwig, C., Kessler, C., Steinfort, A. J. & Ludwig, W. (2001). Versatile high performance digital SQUID electronics, *IEEE Trans. Appl. Supercond.* 11: 1122.

[23] Schurig, T., Drung, D., Bechstein, S., Beyer, J. & Ludwig, F. (2002). High-Tc superconductor dc SQUIDs for unshielded operation and their applications, *Physica C*, 378-381: 1378.

[24] Vrba, J., Betts, K., Burbank, M., Cheung, T., Fife, A. A., Haid, G., Kubik, P. R. & Weinberg, H. (1993). Whole cortex, 64 channel SQUID biomagnetometer system, *IEEE Trans. Appl. Supercond.* 3: 878.

[25] Vrba, J. (1996). SQUID gradiometers in real environments SQUID Sensors: Fundamentals, Fabrication and Applications, (NATO ASI Series E329) ed H. Weinstock (Dordrecht: Kluwer) 117.

[26] Oyama, D., Kobayashi, K., Yoshizawa, M. and Uchikawa, Y. (2006). Development of digital FLL system for SQUID using double counter method, *IEEE Transactions on Magnetics* 42: 3539.

[27] Yoshizawa, M., He, D. F., Nakai, K., Kobayashi, K., Nakamura, Y., Yaegashi, M., Ito, M., Yashiro, H., Daibo, M., Simizu, T., Uchikawa, Y. & Noto, K. (2005). Application of SQUIDs in Iwate CREATE project, *Physica C* 426-431: 1572.

[28] Oyama, D., Harada, Y., Fujine, Y., Kobayashi, K., Uchikawa, Y. & Yoshizawa, M. (2009). Detection of MCG signal by using MgB_2 SQUID on pulse-tube cryocooler, *Intermag Europe* GO-04.

[29] Ono, S., Kimura, M., Sekiya, N., Saito, A., Hirano, S., Ohshima, S., Harada, Y., Takahashi, T., Iriuda, H. & Yoshizawa, M. (2007). Design and fabrication of 5GHz miniaturized bandpass filters using superconducting microstrip quasi-spiral resonators, *IEEE. Trans. Appl. Supercond.* 17: 89.

19

Thermal Behaviour and Refrigeration of High-Temperature Superconducting Fault Current Limiters and Microlimiters

J.A. Veira, M.R. Osorio* and F. Vidal
LBTS, Departamento de Física da Materia Condensada, Universidade de Santiago de Compostela, Santiago de Compostela Spain

1. Introduction

Superconducting fault current limiters (SFCL) based on high temperature superconductors (HTSC) of different types, inductive, resistive or hybrid, have being extendedly studied (Noe & Steurer, 2007; Sokolovsky et al., 2004; Paul et al., 2001) and tested in power plants (Paul et al., 1997; Leung et al., 1997; Zueger et al., 1998). Inductive SFCL, intended to operate at very high powers, are in general based on bulk HTSC whereas resistive or hybrid types are based on both bulk (mainly in the shape of bars) and thin films grown on different types of substrates and mainly patterned in meandered paths. HTSC based limiters represent one of the more promising applications of such materials but both bulk and thin film HTSC present different drawbacks which make difficult the practical implantation of these devices. Important problems concern, in particular, to the thermal behaviour of samples under current faults. In bulk samples, which are quite inhomogeneous, specially the huge elements required for limiters operating at high powers, overcurrents under fault generate hot spots in weak zones which endanger their thermal stability and lead to an irreversible damage or to an unsatisfactory performance of the device (low impedance, long recovery times once the fault is cleared, etc.).

In terms of thermal behaviour, thin films could show a better performance than bulk samples. Their critical current density and electric field above their normal-superconducting transition are around two orders of magnitude higher than those for bulk samples. Accordingly, the heat developed per unit volume under a current fault is also orders of magnitude higher. This dissipation, together with the small thickness of films, should provoke a homogeneous and full transition of the overall sample showing, therefore, a very fast and abrupt increase of the resistance (Antognazza et al., 2002). In addition, the large surface-to-volume ratio of thin film samples, the low boundary thermal impedance film-substrate and the great thermal conductivity of substrates determine a very good heat evacuation once the current fault is cleared. Hence the recovery times observed in limiters based on thin films are much shorter than for limiters based on bulk samples (Osorio et al., 2006). In what concerns the transition to the normal state it may occur, however, that indeed only some parts of the film meander

*Present address: Laboratorio de Bajas Temperaturas, Departamento de Física de la Materia Condensada, Universidad Autónoma de Madrid, Spain.

are triggered to the normal state by the current fault (Paul et al., 2001). In that case dissipation can be so concentrated in some of these parts that films could also be irreversibly damaged. To prevent that some weak parts of the long meander thin film paths be burning out, it has been proposed to fabricate these elements by patterning a set of constrictions alternated with wider portions. With this design, the normal transition would formerly occur in the narrower segments and the heat arising from this overheated sections will rapidly propagate along the whole meander path (i.e. including also the wider parts), hence leading to a complete and more homogeneous transition (Decroux et al., 2003; Antognazza et al., 2005; Duron et al., 2005; 2007; Antognazza et al., 2007; Duron et al., 2007).

For both bulk and thin film based SFCL the thermal behaviour and refrigeration of the superconducting elements determine, therefore, the overall performance of these devices. In this work we will summarize some of our results on this issue including the thermal recovery of inductive SFCL operating with bulk cylinders of different sizes (Osorio et al., 2006) or with artificial weak zones (Osorio et al., 2005; 2008), the improvement obtained by using stacks of bulk rings or thin film washers as secondary (Osorio et al., 2004; 2006), the use of thermoacoustic refrigerators (Osorio et al., 2008) and the possibility of using thin film microbridges as very efficient microlimiters in very low power applications (Lorenzo et al., 2009).

2. Bulk cylinders as secondary of inductive SFCL

Despite the advances in the fabrication of HTSC materials the presence in the samples of inhomogeneities which act as weak zones still represents an important drawback for their use in SFCL. The presence of weak zones provokes an earlier transition to a dissipative state of only a part of the samples that, as indicated above, can cause a great degradation of the performance of the overall SFCL device. In the first part of this section we will summarize our results on the non-homogeneous quench of bulk Bi-2223 cylinders acting as secondaries of inductive SFCL, in particular, on the current limitation capacity under fault of the limiters and their thermal recovery once the current fault is cleared. In the second part we will outline our numerical approaches to account for the effects of the bi-phase character of Bi-2223 samples on the quench under a current fault.

2.1 Non-homogeneous quench of the superconducting secondary of inductive SFCL

2.1.1 Samples' characteristics

In our studies of inductive SFCL based on bulk HTSC we have used samples of nominal composition $Bi_{1.8}Pb_{0.26}Sr_2Ca_2Cu_3O_{10+x}$. This samples were obtained from cylinders of 120 mm in height, 21 mm in internal diameter and 2.4 mm in wall thickness supplied by Can Superconductors (Czech Republic). Commercial samples were cut in shorted cylinders and rings, around 40 mm and 5 mm in height, respectively, to carry out our measurements. Also, other small pieces were properly cut for material characterization: electrical resistivity, characteristic voltage-current curves (CVC), magnetic susceptibility and x-ray diffraction. The results of our characterization tests show a very good reproducibility for different pieces of the same commercial cylinders and also for pieces obtained from different cylinders.

Typical examples of electrical resistivity versus temperature curves, $\rho(T)$, CVC and magnetic susceptibility $\chi(T)$ are displayed in figures 1 to 3. In figure 1 the dotted line represents the linear fir to the normal state resistivity data. In the inset it is shown the width of the

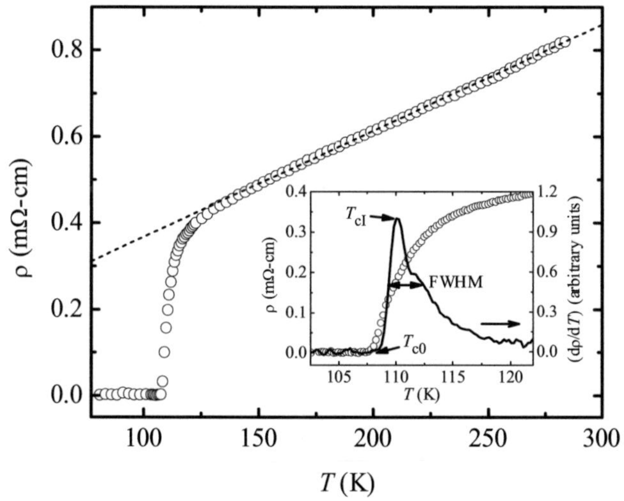

Fig. 1. Resistivity as a function of temperature for a piece of the material employed as secondary of the SFCL. Inset: detailed view of the superconducting transition. The solid line represents the temperature derivative (on arbitrary units). The critical temperatures T_{cI} and T_{c0} are defined as the ones corresponding to the maximum of the derivative and the minimum value for which $\rho \neq 0$, respectively. The full width half maximum (FWHM) is the width of the derivative at half height. Figure from Ref. (Osorio et al., 2004).

transition, defined as the width of the derivative $d\rho/dT$ (solid line) at medium height. T_{cI} the temperature at which $d\rho/dT$ presents its maximum, and T_{c0}, the minimum temperature for which $\rho \neq 0$ are indicated. In the CVC shown in figure 2, symbols represent the experimental data measured at the indicated temperatures and the solid lines the fits to the usual power law function (Prester, 1998; Osorio et al., 2004). In figure 3 circles correspond to the field cooling (FC) of the sample and squares to its zero field cooling (ZFC).

X-ray diffraction analyses (Osorio et al., 2004) show the presence in our samples of two phases: around 15% of Bi2212 and 85% of Bi2223. In $\rho(T)$ or CVC it cannot be distinguished any effect provokated by this inhomogeneity, as may be expected because of the percolative nature of electrical transport properties. It is, however, well visible in the $\chi(T)$ curves. As can be seen in the FC and ZFC measurements of figure 3, the normal-superconducting transition occurs in two steps: one around 108 K (corresponding to Bi2223 phase) and the the other around 98 K (corresponding to Bi2212 phase). We will se below that to numerically account for the electrical resistance and temperature of our samples operating as secondaries in our inductive SFCL under current fault, the bi-phase nature of the material plays a central role.

2.1.2 Samples' behaviour under a current fault

The electrical resistance and temperature evolution under a current fault of cylinders or rings when acting as secondaries in an inductive limiter was measured by attaching to the samples some voltage contacts and thermocouples along their perimeter as indicated in figure 4. Details on the limiter configuration and global experimental set up can be seen in reference (Osorio et al., 2004). In figures 5 and 6 we show typical results for this type of test. In this example data were measured by using a 5 mm in height ring with five couples of voltage

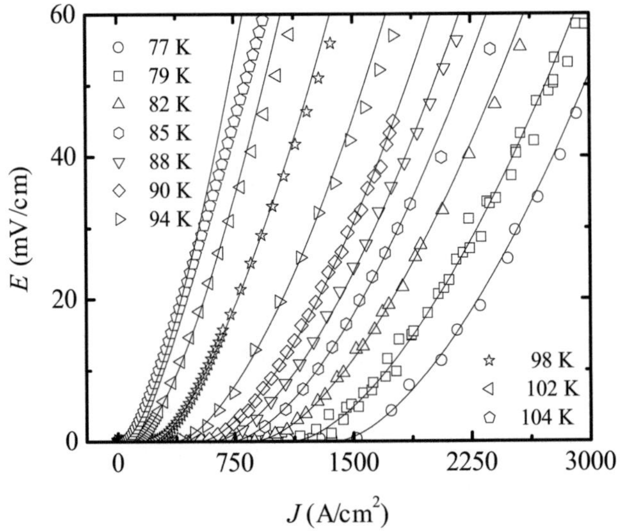

Fig. 2. $E - J$ curves measured in a piece of the superconducting elements employed in the limiter. The solid lines represent the fittings to the usual power law function (Prester, 1998). Figure from Ref. (Osorio et al., 2004).

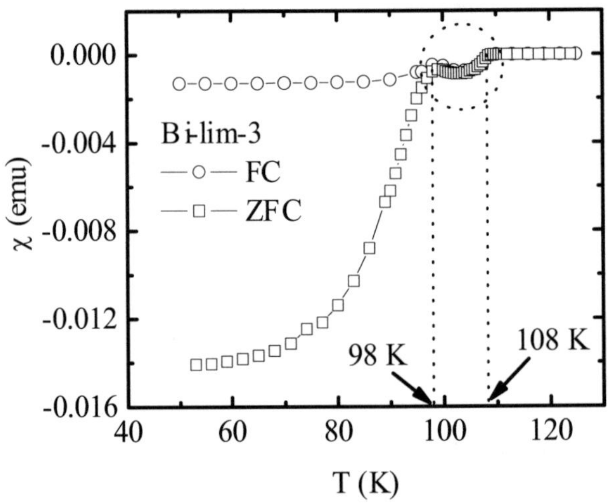

Fig. 3. Variation with temperature of the magnetic susceptibility during field cooling and zero field cooling for a piece of the cylinder used as secondary of the limiter. Notice the presence of two superconducting phases, both identified by their own critical temperatures. Figure from Ref. (Osorio et al., 2004).

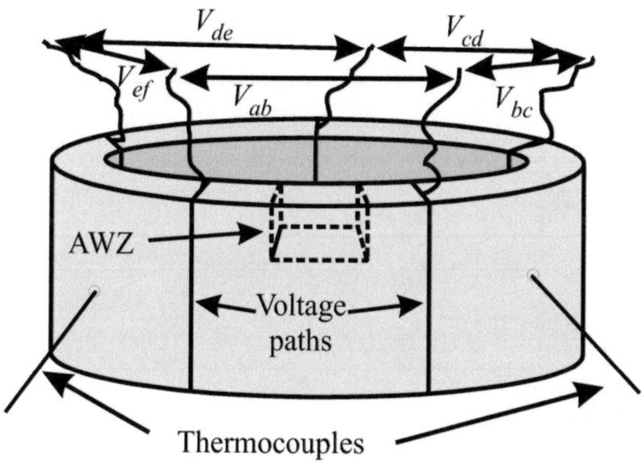

Fig. 4. Scheme of division into regions (five in this example) of a ring with voltage paths and thermocouples attached on each region. The dashed lines indicate the artificial weak zone made by reducing the cross section of the ring in a small portion of its perimeter. Figure from Ref. (Osorio et al., 2004).

contacts and five thermocouples. Figure 5 shows the measured wave-forms. For clarity we use lines instead of symbols and represent only three of the five signals.

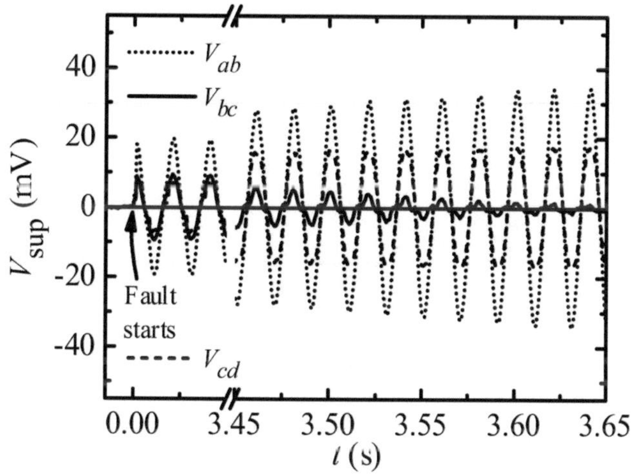

Fig. 5. Voltage measured during a fault with the arrangement depicted in figure 4. The different regions do not behave in the same way, as one of them cools down while others enter in the ohmic state. Figure from Ref. (Osorio et al., 2004).

In figure 6(a) it is shown the rms values of the five voltage signals and in figure 6(b) the corresponding temperature readings. We can observe that the voltage signals of some zones grow continuously during the fault but others experience the inverse tendency, so the voltage is progressively reduced up to typical values of the flux-flow regime and even up to zero indicating that these zones return to the superconducting state. The corresponding temperatures, displayed in figure 6(b), also indicate this tendency: the parts at high voltage heat up to temperatures well above T_{cI} whereas the parts at low voltage scarcely heat up remaining at temperatures below T_{c0}. In this sample, for the region denoted *cd*, the temperature reading is lower that the one that could be expected by inspecting the rms

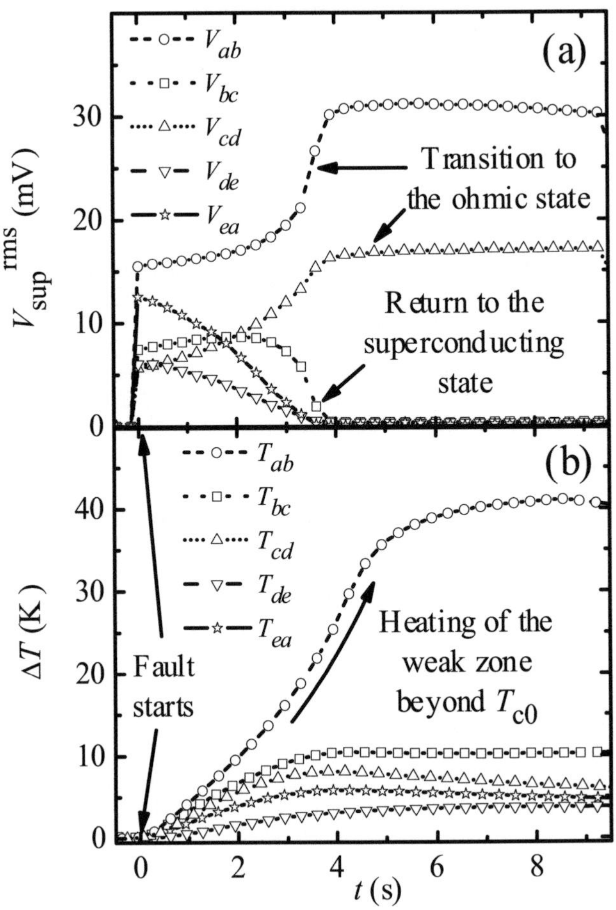

Fig. 6. (a) Rms voltages for all the considered regions of the ring of figure 4. The major part of the sample recovers the superconducting state as the current is limited by the dissipative zones. (b) Temperature excursions obtained for an identical fault. As it can be seen, it exists a remarkable coherence between these results and the voltage measurements. See text for details. Figure from Ref. (Osorio et al., 2004).

voltage. This disagreement could be due to the presence of a relatively small thermal inhomogeneity in this zone, or to an inhomogeneity located in depth, relatively far away from the point at which the thermocouple was attached. In all the studied rings it was found a voltage signal behaviour like the one followed by the voltages denoted *ab*, clearly due to the thermal evolution during the fault of the more dissipative zone in each sample.

The results obtained in this type of test are very similar to those of figure 7 measured by using a ring in which a weak zone was created by reducing its height somewhere (as indicated by dotted lines in figure 4). The symbols represent the temperatures measured with thermocouples attached at the weak zone and at points at ± 1 cm and ± 2 cm far away along the ring perimeter. When the fault is provoked, so the induced current overtakes the critical current, I_c, the whole superconductor enters in the dissipative state, but the temperature

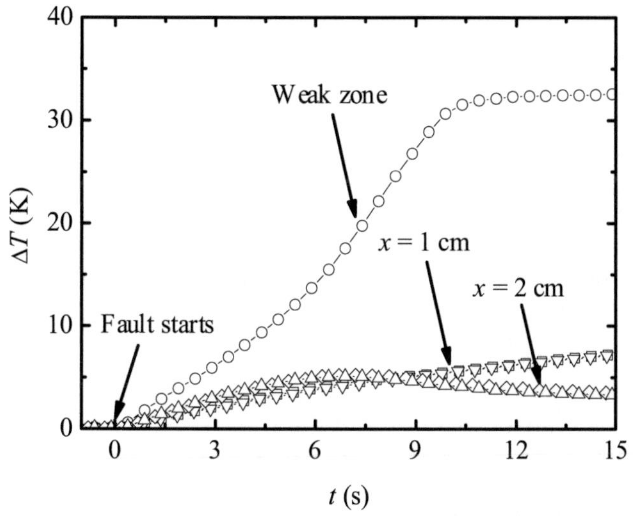

Fig. 7. Temperature excursions when provoking the transition (by exceeding the critical current) of the ring of figure 4 (once the AWZ has been made). The readings of the thermocouples located 1 cm away from the hot spot are denoted by down-triangles and squares, and the ones for 2 cm away by up-triangles and diamonds. Notice that, while the majority of the ring heats up scarcely, and it remains very far from the ohmic state, the dissipative zone (circles) warms up till around the critical temperature. Then it is reached a stable plateau till the end of the fault at $t = 30$ s. The fact that, during the first seconds, the temperature increment is slightly greater at $x = 2$ cm than at $x = 1$ cm can be due to the local inhomogeneities. Figure from Ref. (Osorio et al., 2004).

increment in the majority of the ring is quite low, as the thermocouples indicate, and only in the weak zone it is reached the critical value, T_c. In fact, the temperature increment far away from the weak zone should be due to heat coming from the dissipative zone, instead of being generated in the corresponding parts of the sample. This fact would explain why at 1 cm away from the centre of the weak zone the temperature increases slowly when it has been already reached a stable plateau in the hottest region. Notice that the showed behaviour, with the decreasing temperature in the further parts of the ring, means that the warm zone does not tend to expand itself, but remains essentially still till the end of the fault (at about $t = 30$ s). This fact is in part a consequence of the reduction of the induced current, due to the gain in impedance originated by the very resistive zone. As the current that circulates in the ring diminishes, the dissipation becomes lower, and so the thermal excursion in the majority of the sample. This fact provokes that the resistance of the superconductor, and so the total impedance of the limiter, increase quite smooth.

2.2 The effects of the bi-phase character of samples on the quench

The behaviour of inductive SFCL is mainly determined by the overheating of the weak zone in the superconducting cylinder. In turn, the behavior of this zone seems to be mainly dominated by the minority phase Bi-2212, that has the lower critical temperature $T_{c0} \approx 98$ K, as indicates the change of slope observed in the voltage curves of figure 6(a), which corresponds to $\Delta T \approx 20$ K, i.e. at $T \approx 98$ K, in the temperature versus time curves plotted in figure 6(b).

The role of the minority phase in the thermal evolution of the superconducting element during a fault can be illustrated by using a very simple approach for the temperature distribution in the superconductor. The calculations are done by using an iterative numerical routine in Matlab (The MathWorks, Inc, USA). The magnetic circuit is represented through field variable inductance coefficients (Osorio et al., 2004; MITStaff, 1961; Paul et al., 1995) and the superconducting element is modeled by taking into account the experimental characterization data, as it will be described later. The input variable is the applied voltage, kept as constant, and the fault is simulated by reducing the impedance of the circuit. During the iterative process, the circulating currents (I_p and I_{sup}) are first calculated at a time t, and then the resistivity and the temperature increment of the superconducting element. These steps are repeated till it is reached a set of self-consistent values, so the routine advances to time $t + \delta t$ and so on.

We suppose that the cylinder is divided into two temperature regions, as indicated in figure 8. The region around the weak zone is considered to be at a higher temperature (hot domain) than the remainder part of the cylinder (cold domain). The length, temperature and resistance of each domain are denoted, respectively, by ℓ^h, T^h, $R^h \equiv (\rho^h(I_{sup}, T^h)\ell^h)(A_\perp)^{-1}$ and ℓ^c ($= 2\pi r - \ell^h$, being r the radius of the cylinder), T^c, $R^c \equiv (\rho^c(I_{sup}, T^c)\ell^c)(A_\perp)^{-1}$, where $\rho^{(c,h)}$ represents the resistivity of each domain and $A_\perp$ the cylinder cross section.

The temperature evolution was calculated by applying to both regions the heat balance

$$\varrho A_\perp \ell^{(c,h)} C_p(T^{(c,h)}) \frac{\partial T^{(c,h)}}{\partial t} = \dot{\mathcal{W}} - \dot{\mathcal{Q}} \tag{1}$$

$\dot{\mathcal{W}}$ and $\dot{\mathcal{Q}}$ being the power dissipated inside each superconducting region and the convective flux through their surfaces, respectively, which can be expressed as

$$\dot{\mathcal{W}} = I_{sup}^2 R^{(c,h)}(I_{sup}, T^{(c,h)})$$

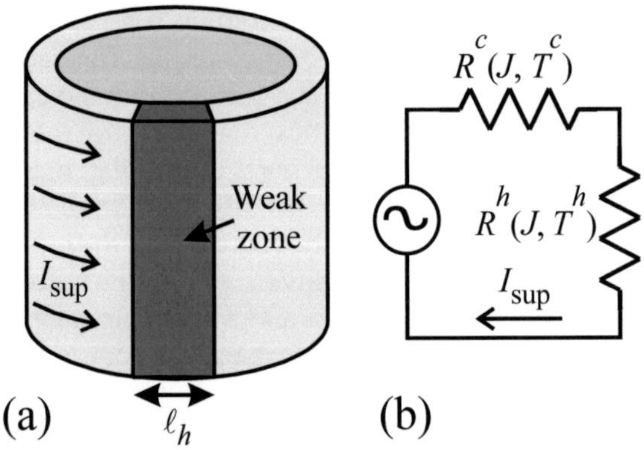

Fig. 8. (a) Scheme of the model supposed for an inhomogeneous superconducting cylinder, which is divided into two regions. The dark one is the hot domain, denoted by the index h (for the cold one stands c), which it is supposed to have a defect that generates a more dissipative zone. Notice that this region occupies only a small portion of the cylinder perimeter, but its whole height, so the current that goes through both parts is the same. (b) Equivalent circuit of the inhomogeneous superconductor. The resistances of both parts are joint in series. Figure from Ref. (Osorio et al., 2004).

and

$$\dot{Q} = hP\ell^{(c,h)}(T^{(c,h)} - T_b),$$

where ϱ is the density of the material, $C_p(T)$ its heat capacity (Xiao et al., 1999), h the convective factor, which can be taken as temperature independent with a small error, T_b is the environmental temperature and P is the cross section perimeter of the cylinder. The circulating current, I_{sup}, depends on the resistances, $R^{(c,h)}$, and so on the length of the hot domain, ℓ^h, which is set as a free parameter in our calculations. It has been also supposed, for the sake of simplicity, that during the fault the heat flow along the cylinder is negligible, in agreement with the low thermal conductivity of this material and with our experimental results.

The lines plotted in figure 9 represent the temperatures determined in this scenario. The experimental data are those corresponding to the ab region in figure 6. The dashed and short-dashed lines were obtained in the framework of the effective-medium-theory (Osorio et al., 2004; Davidson & Tinkhan, 1976), considering that the effective electrical resistivity, ρ_e, of each domain can be approached according to the concentrations of the majority and the minority phases (C_1 and C_2, respectively), the depolarization factor for the minority phase inclusions (X) and the corresponding resistivities, for which it is used the relations

$$\rho(J,T) = \frac{E(J)}{J} \quad \text{for } T < T_{c0} \tag{2}$$

$$\rho_{\text{normal}}(T) \quad \text{for } T \geq T_{c0}, \tag{3}$$

and with $\rho_{\text{normal}}(T) = 103 + 2.3T \ \mu\Omega - cm$ (i.e., the same normal resistivity displayed in figure 1). The value of the critical temperature, T_{c0}, must be chosen depending on the corresponding phase.

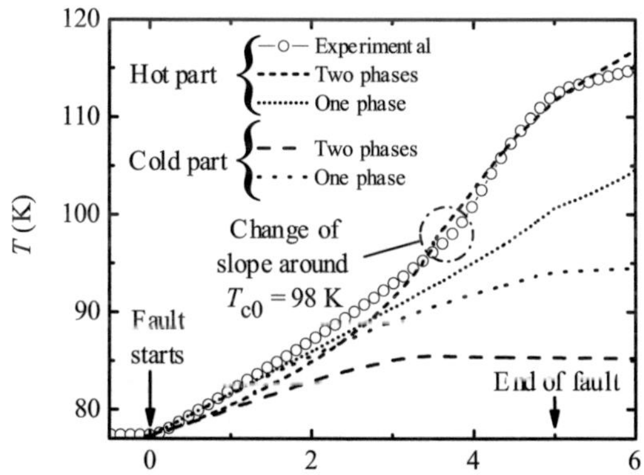

Fig. 9. Temperature excursions simulated when it is supposed that it exists a hot spot created by a slightly different resistivity in a 10% of the perimeter of the cylinder. It is considered two cases: in the first one the superconductor is a pure Bi-2223 phase, while in the second one it is also considered as a Bi-2223 matrix with Bi-2212 inclusions of spherical shape homogeneously distributed in the sample. It is also included the experimental data for the temperature excursion of the ab region of figure 6, which allows fixing the free parameters of the model (i.e. the resistivity and the length of the weak zone). Figure from Ref. (Osorio et al., 2004).

In our calculations (Osorio et al., 2004) we have used the values $X = \frac{1}{3}$ (i.e., we assume that the inclusions are spherical), $C_1 = 0.85$ and $C_2 = 0.15$, in accordance with the x-ray analysis already presented. We have also left the resistivity of the hot domain to be slightly higher than the one of the cold part at each temperature. For that, for ρ_e^h, we have used $\rho_e^h(T^h) = \alpha\rho_e^c(T^h)$ ($\alpha \geq 1$) and carried out the calculations with α, altogether with the length ℓ^h, as free parameters. The dashed and short-dashed lines of figure 9 correspond to the best fit with $\alpha = 1.01$ and $\ell^h = 0.1 \times 2\pi r$. We must note that the precise values of C_1, C_2 and X do not change the main results from our approach. Small variations of these parameters should be balanced by the free ones used in our calculations.

As it can be seen, the general behaviour of the thermal evolution of the hot part is roughly accounted for, in particular the slope changes of the $T(t)$ curve during the fault. The thermal excursion calculated for the cold part of the cylinder is also in good agreement with the data displayed in figures 6 and 7 measured in the parts far away from the weak zones. The dotted lines in figure 9 represent the thermal evolution of the hot (short-dotted) and cold (dotted) parts, calculated by assuming a cylinder composed of only Bi-2223 phase. Clearly, in this scenario the experimental data are not properly accounted for.

Despite the crudeness of our approach, our results indicate that the thermal evolution of the superconducting cylinder during a fault is mainly determined by the minority phase. This behaviour represents one of the most important drawbacks of the inhomogeneous materials to be used for fault current limiter devices, as the limitation can be carried out by only a small part of the superconducting element. This would imply a useless waste of material and, for high values of the critical current density, a huge dissipation in the surroundings of the weakest parts which could lead to a further degradation or even to a local melting. Therefore, work should be done in order to improve the homogeneity of the superconducting samples employed in SFCLs or by designing special configurations which allow to reduce the importance of this problem.

We must finally note that the global resistance of the cylinder, obtained according to the scheme of figure 8(b) by using the expression

$$R(J, T^c, T^h) = \rho_e^c(J, T^c)\left(\frac{\ell^c}{A_\perp}\right) + \rho_e^h(J, T^h)\left(\frac{\ell^h}{A_\perp}\right), \tag{4}$$

roughly reproduces the experimental data (Osorio et al., 2004).

In figure 10, symbols represent the temperature measured in the weak zone during the thermal recovery. The lines represent the temperature determined with our numerical routine, which in this case is, in principle, very simple. The current is set to zero and the temperature is let to vary according to the convective exchange of heat, dashed line, or to a convective plus conductive model (to the neighbouring cold parts), solid line. It is well noticeable that a conductive exchange term between the weak zone and the rest of the cylinder must be included to account for the temperature diminution.

3. Thermal behavior of an inductive SFCL whose secondary has artificial weak zones

The results shown in figure 10 indicate that the convection mechanism is not enough to explain the cooling down of the weak zone once the fault is cleared, but conduction to the cold parts

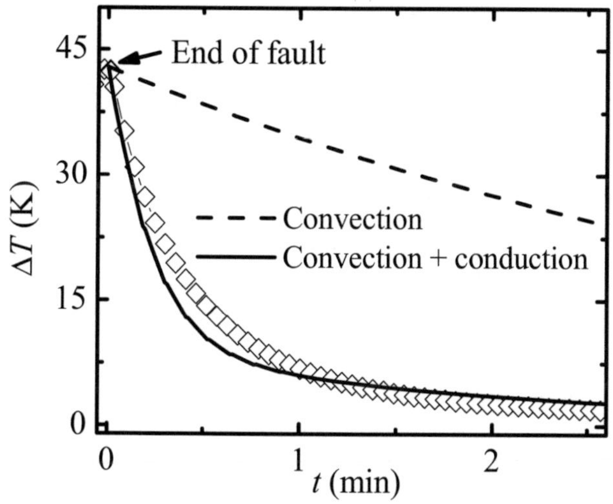

Fig. 10. Temperature measured in the weak zone during the thermal recovery of the bulk cylinder under study. The dashed line shows the calculation for the pure convection model (exponential decrease), while the solid line takes into account not only the convection term, but also a conductive exchange between the weak zone and the rest of the cylinder. Figure from Ref. (Osorio et al., 2005).

of the cylinder plays a major role. Notice that this would entail that this weak zone would refrigerate faster than a homogeneous cylinder, provided that the same final temperature is attained in both cases. This suggests that if the cylinder was made up from weak zones separated by cold segments, after the fault clearance the excess heat could be removed from the hot parts to the colder ones, till the cylinder was thermalized at a low temperature, maybe higher than that of the coolant, but much lower than the maximum value obtained during the fault. It can be argued that, as only the weak parts would enter into the normal state, the impedance of the limiter should be much lower. Nevertheless, and despite this drawback, it could be found a configuration of made-to-measure weak zones, properly spaced, which allow obtaining an acceptable impedance but, in turn, would reduce the recovery time to a competitive value, even in non-liquid environments.

The above ideas were proposed in previous works (Osorio et al., 2005; 2008; 2006), where two possible procedures where stated; in the first one (Osorio et al., 2005), the set of artificial weak zones (AWZ) could be made from a different material, with a lower J_c and a higher resistivity (so somewhat balancing the loss of impedance of the limiter), as depicted in figure 11(a), for a set of n AWZ of length ℓ_n, ℓ^h being the total weak length. Figure 11(b) shows the equivalent electrical circuit for the cylinder, with resistances R^c and R^h for the *cold* and *hot* parts (i.e. the AWZ), respectively.

In the second procedure (Osorio et al., 2008; 2006), each AWZ would be made as a constriction of the cross-section of the cylinder as it is shown in figure 14(a). This configuration allows using the same material for the whole sample, and the reduction in the critical current and the gain in the total impedance is easily yielded by the diminution of the cross-section. For convenience, this idea was initially tested by means of a 1-D numerical simulation routine. This implies assuming that the vertical axis of the cylinder is long enough to neglect

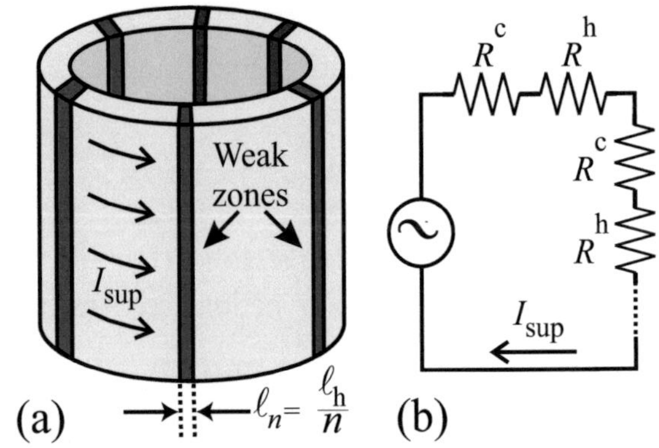

Fig. 11. (a) Scheme of the superconducting cylinder when several thin weak zones are considered instead of a greater size one. The length of each weak zone, ℓ_n, is the total one, ℓ^{h}, divided by the number of slices, n. (b) Equivalent electrical circuit of the inhomogeneous superconducting cylinder. The resistances are both dependent on the current density and the temperature of the corresponding zone: $R^{(\mathrm{c,h})}(J, T^{(\mathrm{c,h})})$. Figure from Ref. (Osorio et al., 2005).

any possible temperature gradient along this direction. In addition, it is assumed that the temperature across the wall thickness is approximately the same no matter the point under consideration, i.e., the surface of the cylinder is not at a lower temperature than the inner parts. This is the rougher approximation of the model, but for cylinders of wall thickness of the order of $2-3$ mm it can be quite acceptable. The AWZ are parametrized by the relative length λ, which is the ratio of the length of the weak zone to the whole perimeter of the cylinder.

Figure 12(a) shows the recovery time calculated for different values of λ normalized to that obtained for a homogeneous cylinder (left axis) as well as the normalized impedance (right axis). Notice two important facts; first of all, the recovery time is not necessarily shorter for a cylinder with a weak zone, unless the AWZ is short enough to warranty a fast evacuation of the heat by conduction. Secondly, the impedance yielded by the AWZ can be lower than that offered by the homogeneous cylinder. This second drawback can be removed, in principle, by using a more resistive material or by making AWZ in the shape of grooves (the lower the cross-section the higher the impedance). Even if the resulting impedance is not as high as that offered by the homogeneous cylinder, this inconvenient could be balanced by a much faster thermal recovery. The solution to the first drawback appears to be splitting the AWZ in short slices, so that the heat can be easily removed by conduction from the cold parts to the colder segments which act as heat reservoirs. This effect is shown in 12(b), where an AWZ of length $\lambda = 2$ (corresponding to the working point chosen in (a), for which the weak zone represents 20% of the perimeter of the cylinder) is split into n slices. Notice that, although the weak zones are individually shorter, the total "weak length" is the same, so the impedance can be considered as invariable with respect to the case of $n = 1$ (i.e. a non-split weak zone). Maybe a better refrigeration of these slices could lead to a slightly lower impedance, but this effect is neglected here.

Figure 13 shows the numerical calculations for the performances of different cylinders acting as secondary of a SFCL. The first one is a homogeneous element, the second a cylinder with a connected weak zone of relative length $\lambda = 2$, and the last one is a cylinder with a split

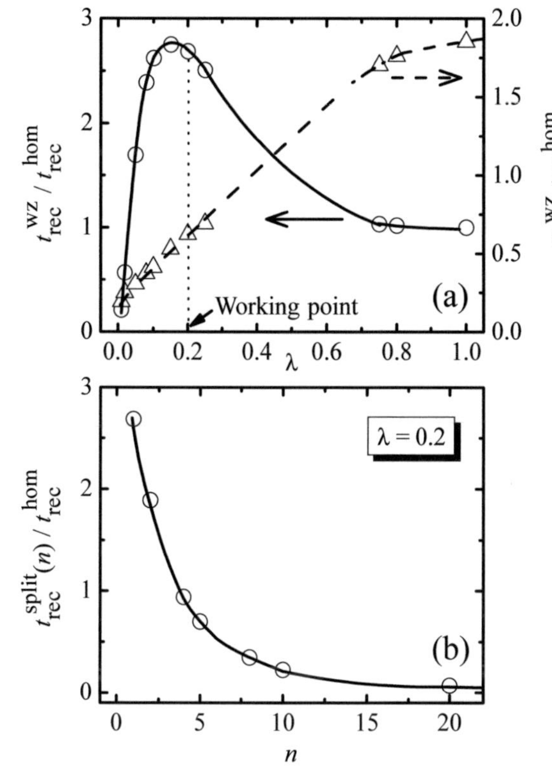

Fig. 12. (a) Recovery time calculated for a set of values of λ normalized to the value obtained for a homogeneous cylinder (circles). The right axis shows the normalized impedance (triangles). Both solid and dashed lines are given as guides for the eyes. (b) Recovery time as a function of the number of slices in which it is divided a weak zone with $\lambda = 0.2$, normalized to the value when the cylinder is fully homogeneous. The solid line is again a guide for the eyes. Figure from Ref. (Osorio et al., 2005).

weak zone characterized by the same relative length and $n = 10$. In figure 13(a) it can be seen the heating and thermal recovery for a fault of about 0.2 s in duration. Notice that the final temperature difference is about 30 K in the case of the homogeneous cylinder, while the AWZ of the second and third cylinders reach the same maximum value, about two times higher. Due to this, the recovery of the connected weak zone is even slower than that of the homogeneous cylinder. However, when the AWZ is split into 10 parts the thermal recovery is noticeably enhanced. Figure 13(b) shows the impedance obtained with these cylinders. The homogeneous cylinder yields a much higher impedance, as the transited length is 80% higher. This drawback can be somewhat balanced by using AWZ made form a material with a higher resistivity (see dashed line, for a material with a normal resistivity two times higher than that of the rest of the superconducting element).

Further tests were realized by means of a finite element routine which allows exploring the effects associated with thick walls. Figure 14 (a) and (b) shows an scheme of the top and the lateral view, respectively, of the cylinder with constrictions equally spaced used in the

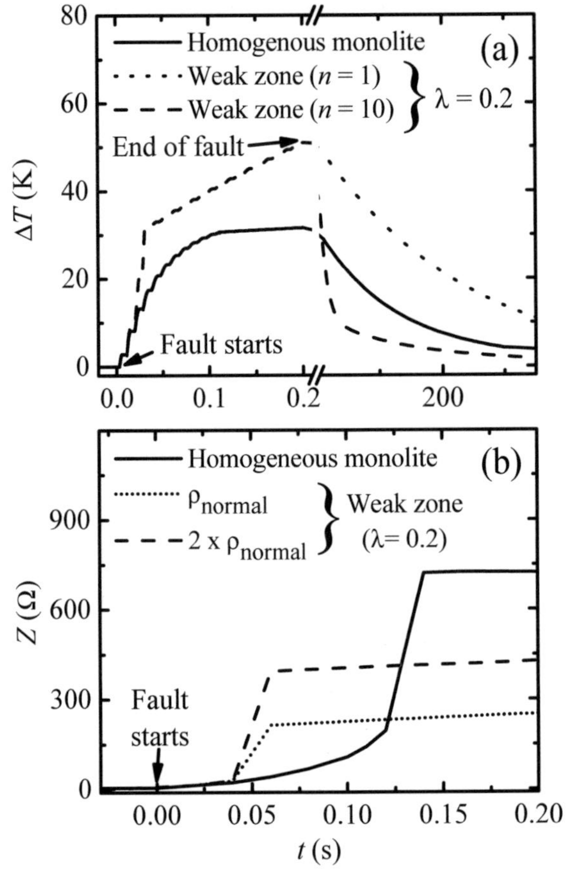

Fig. 13. (a) Temperature excursions calculated for a homogeneous monolith-type cylinder (solid line), a cylinder with a connected weak zone (dotted line) and a cylinder with a weak zone split into 10 small parts (dashed line). The recovery time of the sample with the split weak zone is noticeable shorter than that for the homogeneous sample. In this example, the resistivity of the weak zone is assumed to be 2 times the resistivity of the reminder part. (b) Impedance under a current fault calculated for the device operating with each one of the samples considered in (a). The impedance of the device with the inhomogeneous sample (the same for the connected and the split weak zone) is around a 40% lower than with the homogenous one, except for the first cycles. The dotted line should be the impedance if the resistivity is the same for both the weak zone and the other part of the cylinder. Figure from Ref. (Osorio et al., 2005).

simulation. It is assumed that a constant current circulates along the circumferential perimeter of a cylinder. For the sake of simplicity, the cylinder is considered to be infinite along its axis, hence the problem is 2-D. Due to symmetry, only a quarter of a groove has to be considered. The mesh is denser in the AWZ and its vicinity and coarser far away, where important changes are less likely to occur, as it is shown in figure 14 (c). The details of the numerical routine can be found in reference (Osorio et al., 2008). We just indicate here the main steps of our calculations.

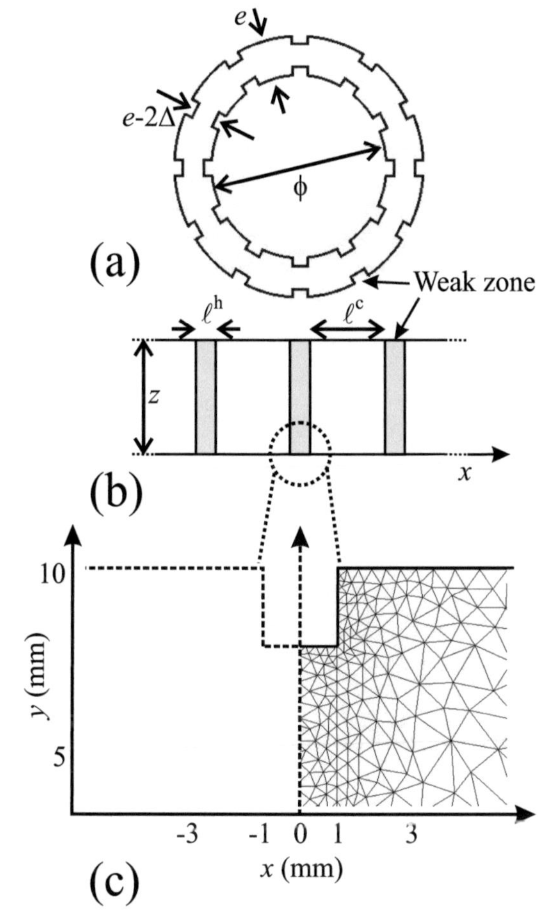

Fig. 14. (a) Schematic top view of the grooved cylinder under study. (b) Lateral view of a segment of the cylinder's perimeter used in the simulation. (c) Top view of the mesh used in FEM calculations. Due to symmetry, only a quarter of a groove is used. Figure from Ref. (Osorio et al., 2008).

When the fault occurs, the circulating current overtakes the critical value I_c and dissipation arises. Following the temperature and current dependent equations for the electric field, the dissipation is calculated and so the temperature at each node of the mesh. Heat exchanges by convection with the coolant and conduction between hot and cold segments are modeled by taking into account experimental values for the temperature dependent convective coefficient for liquid nitrogen (Duron et al., 2007) and helium gas (Chapman, 1984), and the thermal conductivity (Fujishiro et al., 2003; 2006). As we are mainly interested on the thermal recovery, the condition that the circulating current is constant (not realistic for an inductive SFCL) does not preclude at all the validity of our conclusions about the recovery time. More sophisticated models, where the current changes according to the resistance developed in the superconducting element can be found elsewhere (Rettelbach & Schmitz, 2003).

Figure 15 displays the temperature distribution within the cylinder's wall after a fault of 100 ms. We can see that, despite the high temperature attained in the inner of the wall, the heating

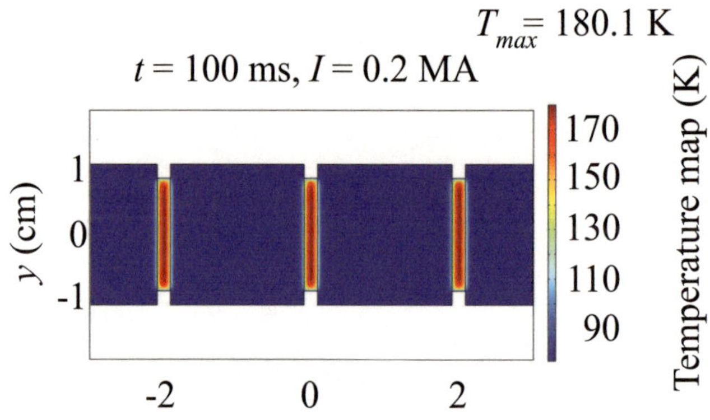

Fig. 15. Temperature map of a superconducting element after a fault of 100 ms in duration. Figure from Ref. (Osorio et al., 2008).

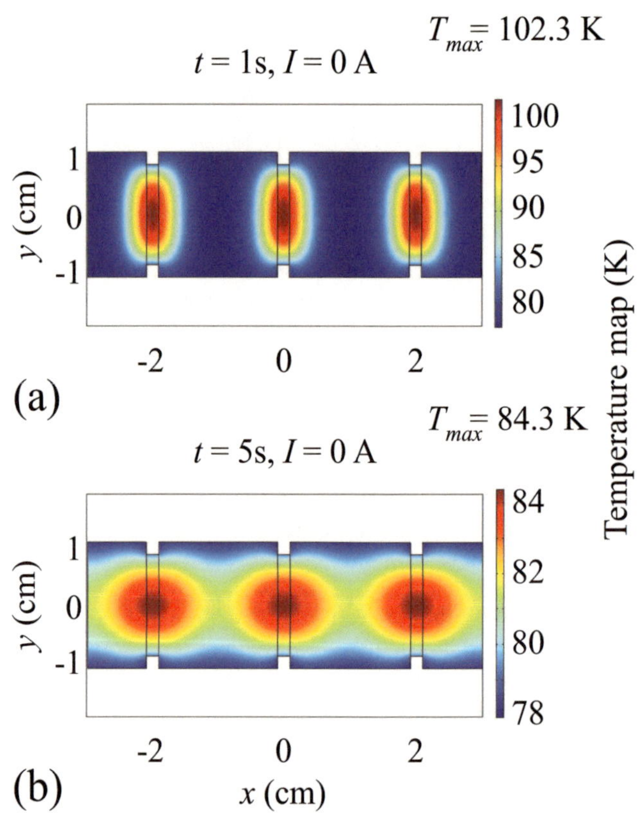

Fig. 16. Thermal recovery for the example of Figure 15 after 1 s of zero-current recovery (a) and after 5 s (b). Figure from Ref. (Osorio et al., 2008).

is quite confined to the AWZ. This effect of "non-propagation" of the hot spot during the fault was one of the main assumptions of the 1-D model, and this more refined model confirms its validity. The evolution of the hot spot once the fault is removed and the current is set to zero is shown in figures 16(a) and 16(b) for $t = 1$ s and $t = 5$ s, respectively. Notice that the redistribution of the excess heat, confined at the beginning in the close vicinity of the weak zone, leads to a temperature map which is not so far above 77 K.

Figure 17 shows an example of the variation of the primary current before and during a fault when the secondary is an ungrooved cylinder (i.e. without AWZ) and a grooved one, UC and GC1, respectively. Both samples are supposed to be made from YBCO, and GC1 is characterized by a ratio of grooved to ungrooved cross-sections $a = 0.8$, and by the lengths of the hot and cold parts being $\ell^h = 0.5$ and $\ell^c = 0.5$ cm. Notice that the limited current is the same no matter the secondary used, so making grooves does not necessarily affect the limitation capacity.

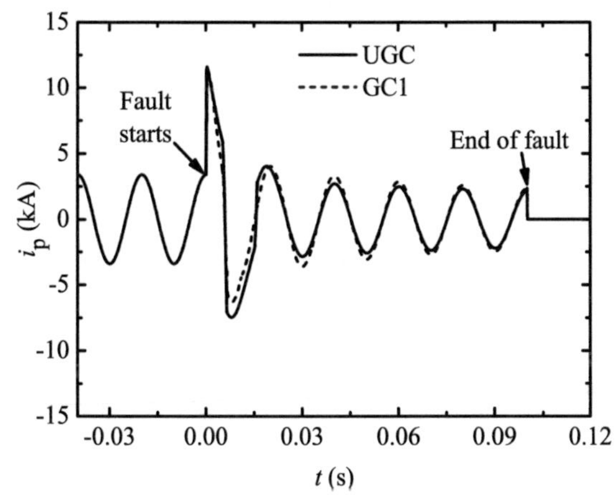

Fig. 17. Variation of the primary current for an ungrooved cylinder (UGC) and a grooved cylinder (GC1). Notice that the limitation performance is identical. Figure from Ref. (Osorio et al., 2008).

As the temperature of the grooves can be very high and the thermal conductivity of YBCO within the *a-b* planes is quite low, about 2.5 $\mathrm{Wm^{-1}K^{-1}}$, it was proposed to use cylinders of YBCO grown in such a way that the current circulates along the *c* axis (Osorio et al., 2008). This entails that the normal state resistivity is 15 times higher (Vanderbemden et al., 1999), and that the thermal conductivity along the perimeter is 10 $\mathrm{Wm^{-1}K^{-1}}$ (and 3 $\mathrm{Wm^{-1}K^{-1}}$ along the cross-section) (Fujishiro et al., 2003; 2006). With these new parameters, the AWZ can be made thinner without reducing the normal state resistance, while the removal of heat is made easier due to the higher thermal conductivity and to the reduction of the path the heat must travel to be exchanged between hot and cold parts. Figure 18 shows our best result for a grooved cylinder characterized by $a = 0.8$, $\ell^c = 0.8$ cm and $\ell^h = 0.2$ cm, when the limiter works in a liquid nitrogen bath (a) and a helium gas atmosphere (b). The thin dotted line shows the temperature at which the recovery ends. Notice that the recovery times for the grooved sample are about 1 s, as demanded in practical applications. The solid lines correspond to the ungrooved cylinder. When a liquid nitrogen bath is used, the recovery for the ungrooved cylinder occurs at about 3.5 s, despite the initial temperature is much lower. For the helium gas atmosphere, its temperature does not even change within the time interval under study (further details of these studies can be found in (Osorio et al., 2008)).

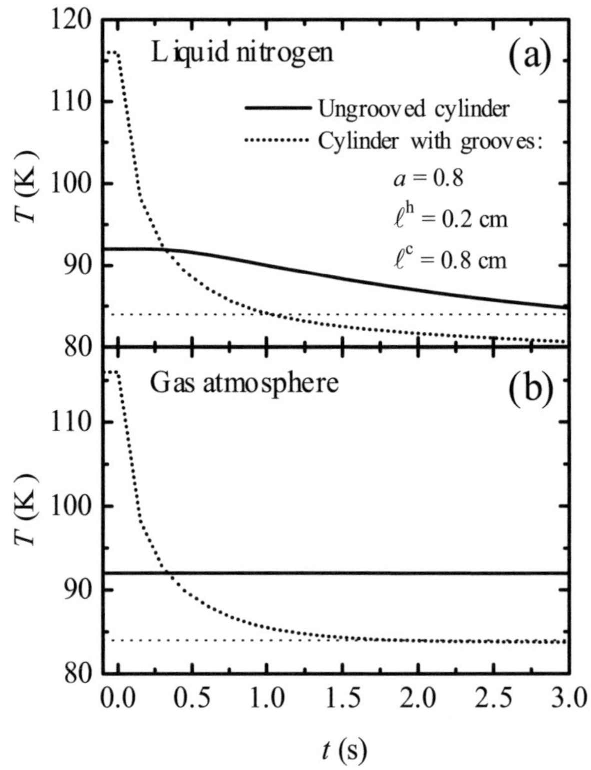

Fig. 18. Thermal recovery for an ungrooved cylinder (solid line) and for a sample with grooves characterized by $a = 0.8$, $\ell^c = 0.8$ cm and $\ell^h = 0.2$ cm in liquid nitrogen (a) and helium gas (b). The thin dotted lines indicate the temperature at which no dissipation arises in the superconducting elements. Figure from Ref. (Osorio et al., 2008).

Our results show that the use of grooves samples can improve the performance of bulk superconductors working in SFCL. However such improvement requires a set of grooves with very close behaviour. We illustrate the importance of this requirement with the example displayed in Fig. 19. When all grooves are identical, which is the case discussed above, the temperature increase of these parts under current fault should reach only up to around 100 K (Fig. 19(a)). If one of the grooves presents a slightly greater ratio a (around 1%) the temperature map changes very little (Fig. 19(b)) and the main efect would be a small decrease of the fault impedance. One major drawback however occur if any of the grooves is smaller than the other: the temperature of the groove with a around 1% smaller should increase up to around 150 K (Fig. 19(c)) which would cause a very long thermal recovery time after fault and therefore a poorer overall performance of the limiter device (a deeper study of this type of drawback can be found in (Osorio et al, 2010)).

4. Stacks of bulk rings and thin film washers as secondary

In the above section we have seen a way to improve the cooling of bulk superconducting samples by making a set of artificial weak zones. Another procedure could be increase the

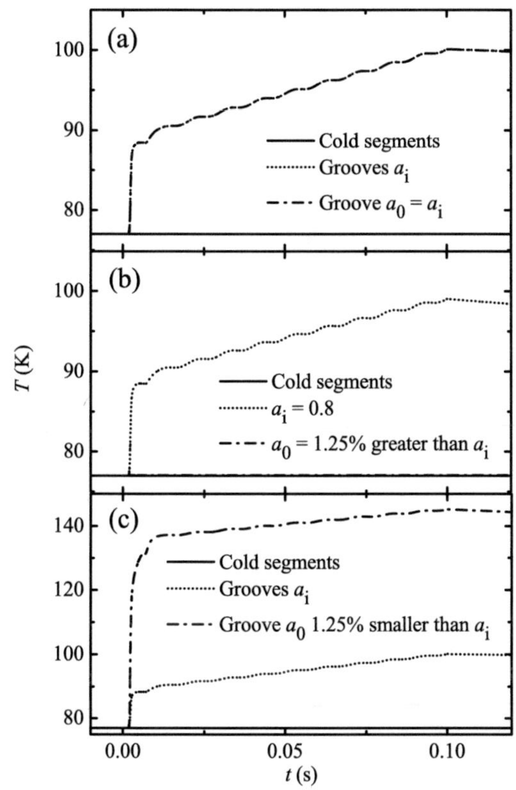

Fig. 19. Temperature excursions under current fault of cold parts and grooves for: (a) All of grooves with the same ratio a; (b) and (c) One of the grooves with greater and smaller a, respectively, than the other. Figure from Ref. (Osorio et al, 2010).

surface-to-volume ratio of the secondary by using short bulk rings or thin film washers. This idea was proposed in a previous work (Osorio et al., 2006), when it was presented a comparison between the performance of both types of secondaries. To balance the reduction in the effective critical current of the limiter (due to the lower cross-section of rings and washers when compared to bulk cylinders), it was proposed to stack these elements up to add up their critical current values.

In figure 20 it is displayed the impedance under current fault for three secondaries: a 40 mm-high Bi-2223 cylinder (labeled 1*R*40) a stack of three 2 mm-high Bi-2223 rings (labeled 3*R*2), and a stack of three 4.5 mm-wide Y123/Au washers (labeled 3*W*4.5). Y123/Au washers were trepanned from wafers made from a $Y_1Ba_2Cu_3O_{7-\delta}$ thin film (300 nm thick) shunted by a gold layer (100 nm thick), deposited on a AlO_2 substrate provided by Theva (Germany).

It can be seen that these secondaries present a very different behaviour under a current fault; with 1*R*40 the increase of impedance is very smooth, whereas it is somewhat abrupter with 3*R*2 and much more with 3*W*4.5. The differences are even more remarkable after the current fault. With 1*R*40 the impedance for the highest voltage continues to increase. On the contrary, with 3*R*2 the thermal recovery (i.e. when the circulating current is the same than before the

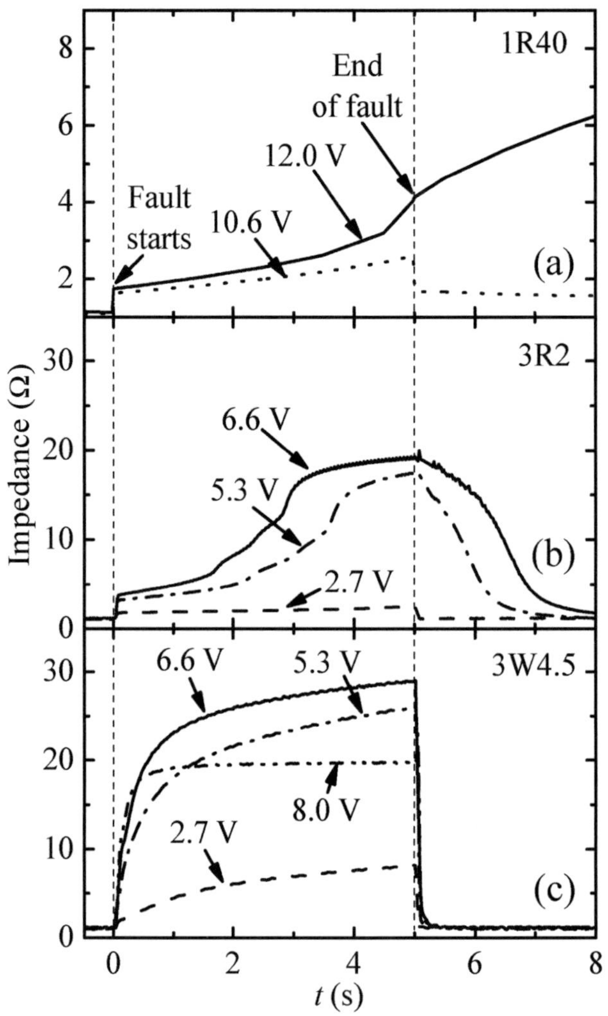

Fig. 20. Impedance at different fault voltages versus time for a limiter working with (a) 1 bulk Bi2223 ring of 40 mm in height, (b) a stack of 3 bulk Bi2223 rings of 2 mm in height and (c) a stack of 3 thin-film Y123/Au washers of 4.5 mm in width. Figure from Ref. (Osorio et al., 2006).

fault) is accomplished within a few seconds, and it is very fast for 3*W*4.5. The recovery times obtained for different faults and these three secondaries are plotted in figure 21. For the secondaries 3*R*2 and 3*W*4.5 the recovery time values plotted were directly obtained from the measured current waveforms, assuming that the limiter was recovered when the amplitude of the current is the same as before the fault. The scattering of the data points around the guide lines, about 10%, represents the precision of our measurements. With this procedure, the recovery times obtained could be somewhat overestimated because the circuit was not opened after each fault. Therefore some slight dissipation in the samples could still remain during the recovery process. Despite these small uncertainties, our results clearly show the better performance of the thiner samples. The secondaries 3*R*2 and 3*W*4.5 show recovery

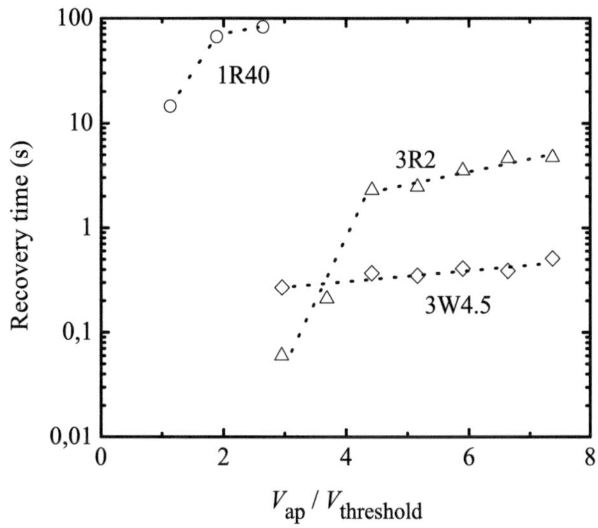

Fig. 21. Recovery time versus applied voltage normalized to the threshold value for the limiter working with a stack of three bulk Bi2223 rings of 2 mm in height (triangles) and a stack of three thin-film Y123/Au washers of 4.5 mm in width (diamonds). It is also included the recovery time when using as secondary a bulk Bi2223 ring of 40 mm in height (circles)($V_{threshold} = 7\ V$ for $1R40$ and $0.9\ V$ for $3R2$ and $3W4.5$). Figure from Ref. (Osorio et al., 2006).

times one and, respectively, two orders of magnitude lower than the one for $1R40$. Note that the lowest times obtained for $3R2$ correspond to trivial examples of low fault voltages for which the samples develop very low resistance.

Furthermore, although $3R2$ and $3W4.5$ operate at a lower power rating than $1R40$, the recovery time will remain unchanged if more rings or washers were added to the stack. For instance, with about 25 of our thin films or short rings one should obtain the same operation currents than for the $1R40$ ring, the secondary being of similar size, with similar or higher impedance but with recovery time strongly shortened, specially with thin films. With this procedure it should be also possible to fabricate limiters able to operate at the high currents withstood, for instance, by many resistive type limiters based on meander thin films, which also present recovery time values in the range of tenths of seconds even when submerged in liquid nitrogen.

In case of a bulk cylinder ($1R40$) of the same diameter as the bulk rings and height 4 cm, a low current (even below the nominal one) circulating after the fault is enough, as discussed above, to provoke appreciable dissipation in the sample. Therefore, the recovery time is extremely enlarged, even for the lowest applied voltages. So, for this secondary the recovery times were obtained by opening the circuit just after the end of the fault, i.e., we measure the *zero current recovery time*. With this procedure, the temperature at which the application of the nominal current does not generate dissipation is measured with the thermocouples attached on the rings. For 40-mm-high rings we found that the nominal current does not produce any appreciable heating for temperatures around 5 K above the working value (77 K). The circles plotted in figure 21 represent the interval between the end of the fault and the moment at which this temperature is reached. The precise value of the temperature threshold and the differences from sample to sample do not change the values of the recovery time in more than 20%.

We must note that using stacks of bulk rings provokes a loss of impedance, as the superconducting elements of the stack would behave as resistances connected in parallel (Osorio et al., 2004). Although this drawback could be balanced by joining together several cores (Osorio et al., 2004), stacks of thin films show a better thermal performance and there is no reduction in the impedance or this is negligible (Lorenzo et al., 2006; 2007). Therefore, thin films offer important advantages concerning their refrigeration and the impedance does not diminish even when many of them are stacked to increase the effective critical current of the limiter.

5. SFCL based on superconducting thin film microbridges and meander paths

Thin film samples present several advantages to be used in current limiting applications, as we have discussed above. We have probed the use of this kind of samples in other two up to now nearly unexplored configurations. One is a hybrid limiting device (i.e inductive-resistive) refrigerated by using a thermoacoustic refrigerator. The other is the microlimiter, a resistive FCL based on "thermally small" (Ferro et al., 2008; 2009) superconducting microbridges (Lorenzo et al., 2009; 2010) intended to operate at very low powers (SQUID based electronics, infrared detectors, etc).

5.1 Hybrid SFCL integrated in a thermoacoustic refrigerator

Thermoacoustic refrigerators, heat engines which are based on the ability of a sound wave to play the role of the mechanical compressor and expander (Swift, 2002), have gained attention in the last decades and the state of art has allowed reaching very low temperatures (Yang & Thummes, 2005; Qiu et al., 2005). Thermoacoustic devices are for the time being slightly less efficient than conventional refrigerators. Nevertheless, their simplicity, low cost and lack of moving parts (which entails a strong diminution of maintenance costs, the main trouble with commercial refrigerators) and contaminating lubricants offer advantages over certain other mechanical devices like mechanical compressors.

A scheme of a thermoacoustic refrigerator is shown in figure 22. It consists of a resonant tube filled with an inert gas (usually Helium at a high pressure, about tens of bars) in which a standing sound wave is generated by using an engine (which can be also of thermoacoustic nature, or just a compressor). Two heat exchangers are located at precise locations inside the tube and they are separated by a porous medium (regenerator) which acts as a heat pump, transferring heat from the cold heat exchanger to the ambient one (kept at room temperature by a water stream), by means of thermoacoustic processes (Swift, 2002). The proper phase between the acoustic pressure and velocity inside the regenerator (where a local traveling wave must exist to verify the heat pumping) is set by an acoustic charge, usually made up from a constriction, a long and thin tube and a huge cavity which behave as a resistor, an inductor and a capacitor in electrical circuits. A third heat exchanger (the "hot" one) is used to prevent the entrance of heat from the acoustic charge into the cold heat exchanger. When the engine is of thermoacoustic nature, the acoustic wave can be generated by a thermal gradient in an appropriately designed porous element and the heating source can be electric power

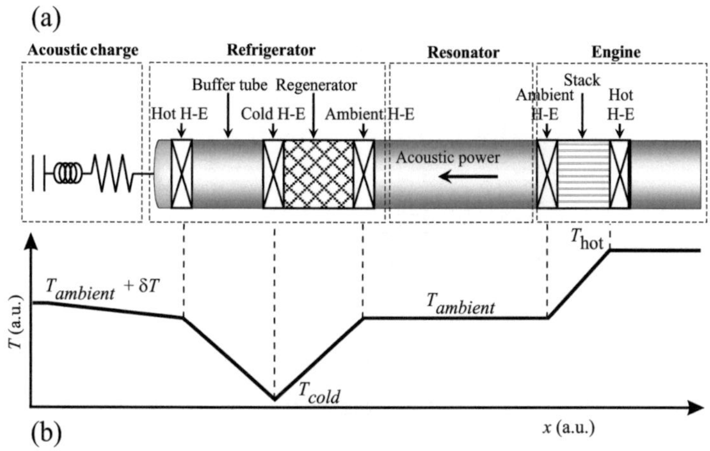

Fig. 22. (a) Schematic view of a thermoacoustic machine. On the right it can be seen the thermoacoustic engine. The thermoacoustic refrigerator, of pulse tube type, is comprised of three heat exchangers (HE), a regenerator and the buffer tube, which limits the entrance in the cold part of the heat generated by the dissipation of acoustic power in the charge. (b) Temperature profile along the TA machine. A strong gradient is imposed on the engine, so the wave can be generated. It is assumed that the dissipation increases the temperature in the acoustic charge. Figure from Ref. (Osorio et al., 2008).

(produced in a photovoltaic system, for instance) or burnt gas, as has already been done in some natural gas plants, where a conventional liquefying system were not cost effective.

In a previous work (Osorio et al., 2008) we proposed a 2.2 kVA prototype of hybrid fault current limiter based on a YBCO film meander path which was directly attached to the cold heat exchanger of a thermoacoustic refrigerator, as it is schematically depicted in figure 23. The main advantage of this configuration is that the close contact between the superconducting path and the cold heat exchanger allows a very efficient removal of the excess heat produced during a fault, thus greatly improving the performance of refrigerators in which the element to be cooled is inside an isolated cavity connected to the cooler through some kind of pipe. The thermoacoustic refrigerator was intented to remove about 50 W at 80 K so the recovery time of the whole SFCL does not exceed 1 s. The final design was tested by using a numerical routine (Osorio et al., 2008) and it was found that the SFCL was able to reduce the fault current in about 3 orders of magnitude, while the superconducting path heated about 130 K during a fault of 100 ms in duration. The total heat deposited in the cold heat exchanger was around 30 J, low enough to be removed by the refrigerator in 1 s.

Resistive configurations were not considered as the length of the superconducting path necessary to get a suitable impedance would be excessive to allow attaching it on the cold heat exchanger of a medium-size thermoacoustic refrigerator. In addition, all the fault power would be dissipated just in the meander path, and so the temperature increase could be very high. A hybrid configuration offers the advantages of a lower dissipation in the superconducting element (part of the energy is required to get the magnetic field inside the core) and also a high impedance which does not demand for a very long meander path. A pure inductive limiter was immediately rejected as the core would occupy too much room inside the refrigerator.

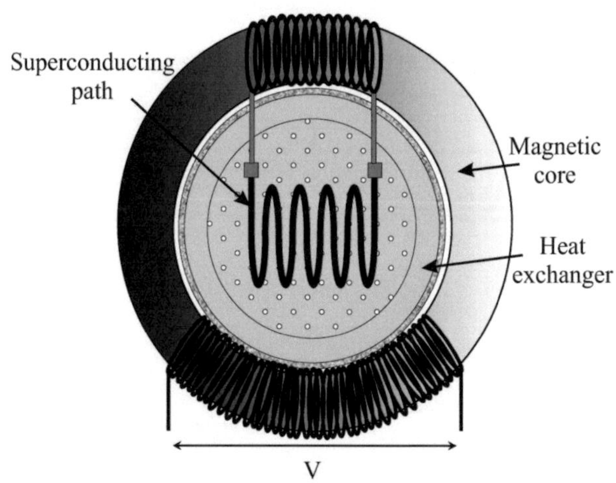

Fig. 23. Axial view of a hybrid limiter attached to the cold heat exchanger. The superconducting element is represented as a meander path. Figure from Ref. (Osorio et al., 2008).

5.2 Superconducting fault current microlimiters

In figure 24 we show a typical electric field versus current density ($E - J$) curve, corresponding to one of the microbridges (denoted BS7) used in our experiments. This microbridge, whose length, width and thickness are, respectively, 385 μm, 28 μm and 300 nm, and with $T_c = 88.6$ K, has been grown on a sapphire substrate. The two characteristic current densities are indicated in this figure: The critical, J_c, at which dissipation first appears, and the so-called supercritical, J^*, at which the microbridge is triggered into highly dissipative states. Let us note here that we have chosen for our studies microbridges of these dimensions to guarantee

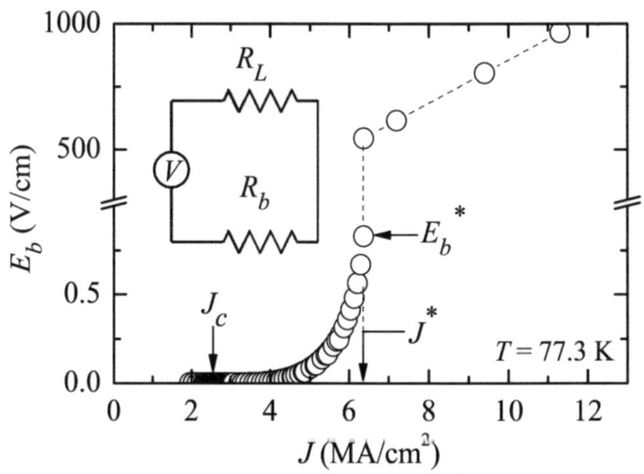

Fig. 24. A typical $E - J$ curve obtained at $T = 77.3$ K in one of the microbridges used in our experiments (BS7). The inset shows a schematic diagram of the circuit used to probe how the microbridge acts as a FCL, protecting the load resistance, R_L, from a voltage fault. Figure from Ref. (Lorenzo et al., 2009).

a good thermal behaviour before, during and after the current fault, as we will see below. In addition, the widths of our microbridges are well above the threshold at which J^* is sample-width dependent (Ruibal et al., 2007). The strong increase of E for current densities around J^* make this type of samples very useful for FCL. However, the current is effectively limited only for electric fields or, equivalently, applied voltages well above that at which the limiter is triggered from normal operation (sample in the superconducting state) to current fault mode. In addition, once J^* is attained, a thermal runaway (Viña et al., 2003; Maza et al., 2008) can be provoked which causes the reduction of the circulating current well below the nominal value (i.e., the current in normal operation without a fault) or very important damage on the microbridge (that can be even burnt out).

The superconducting microbridge exchanges heat mainly with its substrate, because at the operation temperatures, around 90 K, the heat transfer coefficient between YBCO films, for instance, and their substrate is $h_{bs} \approx 10^3$ Wcm^{-2}K, whereas through liquid nitrogen or between the substrate and the copper holder they are 1000 times less (Duron et al., 2007; Lorenzo et al., 2009; Mosqueira et al., 1993). Hence, the conditions for a good refrigeration and, therefore, an optimal operation of the limiting device, depend very much on the relative dimensions of the microbridge and their substrate. The optimal refrigeration should allow us to operate with the minimum microbridge temperature increment relative to that of the bath, ΔT_b. If the Biot number (Lorenzo et al., 2009; Chapman, 1984) of both the microbridge and its substrate are $Bi \ll 1$ and, simultaneously, the thermal diffusion length, L_{th}, is longer than the corresponding thickness, ΔT_b may be crudely approached by

$$\Delta T_b = e_b J E \left(\frac{1}{h_{bs}} + \frac{A_b}{A_s} \frac{1}{h_{sr}} \right), \tag{5}$$

where e_b is the microbridge thickness, A_b and A_s are the surface areas of the superconductor bridge and, respectively, its substrate. As $h_{bs} \gg h_{sr}$, ΔT_b will directly depend on A_b/A_s, the condition of "thermal smallness" being,

$$\frac{A_b}{A_s} < \frac{h_{sr}}{h_{bs}}, \tag{6}$$

under which ΔT_b approaches its "intrinsic" value, which is the temperature increase just associated with the effective thermal resistance of the interface bridge-substrate. For our microbridges similar to BS7 and under faults with characteristic times above 10 ms (of the order of commercial ac current periods), the conditions, $B_i \ll 1$ and $L_{th} > e$ fully apply. In addition, as $A_b/A_s \approx 4 \times 10^{-4}$ whereas $h_{sr}/h_{bs} \approx 10^{-3}$, this microbridge is "thermally small". One may use equation 5 to estimate ΔT_b when the microbridge is in stationary conditions under currents just below J^*. From figure 24, the power density involved is around 5×10^6 W/cm^3 which leads, by also using the appropriate parameter values[1], to $\Delta T_b \approx 0.1$ K. Even for faults involving powers ten times higher than those considered above, ΔT_b will remain below 1 K. One may also use equation 5 to roughly estimate that if the superconducting bridge had A_b/A_s one thousand times larger, as it is the case of those currently proposed for high power applications (Noe & Steurer, 2007; Weinstock, 2000), ΔT_b would take values at least

[1] The substrate thermal conductivities (K_s) and heat capacities (C_s) are of the order of 10 W/cm K and 0.6 J/cm^3 K for sapphire, and 0.2 W/cm K and 1 J/cm^3 K for $SrTiO_3$, whereas for YBCO bridges $K_b \approx 0.05$ W/cm K and $C_b \approx 0.7$ J/cm^3 K. See, e. g., Refs. (Duron et al., 2007; Chapman, 1984; Mosqueira et al., 1993).

two orders of magnitude higher. To confirm these values at a quantitative level, ΔT_b has been calculated by using a finite element method similar to the one described in (Maza et al., 2008). Under the same conditions as above, for the BS7 microbridge we found again $\Delta T_b \approx 0.1$ K.

In the case of $SrTiO_3$ substrates, which have relatively poor thermal conductivities, the condition $Bi \ll 1$ does not apply anymore. Therefore, a term proportional to $1/k_s$ (i.e. to the thermal conductivity of the substrate) must be added in equation 5. For microbridges under the same conditions as before, this leads to $\Delta T_b \approx 2$ K, a value that is confirmed by using the finite element method commented above. Let us stress, finally, that for a bridge on sapphire but having a surface relative to that of its substrate one thousand times larger, the finite element method yields $\Delta T_b \approx 50$ K.

To probe a microlimiter with low thermal dimensions, we have implemented the electrical circuit schematized in the inset of figure 24, with the microbridge BS7 as R_b connected in series to the variable load resistance R_L, this last one representing the impedance of the circuit to be protected. The measurements were made in a cryostat with the sample submerged in a forced flow of helium gas. The temperature of the copper holder of the microlimiter was measured with a platinum thermometer and regulated with an electronic system which ensures a temperature stabilization better than 0.05 K. Two examples of the $I - V$ curves obtained in this $R_L - R_b$ circuit (with $R_L = 4.9\ \Omega$, this value taking already into account the resistance of the circuit electrical wires, of the order of 1.9 Ω) by using the electronic system described elsewhere (Ruibal et al., 2007; Viña et al., 2003) are shown in figure 25. In these curves the voltage was imposed and acquired during pulses of 1 s, a time much longer than the one needed by the microlimiter to reach the stationary state. The bath temperatures were 81.9 K (circles) and 85.0 K (triangles). As below V^* the flux-flow resistance of the microbridge remains much lower than R_L , both curves are almost linear up to V^*. Together with the low heating estimated above, this quasi-ohmic behavior is crucial to allow the microlimiter to work just below V^* under stationary conditions and without disturbing the circuit to be protected.

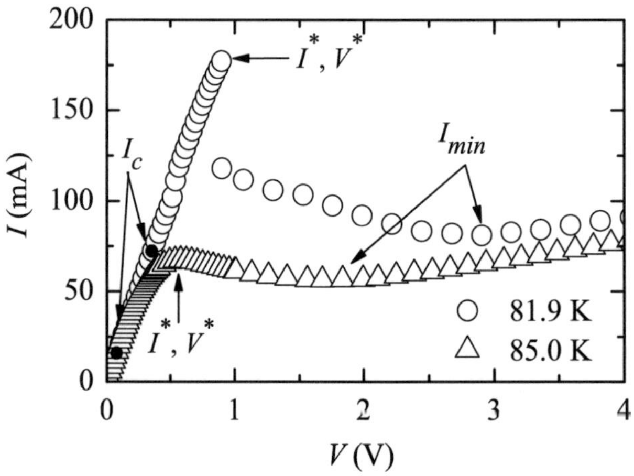

Fig. 25. Two $I - V$ curves, showing their strong bath temperature dependence, of the $R_L - R_b$ circuit schematized in the inset of figure 25 In these examples R_b is the microbridge BS7 and $R_L = 4.9\ \Omega$. Figure from Ref. (Lorenzo et al., 2009).

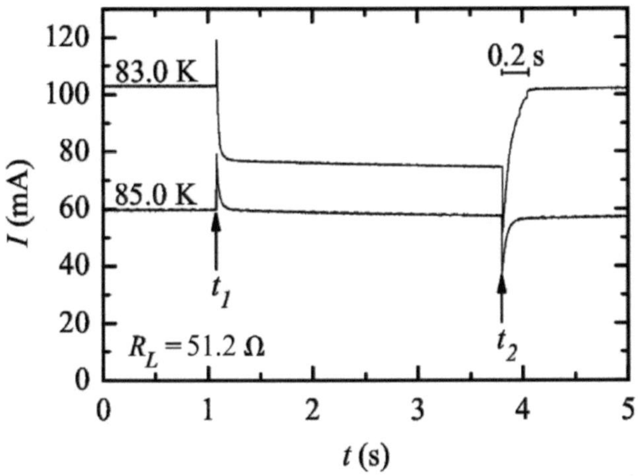

Fig. 26. Two examples of the time evolution of the current measured in presence of voltage faults between t_1 and t_2. Figure from Ref. (Lorenzo et al., 2009).

The results presented in figure 25 also illustrate two other central aspects of the microbridges when working as FCL. Note first that, once the source voltage overcomes V^*, the current in the circuit varies quite slowly even up to voltage faults as important as four times V^*, the current taking a minimum value, I_{min}, at some (temperature-dependent) voltage. Moreover, the sharp drop at V^* of the current is also temperature dependent, being almost absent in the curve at 85.0 K. Both aspects are related and may be explained in terms of the approaches based on the propagation of self-heating hot spots (Skocpol et al., 1974; Gurevich & Mints, 1987; Paulin et al., 1995): Above V^* part of the microbridge becomes normal and then, as the total voltage of the circuit is fixed, the resistance increase originates a current decrease up to the minimum current, I_{min}, capable of sustaining the normal zone. If the fault voltage increases, the hotspot length will grow accordingly, keeping the current roughly constant. At a quantitative level, both aspects may be easily explained by just taking into account the reduced temperature (T/T_c) dependence of I^* and I_{min},

$$I(T) = I(0)\left(1 - \frac{T}{T_c}\right)^n , \tag{7}$$

with $n = 3/2$ for I^* (Ruibal et al., 2007; Viña et al., 2003) and $1/2$ for I_{min} (Skocpol et al., 1974; Gurevich & Mints, 1987; Paulin et al., 1995). Therefore, if the reduced temperature increases both the discontinuity at V^* and the ratio I^*/I_{min} will decrease, in agreement with the results of figure 25 The "optimal" reduced temperature for the microlimiter operation, T_{op}/T_c, will be then given by the condition $I^*(T_{op}) = I_{min}(T_{op})$. By using equation 7, this leads to,

$$\frac{T_{op}}{T_c} = 1 - \frac{I_{min}(0)}{I^*(0)} . \tag{8}$$

At T_{op} the current limited during the voltage fault will be roughly equal to the nominal one. As $I_{min}(0) < I^*(0)$, T_{op} will be near, but below enough, T_c to make $I^*(T_{op})$ adequate for the practical operation of the microlimiter under such an optimal temperature.

The above results, experimentally confirmed with our studies (Lorenzo et al., 2009; 2010), allow us to "thermally tune" a superconducting microlimiter for an optimal protection. The results summarized in figure 26 were measured by using a circuit similar to that of the inset of figure 24. In these examples, R_S is again the BS7 microbridge and $R_L = 51.2\ \Omega$ (taking into account the resistance of the circuit electrical wires). For the curve measured at 83.0 K, the applied voltage was 5.8 V before (t_1) and after (t_2) a fault regime of 9.5 V. For the curve at 85.0 K, which is near T_{op}, these values were 3.5 V and 7.5 V, respectively. As expected, the protection is excellent, whereas there is an overprotection when working well below T_{op}. In both cases the recovery after the fault is achieved under current. This is another considerable practical advantage when compared to the superconducting limiters used in high power applications (Noe & Steurer, 2007; Weinstock, 2000). These results suggest that optimal current limitation in superconducting electronics could be accomplished by the own conductive pathways after a proper design (e. g. by decreasing the width of the pathway at well refrigerated selected locations), thus improving compactness.

6. Conclusion

We have summarized some of our works on the thermal behaviour and the refrigeration of fault current limiters based on HTSC. Our results were experimentally or numerically obtained for the three basic SFCL configurations (resistive, inductive and hybrid types). For inductive devices we report on the effects on the limiter's performance of the inhomogeneous nature of HTSC, on the advantages of using this type of samples with artificially created weak zones or with stacks of elements of great surface-to-volume ratio, in particular thin film washers, which present a extremely large value of this ratio. The good refrigeration conditions of the superconducting elements obtained with these designs allow us to operate with limiters of improved thermal stability under a current fault and shorter recovery times after the fault removal. For bulk cylinders with AWZ, the recovery time values are several times lower than with homogeneous cylinders, while the impedance remains unchanged. By using stacks of bulk rings, with a large surface-volume ratio, the recovery time is about one order of magnitude shorter and without impedance loss. By using stacks of thin film washers, the recovery time is two orders of magnitude lower than with bulk cylinders and the impedance is even better.

For hybrid and resistive configurations, we have presented results for designs scarcely explored. On one hand, a limiter based on a thin film meander path integrated in the cold heat exchanger of a thermoacoustic refrigerator. The hybrid design of this device would allow us to operate at a power about 2.2 kVA with a thermoacoustic refrigerator capable of removing about 50 W at 80 K. Finally, we have probed the use of thin film microbridges to operate in very low power applications. Our results with these samples indicate that, by using the appropriate substrates and the right ratio microbridge-to-substrate area, it is possible to operate in a regime where a relative small fault provokes their transition to highly dissipative states that strongly increase the limitation efficiency.

7. Acknowledgements

The earlier phases of the research summarized here were partially supported by the electrical company UNION-FENOSA under Grants No. 0666-98 and 0220-0085. This work has been supported by Ministerio de Ciencia e Innovación ERDF FI2010-19807 and by Xunta de Galicia ERDF 2010/XA043 and 10TMT206012PR.

8. References

Noe, M. & Steurer, M. (2007). High-temperature superconductor fault current limiters: concepts, applications, and development status. *Superconductor Science and Technology*, Vol. 20, No. 3, January 2007, R15-R29, ISSN 0953-2048.

Sokolovsky, V.; Meerovich, V.; Vajda, I.; Beilin, V. (2004). Superconducting FCL: Design and Application. *IEEE Transactions on Applied Superconductivity*, Vol. 14, No. 3, September 2004, 1990-2000, ISSN 1051-8223.

Paul, W.; Chen, M.; Lakner, M.; Rhyner, J.; Braun, D.; Lanz, W. (2001). Fault current limiter based on high temperature superconductors - different concepts, test results, simulations, applications. *Physica C*, Vol. 354, No. 1-4, May 2001, 27-33, ISSN 0921-4534.

Paul, W.; Lakner, M.; Rhyner, J.; Unternährer, P.; Baumann, Th.; Chen, M.; Widenhorn, L.; Guérig, A. (1997). Test of 1.2 MVA high-Tc superconducting fault current limiter. *Superconductor Science and Technology*, Vol. 10, No. 12, December 1997, 914-918, ISSN 0953-2048.

Leung, E.M.; Rodriguez, I.; Albert, G.W.; Burley, B.; De, M.; Gurrola, P.: Madura, D.; Miyata, G.; Muehleman, K.; Nguyen, L.; Pidcoe, S.; Ahmed, S.; Dishaw, G.; Nieto, C.; Kersenbaum, I.; Gamble, B.; Russo, C.; Boenig, H.; Peterson, D.; Motowildo, L.; Haldar, P. (1997). High temperature superconducting fault current limiter development. *IEEE Transactions on Applied Superconductivity*, Vol. 7, No. 2, June 1997, 985-988, ISSN 1051-8223.

Berends, F.A.; Papadopoulos, C.G.; Pittau, R.; Zueger, H. (1998). 630 kVA high temperature superconducting transformer. *Cryogenics*, Vol. 38, No. 11, November 1998, 1169-1172, ISSN 0011-2275.

Antognazza, L.; Decroux, M.; Reymond, S.; de Chambrier, E.; Triscone, J.M.; Paul, W.; Chen, M.; Fischer, Ø. (2002). Simulation of the behaviour of superconducting YBCO lines at high current densities. *Physica C*, Vol. 372-376, No. 3, August 2002, 1684-16877, ISSN 0921-4534.

Osorio, M.R.; Lorenzo, J.A.; Toimil, P.; Ferro, G.; Veira, J.A.; Vidal, F. (2006). Inductive superconducting fault current limiters with Y123 thin-film washers versus Bi2223 bulk rings as secondaries. *IEEE Transactions on Applied Superconductivity*, Vol. 16, No. 3, September 2006, 1937-1942, ISSN 1051-8223.

Decroux, M.; Antognazza, L.; Reymond, S.; Paul, W.; Chen, M.; Fischer, Ø. (2003). Studies of YBCO strip lines under voltage pulses: optimization of the design of fault current limiters. *IEEE Transactions on Applied Superconductivity*, Vol. 13, No. 2, June 2003, 1988-1991, ISSN 1051-8223.

Antognazza, L.; Decroux, M.; Therasse, M.; M. Abplanalp, M.; Fischer, Ø. (2005). Test of YBCO thin films based fault current limiters with a newly designed meander. *IEEE Transactions on Applied Superconductivity*, Vol. 15, No. 2, June 2005, 1990-1993, ISSN 1051-8223.

Duron, J.; Antognazza, L.; Decroux, M,; Grilli, F.; Stavrev, S.; Dutoit, B.; Fischer, Ø. (2005). 3-D finite element simulations of strip lines in a YBCO/Au fault current limiter. *IEEE Transactions on Applied Superconductivity*, Vol. 15, No. 2, June 2005, 1998-2002, ISSN 1051-8223.

Duron, J.; Dutoit, B.; Grilli, F.; Decroux, M.; Antognazza, L.; Fischer, Ø. (2007). Computer Modeling of YBCO Fault Current Limiter Strips Lines in Over-Critical Regime With

Temperature Dependent Parameters. *IEEE Transactions on Applied Superconductivity*, Vol. 17, No. 2, June 2007, 1839-1842, ISSN 1051-8223.

Antognazza, L.; Decroux, M.; Therasse, M.; Abplanalp, M.; Duron, J.; B. Dutoit, B.; Fischer, Ø. (2007). Thermally Assisted Transition in Thin Film Based FCL: A Way to Speed Up the Normal Transition Across the Wafer. *IEEE Transactions on Applied Superconductivity*, Vol. 17, No. 2, June 2007, 3463-3466, ISSN 1051-8223.

Duron, J.; F. Grilli, F.; L. Antognazza, L.; M. Decroux, M.; B. Dutoit B.; Fischer, Ø. (2007). Finite-element modelling of YBCO fault current limiter with temperature dependent parameters. *Superconductor Science and Technology*, Vol. 20, No. 4, April 2007, 338-344, ISSN 0953-2048.

Osorio, M. R.; Ruibal, M.; Veira, J.A.; Vidal, F. (2005). Thermal recovery of inductive superconducting fault current limiters with weak zones in the secondary. *Superconductor Science and Technology*, Vol. 18, No. 5, April 2005, 739-746, ISSN 0953-2048.

Osorio, M. R.; Lorenzo Fernández, J. A.; Veira, J. A.; Vidal, F. (2008). Optimal refrigeration of bulk superconducting elements in fault current limiters by using artificial weak zones. *Superconductor Science and Technology*, Vol. 21, No. 9, July 2008, 095011(8pp), ISSN 0953-2048.

Osorio, M. R.; Cabo, L.; Veira J. A.; Vidal, F. (2004). Inductive fault current limiter based on multiple superconducting rings of small diameter. *Superconductor Science and Technology*, Vol. 17, No. 1, November 2003, 98-102, ISSN 0953-2048.

Osorio, M. R.; Bétrancourt, A.; François, M. X.; Veira, J. A.; Vidal, F. (2008). A superconducting fault current limiter integrated in the cold heat exchanger of a thermoacoustic refrigerator. *Superconductor Science and Technology*, Vol. 21, No. 9, July 2008, 095013(7pp), ISSN 0953-2048.

Ferro, G.; Ruibal, M.; Veira J. A.; Maza, J.; Vidal, F. (2008). Thermal effects in the flux-creep regime of $YBa_2Cu_3O_{7-\delta}$ thin film microbridges under high current densities in self field. *Journal of Physics: Conference Series*, Vol. 97, No. 1, March 2008, 012016(5pp), ISSN 1742-6596.

Ferro, G.; Maza, J.; Ramallo, M.V.; Veira J. A.; Vidal, F. (2009). Influence of thermal, morphological and measuring variables on the thermal runaway of superconducting thin films under high current densities. *Journal of Physics: Conference Series*, Vol. 150, No. 5, March 2009, 052053(5pp), ISSN 1742-6596.

Lorenzo, J. A.; Osorio, M. R.; Veira J. A.; Vidal, F. (2009). High-temperature superconducting fault current microlimiters. *Superconductor Science and Technology*, Vol. 22, No. 2, January 2009, 025009(4pp), ISSN 0953-2048.

Lorenzo Fernández, J. A.; Ferro, G.; Osorio, M. R.; Veira J. A.; Tello, M.J.; Vidal, F. (2010). Thermal behaviour of high-temperature superconducting fault current limiters: Application to *microlimiters. Journal of Physics: Conference Series*, Vol. 234, No. 4, July 2010, 042019(6pp), ISSN 1742-6596.

Osorio, M. R.; Veira, J. A.; Vidal, F. (2010). Behaviour of different artificial weak zones in superconducting elements working in inductive fault current limiters. *Journal of Physics: Conference Series*, Vol. 234, No. 3, June 2010, 032045(7pp), ISSN 1742-6596.

Prester, M. (1998). Current transfer and initial dissipation in high-Tc superconductors. *Superconductor Science and Technology*, Vol. 11, No. , November 1998, 333-357, ISSN 0953-2048.

Osorio, M. R.; Cabo, L.; Veira, J. A.; Vidal, F. (2004). Non-homogeneous quench of the superconducting secondary of an inductive fault current limiter: implications for current limitation. *Superconductor Science and Technology*, Vol. 17, No. 7, May 2004, 868-875, ISSN 0953-2048.

M.I.T. Staff. (1961). *Magnetic Circuits and Transformers*, John Wiley & Sons, ISBN 10-0-262-63063-4, New York.

Paul, W.; Baumann, T.; Rhyner, J.; Platter, F. (1995). Tests of 100 kW High-Tc superconducting fault current limiter. *IEEE Transactions on Applied Superconductivity*, Vol. 5, No. 2, June 1995, 1059-1062, ISSN 1051-8223.

Xiao, L. Y.; Kiyoshi, T.; Ozaki, O.; Wada, H. (1999). Case Study on the Quench Evolution and Passive Protection of High Tc Superconducting Pancake Coil. *Cryogenics*, Vol. 39, No. 4, Paril 1999, 293-298, ISSN 0011-2275.

Davidson, & M. Tinkham, A. (1976). Phenomenological equations for the electrical conductivity of microscopically inhomogeneous materials. *Physical Review B*, Vol. 13, No. 1, April 1976, 3261-3267, ISSN 0163-1829.

Osorio, M. R.; Toimil, P.; Lorenzo, J. A.; Ruibal, M.; Ferro, G.; Veira, J. A.; Vidal, F.; (2006). Quench and thermal recovery in a superconducting fault current limiter with artificial weak zones. Journal of Physics: Conference Series, Vol. 43, No. 1, June 2006, 929-932, , ISSN 1742-6596.

Chapman, A. J. (1984). *Heat Transfer*, MacMillan Publishing Company, ISBN 0-62-946080-8, New York.

Fujishiro, H.; Teshima, H.; Ikebe, M.; Noto, K. (2003). Thermal conductivity of YBaCuO bulk superconductors under applied field: effect of content and size of Y211 phase. *Physica C*, Vol. 392-396, No. 1, October 2003, 171-174, ISSN 0921-4534.

Fujishiro, H.; Ikebe, M.; Teshima, H.; Hirano, H. (2006). Low-Thermal-Conductive DyBaCuO Bulk Superconductor for Current Lead Application. *IEEE Transactions on Applied Superconductivity*, Vol. 16, No. 2, June 2006, 1007-1010, ISSN 1051-8223.

Rettelbach T. & Schmitz, G. J. (2003). 3D simulation of temperature, electric field and current density evolution in superconducting components. *Superconductor Science and Technology*, Vol. 16, No. 5, May 2003, 645-653, ISSN 0953-2048.

Vanderbemden, Ph.; Cloots, R.; Ausloos, M.; Doyle, R. A.; Bradley, A. D.; Lo, W.; Cardwell, D. A.; Campbell, A. M.; (1999). Intragranular and intergranular superconducting properties of bulk melt-textured YBCO. *IEEE Transactions on Applied Superconductivity*, Vol. 9, No. 2, 2308-2311, ISSN 1051-8223.

Lorenzo Fernández, J. A.; Osorio, M. R.; Toimil, P.; Ferro, G.; Blanch, M.; Veira J. A.; Vidal, F. (2006). The impedance of inductive superconducting fault current limiters operating with stacks of thin film Y123/Au washers or bulk Bi2223 rings as secondaries. *Superconductor Science and Technology*, Vol. 19, No. 12, December 2006, 1274-1277, ISSN 0953-2048.

Lorenzo Fernández, J. A.; Osorio, M. R.; Toimil, P.; Ferro, G.; Blanch, M.; Veira, J. A.; Tello, M.; Vidal, F. (2007). Improvement of an inductive fault current limiter with the use of stacks of thin film washers as secondary. *Physica C*, Vol. 460-462, No. 2, September 2007, 1453-1454, ISSN 0921-4534.

Swift, G. (2002). *Thermoacoustics: A unifying perspective for some engines and refrigerators*, Acoustic Society of America, ISBN 0-7354-0065-2, New York.

Yang L. W. & Thummes, G. (2005). High frequency two-stage pulse tube cryocooler with base temperature below 20 K. *Cryogenics*, Vol. 45, No. 2, February 2005, 155-159, ISSN 0011-2275.

Qiu, L. M.; Sun, D.M.; Yan, W.L.; Chen, P.; Gan, Z.H.; Zhang, X.J.; Chen, G.B. (2005). Investigation on a thermoacoustically driven pulse tube cooler working at 80 K. *Cryogenics*, Vol. 45, No. 5, May 2005, 380-385, ISSN 0011-2275.

Ruibal, M.; Ferro, G.; Osorio, M. R.; Maza, J.; Veira, J. A.; Vidal, F. (2007). Size effects on the quenching to the normal state of $YBa_2Cu_3O_{7-\delta}$ thin-film superconductors. *Physical Review B*, Vol. 75, No. 1, January 2007, 012504(4pp), ISSN 0163-1829.

Viña, J; González, M. T.; Ruibal, M.; Currás, S. R.; Veira, J. A.; Maza, J.; Vidal, F. (2003). Self-heating effects on the transition to a highly dissipative state at high current density in superconducting $YBa_2Cu_3O_{7-\delta}$ thin films. *Physical Review B*, Vol. 68, No. 22, December 2003, 224506(10pp), ISSN 0163-1829, and references therein.

Maza, J.; Ferro, G.; Veira J. A.; Vidal, F. (2008). Transition to the normal state induced by high current densities in $YBa_2Cu_3O_{7-\delta}$ thin films: A thermal runaway account. *Physical Review B*, Vol. 78, No. 9, September 2008, 094512(9pp), ISSN 0163-1829.

Mosqueira, J.; Cabeza, O.; François, M. X.; Torrón, C.; Vidal, F. (1993). Measurements of pool boiling heat transfer from ceramic $YBa_2Cu_3O_{7-delta}$ superconductors to liquid nitrogen. *Superconductor Science and Technology*, Vol. 6, No. 8, August 1993, 584-588, ISSN 0953-2048.

Weinstock, H. (Editor). (2000). *Applications of Superconductivity* (NATO ASI Series E: Applied Sciences - Vol. 365), pp. 247-415, *Proceedings of the NATO Advanced Study Institute on Applications of Superconductivity* , Loen (Norway), June 1997, Kluwer Academic Publishers, ISBN 0-7923-6113-X, Dordrecht (The Netherlands).

Skocpol, W. J.; Beasley, M. R.; Tinkham, M. (1974). Self-heating hotspots in superconducting thin-film microbridges. *Journal of Applied Physics*, Vol. 45, No. 9, September 1974, 4054-4066, ISSN 0021-8979.

See e. g., Gurevich, A. VI. & Mints, R. G. (1987). Self-heating in normal metals and superconductors. *Reviews of Modern Physics* , Vol. 59, No. 4, December 1987, 941-999, ISSN 0034-6861.

Poulin, G. D.; Lachapelle, J.; Moffat, S. H.; Hegmann, F. A.; Preston, J. S. (1995). Current-voltage characteristics of dc voltage biased high temperature superconducting microbridges. *Applied Physics Letters*, Vol. 66, No. 19, May 1995, 2576-2578, ISSN 0003-6951.